# 20th International Conference on Condensed Matter Nuclear Science 2016



Published in the Journal of Condensed Matter Nuclear Science (JCMNS) Volume 24

Sendai, Japan
2 – 7 October 2016

**Editor:**

**Jean-Paul Biberian**

ISBN: 978-1-5108-9473-0

**Printed from e-media with permission by:**

Curran Associates, Inc.
57 Morehouse Lane
Red Hook, NY 12571

**Some format issues inherent in the e-media version may also appear in this print version.**



Printed with permission by Curran Associates, Inc. (2021)

For permission requests, please contact
International Society for Condensed Matter Nuclear Science (ISCMNS)
at the address below.

International Society for Condensed Matter Nuclear Science (ISCMNS)
The Willows, Hobro,
Wolverley, Kidderminster
Worcs DY11 5ST
England

www.iscmns.org
webmaster@iscmns.org

**Additional copies of this publication are available from:**

Curran Associates, Inc.
57 Morehouse Lane
Red Hook, NY 12571 USA
Phone: 845-758-0400
Fax: 845-758-2633
Email: curran@proceedings.com
Web: www.proceedings.com

# JOURNAL OF CONDENSED MATTER NUCLEAR SCIENCE

## Experiments and Methods in Cold Fusion

## Proceedings of the 20th International Conference on Condensed Matter Nuclear Science, Sendai, Japan, October 02–07, 2016

**VOLUME 24, October 2017**

# JOURNAL OF CONDENSED MATTER NUCLEAR SCIENCE

Experiments and Methods in Cold Fusion

JOURNAL OF CONDENSED MATTER NUCLEAR SCIENCE

Volume 24, October 2017

 ISSN 2227-3123

ICCF20
ICCF20

J. Condensed Matter Nucl. Sci. 24 (2017) 1–311
 ISSN 2227-3123

# JOURNAL OF CONDENSED MATTER NUCLEAR SCIENCE

Volume 24 2017

## CONTENTS

# Opening Address

ICCF 20 Conference 02–07 October 2016, Sendai, Japan

Ladies and Gentlemen, On behalf of the Organizing Committee, I would like to welcome all of you to the International Conference on Condensed Matter Nuclear Science, ICCF20. This is the 20th conference of the ICCF series, which started in 1990 in Salt Lake City in the U.S., and hence, continues 26 years. The ICCF conferences were held in Japan three times already so far: ICCF3 in Nagoya, ICCF6 in Lake Toya, Hokkaido and ICCF12 in Yokohama. We are very happy to have ICCF20 here in Sendai, Eleven years after the last one in Japan.

In these 11 years, the study of the anomalous heat effect has been greatly enhanced. Based on my personal opinion, I can give two examples as follows. First: the introduction of a device in which deuterium gas with particular nm-scale palladium particles by Prof. Arata in Osaka University: There was no electrical form of input, and the associated measurements of excess heat were performed by monitoring the change in temperature and gas pressure. Second: the revival of the nickel–hydrogen system by using nano-sized Ni particles with hydrogen gas and performing at higher temperature regions: I would say this is triggered by Rossi, although the evaluation of his device is quite uncertain now.

The study of anomalous heat generation, such as represented by these, has been making steady progress. Now, it has become very important to specify the material as well as to investigate its physical properties under a variety of experimental conditions. This may be a necessary step toward practical use, although it will take much more time, man-power, and money.

On the other hand, evidence linking the anomalous heat generation and the nuclear reaction is still sparse. First of all, detection of nuclear radiation is impossible if it is a special reaction process in which the reaction product is always in its ground state, as it would be with two deuterons only going to the 4He ground state. However, one can expect nuclear radiation, i.e., gamma-rays, beta-rays associated with a decay of a radioactive nucleus when it is created as a result of the nuclear reaction. Other radiation including X-rays, light, electric waves, etc. are also expected since the system should deliver the whole energy. Much more effort toward detecting these types of radiation simultaneously with the heat generation are highly desirable, now, in order to overcome the present condition where we have mostly circumstantial evidence. In the present ICCF, these subjects are to be discussed more specifically based on the newest experimental results. In addition, many experimental ideas, trials and data related to the study of CMNS are also to be discussed as well as a large variety of theoretical studies including interesting unconventional ideas.

I do believe that extensive discussions in this conference may serve to deepen the basic understanding of CMNS phenomena as well as to develop wide-ranging applications.

We have about 130 participants already registered in this conference. About two thirds of them are from overseas: the participants are distributed over 17 countries. On this occasion of the international exchange, I would like you to know more about Masamune Date. As you see, his statue on horseback is an emblem of ICCF20. He was born in 1567 as a military commander of the period of wars before the Tokugawa era, and he created the foundation of Sendai City as a feudal load of the Tokugawa government. He was not a local person but his one eye had been open to the world. Once Masamune obtained the approval of the Tokugawa, he sent a diplomatic mission to Spain and Rome in 1613: his retainer Tsunenaga Hasekura was the delegate. It was known as the Keicho era mission to Europe. Furthermore, Masamune had been well known as a person who excelled in both literary and military arts. Actually, Masamune was very good at Japanese poetry. His last poem is 「曇りなき心の月を先立てて、浮世の闇を照らしてぞ行く」

I interpret this as follows: “I was just going the way I believed the best as if going with the clear moon light in my heart that illuminated this unpredictable warring world”. I am very pleased that the 20th ICCF is being held in this place, Sendai, where Masamune led a way of life which resembles an investigator’s thought.

Now, let us enjoy new findings and developments in this extremely interesting field of science and technology during these four and a half days, including a half-day sightseeing to Matsushima Bay which is famous for its scenic views.

Thank you very much.

Sincerely,

*Dr. Jirohta Kasagi*
*(Co-Chair of ICCF20 )*

# Mayor's Speech

Sendai City Mayor Emiko Okuyama

ICCF 20 Conference 02–07 October 2016, Sendai, Japan

On behalf of Sendai City, it is with great pleasure that I welcome you and all of our other guests here in Sendai for the 20th International Conference on Condensed Matter Nuclear Science.

I have heard that Condensed Matter Nuclear Science has a great possibility of changing the entire industrial structure if its clean nuclear energy can be realized. I am also glad to know a number of researchers have been involved in this filled with active discussion and intensive study all over the world. Tohoku University located in Sendai is the first university in Japan to have a special division for CMNS, collaborating with a private company. Its aim is to obtain basic data on nuclear reactions in anomalous heat generation phenomena, as well as in nuclear transmutation phenomena. We have high expectations that the division at Tohoku University will achieve revolutionary progress in nuclear science and practical prospects for commercializing new clean energy.

As a representative of Sendai citizens, I would like to deeply thank all of the people who have supported us in our recovery from the Great East Japan Earthquake of 11 March 2011. We had promoted our reconstruction plans since the tragedy and completed the first step for restoration in 2016. It is our pleasure to let you know our current situation on this occasion. Thanks to support from all over the world, we have managed to overcome the disaster and proceed to a better life. For these reasons, it might not be difficult for you to imagine just how pleased we are to welcome you for ICCF20 as you open this conference here in Sendai after 11-year-absence from ICCF12 held in Yokohama, Japan. Sendai city, known as the 'city of trees', has abundant nature, beautiful scenery through four seasons and a variety of great food and hot springs. You can enjoy unique Japanese culture and hospitality as well. I am sure your stay in Sendai will be helpful to activate your discussion and interactions with each other.

It is delightful to have all of you here today. As Sendai City Mayor, I hope for further progress in Condensed Matter Nuclear Science and your success and prosperity.

Thank you very much.

# Welcome Address

ICCF 20 Conference 02–07 October 2016, Sendai, Japan

Good morning Ladies and Gentlemen. It is my great pleasure to have the opportunity to deliver a welcome address at the International Conference on Condensed Matter Nuclear Science. On behalf of Tohoku University, I warmly welcome all of you participating in this conference which is hosted by the Research Center of Electron Photon Science of our university.

As an Executive Vice President of the University, I first wish to take this opportunity to introduce Tohoku University. The University was established under the name of Tohoku Imperial University in 1907 as the third oldest university in Japan, following Tokyo Imperial University in 1877 and Kyoto Imperial University in 1897. Today, the university has 10 Faculties, 18 Graduate Schools, three Professional Graduate Schools, six Research Institutes, and 12 Inter-department Institutes for Education and Research. The total number of students is about 18,000 including about 7000 graduate students, while the total staff is about 6500 including 3200 academic teachers and professors.

The traditional mottoes of our university are 'Open Doors', 'Research First' and 'Practice Oriented Research and Education'. 'Open Doors' originates from having accepted the first female students at a university in Japan, in 1913. The 'Research First' reflects the situation of the time the university was founded. Then, bright young professors produced original research results one after another, which were used directly for student's education. Furthermore, the university has cherished a tradition of "Practice Oriented Research and Education," in which the results of cutting-edge research are put to use for the good of society and the improvement of living standards. Based on these traditions, Tohoku University has been a university which promotes high education and original research of the world's highest standard. Moreover, it contributes to the realization of human society in peace and fairness by using the results of research to solve problems which society faces, as well as to raise talented people to be leaders.

Last year, Tohoku University established a new joint research division with Clean Planet Inc. to study condensed matter nuclear reactions. I think that this unique research division reflects these mottoes of our university. I am very much pleased that the division is hosting the pre-eminent conference of Condensed Matter Nuclear Science only one and a half years after its establishment.

I heard from my friend Prof. Kasagi that this ICCF conference series started as a conference on Cold Fusion. The important and fundamental problem which is not yet solved exists in the subject of the conference. I remember so-called Cold Fusion. Although many researchers did challenging research after the famous press release of Fleischman and Pons, it seems that very many research efforts are still required in order to explain the mystery. On the other hand, if the mystery is solved and if completely different clean energy source becomes clear, then our life would change drastically.

I am very concerned about environmental problems as a researcher in the field of Physical Oceanography. It seems to me that we stand at the crossroads with the current use of nuclear power after the Great East Japan Earthquake on 11 March 2011, followed by one of the worst nuclear accidents on record at the Fukushima nuclear power plant. Thus, it is very meaningful that discussions on the possibility of future clean nuclear energy will be conducted here over the next several days by many scientists, engineers and specialists from the world.

I welcome all of you and sincerely hope that this conference will contribute to further progress in research on Condensed Matter Nuclear Science. Thank you very much.

Sincerely,

*Dr. Kimio Hanawa*
*(Executive Vice President, Tohoku University )*

# Welcome Address

ICCF 20 Conference 02–07 October 2016, Sendai, Japan

Good morning to everybody. Welcome to the City of Sendai, one of the most beautiful cities in Japan.

I am really proud you are here. I am Hiroyuki Hama, the Director of the Research Center for Electron Photon Science, Tohoku University. Research in our lab., which was established 50 years ago, is centered on high-energy electron accelerators. Also major scientific fields are nuclear physics, nuclear chemistry and photon science. In the long history of the lab., I think the group of Profs. Iwamura and Kasagi, who chair the conference, is addressing pretty unique activity.

We had suffered from a big earthquake and Tsunami in 11 March 2011. Yes, it was really a tragedy for us. We have still not recovered completely, for example, some cities near seaside are now under re-construction. For us, the accelerators were heavily damaged and it took for more than 2 years to recover the research activity in the lab.

In addition, all of the people living in Japan were scared by the incredible accident of the nuclear power reactor. Now I do not want to discuss the issue of whether nuclear power is safe or not. However, the issue of energy for sustainable human life is getting to be a big subject.

It is a common problem in all of the countries of the world. Very recently, the Japanese government has been discussing about the future of a fast breeder reactor called MONJU. The fast breeder reactor was introduced as the ultimate energy source. However, the MONJU may possibly shut down. The conclusion at the moment, I have thought is that the technology of recycling nuclear power seems to be not organized perfectly by human science and technology.

I do not know that is a final conclusion, but I definitely understand that the problem of energy has to be solved by the science, not emotion. In this sense, this international conference, ICCF20 will open the door for new solution of the energy problem. I really hope it. So please have many fruitful discussions to grasp the prospect for the human future. Also enjoy the Sendai city in this beautiful autumn season.

Thanks a lot for coming. In closing, I have a small question. The name of ICCF means International Conference on Condensed Matter Nuclear Science. But there is no F. Personally I can suppose what it is. Thanks, enjoy your stay, and have a nice conference.

Sincerely,

*Dr. Hiroyuki Hama*
*(Director )*
*Research Center for Electron Photon Science (ELPH) Tohoku University*

J. Condensed Matter Nucl. Sci. 24 (2017) 1–14

Research Article

# The Fleischmann–Pons Calorimetric Methods, Equations and New Applications

Melvin H. Miles*

*University of LaVerne, LaVerne, CA 91750, USA*

**Abstract**

The Fleischmann–Pons Dewar isoperibolic calorimetry remains the most accurate system for measuring excess power in cold fusion experiments. The use of mathematical modelling and numerical integration of the experiment data, along with appropriate averaging methods, can achieve a calorimetric accuracy of $\pm 0.1$ mW. The Fleischmann–Pons "Lower Bound Method" can be used to determine the heat-transfer co-efficient during an experiment and yields accurate results. Important information can be obtained from time periods where the cell is first turned on and when the cell is shut off. Periods of large cell temperature changes are accurately modelled by the differential equations. These Fleischmann–Pons calorimetric methods and equations are applied to a new experiment where excess power is produced in the $Pd/D_2O$ + 0.1 M $KNO_3$ systems. A heat conduction calorimeter was used in this study, but the same methods and equations apply.


*Keywords:* Cell cooling, Cell heating, Deuterium oxide, Differential equations, Excess power, Lower bound method, Palladium

**Remarks from Dr. Jean-Paul Biberian (Editor-in-Chief)**

This paper by Melvin Miles is an important contribution to the understanding of the calorimetry developed by Martin Fleischmann and Stanley Pons in their original work. As in many major discoveries, an inventor creates his own instrument. Fleischmann and Pons built a new type of very simple and accurate calorimeter with unusually high precision. The quality of their apparatus allowed them to make their discovery. Both in the past and present their method has been challenged. It is the purpose of this paper to give a clear view of this unique device. As I do for all papers published in the journal, I have asked a first referee to analyze the paper. As the reviewer raised a number of questions, I asked the advice of a second reviewer. In the end, I decided to publish the paper together with the comments of the reviewers and the answers of the author. This is unusual, but not unprecedented. This way of doing things has the advantage for the reader, who will have in hand all the pros and cons to make his own evaluation of the work (see the Comments of Referees at the end of this article).

*E-mail: melmiles1@juno.com.

© 2017 ISCMNS. All rights reserved. ISSN 2227-3123

## 1. Introduction

Martin Fleischmann had remarkable skills in mathematical modelling and had made many notable contributions to science prior to the Fleischmann–Pons discovery of cold fusion [1]. Stanley Pons was also recognized as a top electrochemist with considerable computer skills. Together they developed the accurate Fleischmann–Pons Dewar isoperibolic calorimetry with computerized data acquisition [2–4]. However, the vastly inferior calorimetric experiments by CalTech, MIT and Harwell, along with personal ridicule, blocked the scientific acceptance of cold fusion in 1989 [5]. An important new scientific discovery soon became a scientific tragedy. Many cold fusion scientists who had observed cold fusion effects in their experiments and continued this research suffered severe career damage. This stigmatizing of cold fusion research for the past 28 years is the real scientific fiasco. Fleischmann and Pons were correct. Sometimes scientists get it wrong, but important discoveries have often been rejected for many years (Galileo, Joule, Semmelweis, Arrhenius, Wegener and many others).

There are numerous reasons why this Fleischmann–Pons calorimetry is still important today – nearly 28 years later. The accuracy of the Fleischmann–Pons calorimetry was the major topic of the 1989 cold fusion controversy, and the field has never recovered from the CalTech, MIT and Harwell rejections of this calorimetry. This mathematically based calorimetry remains poorly understood by most scientists, and this includes many cold fusion researchers. This calorimetry continues as the most accurate for electrochemical reactions and has a very small error of only $\pm$ 0.1 mW when correctly applied [6]. This calorimetry also has possible applications for many other electrochemical studies in addition to cold fusion.

The advantages of the Fleischmann–Pons Dewar cell are that: (1) it provides a direct view inside the cell during an experiment, (2) it has no memory effect because of the direct heat transfer by radiation photons, (3) the Stefan–Boltzmann constant ($5.670373 \times 10^{-8}$ $Wm^{-2}K^{-4}$) along with the cell surface area provides an estimate of the radiative heat transfer co-efficient, (4) it provides a wide dynamic range for the cell temperature (100 K) and cell voltage (10 V), (5) the small cell diameter (2.5 cm) provides good mixing by the electrolysis gases, (6) the inherent safety and self-purifying effects provided by the gases exiting the cell, and (7) the high accuracy achieved by computer data acquisition and mathematical modelling.

These cold fusion calorimetric experiments consist of four measurements: the cell temperature, the bath temperature, the cell current, and the cell voltage. The temperature measurements are the least accurate. Because of this, Fleischmann and Pons measured temperatures to within $\pm$0.001 K by using calibrated thermistors. In contrast, CalTech measured temperatures to within $\pm$0.03 K while MIT and Harwell measured temperatures to only within $\pm$0.1 K. For a typical difference of 10.0 K between the cell and bath temperature and a 1.0 W cell input power, the minimum possible error would be $\pm$0.1 mW for the Fleischmann–Pons calorimetry, $\pm$3 mW for the CalTech calorimetry, and $\pm$10 mW for the MIT and Harwell calorimetry. The averaging of data and the use of mathematical modelling allowed Fleischmann and Pons to reach their limit of $\pm$ 0.1 mW while the MIT calorimetric error was reported as $\pm$50 mW [7].

## 2. Experimental

A new experiment was designed to test some of the Fleischmann–Pons calorimetric methods and equations. However, this experiment was done in a copper calorimeter with heat transfer by conduction [8]. Furthermore, the electrolyte of $KNO_3$ was lithium-free to test proposals of lithium being involved in the production of excess heat [9]. Nevertheless, the same equations and methods, as used by Fleischmann–Pons calorimetry, were applied.

The palladium wire cathode (1.0 mm $\times$ 23 mm) was from a Johnson–Matthey) source that had previously provided cathodes which produced excess heat at China Lake as well as at NHE in Japan [10,11]. This $Pd/D_2O$ + 0.1 M $KNO_3$ cell was prepared using Cambridge Isotope Laboratories $D_2O$ (99.9 atom% D). Previous control experiments produced no excess heat with $Pd/H_2O$ + 0.1 M $KNO_3$ using this calorimeter [8].

This calorimetric experiment involved two thermistors fixed in position on opposite sides of the cell wall and immersed in a heat transfer fluid (Mobil-1 oil) as reported previously [8,12,13]. The cell and thermistors fit tightly into the inner calorimetric chamber. This cell was equipped with a centering ring at the cell top which prevented any shift in the cell position. This same calorimeter was used in co-deposition studies of palladium, ruthenium and other transition metals [12,13].

The results for this experiment are based on one thermistors used for the cell ($T_2$) and one used for the bath ($T_3$). These two thermistors always showed excellent stability ($\pm 0.005$ K) and passed all calibration tests both before and following this experiment. A third thermistor ($T_5$) on the opposite side of the cell wall occasionally showed erratic behavior and failed calibration tests following this experiment. However, there were periods where both thermistor gave good agreement as found in previous experiments [8,12,13].

## 3. Lower Bound Method

This method assumes that the excess power is zero and then calculates the lower bound heat transfer co-efficient ($k'$) resulting from this assumption [3]. The calorimetric equations based on the First Law of Thermodynamics expressed as power is

$$C_\mathrm{p}M\,\mathrm{d}T/\mathrm{d}t = P_\mathrm{EI} + P_\mathrm{H} + P_\mathrm{X} + P_\mathrm{R} + P_\mathrm{C} + P_\mathrm{gas} + P_\mathrm{W}. \quad (1)$$

This differential equation is always applied in the Fleischmann–Pons calorimetry. Positive power terms involving energy supplied to the cell are the electrochemical power ($P_\mathrm{EI}$), internal heater power ($P_\mathrm{H}$), if used, and any form of excess power ($P_\mathrm{X}$). The negative power terms for energy escaping the cell are heat radiation ($P_\mathrm{R}$), heat conduction ($P_\mathrm{C}$), heated gases ($P_\mathrm{gas}$), and the pressure–volume work done by the generated gases on the surroundings ($P_\mathrm{W}$). As a first approximation, the small $P_\mathrm{gas}$ and $P_\mathrm{W}$ terms can often be neglected. The first term involves the heat capacity of the calorimeter ($\mathrm{JK}^{-1}$) expressed in terms of equivalent moles of $D_2O$ (M) and the molar heat capacity ($C_\mathrm{p}$) of $D_2O$ (84.35 J $\mathrm{mol}^{-1}\mathrm{K}^{-1}$). The rate of change of the cell temperature ($\mathrm{d}T/\mathrm{d}t$) depends on the relative rates of energy added to the cell and leaving the cell. However, $\mathrm{d}T/dt$ is never exactly zero (no steady state) because of the changing electrolyte level in an open cell resulting in a changing heat transfer co-efficient [14]. Mathematical equations for these various power terms are given in Appendix A.

For heat conduction calorimetry, Eq. (1) can be expressed as

$$C_\mathrm{p}M\mathrm{d}T/\mathrm{d}t = P_\mathrm{EI} + P_\mathrm{H} + P_\mathrm{X} - k\,(T - T_\mathrm{b}) + P_\mathrm{R} + P_\mathrm{g} + P_\mathrm{W}. \quad (2)$$

Using the lower bound assumption of $P_\mathrm{X} = 0$ yields

$$C_\mathrm{p}M\,\mathrm{d}T/\mathrm{d}t = P_\mathrm{EI} + P_\mathrm{H} + 0 - k'(T - T_\mathrm{b}) + P_\mathrm{R} + P_\mathrm{g} + P_\mathrm{W}, \quad (3)$$

where $k'$ is the resulting lower bound cell constant. Equation (2) minus Eq. (3) yields

$$P_\mathrm{X} = (k - k')(T - T_\mathrm{b}). \quad (4)$$

Note that the lower bound $k'$ will always be less than or equal to the true heat transfer co-efficient, $k$. Because $P_\mathrm{X}$ will likely approach zero sometime during an experiment, the highest value obtained for $k'$ can be used to approximate the true $k$. If the true heat transfer co-efficient is known, then Eq. (4) provides a simple method for accurately calculating excess power. The lower bound heat transfer co-efficient can be calculated for each experimental point using

$$k' = (P_\mathrm{EI} + P_\mathrm{H} + P_\mathrm{C} + P_\mathrm{g} + P_\mathrm{w} - C_\mathrm{p}M\,\mathrm{d}T/\mathrm{d}t)/(T - T_\mathrm{b}), \quad (5)$$

where $P_C \gg P_R$ for a heat conduction calorimeter.

Rearranging Eq. (4) gives

$$k'_C = k_C - P_X/(T - T_b). \tag{6}$$

For $Pd/D_2O$ or $Pd/H_2O$, the exothermic loading of palladium with deuterium or hydrogen will generally result in a small initial excess power effect. Therefore, $k'$ can be near zero or even negative initially when $T - T_b$ is very small [3]. Later, as $T - T_b$ increases and as the exothermic loading is completed, then $k'$ may approach $k$ in value. Equation (6) is similar to the equation $y = a - b/x$ as $x$ ranges from small to large values.

Initial experimental results for the lower bound constant, $k'$, versus time, are shown in Fig. 1 for the new $Pd/D_2O$ + 0.1 M $KNO_3$ experiment. The value for $k'$ is near zero at 2 min and approaches the true $k = 0.1350$ W/K as the time increases. Previous control experiments using this same calorimeter have always given cell constants ranging from 0.1324 to 0.1340 W/K [8,12,13]. From the initial values for $k'$, it can be estimated that the excess power due to loading was about 5 mW. A sharp increase in $k'$ was noted near 50 min where the loading reaches about $x = 0.6$ for $PdD_X$ and the rate of deuterium loading decreases significantly. The results shown in Fig. 1 indicate that this palladium cathode loaded readily with deuterium. Also, there was an early excess power effect in addition to the exothermic loading of deuterium into the palladium which causes changes in $k'$ as shown in Fig. 1.

Equations derived by Fleischmann [14] show that the cell temperature measured inside the cell should initially increase linearly with time. This was approximately the behavior in this new $Pd/D_2O$ + 0.1 M $KNO_3$ experiment where $dT/dt = 1.53 \times 10^{-4}$ K/s over the first 40 min. However, the cell temperature in this experiment was measured on the outside cell surface in a secondary compartment containing a heat transfer fluid (Mobil-1 oil) as in previous experiments [8,12,13]. This gave an episode from 6 to 12 min of nonlinear temperature changes due to the sluggish mixing of this fluid by convection. The theoretical cell heating rate is given by the Fleischmann–Pons equation

$$dT/dt = [(E - E_H)I + P_X - k(T - T_b)]/C_pM, \tag{7}$$

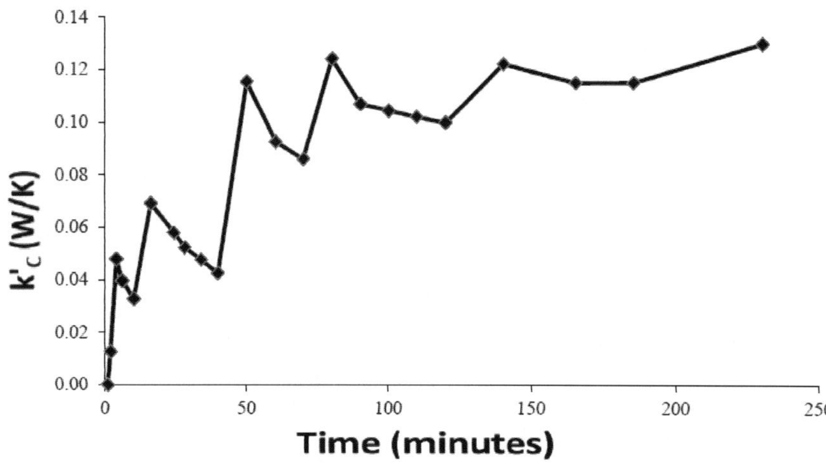


**Figure 1.** The lower bound cell constant versus time for the first 240 h ($I = 0.0365$ A).

where $C_pM = 450$ JK$^{-1}$, $k = 0.1350$ WK$^{-1}$ and $P_g$ and $P_W$ are small and can be neglected due to the initial low cell current ($I = 0.0365$ A). Table 1 gives the theoretical $dT/dt$ values for selected experimental measurements using Eq. (7).

The theoretical cell heating rate ($dT/dt$ calculated using Eq. (7) is nearly constant over the first 40 min with a mean value of $dT/dt = 1.59 \pm 0.04 \times 10^{-4}$ K s$^{-1}$. This is in good agreement with the mean experimental rate of $1.53 \times 10^{-4}$ K s$^{-1}$ and supports the $C_pM$ and $k$ values used as well as the measured excess power. The increases in excess power apparently extends the linear $dT/dt$ region for a longer time period than the expected 15–20 min [14]. The experimental rate of the cell heating between 45 and 65 min was measured as $0.83 \times 10^{-4}$ K s$^{-1}$ and again this is in good agreement with the theoretical rates at 50 and 60 min shown in Table 1. The agreement between the theoretical and experimental values for $dT/dt$ shows that Eq. (7) provides a good mathematical model for the cell behavior.

## 4. Excess Power Measurements

The excess power for this Pd/$D_2O$ + 0.1 M $KNO_3$ experiment is shown in Fig. 2. These values for excess power were calculated using the Lower Bound Method as expressed by Eqs. (4) and (5). The mean excess power value for each day is displayed or split excess powers on days when the current is changed. An excess power effect in addition to the 5 mW exothermic deuterium loading was observed for the first day of this experiment as shown in Table 1. Larger excess power effects were observed when the cell current increases from 36.5 to 146 mA on Day 4, to 219 mA on Day 7, and to 292 mA on Day 8. The additions of make-up $D_2O$ caused the excess power to decrease (5.2 mL on Day 6, 6.6 mL on Day 8, and 5.2 mL on Day 12). This decreases in the excess power is generally explained by the resulting cell cooling or by the introduction of $H_2O$ contamination. The largest excess power was 73 mW on Day 8 at a cell current of 292 mA (0.400 A cm$^{-2}$). This yields 3.65 W/cm$^3$ for this small palladium cathode ($0.1 \times 2.3$ cm, $V = 0.020$ cm$^3$). This is the normally expected value of 1–5 W/cm$^3$ for these electrolysis experiments [5].

These excess power calculations used $k = 0.1350$ W K$^{-1}$ for a cell filled with 50 mL of $D_2O$ giving $C_pM = 450$ J K$^{-1}$. The changes in the cell electrolyte volume were calculated by Faraday's law (–0.812 mL of $D_2O$ per day at a cell current of 100 mA). From previous experiments, the cell constant varies by 0.0002 W K$^{-1}$ per mL, and the heat capacity changes by 4.65 JK$^{-1}$ per mL of $D_2O$ [12,13]. These changes were applied each day in the calculations of the excess power.

There is a good explanation for the zero excess power results for Day 13. At the end of Day 12, the cell was turned off to obtain results for the rate of the cell cooling. This cell cooling was measured with the cell turned off for 133 min allowing the deuterium inside the palladium to escape (deload). Furthermore, 5.2 mL of $D_2O$ was then added followed by turning the cell back on at a higher current of 365 mA (500 mA/cm$^2$). The next day (Day 13), the mean excess power was –23.0 mW. However, the $P_g$ and $P_W$ terms had been neglected for simplicity. Because of the high cell current and high cell temperature (about 43°C), it was calculated that $P_g = -16.2$ mW and $P_W = -7.5$ mW. Adding these two neglected terms gave a mean excess power of 0.7 mW for Day 13. Because of this result, the mean

**Table 1.** Theoretical cell heating rates ($dT/dt$).

| Time (min) | $E$ (V)[a] | $P_X$ (W) | $dT/dt$ (K s$^{-1}$) |
|---|---|---|---|
| 10 | 3.625 | 0.0235 | $1.61 \times 10^{-4}$ |
| 20 | 3.850 | 0.0163 | $1.55 \times 10^{-4}$ |
| 30 | 3.899 | 0.0290 | $1.62 \times 10^{-4}$ |
| 40 | 3.922 | 0.0380 | $1.55 \times 10^{-4}$ |
| 50 | 3.944 | 0.0094 | $0.73 \times 10^{-4}$ |
| 60 | 3.956 | 0.0229 | $0.86 \times 10^{-4}$ |

[a] Cell current ($I$) was 0.0365 A.

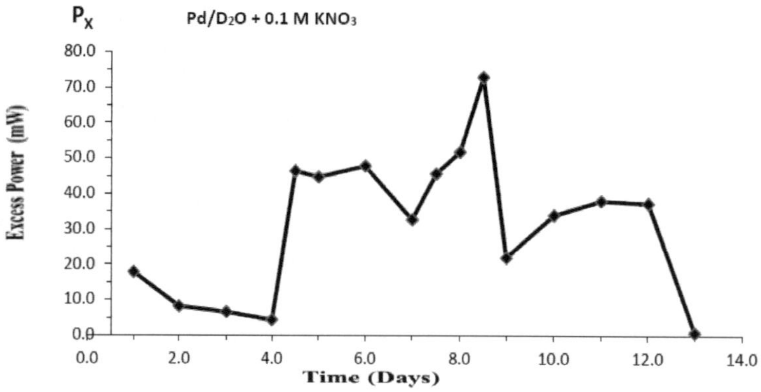


**Figure 2.** Mean values for the excess power measurements for each time period.

$P_X$ values for all other days were corrected for the small $P_g + P_W$ sum on each day. These results for Day 13 indicate the accurate modelling of this calorimeter and show the importance of not neglecting the $P_g$ and $P_W$ terms when the excess power is close to zero.

## 5. Cell Cooling Study

For cell cooling with $I = 0$, most calorimetric terms are zero, leaving the simple differential equation (assuming $P_X = 0$)

$$C_pM\,dT/dt = -k(T - T_b). \tag{8}$$

This equation is readily integrated to give

$$\ln(T_o - T_b)/(T - T_b) = (k/C_pM)t. \tag{9}$$

This integrated equation can also be written as

$$\Delta T = \Delta T_o\,\exp(-kt/C_pM), \tag{10}$$

where $\Delta T = T - T_b$ and $\Delta T_o = T_o - T_b$. Related equations for radiative heat transfer during the cell cooling, as in the Fleischmann–Pons Dewar cell, are given in Appendix B. The cooling curve equations are actually quite similar for conductive and radiative heat transfer with only additional small terms for radiative cooling (see Eq. (B.2)).

The expected exponential decreases in $\Delta T$ is shown in Fig. 3 for this Pd/$D_2O + 0.1$ M $KNO_3$ experiment. Initially, the decrease in the cell temperature is almost linear with time as shown both in Figs. 3 and 4. Except for the first two points in Fig. 4, there is almost a linear decrease in $\Delta T$ from 2 to 14 min. This is explained by Eq. (10) for small values for time ($t$) where

$$\Delta T = \Delta T_o(l - -kt/C_pM). \tag{11}$$

The theoretical cell cooling rate can also be expressed by the differential equation

$$dT/dt = (-k\Delta T + P_X - P'_X)/C_pM, \tag{12}$$

where $P_X$ is any lingering excess power when the cell is off and $P'_X$ represents the endothermic deloading of deuterium from the palladium. This deloading will be very fast when the cell is first turned off. Any excess power

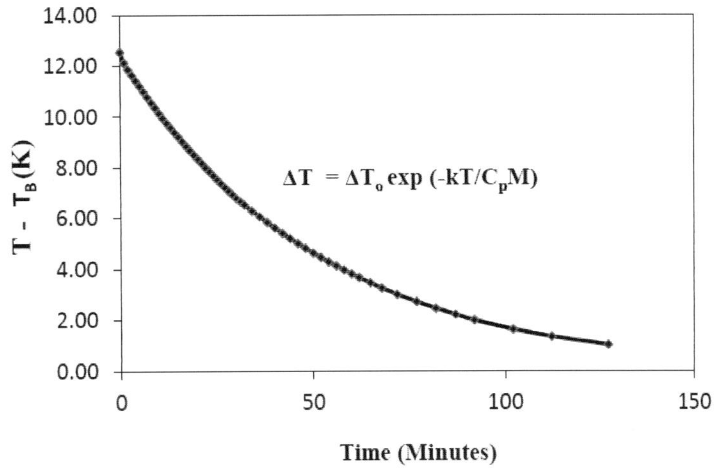


**Figure 3.** Experimental cell cooling at zero current for 130 min.

(heat-after-death) will slow down the cell cooling rate while the endothermic deloading will increase the cell cooling rate. However, $P_X$ and $P'_X$ are unknown, and we can only look for any possible effects that they may exert on the experimental cooling rates. Assuming $P_X - P'_X = 0$, the experimental and theoretical cooling rates are compared in Table 2. Due to the low level of $D_2O$ during this cell cooling experiment (44 mL), $k = 0.1338$ $WK^{-1}$ and $C_pM = 410$ $JK^{-1}$.

The initial high experimental rate of cell cooling could be explained by the rapid deloading of deuterium. Oth-

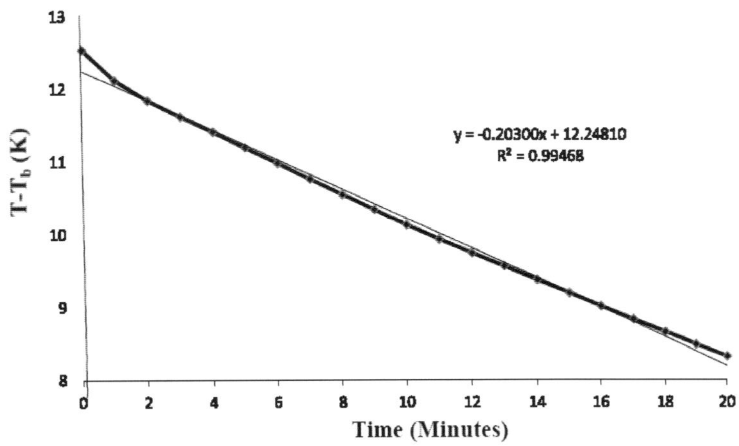


**Figure 4.** Initial cell cooling for the first 20 min.

**Table 2.** Experimental versus theoretical cooling rates for the $Pd/D_2O + 0.1$ M $KNO_3$ cell.

| Time (min) | $\Delta T$ (K) | Experimental (K/min) | Theoretical (K/min)[a] |
|---|---|---|---|
| 2 | 11.84 | –0.28 | –0.232 |
| 5 | 11.19 | –0.22 | –0.219 |
| 10 | 10.31 | –0.21 | –0.202 |
| 15 | 9.19 | –0.18 | –0.180 |
| 20 | 8.32 | –0.17 | –0.163 |
| 30 | 6.77 | –0.14 | –0.133 |
| 40 | 5.61 | –0.11 | –0.110 |
| 50 | 4.63 | –0.09 | –0.091 |

[a] From Eq. (12) with $P_X - P'_X = 0$, $k = 0.1338$ WK$^{-1}$, and $C_pM = 410$ JK$^{-1}$.

erwise, the experimental and theoretical cell cooling rates are in good agreement. If the excess power of 0.038 W continued with the cell off, this would only slow the cooling rate by 0.038 W/410 JK$^{-1}$ $= 9.3 \times 10^{-5}$ K/s or by 0.0056 K/min and would be difficult to detect.

The cell cooling expressed by the integrated equation (Eq. (9)) is shown in Fig. 5. An excellent fit ($R^2 = 0.9999$) to the straight-line behavior is observed with the line slope equal to 0.01974 min$^{-1}$ or $3.290 \times 10^{-4}$ s$^{-1}$. This agrees reasonably well with the expected slope of $3.2963 \times 10^{-4}$ s$^{-1}$ given by $k/C_pM$ where $k = 0.1338$ WK$^{-1}$ and $C_pM = 410$ JK$^{-1}$. Fleischmann and Pons have reported large deviations from the expected straight-line slope for the integrated equation when there is a large and lingering excess power effect [4]. The excess power present (0.038 W) prior to the cooling shown in Fig. 5 would likely be too small to detect such effects.

A dark film on the palladium cathode is generally observed for experiments conducted in $D_2O$ + LiOD solutions. However, for this lithium-free experiment conducted in $D_2O + KNO_3$ the palladium cathode following this experiment showed the bright luster of palladium metal. This suggests that the black colorations on the cathode in lithium based electrolytes may be due to Li–Pd nitride compounds formed by the deposition of lithium metal onto the palladium cathode followed by reactions with nitrogen from the air introduced during the $D_2O$ additions.

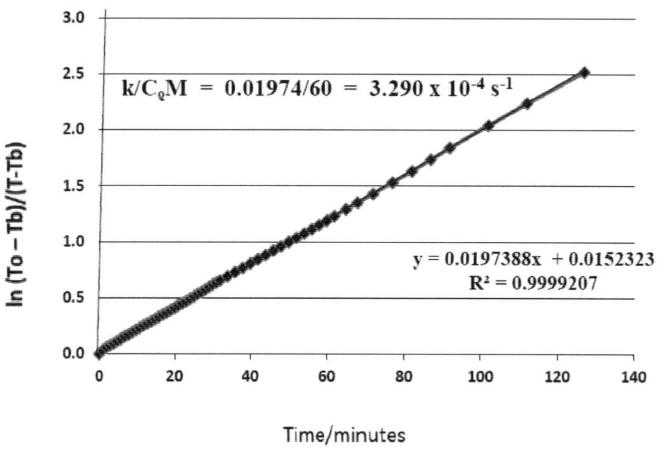


**Figure 5.** The experimental cell cooling expressed by the integrated equation (Eq. (9)).

## 6. Conclusions

The Fleischmann–Pons lower bound method provides a means of determining the heat transfer co-efficient for a calorimetric cell during an experiment while yielding accurate calorimetric results. Periods of large cell temperature changes with time are correctly modelled by the use of the Fleischmann–Pons differential equations. These calorimetric equations modified for heat transfer by conduction accurately describes the calorimetric experimental behavior of this $Pd/D_2O+0.1$ M $KNO_3$ experiment. This experiment showed that normal amounts of excess power (3.65 W/cm$^3$) can be obtained in a lithium-free electrolyte for the $Pd/D_2O$ electrolysis system.

## Acknowledgements

This work was supported from an anonymous fund at the Denver Foundation via Dixie State University and administered by Kalynn Larson at the Dixie Foundation. An adjunct faculty position at the University of LaVerne is also acknowledged.

## References

[1] M.H. Miles and M.C.K. McKubre, Cold fusion after a quarter century: The Pd–D system, Chapter 13, in *Developments In Electrochemistry: Science Inspired*, M. Fleischmann, D. Pletcher, Z.-Q. Tian and D.E. Williams (Eds.), Wiley, West Sussex, UK, 2014.

[2] M. Fleischmann, S. Pons, M.W. Anderson, L.J. Li and M. Hawkins, Calorimetry of the palladium–deuterium–heavy water system, *J. Electroanal. Chem.* **287** (1990) 293–348.

[3] M. Fleischmann and S. Pons, Calorimetry of the Pd–$D_2O$ systems from simplicity via complications to simplicity, *Phys. Lett.* A **176** (1993) 118–129.

[4] M. Fleischmann and S. Pons, Heat after death, *Trans. Fusion Technol.* **26**(4T) (1994) 97–103.

[5] M.H. Miles, B.J. Bush and D.E. Stilwell, Calorimetric principles and problems in measurements of excess power during Pd–$D_2O$ electrolysis, *J. Phys. Chem.* **98** (1994) 1948–1952.

[6] M. Fleischmann and M.H. Miles, The instrument function of isoperibolic calorimeters: excess enthalpy generation due to parasitic reduction of oxygen, *Condensed Matter Nuclear Science*, World Scientific, New Jersey, 2006, pp. 247–268.

[7] M.H. Miles and P.L. Hagelstein, New analysis of MIT calorimetric errors, *J. Condensed Matter Nucl. Sci.* **8** (2012) 132–138.

[8] M.H. Miles and M. Fleischmann, New approaches to isoperibolic calorimetry, in *ICCF-15 Proc.*, V. Violante and F. Sarto (Eds.), Rome, Italy, October 5–9, 2009, 22–26.

[9] C.L. Liang, Z.M. Dong, Y.P. Fu and X.Z. Li, Lithium – an important additive in condensed matter nuclear science, *J. Condensed Matter Nucl. Sci.* **19** (2016) 164–172.

[10] M.H. Miles, B.F. Bush, K.B. Johnson, Anomalous Effects in Deuterated Systems, NAWCWPNS TP 8302, September 1996.

[11] M.H. Miles, Calorimetric studies of $Pd/D_2O$+LiOD electrolysis cells, *J. Electroanal. Chem.* **482** (200) 56–65.

[12] M.H. Miles, Electrochemistry and calorimetry of ruthenium co-deposition, *J. Condensed Matter Nucl. Sci.* **8** 2012() 115–123.

[13] M.H. Miles, Co-deposition of palladium and other transition metals in $H_2O$ and $D_2O$ solutions, *J. Condensed Matter Nucl. Sci.* **13** (2014) 401–410.

[14] M.H. Miles, Excerpts from Martin Fleischmann letters, *J. Condensed Matter Nucl. Sci.* **19** (2016) 210–218.

## Appendix A. Power Terms

*Electrochemical power*

$$P_{EI} = (E - E_H)I. \tag{A.1}$$

*Heat radiation power*

$$P_R = -k_R(T^4 - T_b^4). \tag{A.2}$$

*Heat conduction power*

$$P_C = -k_C(T - T_b). \tag{A.3}$$

Heated gases ($D_2$, $O_2$, $D_2O$ vapor)

$$P_g = -(I/F)\,[0.5C_{p,D_2} + 0.25C_{P,O_2} + 0.75P'C_{p,D_2O}]\,\Delta T - 0.75(I/F)P'L, \tag{A.4}$$

where

$$P' = P/(P* - P) \quad \text{and} \quad P* = P_{D_2} + P_{O_2} + P_{D_2O} \approx P_{\text{atm}}.$$

*Pressure–Volume Work*

$$P_W = -0.75(I/F)RT \tag{A.5}$$

(see also M.H. Miles, *J. Condensed Matter Nucl. Sci.* **13** (2014) 392–400).

## Appendix B. Cell Cooling For Radiative Heat Transfer

$$C_pM\,\mathrm{d}T/\mathrm{d}t = -k_R(T^4 - T_b^4), \tag{B.1}$$

$$\ln(T_0 - T_b)/(T - T_b) - \ln(T_0 + T_b)/(T + T_b) + 2\left[\tan^{-1}(T/T_b) - \tan^{-1}(T_0/T_b)\right] = 4T_b^3k_Rt/C_pM. \tag{B.2}$$

*Note*:

$$P_R = -k_R(T^4 - T_b^4) = -k_R\left[4T_b^3\Delta T + 6T_b^2\Delta T^2 + 4T_b\Delta T^3 + \Delta T^4\right],$$

where $\Delta T = T - T_b$.

## Comments of Referee 1

This paper is confusing because it describes the Fleischmann calorimeter and the equations applied by Fleischmann while using a different calorimeter to make measurements.

As several reviews of the Fleischmann method have shown, his method contains greater potential errors than he claimed. Nevertheless, the reviewers generally agree that the method was sufficiently accurate to support his claim for production of excess energy. Over the years, this method has been described and explained often enough that another description is not required. What is required in this paper is a detailed description of the method used here to obtain the claimed excess power, independent of what Fleischmann did 28 years ago.

Therefore, the Introduction should be rewritten and concentrate on what was done in this study without rearguing the history and merits of the Fleischmann method.

Presumably the calorimeter used here was the same one described in M. Miles, M. Fleischmann, Measurements of Excess Power Effects In Pd/$D_2O$ Systems Using a New Isoperibolic Calorimeter, *J. Condensed Matter Nucl. Sci.*

**4** (2011) 45–55 and M. Miles, Calorimetry, ARL Workshop, Adelphi, MD, 2010. If so, the physical shape and arrangement of the various internal components need to be described in the paper. Where exactly is the temperature measured? How stable is the temperature measurement? The author confuses the absolute accuracy (i.e. value relative to an universal standard), which is not important, with the stability of the temperature calibration over a period of time. What is the amount of noise on the value and the amount of drift over time. Thermistors are known to suffer from both sources of error.

Equation (1), as written, is very confusing. The goal of the study is to measure Px. Therefore, the equation should be written as equal to $P_x$. The heat capacity ONLY comes into play when temperature of the cell changes. During this time, the $\Delta T$ term would be slightly too small when the temperature is increasing and slightly too large when the temperature is decreasing. The $C_p$ term can be used to correct for this variation. On the other hand, less error is introduced if measurements of $P_x$ are only made when the temperature is constant. After all, excess power only has value if and when it is sustained as a steady source of power.

Assuming constant temperature: $P_x = k(T - T_b) + A$, where $k =$ (total power applied to the calorimeter)/$(T - T_b)$ and $A$ is a constant obtained by fitting $k$ to an equation over a range of applied power. Sine this is an open cell, the energy escaping as $D_2$ and $O_2$ gas needs to be added using the neutral potential. The terms $P_{gas}$ and $P_w$ are constants, hence are automatically corrected for by the calibration.

Because the calorimeter cannot measure the absolute amount of power, it must be calibrated as the author describes. Apparently, this calibration is achieved by calculating $k$ when no excess power being made by the sample. This value of $k$ is then used to determine the presence of excess power as being any apparent increase in power over what is called "the lower bound" value.

The description of charges in $k$ being caused by changes in the sources of power, such as that produced during loading, is confusing. The calibration constant is and must be a constant. It does not change. Once its value is obtained by calibration, it is then used to calculate any creation of power in excess that present during calibration.

In Fig. 1, the measurements are made so seldom, the time variation has no meaning. Should the points be connected as the author has done or should a line be drawn through the average, thereby assuming the variation results from random scatter in the measurement?

The description below Fig. 1, where the sluggish response of the oil is described, negates the heat capacity correction. This behavior, which is not typical of the Fleischmann calorimeter nor of any other needs to be acknowledged up front rather than confusing the reader initially with a description having no relationship to this behavior.

In fact, the heat capacity correction is very complex in this system. The energy originating at the cathode first transfers heat to the water, then energy reaches the glass container and the metal containing the oil before the proper temperature is achieved in the oil. Each of these heat sinks results in a different effective correction. Consequently, $C_p$ and $M$ are not known but have to be determined as variables based in the behavior of the system. This reviewer does not see how this determination is made.

The claim for heat production using $KNO_3$ instead of LiOD is important provided the claim can be believed. Unfortunately, the description does not give much confidence in the claim. Therefore, the paper needs extensive revision before it should be published.

### Response to Referee 1

The comments of the first reviewer show the need for my discussion of the Fleischmann–Pons (F–P) calorimetry. This mathematically based calorimetry remains poorly understood by most scientists, and this includes many people involved with cold fusion such as this reviewer. Almost every comment by this reviewer illustrates the need to explain this F–P electrochemical calorimetry to those who have not taken the time to carefully study the F–P methods and equations. For starters, there are no valid reviews of this calorimetry that have found "greater potential errors" than claimed. However, there are many erroneous comments and reviews of the F–P calorimetry.

First and foremost, the F–P calorimetry always introduces a differential equation based on the change of the cell temperature, $\Delta T = T - T_b$, with time $(d\Delta T/dt)$. The F–P method generally involves the application of this differential equation over two-day periods which include the use of an in-cell heater resulting in significant changes in the cell temperature with time. As shown by the differential equations of chemical kinetics, much more accurate results can be obtained by the mathematical integration of the differential equations. The F–P calorimetric differential equation is too complex to be directly integrated, thus numerical integration methods are applied (see my Ref. [6]). However, integration is possible when the cell is turned off giving a simpler differential equation, and the cell cooling closely follows the mathematical equations as shown in this paper.

There is never a time in cold fusion experiments using an open isoperibolic calorimeter where the cell temperature remains exactly constant (see my Ref. [14]). If the cell temperature actually remained constant, as proposed by this reviewer, then there would be no differential equations, no mathematical integrations and no F–P methods and equations. Two otherwise useful books by Ed Storms have promoted the error expressed by this reviewer regarding the need for the assumption of a constant cell temperature for isoperibolic calorimetry. This has led to false statements in these books about isoperibolic calorimetry such as "this method is hard to make accurate to better than $\pm 250$ mW" and "correct power is only obtained when the temperature is constant".

The correct thermodynamic format for Eq. (1) is to place the power for the system $(C_p M \, d\Delta T/dt)$ on the left side and the sources of power into or out of the system on the right side as was always done by F–P. Anyone can algebraically rearrange Eq. (1) as desired.

This reviewer completely fails to understand the F–P "Lower Bound Method". This method always assumes that the excess power is zero, and the resulting "lower bound" cell constant, $k'$, is calculated. At the beginning of a $Pd/D_2O$ experiment, this $k'$ may be near zero or even negative. The changes in $k'$, such as shown in Fig. 1, are explained by changes in the actual excess power as well as changes for $T - T_b$ (see my Eq. (6)) and not from "random scatter in the measurements".

The heat capacity for any isoperibolic calorimeter consists of all material within the cell up to the highest thermal barrier which controls the rate of enthalpy transfer into the constant temperature bath. For my heat conduction calorimetry, this includes the $D_2O$, the glass cell and electrodes, the heat transfer oil, and the inner copper tube. My experimental and calculated values for the heat capacity always show good agreement. The value for $C_p M$ will gradually decrease during electrolysis due to the loss of $D_2O$. The initial sluggish response of the oil is not unexpected because the heat transfer in the oil involves both conduction and convection, and convection may be somewhat slow in getting started when the experiment begins.

Finally, both the F–P Dewar calorimetry and my system are both isoperibolic calorimeters, and the same equations apply. The only difference is the main method of heat transfer. My calorimeter transfers most of the heat by conduction while the F–P calorimetry transfer most of the heat by radiation. All other calorimetric terms are exactly the same, and the same calorimetric methods and equations apply.

## Referee 2 Comments

The Fleischman–Pons isoperibolic base calorimeter design has been used for critical, extreme-precision, real-time energy balance studies on new battery Japanese technology for over 20 years. The base instrument function equations reiterated in the present paper are, in this reviewers view, unequivocal, and as well-tested and proven (under far more diverse conditions than those described in this paper) as any other known chemical reaction measurement technique. To this group's knowledge, there has never been any serious scientific challenge to the base master equations; there has been some serious misunderstanding of same, however, but this widely used calorimetric technique (and commercially available instrumentation) is well understood and remains above reproach. The author also seems to be the most experienced researcher in this area, and is a known and respected thermodynamics expert.

As regards the verisimilitude of the claims of the experiment, this reviewer would ask what is to be gained in raising the controversy of an excess energy generation phenomena with a calorimeter that is modified in function *apropos* that of the Fleischman–Pons device that was used originally and which is already well-characterized? Specifically, the conduction pathway for heat transfer is indeed used in traditional commercial isoperibolic calorimeters with known reliability, but this pathway was evidently purposely avoided in the Fleischmann–Pons device. It then remains to resolve this question along with the fact that according to all available records, there was never any detection of excess thermal energy by Fleischman–Pons using the isoperibolic calorimetric technique with (a) palladium anode, platinum cathode; (b) any other anion other than deuteroxide; (c) any other cation other than lithium-6 or -7, and (d) any other solvent other than anhydrous (H) deuterium oxide; these were evidently often used as control variables by Fleischman and Pons

Besides providing evidence for a new chemical system under electrolysis that exhibits excess enthalpy, in contradiction to previous results in (b) and (c), we have to consider same is provided by use of a calorimeter that is apparently modified in thermal transfer mechanism to that used to provide the original and contradicting results. In addition, simultaneous (null) calibration results, while referred to in the text for a previous point in time, are not conveniently described or presented for comparison.

While the mathematical description is clear and accurate, the plausibility of the experiment is certainly believable, and there is no apparent reason to doubt the accuracy or integrity of the interesting results, there is reluctance to accept the conclusions of a new excess energy system made with a new hybrid energy transfer path calorimeter without at least providing (1) comparative null results under identical conditions as the present results in the same calorimeter, OR (better) (2) providing the results of the new system in a pure radiative isoperibolic Fleischman–Pons calorimeter.

**Response to Referee 2**

I have no disagreements with this reviewer, but I have a few comments. I worked with three F–P Dewar calorimeters at NHE in Japan in 1997–1998, and I agree that this calorimetry is superior to any other calorimetry that I have used. However, I had very little government support for my cold fusion research following the ending of the U.S. Navy program in 1995. Furthermore, I did not know of any group capable of producing the F–P Dewar glass cells with silvering at the top. Therefore, I found a simpler and less expensive calorimetry design which involved heat conduction across inner and outer copper tubes completely separated by insulation. Incidentally, I asked for the three F–P Dewar calorimeters that I used in Japan when the NHE program ended in 1998, but permission was refused for my taking these calorimeters back to the U.S.

My F–P Dewar calorimetry used in Japan included three co-deposition experiments using the $ND_4Cl$–$ND_4OD$–$PdCl_2 + D_2O$ electrolyte. Significant excess power was observed in all three cells. There were no lithium-6 or lithium-7 cations in these palladium co-deposition experiments. Also, $Cl^-$ was the major anion and not $OD^-$ considering that the deposition of $Pd^{++}$ results in an acidic solution. Therefore, the observation of excess power using the F–P Dewar calorimetry apparently is not restricted to the use of $Li^+$ cations and $OD^-$ anions in the $D_2O$ electrolyte.

I discussed the design of my copper calorimeters with Martin Fleischmann, and his main reason for favoring the Dewar design was that there were no memory effects. Photons directly radiate heat between the Dewar cell and the water bath. For heat conduction calorimeters, the heat may linger inside the insulation. Also heat convection across the oil may vary somewhat with time. I have added to the experimental section to explain this calorimeter and to reference previous experiments where this same calorimeter was used. The main reason for my use of $KNO_3$ as an electrolyte was the opinion of Navy scientists at NRL that my excess power in the co-deposition experiments at NHE could be explained by the formation and reactions of nitrates. My report of no excess power in a $Pd/H_2O+KNO_3$

control experiment served as my experimental rebuttal. Furthermore, I had done extensive research on the $LiNO_3$–$KNO_3$ molten salt electrolyte which showed the excellent electrochemical stability of nitrate systems even at high temperatures.

I found this reviewer's report of the F–P Dewar calorimetry being used for accurate studies of new Japanese battery technology extremely interesting. If this became widely known, it could have a belated positive impact on the acceptance of the F–P cold fusion calorimetry.

J. Condensed Matter Nucl. Sci. 24 (2017) 15–24

Research Article

# CMNS Research – Past, Present and Future

Michael C.H. McKubre*

*Energy Research Center, SRI International, Menlo Park, California, USA*

**Abstract**

As a community, we have invested a great deal of time and money in investigating claims of anomalous heat first asserted by Martin Fleischmann and Stanley Pons in 1989. Despite this effort, we remain unable to specify the phenomenon or phenomena revealed. Although it is clear within the LENR community that the effect is of nuclear origin and is exo-energetic, we have yet to define with confidence the pathway to practical technology that seemed implicit in the original announcement. Bringing this community together largely is the belief that such a path is possible, or even imminent. What is taking us so long?


*Keywords:* CMNS, Cold fusion, Excess heat, Helium, LENR, Tritium

## 1. Introduction

This paper deals only indirectly with scientific issues. I do not believe that the problems of Cold Fusion, LENR, CMNS, whatever name we choose, are primarily scientific, political or even financial. I am aware of technologies – or pre-technology demonstrations – that have the potential to change the basic infrastructure of power generation, worldwide. In a very real and somewhat urgent sense, sustainable, large scale, primary energy production that does not contaminate our world or harm species is the primary responsibility that we face as scientists, humans and self-appointed stewards of the planet. The understandable, but regrettably human, response to such an opportunity has been in far too many instances the search for personal, regional or corporate glory, fame, wealth or power. This condition has petrified our progress by stultifying collaboration. I am no less responsible for – or less guilty than any other – of concealing secrets. The barbarians are not outside the gates – they are inside – they are us.

I spoke about this problem last year at Tony La Gatta's ICCF19 in Padua, Italy, and would like to elaborate here using a few personal examples. The primary, perhaps only, problems I see to progress in understanding, amplification and application are those of communication both within and outside the community. Hopefully, we can use these conference proceedings to make progress in this arena. The science is (almost) there and there is no shortage of planetary or financial incentive.

*Retired. E-mail: mmckubre@gmail.com

© 2017 ISCMNS. All rights reserved. ISSN 2227-3123

## 2. Background

First some mythology and mythological metaphors: Martin Fleischmann – father of this field and one of my earliest mentors – drew parallels between Cold Fusion and Icarus. He tortuously crafted the acronym I.C.A.R.U.S. (Isoperibolic Calorimetry Research and Utility System) for his boiling water calorimeter, first publicly displayed at ICCF3 in Nagoya, Japan, in 1992.

Martin selected the metaphor of ICARUS to bring attention to the perils of flying close to the sun with deficient apparatus and support. Martin told Chris Tinsley [1] and others including me that "*I should have called it Daedalus but I couldn't think of a good acronym.*" Daedalus was the father of ICARUS. Both were imprisoned in the labyrinth that Daedalus had constructed for King Minos to cage the Minotaur – a long story. Daedalus was a technical genius and constructed wings facilitating his and his son ICARUS to escape. Daedalus warned ICARUS that the materials of construction were unsuitable for near-solar excursion. Being young and tempted by the brightest object perceivable in his universe Icarus paid no heed, or could not resist, the temptation – and perished.

If Martin imagined himself a latter-day Daedalus I will avoid the temptation to speculate who the modern ICARUS, or ICARI, are or were. Martin gave us the tools to fly – or at least ample motivation to try. Many have failed to heed his warnings and have fallen to earth. But perhaps the story of Prometheus is also apt and cautionary. Are we self-censoring the delivery of new fire to mankind because we are afraid of the wrath of contemporary Zeus? Are we guarding our secrets for fear that others will take credit? Whichever or both, it needs to stop. Anyone who believes that what we are working on has the potential to relieve or solve mankind's most fundamental problem of sustainable, benign primary energy production should collaborate, cooperate and communicate. Having spent a career in confidential research I understand the motives for reticence but it is time for a new axiom. An alternative possibility suggested in review is that some scientists may have left details unexpressed simply because they did not know they were important or they thought that they were obvious. It is critical that scientists communicate exactly what they have done; without this replication is impossible and progress incapacitated.

Why are we here and how did we get to where we are? Through the work of many more able than I, of course, much of whose work was not immediately recognized and is not well remembered today. I would like to focus in a very personal way on the individuals and deeds that motivated me most and inspired what is now more than 27 years of almost constant single focus on demonstrating and thence understanding what we know as the Fleischmann Pons Heat Effect (FPHE) and its much more important generalization Condensed Matter Nuclear Science (CMNS) in which the nature, rates and consequences of nuclear effects in solid and liquid matter differ importantly from those in free space or rarified conditions.

My first hero of course is Martin Fleischmann, closely followed by John Bockris. I met Martin in person first at Southampton in early 1977, John somewhat earlier. But I knew them both as Titans of Electrochemistry from my very first interest in the subject. I have often used Fig. 1 as the convincing argument for a heat effect in the deuterium-palladium system that is incommensurate with any conceivable chemical or lattice energy storage effect.

As you recall, when operated at constant current in their electrochemical cell, the sloping baseline temperature develops with decreasing conductance of the electrolyte, and increasing impedance and thermodynamic voltage at the two electrodes. The downward temperature excursions in Fig. 1 were caused by additions of $D_2O$ every 12 h, to make up for electrolysis loss in their thermodynamically open cell. With one such addition, the temperature rose unexpectedly and apparently spontaneously, the first derivative of temperature increased then increased again also apparently spontaneously, and the temperature of the electrolyte rose to boiling. These increases were not due to increasing cell voltage or resistance, and the increase in temperature could be explained only by the appearance of a new added heat source. Approximately four and a half days later, with another heavy water addition, this new heat source disappeared and the cell resumed its baseline trajectory.

Knowing the power required to maintain an elevated cell temperature, even eyeball integration of the positive

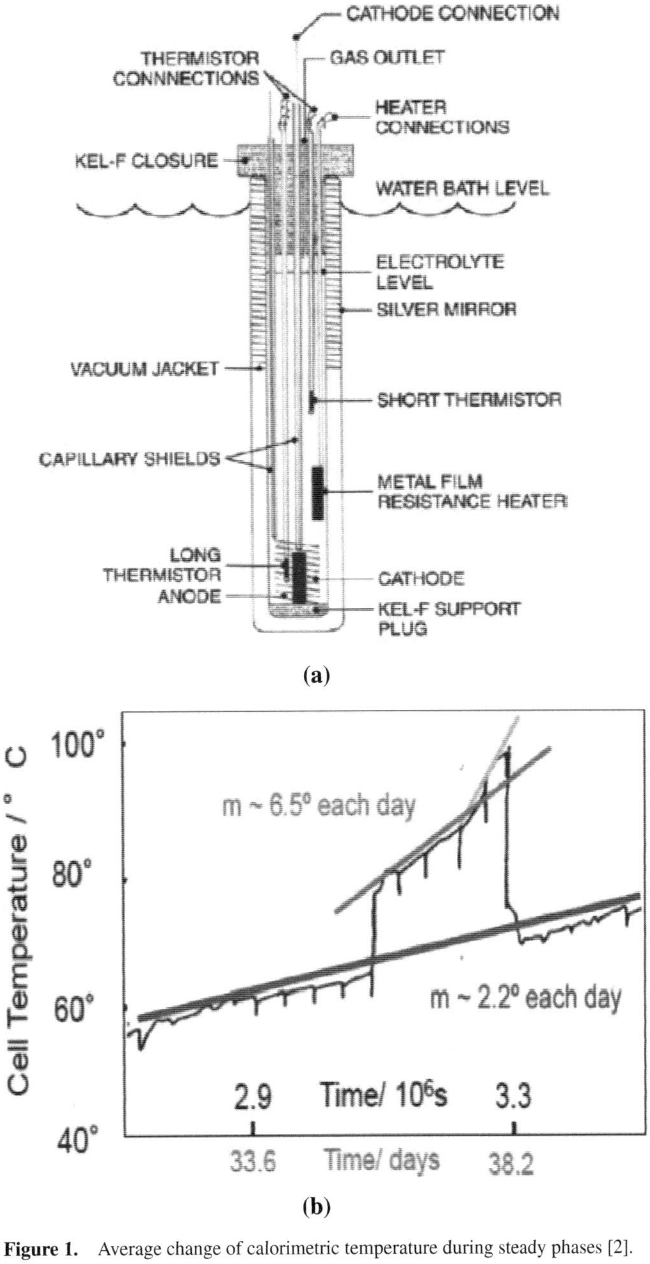


**Figure 1.** Average change of calorimetric temperature during steady phases [2].

heat excursion indicates a heat source larger than known or conceivable chemistry by several orders of magnitude. Normalized to the mass of palladium this suggested a nuclear-level, or at least new and very interesting, energy effect. That statement is still true so long as Fleischmann and Pons were not incompetent, delusional or mendacious. I knew them personally and knew them immediately to be expert scientists and experimentalists – in fact incomparably – as well as honest. With my inside knowledge, I knew that this was not a report of bad science or deliberate deceit. But with the ever-present, now-regretted, spirit of non-cooperation, it was not until I saw the Fleischmann Pons Heat Effect with my own eyes, in my own laboratory, that I knew with certainty that they were not deluded either. And this doubt was for a man whose technical prowess I knew and respected more than any other – so great was the stretch of their claim and the damnation distributed generously (if not completely honestly) by doubters and detractors.

Despite 27+ years, many thousands of person-years of work, and thousands of published papers the world today contains two significant groups: those who have seen the FPHE with their own eyes or who are closely associated with such individuals and therefore believe the effect to be real (potentially with real consequences); and those who have not and do not accept the claim.

The former group number in the hundreds or thousands – I count myself as one. We suffer the weakness of having communicated our results incompletely or inadequately. This failure is both accidental and deliberate. By far the worst paper Martin Fleischmann ever published was his first cold fusion paper [3]. The political pressures on Fleischmann and Pons and the University of Utah present in and leading up to 1989 make the technical insufficiencies of reference [3] somewhat understandable, even forgivable. But other factors already were at work in 1989 and remain at work today to suppress rather than fully reveal technical details sufficient for replication and rapid technological advance.

A very similar story pertains to the words and work of a man with whom I had spent more time (until that point) – the father of modern physical electrochemistry and experimental genius, John O'Mara Bockris shown below with his Electrochemistry Group at Imperial College London (Fig. 2) at that time – the very beginning of modern Physical Electrochemistry. There Bockris was advisor to both Martin Fleischman (indirectly) and John Tomlinson, my PhD supervisor at the University of Wellington – both are shown in this photo circled in the back row. Bockris was therefore my "academic grandfather". Incidentally, 1948 was the year Martin first published on the Palladium–Deuterium system, and the year of my birth. This work has been going on for a very long time.

When I first heard what Bockris rightly claimed (latterly), as the first nuclear evidence for cold fusion – the production of tritium – I was extremely skeptical. If tritium were being produced then where were the neutrons? Horror stories circulated about the leakiness of molecular tritium and the ubiquity of "Exit signs" and other potential sources containing huge amounts of this radioactive element – and poised to leak into unwary experiments. Again, I did not believe primarily because, by that time, we had made scores of tritium assays from Pd/D electrolysis cells – many with well-loaded cathodes – in which we had never seen any increase in tritium levels. Indeed, we observed the converse. Tritium appeared to be slightly reduced in the liquid phase after long-term Pd electrolysis experiments. If we did not observe tritium production in numerous well-performed experiments, then why did others?

There were also tales of tritium spiking and scandal concocted by Gary Taubes [4]. These allegations were stated as fact but never proven or technically supported, and were impossible to believe if one knew the individuals that were accused. But they were messy and complicating, and provided sufficient reason to stay away especially if one had no affirming results to contribute. But again, the case for tritium production had already been well made, this time by one of my old heroes, and two new ones.

The tritium results of 1990, first reported to the world at ICCF1 (in Salt Lake City, Utah) were remarkable in that they came to us already pre-replicated. Three groups with impeccable credibility reported, essentially simultaneously, the production of tritium from deuterium–metal systems: John Bockris' group at Texas A&M; Ed Storms and Carol Talcott at Los Alamos National Laboratory (LANL, New Mexico); Mahadeva Srinivasan, P.K. Iyengar and others working at Bhabha Atomic Research Centre, India (BARC, Trombay, India). Bockris I knew to be a superb and honest scientist, and a supreme experimentalist. Although I did not previously know Storms and Talcott, they were very

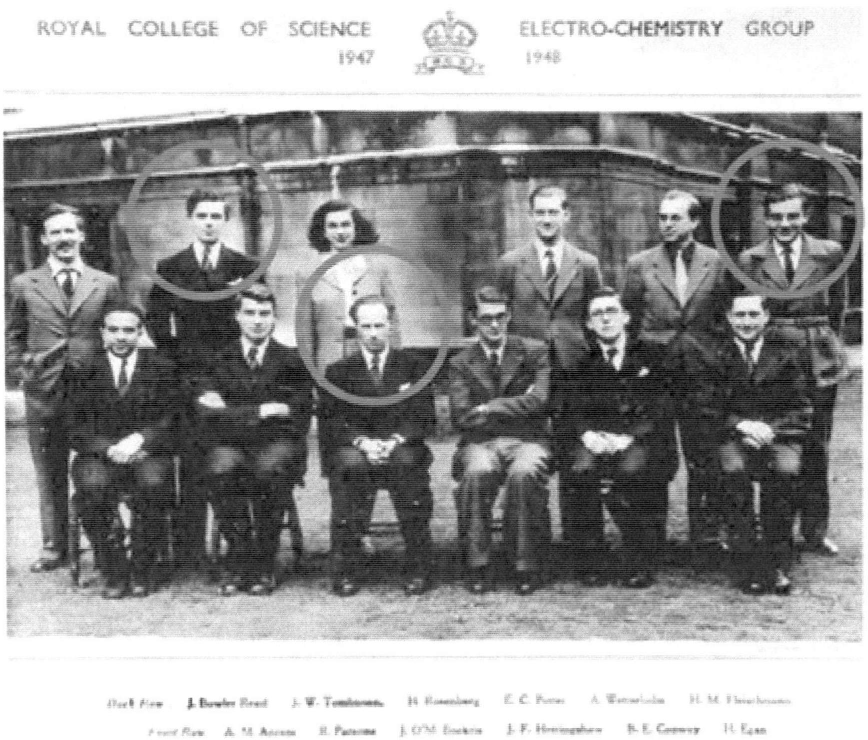


**Figure 2.** Bockris'group, Imperial College, London, 1947–1948.

impressive in person and Los Alamos had a solid reputation, particularly in the area of nuclear materials. Srinivasan and Iyengar were extremely capable individuals, occupying very senior positions at possibly the pre-eminent nuclear research institution in India – one of the best in the world. By its very nature, the evidence presented (tritium) was clearly of nuclear origin. Furthermore the options for doubt listed above for the Fleischmann and Pons Heat Effect (incompetence, delusion, deceit) clearly did not pertain to the tritium results: incompetence could be ruled out by the nature and skills of the individuals and the institutions they represented; delusion and mendacity could be eliminated because they had reproduced each other's observations by different means and in different experiments. Only collusion survives as a possible explanation, and it seems very clear that each was unaware of the other two groups' activities. To this day, I have not heard a single significant or pertinent criticism of any of these groups' activities or results and we must accept the (now) self-evident truth that tritium is produced *de novo* in deuterated metal gas and electrolytic experiments ... and that Bockris was right! This was the first nuclear evidence for cold fusion and it came to us early, already reproduced and – with the wisdom of hindsight – unarguably so.

Why was this not believed: (A) to be significant; (B) at all? The answer to (A) is easy: there was not enough tritium to explain the heat. Heat appeared when tritium did not, and (apparently) *vice versa*; neutrons, while possibly present at the margins of detectability, did not match the rates of tritium production. It should be noted that this last argument is bogus or at least unscientific since if cold fusion were real at all, it occurred by mechanism or mechanisms unknown and very clearly different from hot fusion, while the formation rates of the various products was (and is still) unpredicted.

The answer to (B) is, in my case at least, shameful. More charitably, perhaps, it was due to an excess of caution. But the evidence, when studied and properly assimilated, was clear; the experts and experiments were sound. The fact that tritium could be made at all in what came to be known as "low energy" situations was of vital and critical importance since it challenged what was (then) "known" (or believed) about the "incapacity" of electronic effects to couple to nuclear states and processes. If tritium was produced in an essentially chemical environment – and it clearly had been – it would require a very significant degree of rethinking about the fundamentals of nuclear physics. For many, perhaps most, this was too painful a step and the evidence would be more comfortably ignored. Scientifically, it did not matter that the tritium could not explain the excess heat. What was crucial for science was that tritium was produced at all! In fact, the production of tritium was not only the first solid evidence for cold fusion – it was the first modern evidence of condensed matter nuclear effects.

What did it take for me to believe that tritium atoms can be produced from D, or from D and H, in metals or condensed matter, and thence to believe in cold fusion? The unsurprising but also unacceptable answer is: "I needed to see it with my own yes, in my own laboratory"; in this case, in a completely different experiment. It was not until we (at SRI) observed tritium decay with Brian Clarke [5] by measuring the build-up of helium-3 in our replication of Professor Arata's seminal work with "Double-Structured Cathodes" – in the late 1990s – that I came to accept the tritium evidence as wholly true. I was only then able to incorporate this new fact into my understanding of the mystery that we were studying, and concentrate some effort on trying to comprehend its significance.

Someone said: "*A fool learns only from his own experience*". This is not how science should work, and not how we must make progress. We need to communicate, evaluate, and learn to trust; the literature must be read, understood and respected. Very sadly, the helium-4 story is essentially the same as for tritium. Perhaps even worse, since helium-4 production is claimed to be very closely commensurate with the excess heat, and there is no reason to anticipate the evolution of neutrons (that are not observed), if the primary heat-producing process is, albeit indirectly, $D + D \Rightarrow {}^4He + 24$ MeV (lattice). By way of mitigation, the technical difficulties associated with the measurement of helium-4 are very different from those of helium-3 or hydrogen-3 (tritium). Because of its nuclear stability, Helium-4 is relatively easily measured and resolved from molecular species nominally of the same mass. However, for this very reason, helium-4 is present in the ambient at high levels (5.22 ppm), larger than the expected production in most cold fusion experiments. The primary experimental problem therefore is background interference and suppression.

Knowing this, Melvin Miles designed and implemented an extremely clever deuterium–palladium electrochemical experiment [6], whereby the evolved electrolysis gases, deuterium and oxygen, were used to purge and scavenge helium from his system. Mel's results, reported at ICCF-2 in 1991 in Como, Italy, were stunning. When excess heat was present in the electrochemical experiment, helium-4 appeared approximately commensurately in the evolved electrolysis gas. When excess heat was not present, neither was helium-4. Finally, we had a product that made sense thermodynamically and explained the absence of radiation. This sense was not shared by those with a hot fusion mindset, but if the correlation reported by Miles was real, it proved cold fusion was also, and gave us insight into mechanism. This was a fact that even John Huizenga recognized in his 1992 book [7]. John was counting on non-confirmation – in this, as in other important matters, Huizenga was wrong.

So what was the problem? Why was the significance of Mel Miles' result not well understood or believed at the time? The vessels that Mel used for calorimetry were not helium leak-tight and there was much discussion about the likelihood of helium in-leakage causing or contaminating the result. It took some thinking and calculation to recognize the elegance of the approach that Mel had devised. But most of all, it took the independent replication that Huizenga was counting out, and this all took time. One of the early replicators was SRI so I am quite familiar with the experimental parameters, limitations and consequences of this experiment. In his review of helium-4 and heat correlations in 2015, Abd ul-Rahman Lomax [8] states: "*Miles was amply confirmed, and precision has increased.*

*While there are outliers, there is no experimental evidence contradicting the correlation, and only the exact ratio remains in question. In this, we have direct evidence that the effect is real and is nuclear in nature; the mechanism remains a mystery well worth exploration*". For an experimental result of earth shattering importance, first reported publicly in 1992, why did it take until 2015, 23 years, for this conclusion to be stated with such clarity and conviction? And even now not every researcher in the CMNS world would agree that helium-4 is the primary product, or even a nuclear one!

## 3. Redemption?

What is wrong with this situation, or with us? Is it that experiments are difficult to reproduce, contain uncontrolled critical variables, or lack theory? I have often argued that theory, if not absent in this field, is certainly lagging, but this is not a failure by theorists to be imaginative, productive or rigorous – most are. Instead, I would argue that this is another failure of experimentalists to fully and clearly communicate our results. This failure has denied us the full use of scientific method wherein theory-derived hypotheses are tested experimentally. But absence of theory is not, I believe, a primary or legitimate reason for our failure to advance or convince. I have had very good and productive interactions with enormously capable theorists who became friends. Primarily and initially these included: Giuliano Preparata; Peter Hagelstein; Talbott and Scott Chubb; Akito Takahashi.

Theory is important, in fact crucial, and it will come into its own when we do our experiments better and communicate our results more completely and effectively. But it was the support and interest of two giants in the early days who gave us intellectual cover that maintains today despite their deaths: Julian Schwinger and Edward Teller. Both had a sustaining interest in cold fusion from the very earliest days. If individuals of this caliber believed that solid state, or coherent effects, might modify the trajectory of nuclear interactions sufficiently to generate the heat, tritium, and (later) helium-4 effects, who was I to doubt this possibility? Who was anybody? So the argument that cold fusion was theoretically denied – made mostly by "not-anybodies" – was never serious or important, even though it has had serious and disruptive consequences.

It would take a very long time to mention everyone who has taught me in this field – and how. My principal debts of gratitude for their work and influence are to Martin Fleischmann, of course, being my initial and greatest influence in this field, and to my SRI colleagues – first and most enduring Fran Tanzella, and to Tom Passell – my EPRI Program Manager from 1978 until 1995. I have had and continue to have an extraordinarily fruitful and joyful collaboration with two good friends, Peter Hagelstein and Vittorio Violante – both have done what I could not and taught me about things and places I would never have dreamed exploring. The two last I would like to mention are Ed Storms and Irv Dardik. When each of them see these words they will be annoyed with me for not having mentioned them in my list of "personally influential theorists". But each has influenced me profoundly – Ed with his huge body of work and synthesis of understanding – Irv with his focus on the wider significance and power of multi-frequency resonant effects – a largely untapped field of research that greatly benefited the work at SRI, Energetics and ENEA.

## 4. Where do We Stand?

I would like to conclude with a personal message and a grand challenge. The only place I ever worked was SRI – for 37 years since emigrating to the USA from my native New Zealand *via* postdoctoral studies in England. As many of you know, I retired from SRI in 2016 and returned to New Zealand to be closer to my family. After nearly four decades of contract research for others, I am now a free man (or more nearly so) with significant degrees of freedom.

But my work in the field of CMNS – our work – is not complete. After more than 27 years I still do not know the basic cause of the effect we are studying. I would like to – I would like us all to. To do so in a rational and timely manner, I believe that we need to collaborate and cooperate more effectively as a community than we have. All I can do is exhort. But I think it should and must be done, and I would like to encourage discussion about how we might begin to treat our individual and shared observations of condensed matter nuclear effects as a potentially real solution to the greatest pending problem of humanity, rather than attempting to co-opt it for personal advantage.

Here is a summary of my major observations.

(1) Collectively we have the answer, individually none of us does! We have to stop trying to be heroes and begin playing together as a team. I continue to believe that this is a game worth winning – and that it can be won - and that we can do this if we change the way we approach the solution. I have contributed 27 years to this task – I can give a few more – so long as I see a reasonable trajectory towards resolution.

(2) As a community and individually, we did not pay sufficient attention to the significance of the work of others. Personally, I took a long time to be persuaded by the evidence of: Heat; Tritium; Helium – and only came to accept these as real when I saw evidence with my own eyes and under my experimental control. For whatever reasons, these evidences now fail to persuade, and fresh replications of the best experiments are necessary. We must do the work ourselves.

(3) We need fresh data. We cannot keep pointing out 20-year-old results as the best evidence of an effect.

(4) As a consequence we have limited shared experience. We have not paid sufficient attention to the teaching that does exist, and there exists no consensus around an agreed set of facts. Everyone has a different set of facts. Everyone is taking their own journey. We are not standing on the shoulders of giants, we are standing shoulder to shoulder in the (metaphorical) muck, diving for (real) coins.

(5) The giants that I spoke of before started the job, but they did not complete it. If the likes of: Martin Fleischmann; John Bockris; Fritz Will; Giuliano Preparata; Andrei Lipson, Julian Schwinger – the best I ever knew – could not complete the task that Martin set us, what hope do we have working alone? We must work together!

(6) Despite the negativity expressed both inside and outside our community, there is intense and increasing interest in our field. With the "statute of limitations" expiring on our "glorious transgression" of questioning "established expectation" 27 years ago – talented, young, fresh thinking, imaginative and energetic researchers are entering the field in significant numbers. 30 year-olds do not care what happened 27 years ago – 40 year-olds do not either. These people are our future. Let them learn and do, and watch and help them win.

(7) Our field is not limited by money or constrained by government. Essentially every successful businessman and entrepreneur on the planet is aware that securing a benign primary energy source that is unlimited in power (supply), energy (duration) and availability (location) – is one of the greatest existential planetary questions over which we have any control. Many or most are keen to help. Our job is to make it easy for them. "*Fund me and my idea*" is not helping them or us.

(8) The results we are seeking will benefit all mankind. Likewise the team of talent and consensus needs also to be multi-national. The host nations of ICCF-20 and its satellite, Japan and China, have made significant and conspicuous progress towards fulfilling the potential of Martin's dream, for which we are all extremely grateful. So have the Europeans (especially Russia, Italy, France), and the North Americans (particularly the USA). These groups and nations need to continue to work together – perhaps even better – and we need to expand "the club".

(9) We are presently at the level of building science not technology. In this regime there is no need – or excuse – for competitive instincts or actions. Just as cold fusion or LENR must be a coherent process, so must its solution.

## 5. Grand Challenge

Now allow me to pose a "call to action", what others have called a "Grand Challenge". I posit that our most urgent goals as a community are to:

- Produce fresh experimental results of non-chemical anomalies.
- Replicate these in multiple laboratories.
- Communicate results clearly via technical articles and presentations.

We all know the level of scrutiny this will face, and we must prepare accordingly. As a community, and to make progress on what I (and our funder friends) consider to be a "reasonable timeline", we need to identify the best experiments, rally around them, replicate them, and publish the results. With science established, technology will flow. Although as yet undecided, ICCF21 will occur most probably in the USA, likely in early 2018. If the organizers will allow me, I propose there and then to report on performance against the challenge to:

- Identify what we consider to the best experiments (three or four).
- Recruit multiple laboratories to work on them.
- Write clear scientific papers including multiple authors from the multiple labs. Let us do our own peer review first.
- Publish these papers in our own Journal of Condensed Matter Nuclear Science (JCMNS) or other peer reviewed scientific journal.
- Present the work at ICCF21 in a special session having focus on these replications.

My singular goal, which I invite you to share is to understand for certain the basic cause of the effect we are studying and have assembled here to discuss – by 2018. All I can offer is my help and a threat. In Padua at ICCF19, I delivered a lecture discussing what I believed to be the need and the time for the senior CMNS generation to teach the next. Some listened. Preparing here for ICCF20, my thoughts have circulated around our need to learn what has already been taught, to collaborate broadly and without ego, and to achieve consensus on a limited set of critical experiments that can be performed soon and – if performed successfully – would make a difference. If you accept the proposal of the Grand Challenge, then we have 18 months to produce results that will be evaluated at ICCF21 and illuminated for the world to see!

Do we accept the opportunity given to us – which is huge and critical for mankind? Or do we fail in what I believe is the greatest challenge and opportunity we as scientists will ever face. I promise you, I will be at ICCF21 to help you all to make that call. If by that time we have not coalesced as a community to take full advantage of our collective wisdom then I suggest that the organizers and Chair(s) of ICCF21 convene a panel and discussion on that occasion to identify a best path forward. Our time is limited and the problems that we hope to solve are increasing as our global energy resources diminish and by-product "ashes" increase in the environment. It is time for community action.

## Acknowledgment

With grateful acknowledgment to my esteemed colleagues: Esperanza Alvarez, Yoshiaki Arata, Jianer Bao, Les Case, Emanuele Castagna, Jason Chao, Bindi Chexal, Scott Chubb, Talbott Chubb, Brian Clarke, Dennis Cravens, Steve Crouch-Baker, Irving Dardik, Trevor Dardik, Rob Duncan, Arik El Boher, Alison Godfrey, Ehud Greenspan, Peter Hagelstein, Alan Hauser, Graham Hubler, Nada Jevtic, Antonio La Gatta, Dennis Letts, Shaul Lesin, Jon McCarty, Robert Nowak, Tom Passell, Andrew Riley, Romeu Rocha-Filho, Joseph Santucci, Francesca Sarto, Tara Scarborough, Maria Schreiber, Stuart Smedley, Francis Tanzella, Paolo Tripodi, Matt Trevithick, Vittorio Violante, Dennis van der Vliet, Robert Weaver, Mark Williams, Kevin Wolf, Lowell Wood, Sharon Wing, Tanya Zilov.

## References

[1] C. P. Tinsley, http://www.infinite-energy.com/iemagazine/issue11/fleishmann3.html.
[2] S. Pons, M. Fleischmann, C. Walling and J. Simpson, J. International Patent Publication No. 90/10935 (1990).

[3] M. Fleischmann, S. Pons and M. Hawkins, *J. Electroanal. Chem.* **261** (1989) 301, errata **263** (1990) 187.

[4] G. Taubes, Cold fusion conundrum at Texas A&M, *Science* **248** (4961) (1990) 1299–1304.

[5] M.C.H. McKubre, F.L. Tanzella, P. Tripodi, and P.L. Hagelstein, The emergence of a coherent explanation for anomalies observed in D/Pd and H/Pd system: evidence for 4He and 3He production, in *8th Int. Conf. on Cold Fusion*, Lerici (La Spezia), Italy: Italian Physical Society, Bologna, Italy, 2000. http://lenr-canr.org/acrobat/McKubreMCHtheemergen.pdf.

[6] M. Miles, Heat and helium production in cold fusion experiments, in *2nd Int. Conf. on Cold Fusion, "The Science of Cold Fusion",* Como, Italy: Societa Italiana di Fisica, Bologna, Italy, 1990.

[7] J.R. Huizenga, *Cold Fusion: The Scientific Fiasco of the Century,* University of Rochester Press, Rochester, NY, 1992.

[8] A. Ul-R. Lomax, Replicable cold fusion experiment: heat/helium ratio,
http://www.currentscience.ac.in/Volumes/108/04/0574.pdf.

J. Condensed Matter Nucl. Sci. 24 (2017) 25–31

Research Article

# Fluorescence-based Temperature Sensor for Anomalous Heat from Loaded Palladium Electrodes with Deuterium or Hydrogen

Sangho Bok*, Cherian Mathai, Keshab Gangopadhyay and Shubhra Gangopadhyay†
*Department of Electrical Engineering, University of Missouri, Columbia, MO, USA*

Orchideh Azizi , Jinghao He, Arik El-Boher, Graham Hubler and Dennis Pease
*The Sydney Kimmel Institute for Nuclear Renaissance (SKINR), Department of Physics and Astronomy, University of Missouri, Columbia, MO, USA*

**Abstract**

Anomalous heat generation in palladium-based materials has been studied in various active research groups since M. Fleischmann, and S. Pons demonstrated an anomalous heat in 1989. There have been attempts to explain anomalous heat by deuteron–deuteron nuclear fusion in the Pd lattice while the search for radiation was unsuccessful which indicated that the origin of the excess heat is unknown. Despite the unknown origin of the excess heat, Pd system is in the core of the research for energy production. The excess heat has been demonstrated by using calorimeters that are well characterized. However, these measurements are not able to provide detailed information about a localized heat rather than a heat from a large area. A new method is demonstrated to investigate an excess heat from Pd electrode loaded with deuterium or hydrogen. It is capable of measuring a small amount of heat generated in a localized area with a sub-micrometer resolution by fluorescence imaging with temperature sensitive fluorescence dyes. Considering quantum yield, photostability, and thermal stability, rhodamine 6G (R6G) is selected for temperature sensor along with a polymer, Poly MethylSilsesquioxane (PMSSQ). The thermal quenching of fluorescence resulted in a decreased temperature over time with heat generation. We envisioned that this new method of the temperature measurement provided a novel diagnostic tool for localized excess heat which was not detectable by calorimetry due to intrinsic disability.


*Keywords:* Anomalous heat, Fluorescence, Temperature sensor

## 1. Introduction

Since the demonstration of anomalous heat in 1989 by M. Fleischmann, and S. Pons, anomalous heat generation in palladium-based materials has been studied in various active research groups [1]. Different types of electrodes

*Present address: Department of Engineering and Technology, Southern Utah University, Cedar City, UT, USA.
† Corresponding author. E-mail: gangopadhyays@missouri.edu.

© 2017 ISCMNS. All rights reserved. ISSN 2227-3123

have been tried to reproduce the excess heat. A few exemplary electrodes are Palladium (Pd)-Single Walled Carbon Nanotubes (SWCNT), Pd nanoparticles, and other metals like hydrogen loaded nickel [2]. The origin of the anomalous heat is arguably uncertain because numerous attempts to explain anomalous heat by deuteron-deuteron nuclear fusion in the Pd lattice while the search for radiation was unsuccessful. Although the origin of the excess heat is unknown, Pd electrodes are in the core of the research for excess heat. The demonstration of the excess heat utilizes calorimeters that are well characterized with known errors and limitations as well as beneficial strengths [3]. However, these measurements are not able to provide detailed information about a localized heat rather than a heat from a large area.

Temperature ($T$) is one of the most fundamental parameters in science and daily life. Temperature sensors are commonly used in various fields such as metrology, biology, engineering, climate research, and nanotechnology to name a few in addition to daily life devices like heating and cooling systems, food storage, and production of plants. The important of the temperature sensors are reflected to the market share of about 75% in the world sensor markets [4]. There are different types of temperatures sensors including liquid-filled glass thermometers based on the thermal expansion of materials [5], thermocouples based on the Seebeck effect [6], and optical sensors [7]. Optical temperature sensors have a unique advantage of large-scale temperature measurement with high resolution based on the imaging technique [8]. Among various optical temperature sensors including infrared (IR) thermometers, fluorescence temperature sensor is of our interest because it has a potential of using each fluorophore as a temperature probe.

Fluorescence temperature sensing mechanism have various attractive benefits including the potential of the high resolution temperature monitoring when a single fluorophore acts as a temperature probe. Temperature sensor with fluorescence is almost exclusively based on the fluorophore which can be fluorescent dye molecules, fluorescence nanoparticles or aggregates of fluorophores. The temperature probes can be incorporated in the samples of the study or deposited on its surface in the form of a polymer containing the probes. It has been reported that the measurement can be performed for a single point analysis and the large area by imaging technique [9].

The micro-/nano-scale studies of temperature variation require the use of micro-/nano-size temperature sensors. One of the highlighted studies is nanoparticles such as quantum dots and fluorescence dye molecules [8,9] in combination with spectroscopy or microscopy. To achieve sub-micrometer scale resolution in temperature sensing, the measurement systems need to be capable of measuring such resolution. Recent development on super-resolution microscopy has adequate functionalities of this purpose [10]. However, the super-resolution microscopy is expensive and require special training which leaves a question of practical application in low cost and ease of use. Temperature sensitive fluorescent dyes such as rhodamine 6G (R6G) have drawn great interests due to their real-time temperature measurement capability, large area measurement with high resolution, and ease of incorporation in polymer matrices. The plasmonic grating can significantly enhance the fluorescence intensity by coupling both excitation and emission of fluorophores when satisfying the wave vector matching condition [11,12]. Moreover, it can serve as super-resolution imaging platform to visualize nanostructures [11] to image localized heat generation on the Pd electrodes in sub-micrometer scale.

In this article, we demonstrated a temperature characterization method incorporating R6G/polymer coating matrix on top of the Pd electrodes using a simple epi-fluorescence microscope. We experimentally developed a method for dynamic thermal mapping with this temperature sensitive dyes. First, we prepared Pd electrodes with a temperature sensitive R6G mixed with a polymer matrix, poly methyl silsesquioxane (PMSSQ) on Pd electrodes. Then, we established temperature-fluorescence calibration curve for R6G/PMSSQ systems via fluorescence characteristics by a microscope equipped with a heat stage.

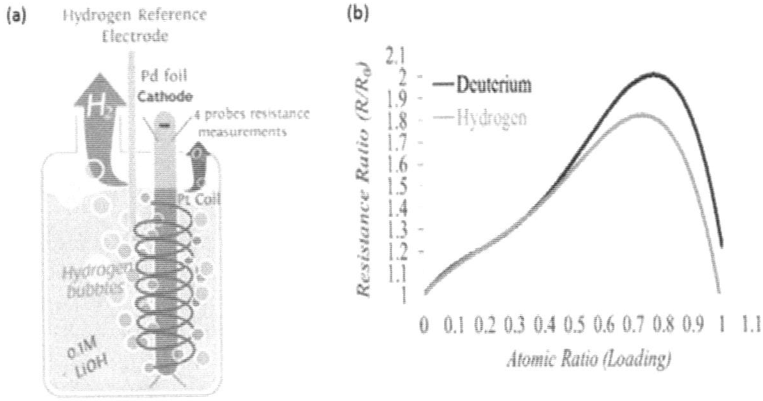


**Figure 1.** (a) Standard three electrode electrochemical cell; (b) Resistance ratio $(R/R_0)$ vs. loading ratio for the Pd/H and Pd/D systems at 25°C [13].

## 2. Experiment

### 2.1. Chemicals and materials

Laser grade R6G was obtained from Exciton (Dayton, OH) and PMSSQ (14% -OH) was purchased from Techneglas (Perrysburg, OH). 200 proof ethanol was purchased from Sigma Aldrich and used as-received. Other analytic grade chemicals such as nitric acid, sulfuric acid, lithium hydroxide were purchased from Sigma Aldrich and used as-received. Deionized (DI) water was used for cleaning and solution preparation.

### 2.2. Preparation of palladium electrode

Palladium foils of 50 $\mu$m thickness were prepared using cold rolling of a 1 mm thick palladium (Holland Moran) followed by thermal annealing at 850°C for 60 min in a protocol to ensure all samples have the same initial conditions. A platinum wire and a standard hydrogen electrode (RHE) were used as a counter and reference electrode, respectively. All potentials are recorded with respect to RHE.

A standard three-electrode electrochemical cell was used to loading hydrogen and deuterium in Pd foils, Fig. 1a. The H(D) loading was determined using in-situ four probe resistance measurements. Figure 1b shows the relation between resistance ratio $(R/R_0)$ versus hydrogen and deuterium atomic ratio in palladium (H/Pd and D/Pd) [13]. The resistance of palladium is expressed as a ratio $R/R_0$, where $R$ is the resistance at a particular loading and $R_0$ is the resistance of the unloaded Pd. The working electrode was a Pd foil of 2.8 cm$^2$ attached to five platinum wires, among which four wires were used for the resistance measurement and one wire for current supply. All experiments were performed at room temperature in 0.1 M LiOH or LiOD solutions. 5 mM $Hg_2SO_4$ was introduced in increments as an impurity into the 0.1 M LiOH (LiOD) solutions in a standard cell to give a concentration between 40 and 400 ppb to enhance H(D) loading in Pd foils and prevent H(D) deloading after the current was tuned off [14].

### 2.3. Temperature sensitive dye/polymer coating

R6G/PMSSQ solution was prepared by dissolving PMSSQ and R6G in pure ethanol with a final concentration of 1 wt.% PMSSQ and 10 $\mu$M R6G (excitation: 530 nm, emission: 550 nm) followed by 10 min sonication. Loaded Pd electrodes described previously and unloaded Pd electrodes were spin-coated at 3000 RPM for 30 s with R6G/PMSSQ solution. This method was expected to produce a conformal, 30 nm-thick layer of fluorophore incorporated PMSSQ on top of the Pd electrodes. The addition of this coating was performed immediately after the loaded samples were out of the loading solution to minimize deloading of H(D) from Pd electrodes. The coated Pd electrodes were imaged with an Olympus BX51WI epifluorescence microscope as described in Section 2.4.

### 2.4. Image acquisition and image analysis

The R6G/PMSSQ coated Pd electrodes were imaged with an Olympus BX51WI epifluorescence microscope equipped an ORCA-flash 2.8 CCD camera, 10/20/40 $\times$ objectives, and illuminated with a fluorescent filter cube. A xenon broadband light source was combined with fluorescent filter cube (an excitation filter (518 nm/20 nm) and a long pass emission filter (541 nm)). The electrode was placed on a pre-cleaned glass slide before the measurement. Same spots of the immobilized electrode were imaged in various times up to several days in different intervals. For analysis, Image J software was utilized. An image taken at 0 min (immediately after the sample was placed under the microscope) was chosen for base fluorescence intensity. Images taken at various time were compared in fluorescence intensity at the same spots.

## 3. Results and Discussion

Since the first claim from Fleischmann and Pons [1] for an anomalous production of heat obtained during an electrochemical experiment, a great number of experiments has been devised and performed, in order to confirm or disprove the effect. The initial disappointment at lack of experimental reproducibility has by now been compensated to a fair degree by realizing special conditions that are usually difficult to achieve [3]. For example, it is now known that a high D/Pd loading ratio is one such difficult necessary condition and that likely there are others. Thus, we carefully prepared each electrode so that it achieved at least D/Pd $>92\%$ because it was reported that no excess heat was observed from the electrodes that had a D/Pd $<90\%$.

Figure 2 shows the resistance ratio $(R/R_0)$ of Pd as a function of time in electrolysis in a pure 0.1 M LiOD solution and then the effect of additions of $Hg_2SO_4$ salt into the electrolyte at different time intervals after the ß phase was formed. The results show that an addition of few ppb level of mercury salt at low current densities results in a decrease of the $R/R_0$ (increase in atomic ratio D/Pd) in a short time. Also, the results show that it is possible to maintain a high loading for more than few weeks by keeping adding mercury salt during electrolysis.

Deuterium loading ratio in a Pd foil was also measured ex-situ using a Rigaku X-ray diffractometer. After a Pd foil was loaded to $R/R_0$ =1.68 (D/Pd~92–95%) in a solution in the presence of $Hg_2SO_4$, the sample was immediately taken out from an electrochemical cell for the X-ray diffraction (XRD) measurement. Lattice parameter calculated by XRD is about $\sim$4.08 AU which corresponds to the atomic ratio of D/Pd $\sim$100% within the depth of a few microns (X-ray penetration depth) in the Pd foil. The XRD measurement showed no significant deloading occurred after the sample was exposed in air for more than 1 h.

The temperature mapping for loaded Pd was conducted without laser/RF excitation to simplify the experimental procedure. Therefore, the temperature change resulted from a slow monoatomic deloading deuterium or hydrogen from the Pd lattice and recombination into molecules of $D_2/H_2$. This exothermic process causes thermal quenching of fluorescence. There are reports on fluorescence thermal quenching processes that are utilized as temperature probes. There are multiple pathways for fluorescence dyes to lose the fluorescence such as oxidation, degradation, quenching, and photobleaching. In the presented experimental condition, oxidation and degradation are not significant comparing to photobleaching which is a major concern for fluorescence dyes [15]. Due to the photobleaching, it is critical to consider the decay of fluorescence over the measurement time as the thermal quenching may be screened. To prevent

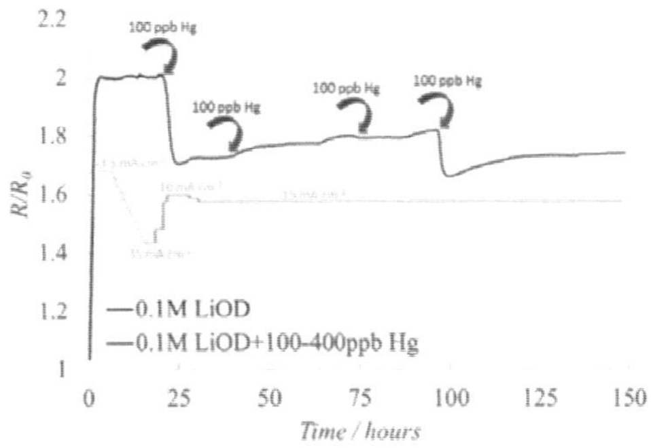


**Figure 2.** Deuterium loading $(R/R_0)$ into Pd in 0.1 M LiOD electrolyte. 100–400 ppb $Hg_2SO_4$ was added to the electrolyte after a-phase was fully transformed to ß-phase and a steady state loading ratio was reached.

the misleading effect of photobleaching, R6G/PMMSQ on unloaded Pd electrodes were examined Unloaded Pd electrodes did not have any exothermic process and therefore, fluorescence decay observed was due to the photobleaching by multiple exposure of the R6G/PMSSQ coating on Pd at the same measurement spots. This calibrated photobleaching data were utilized to compensate the measurement of the loaded Pd electrodes. Once the exothermic process is finished, either the fluorescence recovers or the fluorescence dyes decompose (>250°C) remaining in a low fluorescence. Experimental results undoubtedly showed localized temperature changes after the images were subtracted from the reference fluorescence image (measured at 0 min). Figure 3(a) is obtained by subtracting a fluorescence image at 48 h from a fluorescence image at 0 min. If there is heat generated at specific regions, quenching suppresses fluorescence. By subtracting it from unquenched image, we obtain bright spots in Fig. 3(a) as heat generated regions and dark spots as no heat generation. Figure 3(a) presented three different regions (red circles) with corresponding temperature changes (Fig. 3(b)). The temperature increased over time while region three had fluorescence recovery (from 19 to 48 h) resulting in temperature reduction after subtracting the effect of calibrated photobleaching.

The temperature changes were consistently observed from other loaded Pd electrodes as summarized in Table 1. Although there was no correlation of the behavior between grain boundary and grain in terms of temperature variation,

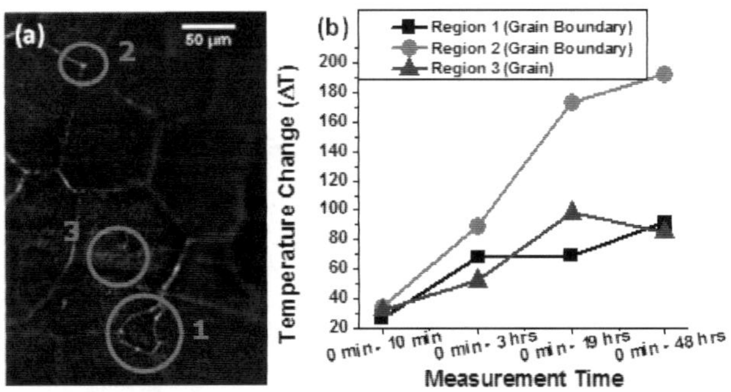


**Figure 3.** (a) Fluorescence image of a loaded Pd coated with a 30nm R6G/PMSSQ. The image was captured in an epi-fluorescence microscope (10×) with an excitation filter (518 nm/20 nm)/long pass emission filter (541 nm). (b) Temperature changes in grain boundaries (regions 1 and 2) and on grain (region 3).

**Table 1.** Fluorescence quenching for loaded Pd electrode.

| Subtracted mages | Intensity difference (thermal quenching) | |
|---|---|---|
| | Grain boundary (AU) | Grain (AU) |
| 0 min–10 min | 433 | 839 |
| 0 min–60 min | 1428 | 960 |
| 0 min–2 h | 1715 | 1107 |
| 0 min–3 h | 1925 | 1531 |
| 0 min–22 h | 2391 | 1848 |

it was evident that most of area showed temperature increase (i.e. fluorescence decrease). The subtraction of fluorescence intensity (e.g. 0 min image – 10 min image) was 433 and 839 AU on grain boundary and grain, respectively. After 22 h, the subtraction became 2391 and 1848 AU on grain boundary and grain, respectively. This means that the electrode generated heat increasingly over time. In 22 h, the photobleaching was estimated to have 600 AU. After considering a photobleaching calibration curve, R6G thermal quenching fluorescence intensity was obtained as below [8]:

$$I = [-0.41\%(T - T_0) + 1]I_0, \quad T_0 = 27°\text{C}. \tag{1}$$

From Eq. (1), the temperature changes ($\Delta T$) were calculated for the grain (1848 AU at 22 h) to be 148°C.

In the current study, up to 40× objective was utilized because higher objectives are available for water or oil immersion. The 40× objective allowed us to observe the Pd electrode in air without damaging the samples with other media. Based on the optical arrangement and the objective, the image had a resolution of 300 nm ×300 nm pixel as shown in Fig. 4. The pixel was a potentially a temperature probe which included multiple fluorophores. Considering the recent advancement in *super-resolution* microscopy, it is expected to have much higher resolution (∼20 nm) [8].

## 4. Conclusion

We have demonstrated a new method as a temperature probe to investigate an anomalous heat from Pd electrode. Current experimental setups utilizing calorimetry are well calibrated for measurement of bulk temperature changes in the experimental cell. The fluorescence based temperature probe is capable of identifying and mapping a very local temperature changes close to 300 nm with a potential for 20 nm combining with the super-resolution analysis. We envisioned that this new method provided a novel diagnostic tool for localized heat which was not detectable by

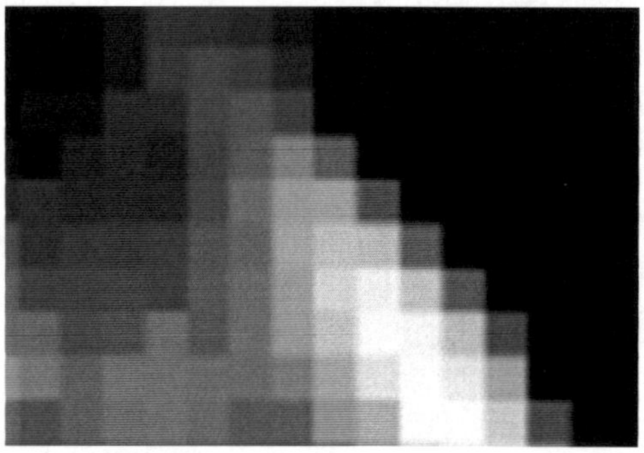

**Figure 4.** Zoom-in of the original image showed pixels. Each pixel was 300 nm ×300 nm which depended on magnification of objective.

calorimetry. Further study includes that excitation of the loaded Pd will be performed with various sources such as lasers and RF to investigate excess heat.

## Acknowledgements

The study is supported by the Sidney Kimmel Foundation, USA.

## References

[1] M. Fleischmann, S. Pons and M. Hawkins, *J. Electroanal. Chem.* 261 (1989) 301 and errata in Vol. 263.
[2] O. Azizi, A. El-Boher, J. H. He, G. K. Hubler, D. Pease, W. Isaacson, V. Violante and S. Gangopadhyay, *Current Sci.* **108** (4) (2015) 565–573.
[3] A. Kitamura, T. Nohmi, Y. Sasaki, A. Takahashi, R. Seto and Y. Fujita, *Phys. Lett.* A **373** (2009) 3109–3112.
[4] P.R.N. Childs, J.R. Greenwood and C.A. Long, *Rev. Sci. Instrum.* **71** (2000) 2959–2978.
[5] L. Michalski, K. Eckersdorf, J. Kucharski and J. McGhee, *Temperature Measurement*, Wiley, Chichester, UK, 2001.
[6] S. Maekawa, *Physics of Transition Metal Oxides*, Springer, 2004.
[7] G. Beheim, Integrated optics, *Microstructures and Sensors* (1995) 285–313.
[8] B. Chen, H. Zheng, J. Yoon, S. Bok, C. Mathai, K. Gangopadhyay, S. Gangopadhyay M.R. Maschmann, Sensors, *IEEE*, 2016.
[9] X. Wang, O.S. Wolfbeis and R.J. Meier, *Chem. Soc. Rev.* **42** (2013) 7834.
[10] A. Wood, B. Chen, C. Mathai, S. Bok, S. Grant, K. Gangopadhyay, P.V. Cornish and S. Gangopadhyay, *Microscopy Today* 25 (1) (2017) 42–47.
[11] B. Chen, A. Wood, A. Pathak, J. Mathai, S. Bok, H. Zheng, S. Hamm, S. Basuray, S. Grant, K. Gangopadhyay, P. V. Cornish and S. Gangopadhyay, *Nanoscale* **8** (2016) 12189–12201.
[12] A. Wood, B. Chen, S. Pathan, S. Bok, C.J. Mathai, K. Gangopadhyay, S.A. Grant and S. Gangopadhyay, *RSC Advances* **5**(96) (2015) 78534–78544.
[13] W. Zhang, Z. Zhang and Z.L. Zhang, *J. Electroanal. Chem.* **528** (2002) 1.
[14] F.L. Tanzella, S. Crouch-Baker, A. McKeown and M.C.H. McKubre, *ICCF6*, Japan, 1996.
[15] S. Bok, V. Korampally, L. Polo-Parada, V. Mamidi, G.A. Baker, K. Gangopadhyay, W.R. Folk, P.K. Dasgupta and S. Gangopadhyay, *Nanotechnology* **23**(17) (2012) 175601.

J. Condensed Matter Nucl. Sci. 24 (2017) 32–41

Research Article

# The *Zitterbewegung* Interpretation of Quantum Mechanics as Theoretical Framework for Ultra-dense Deuterium and Low Energy Nuclear Reactions

Francesco Celani*

*Istituto Nazionale di Fisica Nucleare (INFN-LNF), Via E. Fermi 40, 00044 Frascati, Roma, Italy*

Antonino Oscar Di Tommaso†

*Università degli Studi di Palermo – Department of Energy, Information Engineering and Mathematical Models (DEIM), viale delle Scienze, 90128 Palermo, Italy*

Giorgio Vassallo‡

*Università degli Studi di Palermo – Department of Industrial and Digital Innovation (DIID), viale delle Scienze, 90128 Palermo, Italy*

**Abstract**

This paper introduces a *Zitterbewegung* model of the electron by applying the principle of Occam's razor to the Maxwell's equations and by introducing a scalar component in the electromagnetic field. The aim is to explain, by using simple and intuitive concepts, the origin of the electric charge and the electromagnetic nature of mass and inertia. The *Zitterbewegung* model of the electron is also proposed as the best suited theoretical framework to study the structure of Ultra-Dense Deuterium (UDD), the origin of anomalous heat in metal–hydrogen systems and the possibility of existence of "super-chemical" aggregates at Compton scale.


*Keywords:* Clifford algebra, Compton scale aggregates, Dirac equation, Electric charge, Electron structure, Elementary particles, LENR, Lorenz gauge, Maxwell's equations, Occam's razor, Space–time algebra (STA), Ultra-dense deuterium, Vector potential, Weyl equation, *Zitterbewegung*

*Also at: International Society for Condensed Matter Nuclear Science (ISCMNS)-UK. E-mail: francesco.celani@lnf.infn.it, Mobile: +39 3289859059.

†E-mail: antoninooscar.ditommaso@unipa.it.

‡Also at: International Society for Condensed Matter Nuclear Science (ISCMNS)-UK. E-mail: giorgio.vassallo@unipa.it.

© 2017 ISCMNS. All rights reserved. ISSN 2227-3123

## 1. Introduction

Prof. David Hestenes, emeritus of the Arizona State University, is the most notable and active physicist that advocates the use of geometric algebra in physics [1]. Space–time geometric algebra is a four dimensions real Clifford algebra with Minkowski signature "$+---$" ($Cl_{1,3}(\mathbb{R})$) or with signature "$-+++$" ($Cl_{3,1}(\mathbb{R})$).

Hestenes rewrote Dirac's equation for the electron using space–time algebra $Cl_{1,3}(\mathbb{R})$, eliminating the unnecessary complexities and redundancies arising from the traditional use of matrices and complex algebra. As a matter of fact, the Dirac gamma matrices $\gamma_\mu$ and the associated algebra can be seen as an isomorphism of the four basis vector of space–time geometric algebra. This simple isomorphism allows a full encoding of the geometric properties of the Dirac algebra, and a reformulation of the Dirac equation that does require neither complex numbers nor matrix algebra.

In this context the wave function $\psi$ is characterized by eight real values of the even grade multivectors of space–time algebra (STA). The even grade multivectors of STA can encode ordinary rotations as well as Lorentz transformations in the six planes of space–time. Hestenes associates the rotations encoded by the wave function with an intrinsic characteristic of the electron: the *Zitterbewegung*, a German word for "trembling motion", indicating a rapid rotation that is considered at the origin of both the electron spin and its magnetic moment. Hestenes considers the complex phase of the wave function, solution of the traditional Dirac equation, as the phase of the *Zitterbewegung* (ZBW) rotation, showing "*The inseparable connection between quantum mechanical phase and spin*" [2,3]. Consequently, he rejected the "*conventional wisdom that phase is an essential feature of quantum mechanics, while spin is a mere detail that can often be ignored*" [2].

In this paper we will try to show that the application of Occam's razor principle to the Maxwell's equations advise a particular ZBW interpretation of quantum mechanics proposed by Hestenes [4]. A very simple ansatz based on STA and on a reinterpretation of the "Lorenz gauge" is proposed to explain the origin of the electric charge and the electromagnetic nature of mass and inertia. A ZBW model of the electron is introduced in a preliminary attempt to explain the structure of Ultra-dense Deuterium (UDD) [5,6] and the origin of Anomalous Heat in metal–hydrogen systems. According to this framework, the electron structure consists of a massless charge distribution that rotates at the speed of light along a circumference with a length equal to electron Compton wavelength ($\simeq 2.42631$ pm). The electron mass-energy, expressed in natural units (NU), is equal to the angular frequency of the ZBW rotation and to the inverse of the orbit radius (i.e. 511 keV). The inter-nuclear distance in UDD of $\simeq$ 2.3 pm, found by Holmlid [5], seems to be compatible with proton-electron structures at Compton scale where the ZBW phases of neighbor electrons are synchronized. The existence of structures at an intermediate scale between atomic dimensions/energies and nuclear ones have been proposed by many authors [7–9]. These hypothetical composites are electrically neutral or negatively charged objects that are not repelled by the Coulomb barrier. These structures may generate unusual nuclear reactions and transmutations, considering the different sizes, time-scales and energies of these composites with respect to the dimensions of the particles (like neutrons) normally used in nuclear experiments.

## 2. Maxwell's Equations with Occam's Razor

The STA can be represented by the four dimensions real Clifford algebra $Cl_{3,1}$, with signature "$-+++$", where the standard Euclidean metric is used for spatial coordinates. In this algebra the four unitary basis vectors $\gamma_x, \gamma_y, \gamma_z$ and $\gamma_t$ are isomorphic to Majorana matrices and the following rules apply:

$$\gamma_i\gamma_j = -\gamma_j\gamma_i \quad \text{with} \quad i \neq j \;\; i,j \in \{x,y,z,t\}, \tag{1}$$

$$\gamma_x^2 = \gamma_y^2 = \gamma_z^2 = -\gamma_t^2 = 1. \tag{2}$$

Using this algebra, Maxwell's equations can be rewritten in a compact form starting from the derivatives of the electromagnetic four-potential [1,10]

$$\boldsymbol{A}_\square = \gamma_x A_x + \gamma_y A_y + \gamma_z A_z + \gamma_t A_t. \tag{3}$$

The four vector potential components $A_x$, $A_y$, $A_z$ and $A_t$ are functions of the space–time coordinates $x, y, z, t$ and have dimension in SI units equal to (V s m$^{-1}$), and have the dimension of an energy expressed in eV (electron volts) in natural units (NU), where $\hbar = c = 1$.

$\boldsymbol{A}_\square$ can be seen as a vector field in the space–time continuum and as the unique source of all concepts-entities in Maxwell's equations [11]. We remember that the vector potential should not be viewed only as a mathematical tool but as a real physical entity, as suggested by the Aharonov–Bohm effect [12], a quantum mechanical phenomenon in which a charged particle is affected by the vector potential in regions in which the electromagnetic fields are null.

Using the following definition of the operator $\boldsymbol{\partial}$ in STA

$$\boldsymbol{\partial} = \gamma_x \frac{\partial}{\partial x} + \gamma_y \frac{\partial}{\partial y} + \gamma_z \frac{\partial}{\partial z} + \gamma_t \frac{1}{c}\frac{\partial}{\partial t} = \nabla + \gamma_t \frac{1}{c}\frac{\partial}{\partial t}, \tag{4}$$

the following expression can be written:

$$\boldsymbol{\partial A}_\square = \boldsymbol{\partial} \cdot \boldsymbol{A}_\square + \boldsymbol{\partial} \wedge \boldsymbol{A}_\square = S + \boldsymbol{F} = \boldsymbol{G}, \tag{5}$$

where $S = \boldsymbol{\partial} \cdot \boldsymbol{A}_\square$ and $\boldsymbol{F} = \boldsymbol{\partial} \wedge \boldsymbol{A}_\square$. This means that the electromagnetic field $\boldsymbol{G}$ is a composition of a scalar (grade zero element of $Cl_{3,1}$) field $S$ and a six components bivector (grade two elements of $Cl_{3,1}$) field $\boldsymbol{F}$. Calling $I$ the pseudoscalar unity of $Cl_{3,1}$

$$I = \gamma_x \gamma_y \gamma_z \gamma_t \tag{6}$$

and $c = 1/\sqrt{\epsilon_0 u_0}$ the speed of light, a simple relation between the electric field $\boldsymbol{E}$, the flux density field $\boldsymbol{B}$ and the bivector field $\boldsymbol{F}$ is found:

$$\boldsymbol{F} = \frac{1}{c}\boldsymbol{E}\gamma_t + I\boldsymbol{B}\gamma_t = \frac{1}{c}\left(\boldsymbol{E} + Ic\boldsymbol{B}\right)\gamma_t. \tag{7}$$

Table 1 represents the relation between the fundamental electromagnetic entities and the space–time components of the vector potential $\boldsymbol{A}_\square$.

Using the so called "Lorenz gauge" ($S = \boldsymbol{\partial} \cdot \boldsymbol{A}_\square = 0$) or using the "Coulomb gauge" ($S_i = 0,\ i = 1, 2, 3, 4$) the scalar component $S$ of electromagnetic field $\boldsymbol{G}$ is ignored and its physical meaning is lost. The Occam's razor principle states that "plurality should not be posited without necessity" and according to this rule the number of independent concepts or entities in mathematical models should be reduced as much as possible. For this reason it's important to have a direct derivation of the concepts of "charge density" and "current density" from the electromagnetic

**Table 1.** Relation between electromagnetic entities and the vector potential (electromagnetic tensor).

| $\boldsymbol{\partial A}_\square$ | $\gamma_x A_x$ | $\gamma_y A_y$ | $\gamma_z A_z$ | $\gamma_t A_t$ |
|---|---|---|---|---|
| $\gamma_x \cdot \frac{\partial}{\partial x}$ | $S_1$ | $B_{z1}$ | $-B_{y1}$ | $\frac{1}{c}E_{x1}$ |
| $\gamma_y \frac{\partial}{\partial y}$ | $B_{z2}$ | $S_2$ | $B_{x1}$ | $\frac{1}{c}E_{y1}$ |
| $\gamma_z \frac{\partial}{\partial z}$ | $-B_{y2}$ | $B_{x2}$ | $S_3$ | $\frac{1}{c}E_{z1}$ |
| $\gamma_t \frac{1}{c}\frac{\partial}{\partial t}$ | $\frac{1}{c}E_{x2}$ | $\frac{1}{c}E_{y2}$ | $\frac{1}{c}E_{z2}$ | $S_4$ |

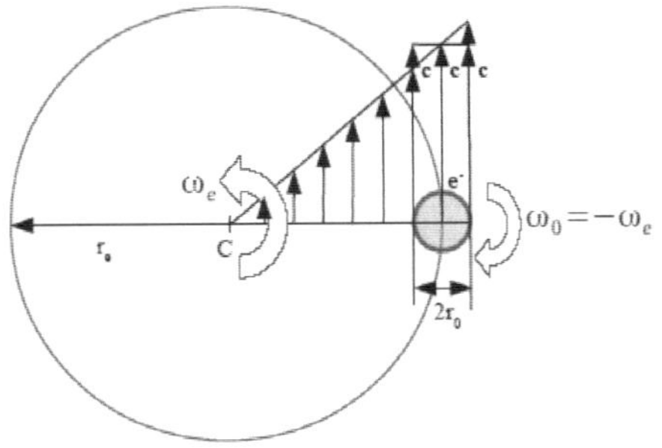


**Figure 1.** *Zitterbewegung* model and speed diagrams of the free electron. All points of the sphere have an absolute speed equal to *c*.

four-potential $\boldsymbol{A}_\square$ reducing, in this manner, the primitive essential concepts used in Maxwell's equations to a unique entity. This aim can be accomplished if we recognize the four space–time derivatives of the scalar field $S$ as the four components of the four current density $\boldsymbol{J}_{\square e}$:

$$\frac{1}{\mu_0}\boldsymbol{\partial} S = \frac{1}{\mu_0}\left(\gamma_x \frac{\partial S}{\partial x} + \gamma_y \frac{\partial S}{\partial y} + \gamma_z \frac{\partial S}{\partial z} + \gamma_t \frac{1}{c}\frac{\partial S}{\partial t}\right) = \boldsymbol{J}_{\square e}, \tag{8}$$

where $\boldsymbol{J}_{\square e} = \gamma_x J_{ex} + \gamma_y J_{ey} + \gamma_z J_{ez} - \gamma_t c\rho = \boldsymbol{J}_\triangle - \gamma_t c\rho = \rho\,(\boldsymbol{v} - \gamma_t c)$ is the four current density vector and $\boldsymbol{v}_\square = \gamma_x v_x + \gamma_y v_y + \gamma_z v_z - \gamma_t c = \boldsymbol{v} - \gamma_t c$ is the four velocity vector. The charge density is related to the time derivative of scalar field $S$

$$\frac{1}{c}\frac{\partial S}{\partial t} = \mu_0 J_{et} = -\mu_0 c \frac{\partial q}{\partial x \partial y \partial z} = -\mu_0 c\rho, \tag{9}$$

Using the seven component electromagnetic field $\boldsymbol{G}$, Maxwell's equations can be rewritten in the following compact form

$$\boldsymbol{\partial G} = \boldsymbol{\partial}^2 \boldsymbol{A}_\square = 0. \tag{10}$$

$\boldsymbol{G}$ is a composition of a scalar and a bivector field and can be viewed as a spinor field in STA.

## 3. Electron *Zitterbewegung* Model

The concept of charge that emerges from the "Lorenz gauge free" Maxwell's equations [11], has a non trivial implication: the analysis of (9) shows that the time derivative of a field $S$ which propagates at the speed of light, must necessarily represent charges that are also moving at the speed of light. Moreover, equation (10) is Lorentz invariant only for charges that move at the speed of light. This observation naturally suggests a pure electromagnetic model of elementary particles based on the ZBW interpretation of quantum mechanics [4,13]. The free electron can be viewed as a charge distribution that rotates at speed of light along a circumference with a length equal to electron Compton wavelength [14] as shown in Fig. 1. The charge may be seen as distributed on a spherical surface with a radius equal to electron classical radius $r_0 \simeq 2.817940 \times 10^{-15}$ m. The electron mass in natural units (NU) is equal to the angular frequency: $\omega_e \simeq 5.11 \times 10^3$ eV (NU) or $\omega_e \simeq 7.763440 \times 10^{20}$ rad/s (SI). This value is also equal to the inverse of

**Table 2.** Parameters of the *Zitterbewegung* model.

| Item | Symbol | Value (SI) | Unit (SI) |
|---|---|---|---|
| Charge | $e$ | $1.602176565 \times 10^{-19}$ | $C$ =A s |
| *Zitterbewegung* orbit radius | $r_e = \lambda_c/2\pi$ | $3.861593 \times 10^{-13}$ | m |
| Intrinsic angular momentum | $\Omega = \hbar = h/2\pi$ | $1.054571726 \times 10^{-34}$ | J s |
| Spin[1] | $\hbar/2$ | $0.527285863 \times 10^{-34}$ | J s |
| Angular speed | $\omega_e$ | $7.763440 \times 10^{20}$ | rad s$^{-1}$ |
| Mass | $m_e$ | $9.109384 \times 10^{-31}$ | kg |
| Current | $I_e$ | 19.796331 | A |
| Magnetic moment (Bohr magneton) | $\mu_B$ | $9.274010 \times 10^{-24}$ | A m$^2$ |
| Vector potential | $A$ | $1.704509 \times 10^{-3}$ | V s m$^{-1}$ |
| Magnetic flux density | $B_e$ | $4.414004 \times 10^{9}$ | V s m$^{-2}$ |
| Magnetic flux | $\phi_e = h/e$ | $4.135667 \times 10^{-15}$ | V s |
| Magnetic energy | $W_m$ | $4.093553 \times 10^{-14}$ | J |
| Electrostatic energy | $W_e$ | $4.093553 \times 10^{-14}$ | J |
| Electron energy at rest | $W_{tot} = m_e c^2$ | $8.187106 \times 10^{-14}$ | J |
| Charge radius | $r_0$ | $2.817940 \times 10^{-15}$ | $m$ |
| Inverse of the FSC | $\alpha^{-1} = r_e/r_0$ | 137.035999 | 1 |
| von Klitzing constant | $R_K = h/e^2 = \mu_0 c/2\alpha$ | 25812.807 | $\Omega$ |

[1]Component of the angular momentum due to Larmor precession along an external magnetic field $B_E$.

ZBW orbit radius $r_e = \lambda_c/2\pi \simeq 1.956\,950 \times 10^{-6}$ eV$^{-1}$ (NU) or $r_e \simeq 3.861593 \times 10^{-13}$ m (SI) and is equal to the reduced Compton wavelength:

$$\left(m_e = \omega_e = \frac{1}{r_e} \approx 0.511 \times 10^6 \text{ eV}\right)_{\text{NU}},$$

$$m_e = \frac{\hbar\omega_e}{c^2} = \frac{\hbar}{c r_e} \approx 9.109384 \times 10^{-31} \text{ kg}.$$

The angular momentum of the free electron according to the proposed model is

$$\Omega = eAr_e = \hbar,$$

where $e$ is the elementary charge and $A \simeq 1.704509 \times 10^{-3}$ V s m$^{-1}$ is the module of the vector potential $\mathbf{A}$ seen by the rotating charge. The value of the vector potential is determined by the current $I_e = e\omega_e/2\pi \simeq 19.796331$ A generated by the rotating charge. The angular momentum $\hbar$ of the free electron may explain the spin value $\hbar/2$ if we consider the electron interaction with an external magnetic flux density field $\boldsymbol{B}_E$, as in the Stern–Gerlach experiment.

We can interpret the electron spin value $\pm\hbar/2$ as the component of the angular momentum vector $\boldsymbol{\Omega} = \hbar$ aligned with the external magnetic flux density field $\boldsymbol{B}_E$. In this case the angle between the $\boldsymbol{B}_E$ vector and the angular momentum has only two possible values, namely $\pi/3$ and $2\pi/3$ while the electron is subjected to Larmor precession.

All parameters that can be deduced from the electron ZBW model are resumed in Table 2, where the first three rows are referred to the model's input parameters. Thanks to the ZBW model it's possible to show a simple and intuitive explanation of the relativistic mass concept. The position in space–time of a charge that moves at the speed of light must be a light-like vector in every reference frame:

$$\lambda_c^2 - c^2T^2 = 0, \tag{11}$$

where $T = \lambda_c/c$ is the ZBW period and $\lambda_c$ the ZBW orbit length for an electron at rest. For an electron moving at constant speed $v_z$ along the $z$ axis orthogonal to the rotation plane, calling $\lambda_v$ and $T_v$ the new orbit length and the new ZBW period respectively, we can write

$$\lambda_v^2 + v_z^2 T^2 - c^2 T^2 = 0, \tag{12}$$

$$v_z^2 T^2 - c^2 T^2 = -c^2 T_v^2, \tag{13}$$

$$\lambda_v^2 - c^2 T_v^2 = 0. \tag{14}$$

Moreover, by substituting (11) into (12) we get

$$\lambda_v^2 = c^2 \frac{\lambda_c^2}{c^2} - v_z^2 \frac{\lambda_c^2}{c^2} = \lambda_c^2 \left(1 - \frac{v_z^2}{c^2}\right) \tag{15}$$

or

$$\lambda_v = \lambda_c \sqrt{1 - \frac{v_z^2}{c^2}}. \tag{16}$$

Finally, by considering that the mass is inversely proportional to $\lambda_v$ and $T_v$, it is possible to write the relativistic expression of the mass as

$$m = \frac{m_e}{\sqrt{1 - \frac{v_z^2}{c^2}}}, \tag{17}$$

where $m_e$ is the electron mass at rest. In Fig. 2 the helical space trajectory of the spinning charge of an electron subjected to an acceleration directed along the positive $z$ axis is represented. Due to the acceleration the radius reduces itself according to (16).

This electron model may highlight some important connections and similarities between Maxwell's and Weyl equation.

By using STA the Dirac equation

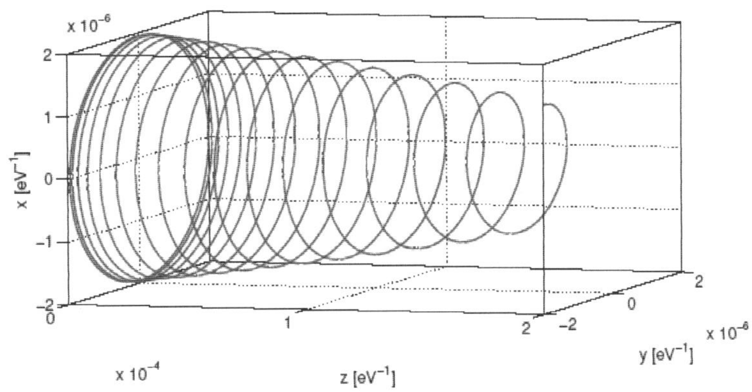


**Figure 2.** *Zitterbewegung* trajectory during an acceleration of the electron in the $z$-direction.

$$i\partial\!\!\!/\psi - m\psi = 0 \tag{18}$$

becomes the Hestenes–Dirac equation [15]

$$\boldsymbol{\partial}\psi - m\psi\gamma_t\gamma_x\gamma_y = 0, \tag{19}$$

where $\boldsymbol{\partial}$ is an operator similar to the one used in Maxwell's equations $\boldsymbol{\partial G} = 0$ (see (4)), but using the space–time Minkowski signature of $Cl_{1,3}$ "$+---$". For a massless particle $m = 0$ and (19) becomes the Weyl equation

$$\boldsymbol{\partial}\psi = 0. \tag{20}$$

Both equations (10) and (20) can be seen, from a mathematical point of view, as an extension in four dimensions of the Cauchy–Riemann conditions for analytic functions of a complex ("two-dimensional") variable [11,16]. The solution is always a *spinor* field. A spinor is a mathematical object that in Clifford algebra is simply a multivector with only even grade components. The movement of a point charge that rotates in the plane $\gamma_x\gamma_y$ and at same time moves up along the $\gamma_z$ axis can be seen as the composition of an ordinary rotation in the plane $\gamma_x\gamma_y$ and a scaled hyperbolic rotation in the plane $\gamma_z\gamma_t$. These two rotations can be encoded with a single spinor of $Cl_{3,1}$.

The momentum $P_\|$ of the rotating charge is equal to the elementary charge times the vector potential generated by the ZBW current:

$$P_\| = eA_\| = \frac{\hbar}{r_\mathrm{e}} = \frac{\hbar\omega_\mathrm{e}}{c} = m_\mathrm{e}c.$$

An electron that moves with velocity $v \ll c$ along an axis orthogonal to the ZBW rotation plane has a momentum $P_\perp$ given by

$$P_\perp = eA_\perp \simeq P_\|\frac{v}{c} = m_\mathrm{e}v,$$

where $P_\|$ is the component of the momentum of rotating charge parallel to the ZBW rotation plane, $P_\perp$ is the orthogonal component and $A_\|$ and $A_\perp$ are the relative components of the vector potential seen by the spinning and moving charge. A variation of speed $a = \mathrm{d}v/\mathrm{d}t$ implies a force

$$f = \frac{\mathrm{d}P_\perp}{\mathrm{d}t} = e\frac{\mathrm{d}A_\perp}{\mathrm{d}t} = m\frac{\mathrm{d}v}{\mathrm{d}t}.$$

## 4. Electromagnetic Composite at Compton Scale

If the electron is a current loop (ring) whose circumference is equal to its Compton wavelength, it is reasonable to assume the possibility of existence of "super chemical" structures of picometric (1 pm = $10^{-12}$ m) dimensions.

A simple ZBW model of the proton consists in a current ring generated by an elementary positive charge that rotates at the speed of light along a circumference with a length equal to the proton Compton wavelength ($\approx 1.32141 \times 10^{-15}$ m). According to this model the proton is much smaller than an electron ($r_\mathrm{e}/r_\mathrm{p} = m_\mathrm{p}/m_\mathrm{e} \approx 1836.153$). A hypothetical very simple structure formed by a proton centered in the middle of the electron orbit would have a potential energy of $-e^2/r_\mathrm{e} \approx -3.728$ keV, corresponding to a photon wavelength of $\lambda_\varphi \approx 3.325 \times 10^{-10}$ m. This structure may be created only in presence of particular catalytic environments. A "resonant cavity" with dimensions comparable to

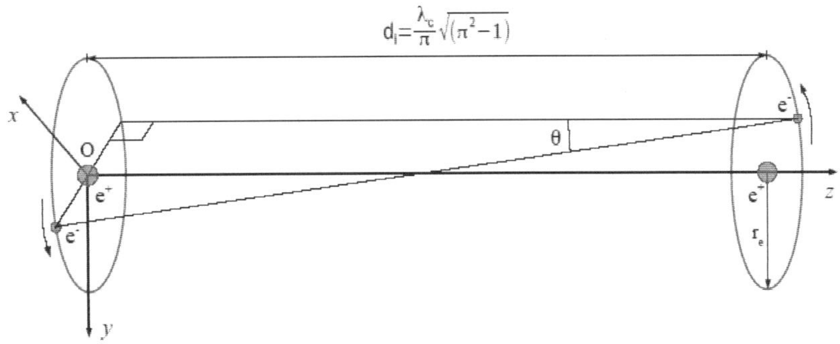


**Figure 3.** UDH protons distance.

$\lambda_\varphi = 332.5$ pm may facilitate the photon emission, acting as an "impedance matcher" with the external environment. Nickel has a lattice constant of 352.2 pm, a value not very far from $\lambda_\varphi$, and each Ni lattice cell may act as a resonant cavity and as an "energy emission catalyzer" in presence of Rydberg State Hydrogen, atomic hydrogen or hydrogen plasma, in systems very far from equilibrium. The generation of atomic hydrogen at relatively low temperatures may be catalyzed by nanostructured materials that catalyze the splitting of molecular hydrogen [17].

The hypothesis of existence of Compton-scale composites (CSC) has been experimental confirmed by prof. Holmlid [5]. The inter-nuclear distance in UDD of $\approx 2.3$ pm, found by Holmlid, seems compatible with deuteron–electron (or proton–electron in UDH) structures where the ZBW phases of adjacent electrons are synchronized. Such distance may be obtained imposing, as a first step, the condition that the space–time distance $d_\square$ between adjacent electrons rotating charges is a light-like vector:

$$d_\square^2 = d_\Delta^2 - c^2\delta t^2 = 0,$$

where $d_\Delta$ is the ordinary Euclidean distance in 3D space. This condition is satisfied if $d_\Delta$ is equal to electron Compton wavelength ($d_\Delta = \lambda_c$), $\delta t = T = \lambda_c/c = 2\pi r_e/c$ is the ZBW period and the phase difference between adjacent electrons is equal to $\pi$. In this case by applying the Pythagorean theorem we can find the internuclear deuteron distance $d_i$ (see Figs. 3 and 4)

$$d_i = \frac{\lambda_c}{\pi}\sqrt{\pi^2 - 1} \simeq 2.3001 \times 10^{-12} \text{ m}.$$

We must remark that the hypothesis of existence of exotic forms of "hydrogen" is not new and has been proposed in different ways by many authors (Mills [7], Dufour [9], Mayer and Reitz [8,18] and many others). An indirect support comes also from the numerous claims of observation of anomalous heat generation in Nickel–hydrogen systems.

**Figure 4.** UDH model.

These anomalies have been reported by many authors (Notoya, Mills, Piantelli, Mizuno, Hagelstein, Godes, Celani and others).

Mayer and Reitz, starting from a ZBW model of the electron, propose a three-body system model at the Compton scale, composed by a proton and two electrons. Prof. Piantelli in patent application WO 2012147045 "Method and apparatus for generating energy by nuclear reactions of hydrogen adsorbed by orbital capture on a nanocrystalline structure of a metal" proposes an orbital capture of "H- ions" by nickel atoms in nano-clusters as a trigger for Low Energy Nuclear Reactions [19]. The orbital capture of the negatively charged structures at pico-metric scale described by Mayer and Reitz may be viewed as an alternative explanation to the capture of the much larger H- ions.

Anomalous heat generation in Ni-H systems has also been reported by prof. Sergio Focardi and Andrea Rossi. Their impressive and revolutionary "energy catalyzer" (or "E-Cat") seems capable of reliable generation of thermal energy at kilowatt levels for many months! All these hypothetical "Compton Scale Composites" are electrical neutral or negatively charged objects that cannot be repelled by the Coulomb barrier and may generate unusual nuclear reactions and transmutations, considering the different sizes, time-scales and energies of this structures with respect to the particles (as neutrons) normally used in nuclear experiments.

## 5. Conclusions

Simplicity is an important and concrete value in scientific research. The application of Occam's razor principle to the Maxwell's equations suggests, as a natural choice, a ZBW interpretation of quantum mechanics. According to this framework, the electron structure consists of a massless charge that rotates at the speed of light along a circumference with a length equal to electron Compton wavelength. Following this interpretation, the electron mass-energy, expressed in natural units, is equal to the angular frequency of the ZBW rotation and to the inverse of orbit radius. The ZBW interpretation of quantum mechanics has been proposed as the best suited theoretical framework to understand the structure of ultra-dense deuterium and the origin of anomalous heat in metal–hydrogen systems and low energy nuclear reactions.

## References

[1] D. Hestenes, *Oersted Medal Lecture 2002: Reforming the Mathematical Language of Physics*, 2002.
[2] D. Hestenes, Mysteries and insights of Dirac theory, *Annales de la Fondation Louis de Broglie* **28** (2003) 3.
[3] D. Hestenes, The Zitterbewegung interpretation of quantum mechanics, *Found. Phys.* **20**(10) (1990) 1213–1232.
[4] D. Hestenes, Quantum mechanics from self-interaction, *Found. Phys.* **15**(1) (1985) 63–87.
[5] S. Badiei, P.U. Andersson and L. Holmlid, High-energy Coulomb explosions in ultra-dense deuterium: time-of-flight-mass spectrometry with variable energy and flight length, *Int. J. Mass Spectrometry* **282**(1–2) (2009) 70–76.
[6] L. Holmlid and S. Olafsson, Spontaneous ejection of high-energy particles from ultra-dense deuterium D(0), *Int. J. Hydrogen Energy* **40**(33) (2015)10559–10567.
[7] R.L. Mills, J.J. Farrell and W.R. Good, *Unification of Space time, the Forces, Matter, and Energy*, Science Press, Ephrata, PA 17522, 1992.
[8] F.J. Mayer and J.R. Reitz, Electromagnetic composites at the Compton scale, *Int. J. Theoret. Phys.* **51**(1) (2012) 322–330.
[9] J. Dufour, An introduction to the pico-chemistry working hypothesis, *J. Condensed Matter Nucl. Sci.* **10** (2013) 40–45.
[10] D. Hestenes, Spacetime physics with geometric algebra, *Amer. J. Phys.* **71** (2003) 691–714.
[11] G. Bettini, *Clifford Algebra, 3- and 4-dimensional Analytic Functions with Applications*, Manuscripts of the Last Century. viXra.org, *Quantum Physics* (2011) 1–63.
[12] Y. Aharonov and D. Bohm, Significance of electromagnetic potentials in the quantum theory, *Phys. Rev.* **115** (1959) 485–491.
[13] D. Hestenes, Zitterbewegung modeling, *Found. Phys.* **23**(3) (1993) 365–387.

[14] F. Santandrea and P. Cirilli, Unificazione elettromagnetica, concezione elettronica dello spazio, dell'energia e della materia, *Atlante di numeri e lettere, Laboratorio di ricerca industriale* (2006) 1–8. http://www.atlantedinumerielettere.it/energia2006/labor.htm.

[15] D. Hestenes, Real spinor fields, *J. Math. Phys.* **8**(4) (1967) 798–808.

[16] D. Hestenes, *Clifford Algebra and the Interpretation of Quantum Mechanics*, Springer, Netherlands, Dordrecht, 1986, pp. 321–346.

[17] F. Celani, E. Purchi, F. Santandrea, S. Fiorilla, A. Nuvoli, M. Nakamura, P. Cirilli, A. Spallone, B. Ortenzi, S. Pella, P. Boccanera and L. Notargiacomo, Observation of macroscopic current and thermal anomalies, at high temperature, by heterostructures on thin and long Constantan wires under $H_2$ gas, *Int. Conf. on Condensed Matter Nucl. Sci., ICCF-19*, Padua, Italy, April 2015, pp. 13–17.

[18] F. J. Mayer and J.R. Reitz, Thermal energy generation in the earth. *Nonlinear Processes Geophys.* **21**(2) (2014) 367–378.

[19] F. Piantelli, Method and apparatus for generating energy by nuclear reactions of hydrogen adsorbed by orbital capture on a nanocrystalline structure of a metal, November 1 2012. WO Patent App. PCT/IB2012/052,100.

J. Condensed Matter Nucl. Sci. 24 (2017) 42–59

Research Article

# Effects of D/Pd Ratio and Cathode Pretreatments on Excess Heat in Closed Pd|$D_2O+D_2SO_4$ Electrolytic Cells*

Jie Gao† and Wu-Shou Zhang‡

*Institute of Chemistry, Chinese Academy of Sciences, P.O. Box 2709, Beijing 100190, China*

Jian-Jun Zhang

*College of Chemistry and Material Science, Hebei Normal University, Shijiazhuang 050024, China*

**Abstract**

The resistance method was used in situ to determine the D/Pd ratio in the cathode of Pd|$D_2O+D_2SO_4$ electrolytic-calorimetric systems. Thirty-three experimental runs with excess heat production indicated that there is no clear correlation between the excess power and the D/Pd ratio. Most excess heat occurs when the ratio is $0.5 < D/Pd < 0.7$. Three different cathode pretreatments were tested. It is found that the excess heat reproducibility rate is 2/11 (18%), 21/32 (66%) and 25/50 (50%), and maximum excess power of $45 \pm 18$ mW, $117 \pm 24$ mW and $128 \pm 19$ mW were achieved for samples etched in aqua regia, heated in $D_2SO_4$ and open electrolyzed in $D_2SO_4$, respectively, based on 93 runs of calorimetry with 25 Pd plates. This indicates that heating and open electrolyzing in $D_2SO_4$ are effective methods of pretreatment for excess heat production in subsequent electrolysis. Surface characterizations with 3D non-contact optical profiling, scanning electron microscopy (SEM) and energy dispersive X-ray spectrometry (EDS) were performed for Pd samples of cold rolled, pretreated and electrolyzed. The results showed that the surface roughness of palladium increases prominently after pretreatment. New element silver was detected on Pd surface after electrolysis.


*Keywords:* Current density, Surface roughness, Temperature, Transmutation

## 1. Introduction

In our previous studies, we observed anomalous heat in Pd|$D_2O$ electrolytic cells by Seebeck Envelope Calorimetry (SEC) [1–5]. Two factors were found to be very important for experimental reproducibility. One is that the temperature increment (i.e. the difference between steady temperature and initial temperature) during electrolysis must be high

*In memory of Prof. John Dash (1933–2016), who's lab at Portland State University has been visited by the author W.-S. Zhang from September 2014 to November 2016.

†Also at: College of Chemistry and Material Science, Hebei Normal University, Shijiazhuang 050024, China.

‡ Corresponding author. E-mail: wszhang@iccas.ac.cn or zhangwushou@hotmail.com; Phone: +86-10-6255 4276; Fax: +86-10-6255 9373.

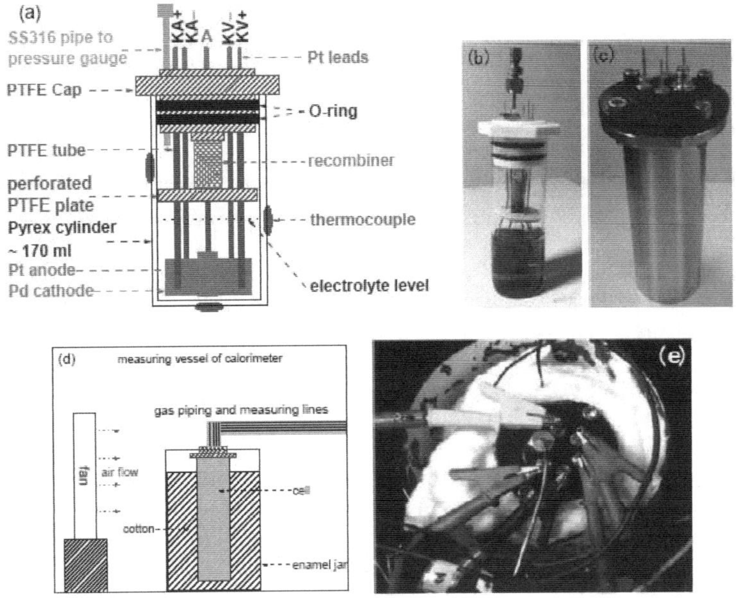


**Figure 1.** (a) Schematic of electrolytic cell (not in scale) used from January 2011 to October 2014. The symbol A represents anode; symbols KA+ (also the cathode), KA−, KV+ and KV− represent two source terminals and two sense terminals, respectively, in a 4-terminal AC resistance meter (HIOKI 3560); (b) photo of electrolytic cell of (a); (c) the cell since November 2014; (d) Schematic of different parts in the calorimeter (not in scale); (e) the cell of (c) with lead wires and thermocouples in calorimeter, the rust of porcelain enamel jar is a result of two explosions that happened before, due to failures of the catalyst.

enough; otherwise no excess heat will be observed [3]. Another is that the electrolysis of a Pd sample in an open cell is an effective way to activate the cathode for excess heat production in subsequent electrolysis in a closed cell [5]. In this paper, deep research on pretreatment methods will be reported.

McKubre et al. [6] and Kunimatsu et al. [7] reported that D/Pd $> 0.8$–$0.9$ is a necessary condition to obtain excess heat in Pd|$D_2O$ electrolytic systems around room temperature. This requirement is in conflict with our experiences [1–5]. From April 2011, we have been measuring in situ the D/Pd ratio in electrolysis and calorimetry. The results showed that high loadings of D into Pd are unnecessary for excess heat production.

In recent years Violante's group found that the surface morphology of the cathode plays a key role in excess heat production in Pd|$D_2O$ electrolytic cells [8,9]. This conclusion is similar to our discovery on pretreatment because both of them are surface effects. In this paper, we will report the characteristics of effective Pd surfaces.

## 2. Experimental Set-up

The electrolytic cell used from January 2011 to October 2014 was an advanced model compared to the previous design [5]. Its schematic and photo are shown in Figs. 1(a) and (b), respectively. The cell is made of borosilicate glass ($\varphi_{\rm in}$ 41 $\times$ $\varphi_{\rm out}$ 45 $\times$ 130 mm$^3$) and it has a PTFE cap with a SS316 pipe connected to a deuterium gas cylinder and a pressure gauge. The bottom part of the male cap has two grooves of 4 mm width and 2.5 mm depth for O-rings ($\varphi_{\rm in}$ $=$ 31.5 mm, width $=$ 3.55 mm, made of nitrile butadiene rubber) to seal the cell. In addition, the cap has five holes for electrode leads as shown in Fig. 1(a). High-temperature sealant (Thomas Technologies Co. Ltd, Chengdu, China), which solidifies at 120°C, is filled into the gaps between the leads and the cap to prevent escape of the off gases from the cell. A Pt cylindrical mesh filled with recombination catalyst (Alfa Aesar, Stock #89106, Lot #C24R036) to recombine excess $D_2$ and $O_2$ is suspended under the PTFE cap as shown in Fig. 1(b). The quantity of catalyst generally varies from 35 to 45 pellets (weight $\sim$ 0.05 g /pellet).

From November 2014 to now, a SS316 flange was used outside the borosilicate cylinder to prevent an explosion, as shown in Fig. 1(c). The cap is also made of SS316, and is covered with Teflon coating to prevent corrosion. Cored screws and small O-rings are used to seal gaps between the leads and cap as partly shown in Fig. 1(c). Four screws are used to fix the cap and the flange. This cell gives better pressure tightness than the prior one. However, the Teflon coating is not so stable and has to be replaced after some runs.

The electrolyte is $D_2O + D_2SO_4$ mixed solution, to which we always added 0.2–0.4 g of titanium powder. Both deuterated reagents are from Beijing Chemical Reagent Company (>99.9 atom % D) with the mass ratio of $D_2O$:$D_2SO_4$ about 5:1. The platinum foil (normally $0.02 \times 31 \times 43$ mm$^3$) anode and leads are from General Research Institute for Nonferrous Metals (GRINM) as before [3].

Palladium samples from four different sources were used during the experiments. The physical properties, pretreatment methods and maximum excess power produced during electrolysis in closed Pd|$D_2O$ cells are listed in Table 1 for Pd #4 to 33 in total of 157 runs from July 27, 2011 to May 26, 2016. Most of samples are from Alfa Aesar, Stock #11514, 99.9% purity. Pd #6, provided by John Dash, is also from Alfa Aesar. Pd #4, 5 and 17, provided by Da-Lun Wang, are from Kunming Institute of Precious Metals (KIPM). Pd #7, 18 and 27 are from GRINM, 99.95% purity. Pd #20 and 21 are from AVIC Beijing Institute of Aeronautical Materials (BIAM). Before the first electrolysis run, a Pd sample from Alfa or KIPM is cold rolled to the desired thickness and then washed with pure alcohol and de-ionized water to remove surface contamination. A scanning electron microscope in combination with an energy dispersive spectrometer (SEM-EDS; HITACHI, S-4300) is used to characterize the Pd surface morphology and elemental composition. A 3D Noncontact Optical Profiler (BRUKER ContourGT) and an Atomic Force Microscope (BRUKER Dimension FastScan AFM) are used to measure the surface roughness.

Five Pt leads ($\varphi$0.8 mm), covered with heat-shrink Teflon tubing for protection, are welded to the Pt anode and a Pd cathode by a spot-welder. The loading ratios of D/Pd are determined by the resistance method using HIOKI 3560 AC m$\Omega$ HiTESTER to avoid DC voltage shifts during electrolysis, as discussed [10].

Before and after each electrolysis, the cell is weighed with a Mettler balance (PM1200, 1 mg readability) to monitor the weight loss. If the catalyst is working well, electrolysis at 3 A for 10 h will induce mass loss $\Delta m$ <0.2 g and the corresponding decomposition energy of heavy water will be less than 3 kJ; otherwise, the calorimetry may be complicated because excess heat is not prominent in these tests.

Three K-type thermocouples are attached outside the cell for temperature measurements, as shown in Fig. 1(a). The first is at the height of the recombination catalyst to monitor the recombining process of deuterium and oxygen gases. The second thermocouple is at the height of electrolyte level, and the third is on the bottom of the cell. The average value of the second and third temperatures is considered to be the electrolyte temperature in this paper. Generally, the temperature around the electrolyte level was greater than that of bottom by 5–10°C during experiments.

The power supply and data logging system are similar to that [5] with some modifications. All parameters are measured every 10 s. The input energy is the integration of input power (electrolytic current $\times$ cell voltage) with electrolysis time. The output energy is the integration of output power with electrolysis time and cooling time (around 10 h) of a heated cell.

Measurement of excess power is made by a decimeter-size heat-flow calorimeter which was fully demonstrated to have stability and accuracy in [4,5,11]. The electrolytic cell is placed in an enamel jar filled with cotton for heat insulation as shown in Figs. 1(d) and (e). The jar is located in the center of the measuring vessel with inner sizes of 26 cm $\times$ 26 cm $\times$ 26 cm and the air in the vessel is mixed with a fan to ensure temperature homogeneity. The

**Table 1.** Summary of excess power for various samples from July 27, 2011 to May 26, 2016.

| Pd # | size | mass | area | $P_{ex,max}$* | Rp[#] | source (Lot #) purity | Exp. # | pretreatment methods |
|---|---|---|---|---|---|---|---|---|
| | $mm^3$ | g | $cm^2$ | mW | | | | |
| 4 | 0.5 × 10 × 30 | 1.7836 | 6.28 | 117 ± 15 | 4/10 | KIPM, unknown | 110727-110907 | open electrolyzing |
| 5A | 0.5 × 11 × 31 | 1.8854 | 7.24 | 73 ± 17 | 2/18 | KIPM, unknown | 120508-121009 | open electrolyzing |
| 5B[&] | 0.33 × 18 × 31 | 1.7682 | 11 | 59 ± 32 | 2/12 | KIPM, unknown | 121218-130524 | open electrolyzing |
| 6 | 0.25 × 12 × 34 | 1.0767 | 8.1 | 61 ± 23 | 5/13 | Alfa Aesar, 99.9% | 121013-121129 | treated in flame for 40 min, open electrolyzing |
| 7 | 0.05 × 14 × 31 | 0.3043 | 8.41 | 28 ± 26 | 1/5 | GRINM (03-120) 99.95% | 130824-130906 | open electrolyzing |
| 8 | 0.26 × 14 × 31 | 1.2798 | 8.60 | 45 ± 18 | 1/8 | Alfa Aesar (E17Y001) 99.9% | 130912-130925 | etching in hot $H_2SO_4$ and aqua regia |
| 9 | 0.2 × 14 × 33 | 1.3969 | 9.86 | – 14 ± 18 | 0/2 | Alfa Aesar (E17Y001) 99.9% | 130927-130929 | etching in aqua regia |
| 10 | 0.26 × 10 × 32 | 0.9886 | 6.62 | – 34 ± 18[$] | 0/2 | Alfa Aesar (E17Y001) 99.9% | 130930-131011 | etching in aqua regia |
| 11 | 0.30 × 11 × 33 | 1.3863 | 7.56 | 16 ± 19 | 0/3 | Alfa Aesar (E17Y001) 99.9% | 131013-131023 | etching in aqua regia, open electrolyzing |
| 12 | 0.26 × 11 × 31 | 1.1296 | 6.97 | 42 ± 19 | 2/5 | Alfa Aesar (E17Y001) 99.9% | 131028-131105 | open electrolyzing |
| 13 | 0.24 × 11 × 31 | 1.0686 | 7.08 | 66 ± 16 | 3/4 | Alfa Aesar (E17Y001) 99.9% | 131109-131115 | open electrolyzing |
| 14 | 0.22 × 10 × 33 | 1.1226 | 6.29 | 101 ± 22 | 1/4 | Alfa Aesar (E17Y001) 99.9% | 131120-131126 | open electrolyzing |
| 15 | 0.26 × 10 × 30 | 1.0890 | 6.39 | 87 ± 24 | 4/4 | Alfa Aesar (E17Y001) 99.9% | 131129-131226 | heating in $D_2SO_4$ |
| 16 | 0.34 × 10 × 31 | 1.3704 | 6.80 | 117 ± 24 | 6/7 | Alfa Aesar (E17Y001) 99.9% | 131231-140123 | heating in $D_2SO_4$ |
| 17 | 0.26 × 10 × 31 | 1.1317 | 6.44 | 27 ± 16 | 1/3 | KIPM, unknown | 140112-140116 | heating in $D_2SO_4$ |
| 18 | 0.05 × 10 × 32 | 0.2280 | 6.56 | 48 ± 16 | 1/2 | GRINM (03-120) 99.95% | 140118-140120 | heating in $D_2SO_4$ |
| 19 | 0.34 × 11 × 31 | 1.5105 | 6.95 | 30 ± 16 | 1/2 | Alfa Aesar (E17Y001) 99.9% | 140211-140214 | heating in $D_2SO_4$ |
| 20 | 0.8 × 1.4 × 31 | 0.2875 | 1.39 | 45 ± 33 | 2/3 | BIAM, unknown | 140217-140222 | annealing, heating in $D_2SO_4$ |
| 21 | ϕ2.4 × 20 | 1.1158 | 1.54 | 23 ± 33 | 0/2 | BIAM, unknown | 140224-140226 | mercerizing, heating in $D_2SO_4$ |
| 22 | 0.36 × 10 × 34 | 1.5710 | 7.37 | 70 ± 33 | 3/3 | Alfa Aesar (E17Y001) 99.9% | 140301-140315 | heating in $D_2SO_4$, electro-deposited Cu or Au |
| 23 | 0.34 × 11 × 31 | 1.4627 | 7.07 | 40 ± 14 | 3/3 | Alfa Aesar (E17Y001) 99.9% | 140319-140325 | heating in $D_2SO_4$, electro-deposited Cu |
| 24 | 0.26 × 11 × 31 | 1.0146 | 6.38 | 26 ± 14 | 1/2 | Alfa Aesar (E26W014) 99.9% | 140327-140410 | heating in $D_2SO_4$ |
| 25 | 0.30 × 11 × 31 | 1.2388 | 6.85 | 31 ± 19 | 1/2 | Alfa Aesar (E26W014) 99.9% | 140412-140528 | open electrolyzing |
| 26 | 0.26 × 10 × 30 | 1.0972 | 6.51 | 70 ± 19 | 1/1 | Alfa Aesar (E26W014) 99.9% | 140605 | open electrolyzing |
| 27 | 0.05 × 15 × 32 | 0.3351 | 9.65 | – 29 ± 19[$] | 0/1 | GRINM (03-120) 99.95% | 141012 | none |
| 28 | 0.26 × 11 × 31 | 1.0604 | 7.14 | 61 ± 19 | 3/4 | Alfa Aesar (J30Z002) 99.9% | 141121-141202 | open electrolyzing |
| 29A | 0.3 × 11 × 31 | 1.0570 | 7.28 | 128 ± 19 | 4/4 | Alfa Aesar (J30Z002) 99.9% | 141204-141214 | open electrolyzing |
| 29B[&] | 0.12 × 26 × 32 | 0.9496 | 16.73 | 24 ± 14 | 1/4 | Alfa Aesar (J30Z002) 99.9% | 150705-160526 | etching in aqua regia |
| 30 | 0.26 × 10 × 31 | 1.0511 | 6.64 | 47 ± 32 | 1/3 | Alfa Aesar (J30Z002) 99.9% | 141218-141224 | open electrolyzing |
| 31 | 0.22 × 14 × 30 | 1.1995 | 8.39 | 108 ± 32 | 3/3 | Alfa Aesar (J30Z002) 99.9% | 141226-141231 | open electrolyzing |
| 32 | 0.27 × 22 × 30 | 2.1150 | 14.04 | 100 ± 32 | 5/10 | Alfa Aesar (J30Z002) 99.9% | 150107-150316 | open electrolyzing |
| 33 | 0.26 × 12 × 31 | 1.1428 | 7.98 | 61 ± 45 | 1/8 | Alfa Aesar (J30Z002) 99.9% | 150328-150531 | open electrolyzing |

* $P_{ex,max}$ is the maximum stable (always > 1 hr) excess power; [#] Rp is reproducibility; [$] Although the excess power is a negative value beyond one standard deviation, it should be positive after including mass loss; [&] Pd #5B and 29B were cold rolled from Pd #5A and 29A, respectively.

calorimeter is calibrated with an electric heater in 2-W steps, up to a power level greater than the electrolysis power, e.g. the maximum calibration power is 12 W if the electrolysis power is planned to be around 10 W. The input power $P$ and output signal $E$ can be simulated by a linear equation. For accuracy, we fit their relationship with a quadratic form:

$$P = D_1 E + D_2 E^2, \tag{1}$$

where $P$ is the input power in Watts and $E$ is the output electromotive force in Volts; $D_1$ is the device constant in W $V^{-1}$; $D_2$ is the coefficient of the quadratic term. The intercept term in Eq. (**??**) is forced to zero to ensure no signal without input power. The device constant$D_1$ and the uncertainty over more than 41 months are shown in Fig. 2. Fluctuations of room temperature in different seasons affect the long-term precision; however, this calorimeter is calibrated every 2–4 weeks when it is used in experiments, and this is accurate enough to give a true signal of excess power with the uncertainty around 30 mW for many experiments. At the same time, we consider a result to be assured production of excess heat only when both of the excess power and the excess energy (output energy minus input energy) are positive values, to avoid misjudgments. If both of them are negative and the mass loss rate of the cell in electrolysis is greater than 20 mg $h^{-1}$, the result has to be considered a failed experiment.

On Oct. 10, 2014 (Exp. #141010), a control experiment was conducted with a Pt foil ($0.02 \times 17 \times 29$ $mm^3$) in place of Pd cathodes. Electrolysis current of 3 A was applied for 12 h, the input and excess power were $8.462 \pm 0.003$ W and $-0.014 \pm 0.019$ W, respectively, under steady state. Another control experiment (Exp. #141012) was done using Pd #27. (We had difficulty producing excess heat with this batch of samples.) The stable excess power was $-0.029 \pm 0.019$ W while the input power was $8.721 \pm 0.004$ W at 3 A. These two results and other negative results with Pd|$D_2O$ all confirmed that our calorimetry is credible.

## 3. Results

### 3.1. Effects of D/Pd ratio and other parameters on excess power

The relationship between the resistance ratio and the D/Pd ratio at room temperature has been well known for many years [10,12,13]; however, the temperature coefficient of resistance of Pd–D, $\alpha_D$, in the miscibility gap has not been determined up to now, so the similar parameter of Pd–H, $\alpha_H$, had to be used by various groups in the past. We found that the application of $\alpha_H$ to the Pd–D system will induce what appears to be a physically impossible result, i.e. the D/Pd ratio will increase with time but not decrease as expected in a short period just after electrolysis is stopped. Therefore, we had to determine $\alpha_D$ by ourselves. We calculated it using the resistance data during the cooling process after the electrolysis is stopped, as shown in Fig. 3(a). Of course, we had to ensure resistance only decreases linearly

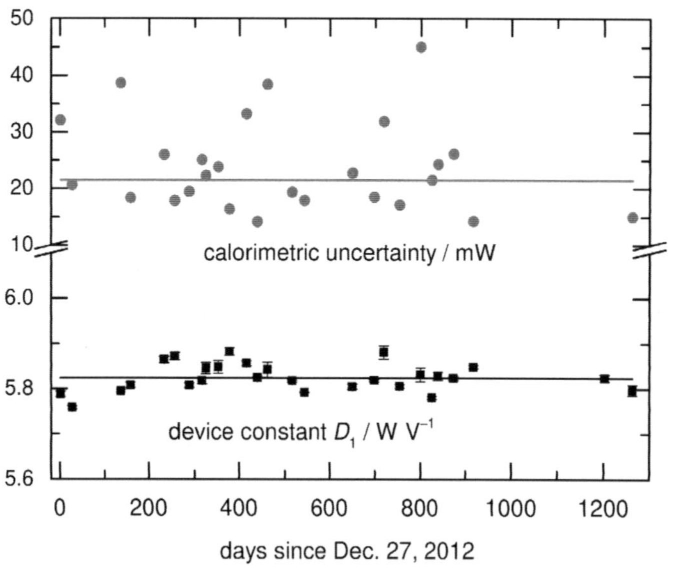


**Figure 2.** Changes of device constant (average value $\langle D_1 \rangle = 5.8246 \pm 0.0312\,(0.54\%)$ W $V^{-1}$) and uncertainty (average value $\langle \sigma \rangle = 24.3 \pm 8.6$ mW) of the calorimeter at 25°C (the set temperature for all results listed in this paper) over 41 months (Dec. 27, 2012 to June 11, 2016).

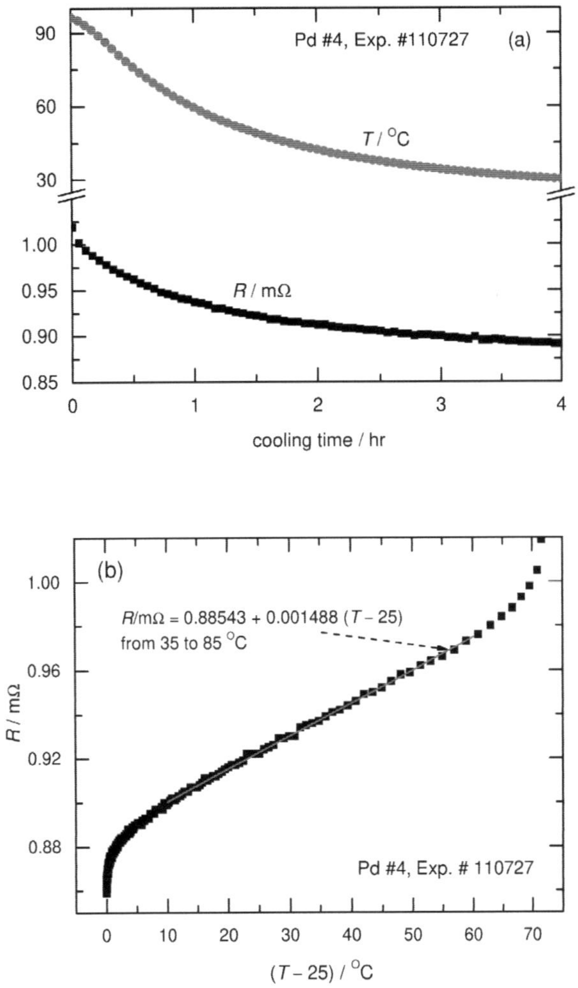


**Figure 3.** (a) Changes in resistance and temperature of a Pd cathode with time during cooling after stopping deuterium absorption by electrolysis; (b) Resistance vs. temperature in (a), only middle part has linear relationship between these two variables as shown by the fitting line and formula (Pd #4, Exp. #110727).

with temperature, and this means only part of data from cooling is chosen, as shown in Fig. 3(b). $\alpha_D$ with 0 <D/Pd <0.7 is obtained by collecting some runs of experiments as shown in Fig. 4 and simulated by a linear equation, the resulted resistance coefficient is

$$\alpha_D/\mathrm{K}^{-1} = 0.003965 - 0.003597x. \quad (2)$$

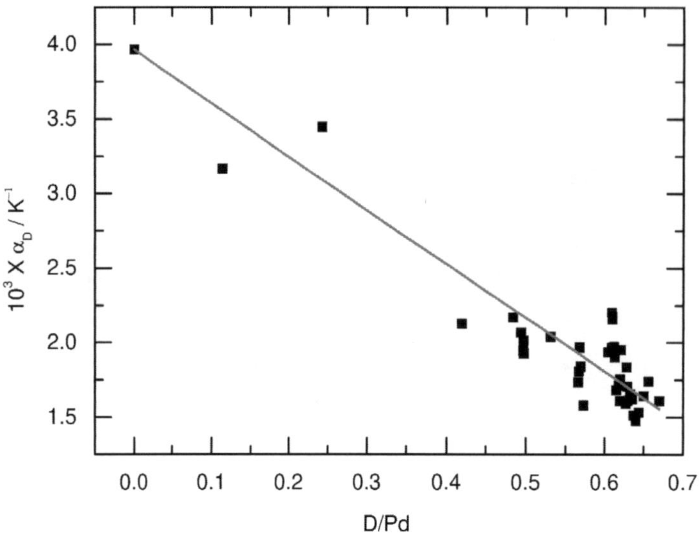


**Figure 4.** Collections of temperature coefficients of resistance of Pd–D in 39 runs of experiment for Pd #1, 4, 5 and 6 from May 10, 2011 to Oct. 24, 2012. The corresponding fitting formula is Eq. (**??**).

In our experiments, the D/Pd ratio at different temperatures was determined in situ by combining this factor and Eqs. (**??**) and (13) in [10]. Of course, $\alpha_D$ here must also be considered as a qualitatively and acceptable substitute of $\alpha_H$, it needs to be measured in carefully designed experiments in future.

Figure 5 shows an example of in situ D/Pd ratio measurement during excess heat production in electrolysis. We find that the cathode resistance and the cell temperature increase with time as shown in Figs. 5(a) and (b), respectively. The D/Pd ratio increases with time due to deuterium absorption at the beginning of electrolysis; however, the cathode begins to desorb deuterium after 12 min of electrolysis, because the temperature is too high as we discussed previously [14]. It is found that the average D/Pd ratio is only 0.60 when excess power is 0.103 W as shown clearly in Fig. 5(c). This is in direct conflict with the previous conclusion that high loading ratios of D/Pd are needed for excess power production [6,7].

Figure 6 is a collection of data from 75 runs of positive results with excess power, with 21 samples. It illustrates how excess power is dependent on different parameters. Because the D/Pd ratio was not measured before July 2011, and it was not successfully measured in every run after that, the data points in Fig. 6(a) are from only 33 runs with 16 samples. We find that there is no clear correlation between the excess power and the D/Pd ratio. Most excess heat occurs when D/Pd is 0.5–0.7 below the threshold value of 0.8–0.9 cited [6,7]. This means that high loading ratios of D/Pd are not necessary for excess heat production. Another effect is the initial D/Pd ratio; we obtained excess heat for Pd samples both with deuterium already absorbed (e.g. Fig. 5(a)) and without any deuterium at the beginning of electrolysis. Therefore, the initial D/Pd ratio has no clear influence on excess heat production. In our experiments, the closed cell was only filled with deuterium gas to increase the D/Pd ratio in some runs; however, we did not find effects of atmospheres on the deuterium absorption during electrolysis. Therefore, we always carried out experiments with

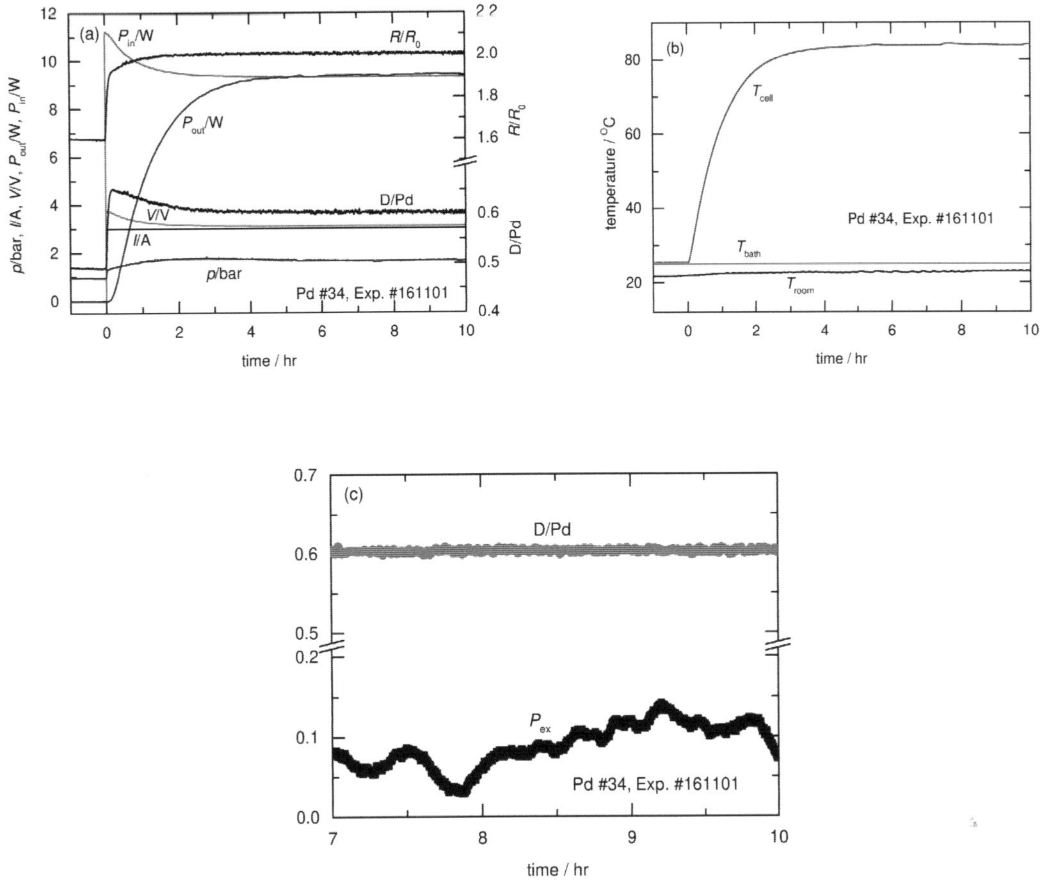


**Figure 5.** Applied current $I$, cell voltage $V$, input power $P_{\text{in}}$, output power $P_{\text{out}}$, gas pressure $p$, cathode resistance ratio $R/R_0$ (instantaneous resistance to the initial value of the free deuterium at 25°C), D/Pd ratio, cell temperature $T_{\text{cell}}$, room temperature $T_{\text{room}}$, bath temperature $T_{\text{bath}}$ (= temperature of outer wall of calorimeter) and excess power $P_{\text{ex}}$ in electrolysis of Exp. #161101 with Pd #34 (0.25 × 12 × 33 mm $^3$, Alfa Aesar, Lot #J30Z002). Average values $\langle P_{\text{ex}} \rangle = 103 \pm 17$ mW and $\langle \text{D/Pd} \rangle = 0.60 \pm 0.02$ in 8–10 h.

the closed systems with only air in the cell.

On the other hand, from Fig. 6(c), we find there is clear correlation between the excess power and the cell temperature as observed by us previously [3] and also reported by Storms recently [15].

In our experiments, we intentionally increase the cell temperature by using more cotton in the jar, although the applied current heat the cell at the same time. This is the reason why there is correlation between the excess power and the current density as shown in Fig. 6(b) but Figs. 6(b) and (c) are unlike each other.

### 3.2. The effects of different pretreatments on excess heat reproducibility

In a previous paper [5], we reported that open electrolyzing of Pd cathode in $D_2SO_4$ heavy water solution until the electrolyte is used up is an effective way to stimulate excess heat in subsequent runs of electrolysis, and Pd #1 maintained its activity after lying in a sample box for more than 6 months. Since then, we used this sample as a cathode in a $H_2O$ cell in two runs (Exp. #091001 and 091002) for blank calorimetry [5]. After that, this sample was dead; it produced no excess heat in subsequent 13 runs of calorimetry with three runs of open electrolyzing for activation. It

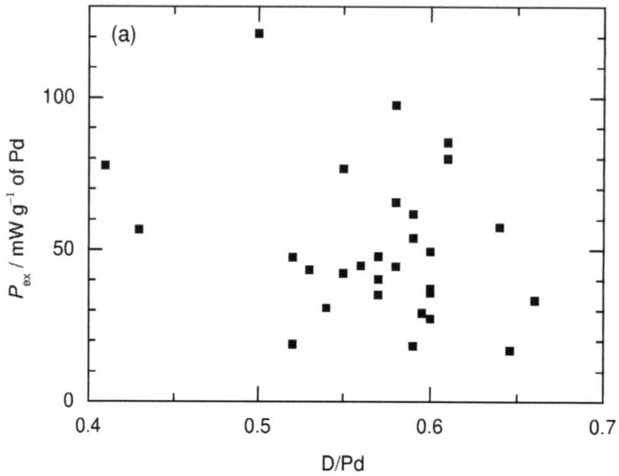


Fig. 6(a)

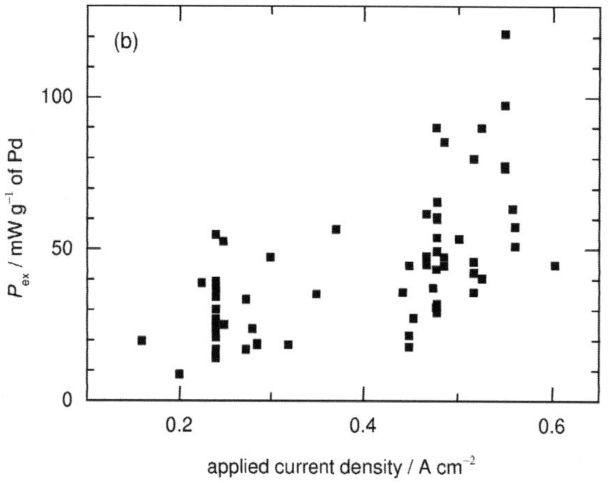


Fig. 6(b)

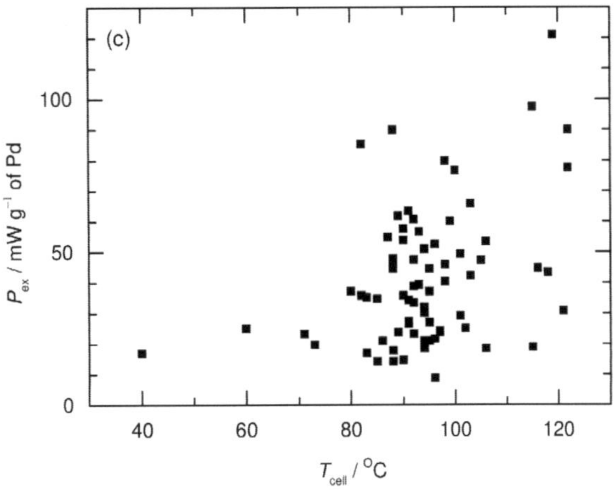


Fig. 6(c)

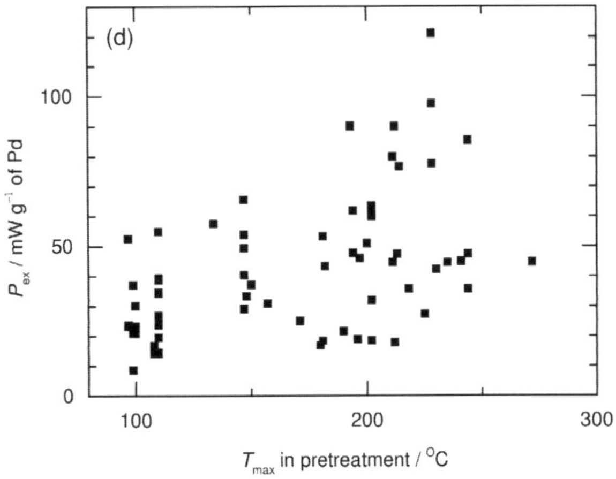


Fig. 6(d)

**Figure 6.** Specific excess power vs. (a) D/Pd ratio, (b) current density, (c) temperature, and (d) maximum temperature reached in pretreatments for 75 runs of positive calorimetric experiments with 21 samples from Dec. 23, 2008 to Mar. 28, 2015.

seems that history of H contamination suppresses excess heat production in Pd–D systems. Only Pd #2 and 4 in [5] were still active or could be activated by open electrolyzing thereafter.

In 2012, we tested Pd #5A from KIPM in 17 runs of calorimetry with four runs of open electrolyzing for activation; however, only one run (Exp. #120612) gave excess power of 73 ± 17 mW just after open electrolyzing (Exp. #120611). On Dec. 17, 2012, Pd #5A was cold rolled to 0.33 mm from original thickness of 0.5 mm and it was designated Pd #5B.

**Table 1.** Summary of excess power for palladium samples pretreated with etching in aqua regia.

| Pd # | Etching times | $P_{ex,max}$*/mW | Reproducibility | Exp. # |
|---|---|---|---|---|
| 8 | 3 | $45 \pm 18$ | 1/4 | See Table 3 |
| 9 | 1 | $-14 \pm 18$ | 0/2 | 130927/29 |
| 10 | 2 | $-34 \pm 18$ | 0/2 | 131008/11 |
| 11 | 1 | $-30 \pm 19$ | 0/1 | 131013 |
| 29B | 1 | $24 \pm 14$ | 1/2 | 150705/07, 160423/26 |

* $P_{ex,max}$ is the maximum stable (always >1 h) excess power.

After soaking in sulfuric acid for 15 min and washing, it was assembled to the cell for electrolysis and calorimetry. It showed a sign of excess power of $32 \pm 32$ mW the next day (Exp. #121218). This sample showed excess heat in two runs over 12 runs of calorimetry with one run of open electrolyzing for activation. The maximum excess power was $59 \pm 32$ mW in Exp. #130110 at applied current of 3 A and a cell temperature of 92°C.

From Aug. 24 to Sep. 6, 2013, Pd #7 from GRINM was tested in five runs of calorimetry with three runs of open electrolyzing for activation. The maximum excess power was only $28 \pm 26$ mW in Exp. #130827 just after activation by open electrolyzing the day before. In our experience, it is difficult to produce excess heat with this batch of samples from GRINM.

As far as open electrolyzing is concerned, three processes take place in this procedure. The first is etching by acid, the second is heating in acid and the third is electrolyzing. To distinguishing the effects of these three processes, we designed three treatment methods with samples Pd #8 to 33, from September 2013, as shown in Table 1 and discussed below.

### 3.2.1. *Etching in aqua regia*

Five samples (Pd #8 to 11, 29B) were pretreated with etching in aqua regia, and 11 runs of calorimetry were carried out. Most of these runs did not produce any excess heat, as shown in Table 2. Now we take Pd #8 as an example to illustrate the calorimetric results shown in Table 3. We find that only a little excess heat was observed in Exp. #130917 when Pd was etched in aqua regia for 15 min and then electrolyzed for 12 h at 3 A, whereas there was no excess power in the other three runs with this sample (Exp. #130919/23/25).

Because mass losses from the cell were prominent in experiments during 2013, we tested this pretreatment method again in 2015. We chose Pd #29A, which had given the maximum excess heat after pretreated with open electrolyzing (see Section 3.2.3), and we cold rolled it to the thickness of 0.12 mm and designated it Pd #29B. It was etched in aqua regia for 72 s and about 6.4 $\mu$m of thickness from the surface was removed, according to its mass loss (from 1.0624 to 0.9496 g). The subsequent electrolysis at 3 A for 12 h only gave excess power of $6 \pm 14$ mW (Exp. #150705). The mass loss of this run was only 28 mg. All experiments listed in Tables 2 and 3 show that etching alone has no significant effect on excess heat production.

**Table 2.** Summary of excess power for Pd #8 pretreated with etching in aqua regia.

| Exp. # | Etching time | $I \times t$ (A h) | Mass loss (mg) | $T_{cel}$(°C) | $P_{in}$(W) | $P_{ex}$(mW) | D/Pd |
|---|---|---|---|---|---|---|---|
| 130917 | 15 min | $3.0 \times 12$ | 257 | 85 | 7.984 | $45 \pm 18$ | 0.57 |
| 130919 | None | $3.2 \times 10$ | 218 | 88 | 9.747 | $-11 \pm 18$ | 0.47 |
| 130923 | 20 min | $3.3 \times 12$ | 272 | 90 | 9.310 | $1 \pm 18$ | 0.52 |
| 130925 | 30 min | $3.3 \times 12$ | 506 | 91 | 8.866 | $-27 \pm 18$ | 0.54 |

**Table 3.** Summary of excess power for palladium samples pretreated with heating in $D_2SO_4$.

| Pd # | Heating times | $P_{ex,max}$*/mW | Reproducibility | Exp. # |
|---|---|---|---|---|
| 15 | 4 | 87 ± 24 | 4/4 | 131129, 131222/24/26 |
| 16 | 5 | 117 ± 24 | 6/7 | See Table 5 |
| 17 | 3 | 27 ± 16 | 1/3 | 140112/14/16 |
| 18 | 2 | 48 ± 16 | 1/2 | 140118/20 |
| 19 | 2 | 30 ± 16 | 1/2 | 140211/14 |
| 20 | 4 | 45 ± 33 | 2/3 | 140218/20/22 |
| 21 | 1 | 23 ± 33 | 0/1 | 140226 |
| 22$ | 3 | 70 ± 33 | 2/3 | 140301/13/15 |
| 23$ | 3 | 40 ± 14 | 1/3 | 140319/21/25 |
| 24 | 1 | 26 ± 14 | 1/1 | 140327 |
| 28 | 2 | 61± 19 | 2/3 | 141121/27/29 |

* $P_{ex,max}$ is the maximum stable excess power.
$ See Table 8.

### 3.2.2. *Heating in $D_2SO_4$*

Thirty-two runs of calorimetry with 11 samples were conducted to study the effects of pretreatment by heating in $D_2SO_4$ on subsequent electrolysis, as shown in Table 4. Details of Pd #16 as an example are listed in Table 5; we find that most runs with this sample show excess heat after activation. It produced maximum excess power of 117 ± 24 mW in Exp. #140107 after being heated for 2 h and reaching the maximum temperature of 244°C in pretreatment. In all 11 samples listed in Table 4, 9 samples produced excess power but only four of them give maximum excess power $P_{ex,max} > 3\sigma$, the rest are not so activated, although reproducibility is high for this method.

### 3.2.3. *Electrolyzing in $D_2SO_4$*

Fifty runs of calorimetry with 12 Pd samples were performed to test the effect of open electrolyzing in pretreatment on excess heat production in subsequent electrolysis, as shown in Table 6. All runs with Pd #29A as an example are listed in Table 7. Before open electrolyzing, PTFE sealing tape is used to wrap the cell cap to protect the O-ring from corrosion by evolution gas, an open bottle is filled with $D_2SO_4$ until the Pd cathode is completely immersed, and the electrolysis current is usually set as 4 A. During 3 h of open electrolyzing in Exp. #141209 with Pd #29A, the maximum temperature reached 228°C at 2.5 h. In subsequent electrolysis in a closed cell at 4 A for 15 h, the stable excess power $P_{ex}$ reached 103 ± 19 mW from 10 to 11 h. The activity of this sample continued in the next two runs as observed before [5]. It gave 128 ± 19 and 82 ± 19 mW in Exp. #141212 and 141214, respectively, without any pretreatment, as shown in Table 7. From Table 6, we find the reproducibility of this method is 25/50 (50%), which is somewhat less than the heating method of 21/32 (66%); however, there are six samples giving excess power of

**Table 4.** Summary of excess power for Pd #16 pretreated with heating in $D_2SO_4$.

| Exp. # | Heating time, maximum temperature | $I \times t$(A h) | Mass loss (mg) | $T_{cell}$(°C) | $P_{in}$(W) | $P_{ex}$(mW) | D/Pd |
|---|---|---|---|---|---|---|---|
| 131231 | 1.5 h, 235°C | 3.3 × 12 | 1372 | 95 | 9.357 | 61 ± 24 | 0.58 |
| 140102 | 1.5 h, 242°C | 3.3 × 8 | 847 | 95 | 9.304 | −12±24 | 0.58 |
| 140104 | 2 h, 244°C | 3.3 × 12 | 572 | 92 | 9.768 | 65 ± 24 | 0.52 |
| 140107 | 2 h, 244°C | 3.3 × 24 | 135 | 82 | 9.813 | 117 ± 24 | 0.61 |
| 140110 | None | 3.3× 1, 3× 11 | 371 | 90 | 9.761 | 49 ± 16 | 0.60 |
| 140122 | 2 h, 241°C | 3.3× 1.5, 3× 10.5 | 1624 | 90 | 9.260 | 29 ± 16 | 0.57 |
| 140123 | None | 3.3× 0.8, 3× 10.3 | | 90 | 9.365 | 19 ± 16 | 0.58 |

**Table 5.** Summary of excess power for palladium samples pretreated with open electrolyzing in $D_2SO_4$.

| Pd # | Times of open electrolyzing | $P_{\mathrm{ex,max}}$*/mW | Reproducibility | Exp. # |
|---|---|---|---|---|
| 11 | 2 | $16 \pm 19$ | 0/2 | 131018/23 |
| 12 | 3 | $42 \pm 19$ | 2/5 | 131028/29, 131101/03/05 |
| 13 | 2 | $66 \pm 16$ | 3/4 | 131109/10/12/15 |
| 14 | 3 | $101 \pm 22$ | 1/4 | 131120/21/23/26 |
| 25 | 1 | $31 \pm 19$ | 1/2 | 140412/0528 |
| 26 | 1 | $70 \pm 19$ | 1/1 | 140605 |
| 28 | 1 | $61 \pm 19$ | 3/4 | 141202 |
| 29A | 2 | $128 \pm 19$ | 4/4 | see Table 7 |
| 30 | 3 | $47 \pm 32$ | 1/3 | 141218/22/24 |
| 31 | 3 | $108 \pm 32$ | 3/3 | 141226/29/31 |
| 32 | 5 | $100 \pm 32$ | 5/10 | 150107/10/12/16/21,150309/11/13/15/26 |
| 33 | 3 | $61\pm 45$ | 1/8 | 150328,150409/14150507/21/31 |

* $P_{\mathrm{ex,max}}$ is the maximum stable (always > 1 h) excess power.

$P_{\mathrm{ex,max}} > 3\sigma$ in all 11 samples tested. In comparison, this parameter is only four in all 11 samples for heating method in Table 4. This means open electrolyzing is apparently the most effective method of activation.

### 3.2.4. *Other activation methods*

It was reported that applying reverse current for 1 h in electrolysis can activate a Pd cathode for excess heat production [16]; however, we tested five samples in eight runs (Exp. #080731, 080804, 081125 & 081130 with Pd #1; Exp. #090527 with Pd #2; Exp. #121118 with Pd #6; Exp. #130914 with Pd #8; Exp. #140319 with Pd #23) and no significant excess heat was stimulated by this method.

We also tested reverse polarization in pretreatment during open electrolyzing (Exp. #090814, 090915 and 090921 with Pd #1), there was no excess heat observed in subsequent calorimetry for closed electrolysis. Only one positive example came from Pd #4, when we applied reverse polarization of 3 A for 20 min at the beginning of pretreatment with open electrolyzing, then normal current of 3.5 A was applied for 2 h and 3.75 A was applied for 1 h successively (Exp. #110728). The subsequent four runs of calorimetry with closed electrolysis gave excess power of $96 \pm 15$ mW in Exp. #110729, $117 \pm 15$ mW in Exp. #110802, $52 \pm 15$ mW in Exp. #110804, and $88 \pm 15$ mW in Exp. #110805. However, we are not sure of the role of reverse electrolysis here.

It has been said that an oxide layer on the palladium surface may be important for excess heat production [17,18]. In ICCF17, Dash et al. also reported that Pd and Ti foils being heated in flame at 700°C for 20–60 min had given excess power in subsequent electrolysis [19]. At our request, John Dash sent us three samples (Pd #6 in Table 1 and two Ti foils). We tested the Pd sample in six runs (Exp. #121013/14/16/17/24, 121115) at current of 3 A. Only three runs produced a little excess heat with the maximum value of $61 \pm 23$ mW in Exp. #121017. We also activated it by electrolyzing two times; however, it did not produce excess heat in the two subsequent runs (Exp. #121117/18) after

**Table 6.** Summary of excess power for Pd #29A pretreated with open electrolyzing in $D_2SO_4$.

| Exp. # | Electrolyzing time, maximum temperature | $I \times t$(A h) | Mass loss (mg) | $T_{\mathrm{cell}}$(°C) | $P$ (W) | $P$ (mW) | D/Pd |
|---|---|---|---|---|---|---|---|
| 141204 | 2 h, 214°C | 4× 15 | 1821 | 100 | 12.846 | $81 \pm 19$ | 0.55 |
| 141209 | 3 h, 228°C | 4× 15 | 127 | 115 | 12.545 | $103 \pm 19$ | 0.58 |
| 141212 | None | 4× 15 | 982 | 119 | 13.105 | $128 \pm 19$ | 0.50 |
| 141214 | None | 4× 15 | 662 | 122 | 12.397 | $82 \pm 19$ | 0.41 |

**Table 7.** Summary of excess power for Pd samples plated with Au and Cu.

| Pd # | Exp # | D/Pd | Anode | $I \times t$ (A min) | Plating thickness ($\mu$m) | $P_{ex}$ (mW) |
|---|---|---|---|---|---|---|
| 22 | 140301 | 0.74 | Au | 2.7 A × 45 min | – | 70 ± 33 |
| 22 | 140313 | 0.73 | Cu | 1.0 A × 30 min | 12.1 | 28 ± 14 |
| 22 | 140315 | 0.74 | Cu | 1.0 A × 1 min | 3.37 | 34 ± 14 |
| 23 | 140319 | 0.56 | Cu | 0.1 A × 1 min | 0.43 | 19 ± 14 |
| 23 | 140321 | 0.69 | Cu | 0.1 A × 1 min | 0.45 | 25 ± 14 |

the first open electrolyzing (Exp. #121116). It only gave small excess power of 40 ± 26 mW in the first two of total four runs (Exp. #121120/21/22/29) after the second open electrolyzing (Exp. #121119). One Ti foil (0.6596 g, 0.4 × 10 × 39 mm$^3$, Alfa Aesar, 99.99% metals basis) did not give excess heat in four runs of electrolysis (Exp. #130529/30/31 and 130602) too.

The situation was similar for Pd #21. It was soaked in fused potassium hydroxide (a.k.a. mercerizing) about 20 years ago and the surface is black. It did not produce excess heat in the first run of electrolysis (4 ± 33 mW in Exp. #140224). After heating in hot $D_2SO_4$, the excess power in subsequent electrolysis was still within the range of uncertainty (23 ± 33 mW in Exp. #140226).

We tried to deposit inert metals on the surface of palladium after deuterium was absorbed in order to seal the deuterium into the sample, thereby testing if this method could trigger excess heat. Two samples were tested in 5 runs, as shown in Table 8. Au or Cu was deposited on the Pd surface by means of electroplating. Firstly, one of the samples was heated in $D_2SO_4$ for 1–2 h, and then electrolysis was carried out for the Pd cathode in the cell used for calorimetry. After the D/Pd ratio reaching saturation value, the Pt anode shown in Fig. 1 was switched to Au or Cu (Alfa Acsar, 99.99%) for some minutes. Cu was easily plated; however, it was difficult to form an Au deposit because of the insolubility of Au in sulfuric acid. Both current and time affect the plating thickness, as shown in Table 8. There is no clear evidence that this method can improve excess heat production.

### 3.2.5. *Comparison of different treatment methods*

Table 9 compares the effects of different pretreatments on excess heat production for different Pd samples. We find that etching in aqua regia only gives the maximum excess power of 45 ± 18 mW and 18% reproducibility; heating in $D_2SO_4$ induces maximum excess power of 117 ± 24 mW and 66% reproducibility; open electrolyzing in $D_2SO_4$ leads to the maximum excess power of 128 ± 19 mW and 50% reproducibility. Although heating produced higher reproducibility than open electrolyzing, the latter method has a sample rate of 3/11 with $P_{ex}$ >100 mW while the former has only 1/11. This means that open electrolyzing is the most effective method of activation. Figure 6(d) shows effects of maximum temperature reached in pretreatment (heating or open electrolyzing of Pd in $D_2SO_4$) on excess heat, higher temperature of pretreatment will induce higher excess power as expected. We feel that we need to achieve much higher temperature in future research.

Of course, reproducibility is also source and sample dependent. Pd plates from GRINM were difficult to activate; four samples (Pd #3 in [5], Pd #7, 18 and 27 here) only gave maximum excess power of 48 ± 16 mW with reproducibility of 2/11. Generally, we only use this batch without any pretreatment in control experiments to test the

**Table 8.** Comparison of different pretreatments on excess power for samples Pd #8 to 33.

| Pd # | $P_{ex,max}$/mW | Reproducibility | Pretreatment method |
|---|---|---|---|
| 8–11, 29B | 45 ± 18 | 2/11 (18%) | Etching in aqua regia |
| 15–24, 28 | 117 ± 24 | 21/32 (66%) | Heating in $D_2SO_4$ |
| 12–14, 25, 26, 28, 29A, 30–33 | 128 ± 19 | 25/50 (50%) | Electrolysis in hot $D_2SO_4$ |

accuracy and validity of our calorimetry. On the other hand, samples from one batch of Alfa Aesar showed different levels of excess heat production from one to another. For example, Pd #28 to 33 were all cut from a plate of Lot #J30Z002 and were all pretreated with open electrolyzing; however, Pd #33 only gave maximum excess power of 61 $\pm$ 45 mW (Exp. #150328) and reproducibility of 1/8 while both #31 and 32 gave excess power greater than 100 mW The reasons of this sample dependent on excess heat production are unknown at present.

### 3.3. Surface characteristics of Pd samples

As reported by the Violante group [8,9], the surface morphology of a Pd sample is very important for its excess heat production in electrolysis. For this purpose, we cut a small part from some typical Pd samples in every phase and studied them with light and electron microscopes. Now we take Pd #16 as an example to demonstrate our primary results. This sample was pretreated by heating in $D_2SO_4$ and gave excess power up to 117 $\pm$ 24 mW, as shown in Tables 1 and 5. Its photos in three different phases are shown in Fig. 7; we find its surface becomes rough after heating and electrolysis in $D_2SO_4$, as expected.

#### 3.3.1. *Changes of surface roughness*

To quantitatively characterize the surface morphology, we used a non-contact 3D optical profiler (BRUKER ContourGT) to measure roughness, as shown in Fig. 8 and listed in Table 10. It is clear that all roughness parameters under different definitions increase by around one order of magnitude after heating in $D_2SO_4$, and they increase again by 1/10 to 1/2 after electrolysis for 12 h. AFM study on Pd #14 also support this conclusion, i.e. open electrolyzing also roughens the surface. These results agree with the discovery of the Violante group [8,9]. Of course, more study on comparison of different pretreatments on surface properties is needed. Without this we cannot determine the detailed requirements for excess heat production.

#### 3.3.2. *SEM and EDS characteristics*

Similar to our previous work [2,3], we also studied SEM as shown in Fig. 9, and element composition with EDS, as shown in Tables 11 and 12 for different phases of Pd #16. Figure 9(a) is an SEM of the fresh sample and there is only base metal Pd and some C and O on the surface (see Table 11). After heating in $D_2SO_4$ for 2 h, many microstructures appear on the Pd surface as shown in Fig. 9(b). After electrolysis, new elements S, Ag and Pt occur on the surface as shown in Fig. 9(c) and listed in Table 12. For these new elements, S comes from the electrolyte; Pt comes from

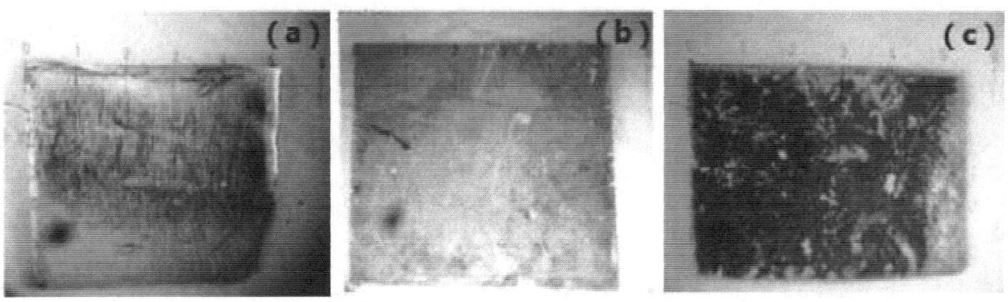


**Figure 7.** Optical photos of Pd #16 just cold rolled (a), after heating in $D_2SO_4$ for 2 h (b) and after electrolysis for 12 h (c). The units of scale are in mm.

**Table 9.** Roughness parameters of surface shown in Fig. 8.

| Figure # | $R_a$ ($\mu$m) | $R_p$ ($\mu$m) | $R_q$ ($\mu$m) | $R_v$ ($\mu$m) |
|---|---|---|---|---|
| a | 0.235 | 1.807 | 0.307 | −2.014 |
| b | 2.426 | 14.046 | 3.055 | −14.084 |
| c | 3.545 | 15.382 | 4.307 | −16.492 |

*Note*: $R_a$, arithmetic average; $R_p$, maximum peak height; $R_q$, root mean squared; $R_V$, maximum valley depth.

**Table 10.** Relative atomic percent concentrations at spots shown in Fig. 9(a).

| Element | #1 | #2 | #3 |
|---|---|---|---|
| C | 37.3 | 68.1 | 22.9 |
| O | 0.0 | 15.2 | 0.0 |
| Pd | 62.7 | 16.7 | 77.1 |

the anode and Ag may be the product of transmutation as reported before [2,3]. Spot #1 in Fig. 9(c) is the center of a crater, where should be the place of deuterium bubbles evolved from and silver is prominent. This situation is the same as Fig. 9 in [3]. This means that silver is transmuted in the deuterium electrode reaction.

## 4. Discussions and Conclusion

Although we can observe excess heat with some reproducibility, the amplitude of excess power is still very small now and we need to increase it by 2 or 3 times at least. Otherwise, the scientific community may have doubts about the calorimetry. Regarding the Pd surface pretreatment, we need more research to clarify the surface properties and their physical effects. This may be the key step to amplifying excess power.

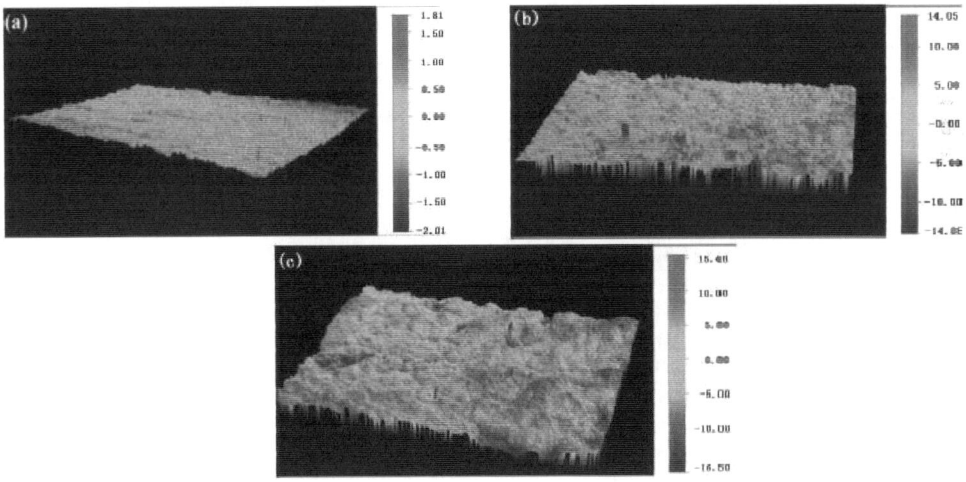


**Figure 8.** Typical 3D morphology of Pd #16. Regions of (a)–(c) are chosen from Figs. 7(a)–(c), respectively. The areas are 0.234 × 0.310 mm$^2$ for (a) and (c), and 0.223 × 0.299 mm$^2$ for (b); the units of height on the right side are in $\mu$m.

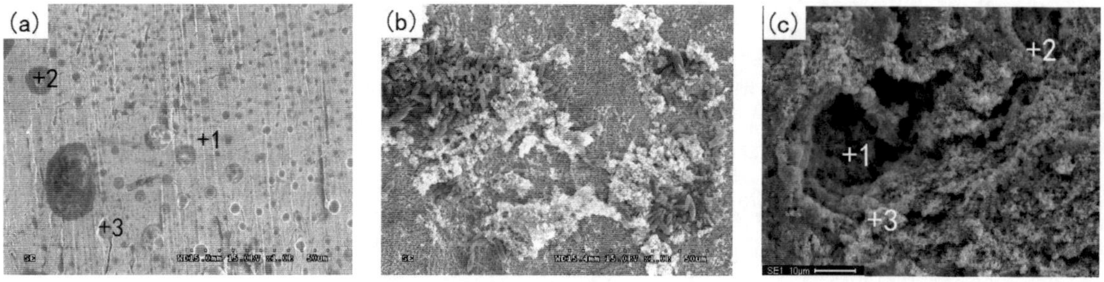


**Figure 9.** SEM of Pd #16, (a)–(c) are chosen from Figs. 7(a)–(c), respectively.

**Table 11.** Relative atomic percent concentrations at spots shown in Fig. 9(c).

| Element | #1 | #2 | #3 |
|---|---|---|---|
| C | 20.1 | 0.0 | 24.0 |
| O | 4.0 | 0.0 | 0.0 |
| S | 6.9 | 0.0 | 0.0 |
| Pd | 45.5 | 8.1 | 46.7 |
| Ag | 3.0 | 0.0 | 0.6 |
| Pt | 20.5 | 91.9 | 28.7 |

In conclusion, we find that high loading ratios of D/Pd are unnecessary for excess heat production in Pd|$D_2O$ electrolytic cells. Additionally, the temperature is more important than the D/Pd ratio as demonstrated by 75 runs of calorimetry. On the other hand, pretreatments of surface are another necessary step to activate Pd cathodes. Comparing three methods, i.e. etching in aqua regia, heating in $D_2SO_4$ and open electrolyzing in $D_2SO_4$, we found that the latter two are effective ways to activate a Pd sample to produce excess heat in subsequent electrolysis. Light and electron microscopy indicate that pretreatments roughen the surface and this may be responsible for the excess heat production.

## Acknowledgments

This work was supported by NSFC (21153003 and 20973185), 973 Program of MOST in China (2009CB226113).

## References

[1] Z.-L. Zhang, W.-S. Zhang, M.-H. Zhong and F. Tan, *Proc. ICCF8*, Bologna, Italy, May 21 to 26, 2000, p. 91.
[2] W.-S. Zhang, J. Dash and Q. Wang, *Proc. ICCF12*, Yokohama, Japan, Nov. 27 to Dec. 2, 2005, p. 86.
[3] W.-S. Zhang and J. Dash, *Proc. ICCF13*, Dagomys, Sochi, Russia, June 25 to July 1, 2007, p. 202.
[4] W.-S. Zhang, J. Dash and Z.-L. Zhang, *Proc. ICCF14*, Washington DC, USA, Aug. 8–10, 2008, p. 26.
[5] W.-S. Zhang, *Proc. ICCF15*, Roma, Italy, Oct. 5–9, 2009, p. 27.
[6] M.C.H. McKubre, S. Crouch-Baker, A.M. Riley, S.I. Smedley and F.L. Tanzella, *Proc. ICCF3*. Nagoya, Japan, Oct. 21–25, 1992, p. 5.
[7] K. Kunimatsu, N. Hasegawa, A. Kubota, N. Imai, M. Ishikawa, H. Akita and Y. Tsuchida, *Proc. ICCF3*, Nagoya, Japan, Oct. 21–25, 1992, p. 31.
[8] V. Violante, F. Sarto, E. Castagna, M. Sansovini, S. Lecci, D.L. Knies, K.S. Grabowski and G.K. Hubler, *Proc. ICCF14*, Washington DC, USA, Aug. 8–10, 2008, p. 429.
[9] E. Castagna, S. Lecci, M. Sansovini, F. Sarto and V. Violante, *J. Condensed Matter Nucl. Sci.* **8** (2012) 49.

[10] W.-S. Zhang, Z.-F. Zhang and Z.-L. Zhang, *J. Electroanal. Chem.* **528** (2002) 1.
[11] W.-S. Zhang, *Thermochim. Acta* **499** (2010) 128.
[12] M.C.H. McKubre and F.L. Tanzella, *Proc. ICCF12*, Yokohama, Japan, Nov. 27 to Dec. 2, 2005, p. 392.
[13] A. Spallone, F. Celani, P. Marini and V. Di Stefano, *Proc. ICCF12*, Yokohama, Japan, Nov. 27 to Dec. 2, 2005, p. 404.
[14] W.-S. Zhang, Z.-L. Zhang and X.-W. Zhang, *J. Electroanal. Chem.* **481** (2000) 13.
[15] E. Storms, *J. Condensed Matter Nucl. Sci.* **20** (2016) 81.
[16] J. Dash and A. Ambadkar, *Proc. ICCF11*, Marseille, France, Oct. 31 to Nov. 5, 2004, p. 477.
[17] D.D. Dominguez, D.A. Kidwell, D.L. Knies, K.S. Grabowski, G.K. Hubler, J.H. He, V. Violante, *J. Condensed Matter Nucl. Sci.* **8** (2012) 219.
[18] J.-P. Biberian, I. Parchamazad and M.H. Miles, *J. Condensed Matter Nucl. Sci.* **13** (2014) 38
[19] J. Dash, J. Solomon and M. Zhu, *J. Condensed Matter Nucl. Sci.* **13** (2014) 80.

J. Condensed Matter Nucl. Sci. 24 (2017) 60–74

Research Article

# LENR Theory Requires a Proper Understanding of Nuclear Structure

Norman D. Cook*

*Department of Informatics, Kansai University, Osaka, Japan*

**Abstract**

Deciphering the many puzzles of low-energy nuclear reactions (LENR) will require not only knowledge of the electromagnetic and chemical environment of the nuclear species that are susceptible to low-energy excitation, but also a detailed understanding of the many-body problem of the atomic nucleus itself. Today most LENR research focuses on the former issues, while nuclear structure theory remains the same conundrum that it has been for 80 years. The essence of the problem is that there has been no self-consistent *spatiotemporal* description of nuclear structure – and, as a consequence, no reliable framework within which to address the oldest and most outstanding problem in nuclear physics: the nuclear force. The unanswered question in traditional nuclear physics – and the new challenge in LENR – is how to reconcile the short-range nuclear force *known* from nucleon scattering experiments with the wealth of empirical data on nuclear structure. Despite problems due to the early rejection of spatiotemporal explanations in nuclear physics, 21st century computational power has opened the door to realistic "molecular dynamics"-style simulations of nuclear structure. Some implications both for LENR and for nuclear physics are discussed.


*Keywords:* Lattice simulations, Low-energy nuclear reactions, Nuclear structure theory, Realistic nuclear potentials

## 1. Introduction

The philosophical questions raised by quantum theory in the 1920s are issues that virtually nobody interested in empirical research wants to revisit. Various positions have been developed, debated, "proven" and "disproven" countless times – to the full satisfaction of no one (see the 16 "common" interpretations of quantum mechanics, as well as the 12 "minority" interpretations in Wikipedia). There is, consequently, little merit in discussing the phenomena of LENR on the basis of inherently controversial preconceptions about quantum philosophy.

It is nevertheless worth stating the historical fact that, *prior* to the beginnings of nuclear theory, uncompromising philosophical positions concerning "the correct" interpretation of quantum theory were already in place. For example, in 1925 Niels Bohr wrote that "In contrast to ordinary mechanics, the new *quantum mechanics does not deal with a space–time description* of the motion of atomic particles" (emphasis added) [1], see also [2]. Physics without space–time!?! It is no surprise that vociferous debates ensued – focusing on whether or not there is a meaningful distinction between classical mechanics and the newer quantum mechanics of the atom.

*E-mail: cook@res.kutc.kansai-u.ac.jp

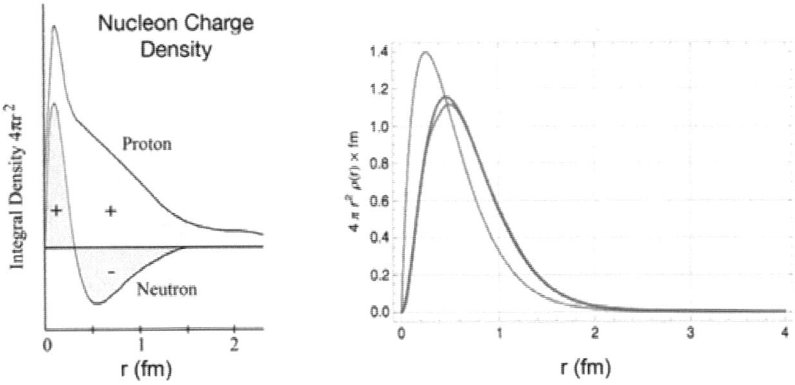


**Figure 1.** (*Left*) The charge densities of the proton and neutron [3]. Most of the proton charge is found within $r < 1.0$ fm; the neutron has a positively charged core ($r < 0.3$ fm) with a negatively charged surround ($r = 0.3$ to $1.0$ fm). (*Right*) Various proton charge distributions obtained from multi-parameter models of the proton. Again, most of the proton charge is found at distances less than 1.0 fm from the nucleon center.

Bohr's intuitions have influenced many theorists over the past century, but it is relevant for modern researchers in LENR to realize that he reached his conclusions well *before* the discovery of the neutron (1932) and with *no knowledge* of the many developments in *nuclear* physics subsequent to 1935. Without casting doubts on Bohr's contributions to atomic physics (i.e., the *electron* structure of the atom), his deeply counter-intuitive philosophy of the quantum world was formulated prior to the advent of nuclear physics. Following Bohr, others soon added their own versions of non-spatiotemporal dynamics (Pauli's exclusion principle in 1925 and Heisenberg's uncertainty principle in 1927). Although Planck, Einstein, Schrödinger, and many others maintained that less radical interpretations were possible, today the rejection of *spatiotemporal* explanations in quantum physics, in general, and in nuclear structure theory, in particular, remains a popular, if dubious, stance.

## 2. The Spatiotemporal Dimensions of the Atomic Nucleus

### 2.1. Nucleons

On the basis of scattering experiments done at relatively low energies (<10 MeV), protons and neutrons are known to have properties indicative of particles of finite size. Their electrostatic and magnetic field radii have been measured many times since the 1950s, starting with the work of Hofstadter (Nobel Prize, 1961) and indicating an RMS charge radius for the proton of less than one fermi (Figs. 1 and 2, Tables 1 and 2). Today, the internal (quark? parton?) substructure of the lone nucleon is far from clear, but an electromagnetic radius of 0.8–0.9 fm for both protons and neutrons is well established. Note that the recent controversy concerning the 5% difference between the charge radius of the proton when measured using muons (0.84 fm) vs. electrons (0.88 fm) does not affect the conclusion that, at the relatively low energies of nuclear structure physics, the nucleon is a space-occupying particle with a rather large, impenetrable core ($R \sim 0.5$ fm) and a cut-off radius at about 1–2 fm.

### 2.2. Nuclear radii

Also dating from Hofstadter's work in the 1950s, the nuclear charge radii, the density of nucleons in the nuclear core, and the thickness of the so-called nuclear skin are well known [6]. Density values that are typically cited in nuclear

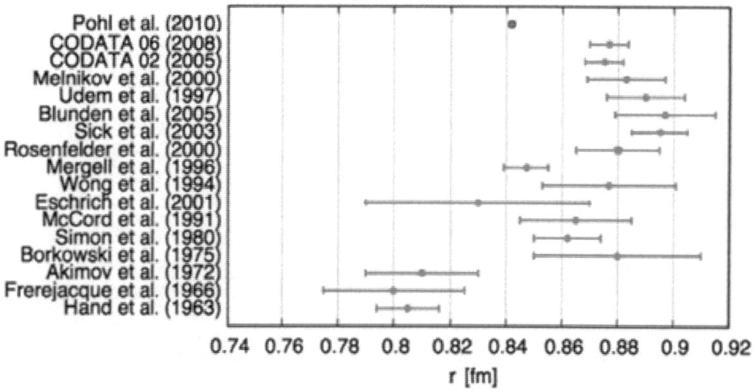


**Figure 2.** A summary of proton charge radii from muon (*red*) and electron (*blue*) scattering experiments [4].

**Table 1.** Summary of modern measurements of the proton electric and magnetic radii [4].

| Electric and magnetic radii of protons | $r_E$ (fm) | $r_M$ (fm) |
|---|---|---|
| Mainz updated | 0.8750(150) | 0.799(28) |
| World updated | 0.8810(110) | 0.867(20) |
| Naïve global average | 0.8790(90) | 0.844(16) |
| Suggested global average | 0.8790(110) | 0.844(38) |

textbooks range from 0.13 to 0.17 nucleons/fm$^3$, and are illustrated in Fig. 3. With the exception of $^4$He, all nuclei exhibit a more-or-less constant core nucleon density, which is indication of the dense "packing" of nucleons into a small volume under the influence of the strong force. The slightly lower core density of the larger nuclei is presumably a consequence of the mutual repulsion of the many positive charges packed into the nuclear volume. However that may be interpreted, the mean charge ($Z$) density ($\sim$ 0.08 protons/fm$^3$) can be used to calculate the mean nucleon ($A = Z + N$) mass density of $\sim$ 0.16 nucleons/fm$^3$ under the assumption that most nuclei ($Z \sim N$) have similar distributions of protons and neutrons.

Two especially interesting aspects of nuclear density are illustrated in Fig. 3. The first is the anomalously high density of the alpha particle itself. It contains one spin-up and one spin-down nucleon for both the so-called isospin varieties, protons and neutrons – and, as such, should be a good representative of so-called "nuclear matter" ($Z = N$). Also yet, the nuclear matter of all other nuclei has a core density that is only half that of the alpha particle (Fig. 3a). This structural feature is precisely what is implied in a "close-packed" lattice model with an extremely high-density tetrahedron of nucleons forming the lowest energy $^4$He isotope at the nuclear center (see below), but remains

**Table 2.** Summary of the electric and magnetic radii of both protons and neutrons [5].

| Particle | Electric rms radii (fm) | Magnetic rms radii (fm) |
|---|---|---|
| Proton | 0.895±0.018 | 1.086±0.012 |
| Neutron | –0.113±0.003 | 0.873±0.015 |

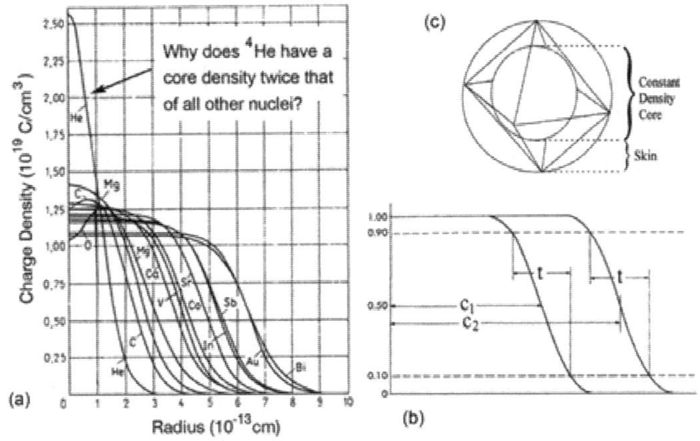


**Figure 3.** (a) The nuclear core density is more-or-less constant across the periodic chart. (b) The skin thickness (t) is also constant. (c) The approximately octahedral shape of most nuclei ($A > 16$) implies a "skin" region where the mean density gradually decreases from the core (data from Hofstadter [6]; reproduced from [7]).

anomalous in a simple-cubic packing (scp) lattice, and is totally unexplained in both liquid-phase and gaseous-phase nuclear models.

The second "anomaly" is the nuclear "skin" region (Fig. 3b) – which, regardless of nuclear radius, exhibits a gradual fall off of density from the core to the nuclear surface. Clearly, a billiard ball-like constant-density sphere (solid, liquid or gas) would exhibit a sharp drop in density from the core to the surface (i.e., no skin), but it is noteworthy that a constant-density, triaxially symmetrical nucleus (i.e., an octahedron with $x = y = z$) (Fig. 3c) would exhibit a gradual decrease from the constant-density core to the apices of the octahedron. In other words, the low-density nuclear skin is consistent with a solid-phase model of nuclear structure, but is anomalous in a spherical liquid-drop model and requires additional assumptions in a gaseous-phase model.

Of importance for understanding the spatiotemporal characteristics of nuclei are the experimental radial measurements that have been obtained from 908 isotopes [8]. As shown in Fig. 4, there is a linear increase in nuclear volume

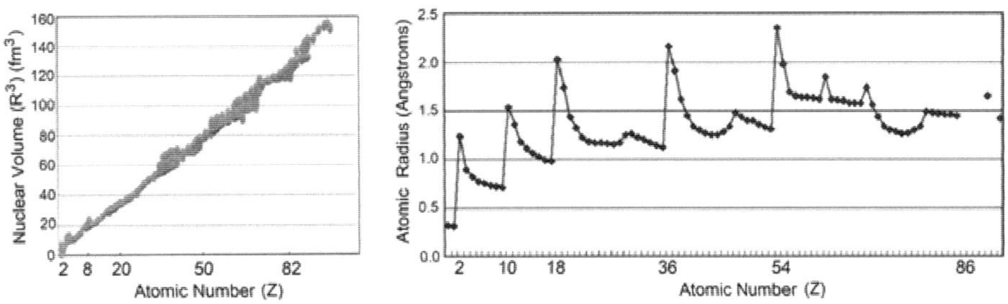


**Figure 4.** A comparison of nuclear and atomic radii. (*Left*) A linear relation is found when the nuclear volume is plotted against the number of protons ($Z$), with only minimal indication of shell structure. *Nuclear* size is strongly dependent on the number of nucleons present – a result to be expected from a constant-density substance (data on 908 isotopes from [8]). (*Right*) In contrast, the covalent radius of atoms shows marked changes – with minima evident at shell closures, and huge jumps when the external valence shell contains one or a few extra electrons [7].

that correlates strongly with the number of protons ($R = 0.995$). In detail, there are interesting isotopic effects that indicate slightly different contributions of protons and neutrons, as well as small shell and subshell effects. The linear trend for *nuclear* radii, however, contrasts sharply with the huge variation for covalent (or ionic) *atomic* radii when plotted against $Z$. That is, the linear increase in *nuclear* charge radii with increasing $Z$ is precisely what would be expected from a constant-density nuclear liquid or solid, whereas the large variations in *atomic* radii are dependent on the closure of electron shells surrounded by external valence electrons – a rather malleable electron "gas" that determines the covalent (or ionic) radius. It is worth emphasizing that the simple fact that both systems are well described by quantum mechanics does *not* imply that nucleons "orbit" within the nuclear interior. On the contrary, nuclei are structurally quite different from atoms.

What do these numbers concerning the nucleon radius and the nuclear texture imply about the local environment of nucleons in the nuclear interior? The long-and-short conclusion is that nuclei are quite literally filled with nucleon matter. Unlike atomic structure – where electron matter occupies only a small percentage of the *atomic* volume, nucleon matter occupies well over 70% of the *nuclear* volume. In other words, unlike atoms, nuclei contain little unoccupied "free space" within which nucleons might randomly roam or "orbit". This empirical fact was already known at the advent of the so-called shell model in the early 1950s, but – in direct contradiction with the structural data – the shell model was built upon the assumption that nucleons are "point particles" that orbit for several intranuclear revolutions before they collide with other nucleons. Understandably, the debates on nuclear structure theory in the 1950s and 1960s were intense! The shell model was deeply unrealistic with regard to the spatiotemporal structure of nuclei, but it was also highly successful in predicting the $J$-values of virtually all ground-state isotopes and more than 500,000 excited states. As a consequence, even today textbooks struggle to explain the dilemma of "successful", but mutually contradictory, nuclear models.

Resolution of the low-density/high-density paradox requires a model that predicts the empirical $J$-values (like the shell model), but is based upon a realistic high-density nuclear texture (like the liquid-drop model) – and, moreover, can explain the alpha-particle substructure of many nuclei. It turns out that all of the $n$-shells and $J$-subshells of the shell model are reproduced in a very specific lattice (an antiferromagnetic face-centered-cubic [fcc] lattice with alternating proton and neutrons layers and built from a central tetrahedron). Two lattice configurations (fcc and hexagonal-close-packing, hcp) exhibit inherent tetrahedral subgrouping (alpha-particle substructure), but only the fcc lattice contains $n$-shells and $J$-subshells consistent with those of the shell model [7] (Section 5). The SCP lattice exhibits neither tetrahedral subgrouping nor any correspondence with the well-known quantum number symmetries of nucleons.

## 3. The Nuclear Force

The starting point for realistic "bottom-up" computations of nuclear structure is the nuclear potential obtained from nucleon–nucleon scattering experiments (Fig. 5). Of importance is the fact that the force is attractive at 0.7–3.0 fm, but strongly repulsive at 0.6 fm or less. The so-called "repulsive core" is a computational headache for theorists (and often unrealistically "softened" for computational convenience), but it bears emphasis that there is no empirical evidence suggesting that nucleons can spatially overlap with one another. Both protons and neutrons are hard-core particles (undoubtedly with internal quark substructure) whose interactions at separations of 1–3 fm (nucleon-center to nucleon-center) utterly dominate the structure of many-nucleon nuclei [9].

Theoretical work on the nuclear potential has been ongoing since the 1950s. As shown in Fig. 5, it consists of several independent components (corresponding to spin and isospin permutations), each of which requires $\sim 10$ variables for a total of 40–50 adjustable parameters [10] in order to reproduce "the" nuclear potential. Needless to say, a 40–50 parameter model allows for a huge range of possible nucleon-nucleon effects that cannot be decided *a priori*, but the overall shape of the nuclear potential (short-range repulsion, mid-range attraction and long-range absence of effect) must be maintained for agreement with experimental facts. Once a specific, multi-parameter model of the

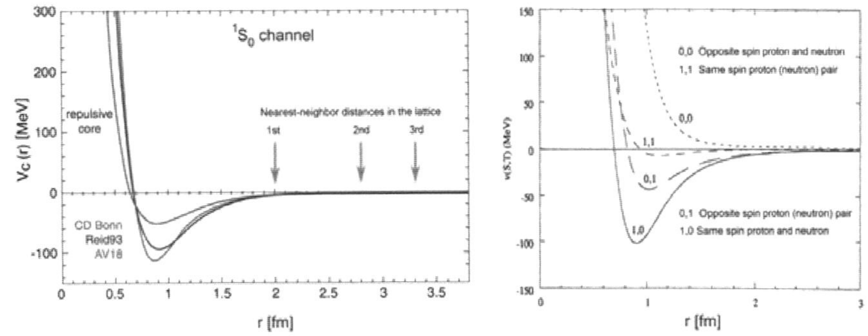


**Figure 5.** (*Left*) Three modern (2014) parametrizations of the nuclear potential well [10]. (*Right*) The spin and isospin components of the AV14 potential (1993) [9]. Most theoretical studies focus on the high-energy, short-range effects (>100 MeV) at distances less than 1.0 fm, but the *known* nuclear force at work in *stable* nuclei produces notably low-energy effects ($< \pm 5$ MeV) at the realistic nucleon–nucleon distances of 2.0–3.0 fm.

nuclear force has been decided on, its application to multinucleon nuclei in so-called "no-core", *ab initio* calculations becomes feasible.

Also herein lies the computational revolution of 21st century nuclear structure theory. Although the nuclear force itself remains an analytically complex mystery, the overall effect of the nuclear potential can be simulated with great precision and implemented within various nuclear models. The possibility for numerical study of nuclear binding energies and excited states, even without a rigorous analytical solution to the nuclear force itself, was of course already discussed in the early days of nuclear structure theory, but the computational reality has remained difficult for all but the smallest ($A \leq 4$) nuclei due to the combinatorial explosion involved in calculating nucleon interactions for larger nuclei.

What now promises to revolutionize nuclear structure theory is the introduction of femto-level nuclear substructure to computational techniques. Ground-breaking work was done by Bauer [11,12] in the 1980s and subsequently by Lee [13] and others, but brought to a new level of rigor by Epelbaum, Meißner and colleagues [10,14–24] with the introduction of what they refer to as "nuclear lattice effective field theory". In the case of both liquid- and gaseous-phase models, the computational bottleneck remains insurmountable insofar as nucleon interactions must be calculated for all nucleon–nucleon permutations in all possible spatiotemporal configurations ($A^{A-1}$) for whatever level of sub-fermi resolution that is deemed necessary. Fortunately, that bottleneck is greatly alleviated when using a solid-phase lattice because (i) the lattice provides a finite number of first-order nuclear configurations (at the $\sim$2 fm lattice-spacing needed to reproduce nuclear core densities), and (ii) the local environment of each nucleon is necessarily determined by the lattice geometry. By restricting nuclear force calculations to nearest and next-nearest neighbors (neglecting more distant neighbors whose influence is, on theoretical grounds, thought to be minimal), the lattice provides a computationally simplified, but rigorous version of the many-body problem of nuclear structure theory.

For those skeptical about the significance of computer simulations, it is worth recalling that molecular dynamics simulations at the atomic/molecular level have already attained a practical level of sophistication in computational chemistry. For example, starting with the coordinates of the atoms contained within large membrane proteins (and the membrane lipids themselves), the mechanisms underlying ligand binding, pore opening/closing, and the flux of ions into and out of the cell has been demonstrated with atomic resolution. Such simulations are fully realistic in terms of the electromagnetic interactions among the atomic constituents. In spite of the fact that direct observation of those molecular events remains impossible (due to the constraints of the uncertainty principle), the simulated spatiotemporal dynamics accurately predict measurable properties (of, for example, ion flow through membrane channels). Note that the distinct advantage of molecular dynamics simulations in computational chemistry, relative to the nuclear case, is *not* due to the scale (dimensions in angstroms vs. femtometers), but simply due to the fact that the *known* electromagnetic force is implemented in the molecular case, whereas the parametrization of the nuclear force is still a contentious issue.

The upshot of the basic structural findings on both nucleons and nuclei (Section 2), and the unambiguous findings on the nuclear force (Section 3) is that the overwhelmingly dominant realm of nuclear force effects in stable nuclei occurs at distances of 1–3 fermi. Again, these short distances from nucleon-center to nucleon-center are the experimental values. Based on theoretical assumptions concerning the much larger distances over which nucleons might "effectively" interact (despite empirical indications to the contrary), a huge "effective field theory" literature has come to dominate nuclear structure theory since the 1950s. Most nuclear theorists would defend effective field theory as a reasonable approach to the difficult many-body problem of the nucleus, but there is a self-fulfilling prophecy at work here. If one assumes that long-range and weak "effective" nucleon-nucleon interactions are important, then a nuclear "gas" (as assumed in the nuclear shell model, ca. 1950) is justified. If, however, one assumes a nuclear liquid (Bohr, 1935) or solid (Wigner, 1937), then the experimentally *known* short-range and strong nuclear force is justified. Both approaches appear to be justifiable, but clearly both cannot be simultaneously correct.

For connoisseurs of the history of nuclear theory, it is worth pointing out that the acceptance of the gaseous texture of the nuclear interior (seemingly an inevitable implication of the shell model) in the early 1950s was the beginning of unbridled theoretical bizarreness in microphysics – first, regarding nucleons (hard-core particles, but transparent to one another, later justified by the invention of a previously unknown "force" implied by the Pauli exclusion principle), secondly, regarding the nuclear force (experimentally short-range, but theoretically long-range, later justified by separating physical parameters into "real" and "imaginary" values), followed by the paradox of quarks (particles that cannot, in principle, be isolated) and onward into string theory and beyond. The well-established and once indubitable facts of nucleon/nuclear size and the nuclear force (Sections 2 and 3) were re-parameterized to allow for a theoretically gaseous nuclear interior, leading to a free-for-all that continues unabated into the 21st century. In contrast, as discussed in Section 5, by deriving the entire shell/subshell structure of the shell model from a particular solid-phase lattice, the necessity of unrealistic, counter-intuitive nuclear modeling disappears. Questions concerning quark substructure remain unanswered, but are safely ignored at the energy levels typical of nuclear structure theory (several MeV per nucleon).

Note that, (i) if nucleons were "point" particles – as assumed in the early shell model; or (ii) if nucleons were somehow transparent to one another – in contradiction to the extremely high-energy "impenetrable core" region of both protons and neutrons; or (iii) if the size of *nucleons* were small relative to the size of *nuclei* (in analogy with atomic physics); or (iv) if the nuclear force acted weakly up to relatively large distances ($\sim$10 fm), then nuclear matter might be accurately characterized as a low-density gaseous substance, with nucleons roaming freely with lengthy "mean free paths" within the nucleus. Molecular dynamics-style simulations would then remain computationally infeasible. The high-density structural reality of the nucleus is, however, quite the converse of a "nucleon gas", and high-resolution, computationally-intensive, spatiotemporal simulations of nuclear structure have become possible.

## 4. Which Lattice?

The one remaining roadblock to realistic nuclear-level simulations concerns the choice of the type of nuclear lattice. The chronic pessimist might argue that there are simply too many possibilities to embark on a coherent research project, but the results of theoretical studies in the 1960s and 1970s on the "condensation" of nuclear matter in neutron stars paint a much more optimistic picture. Specifically, under the assumption of "infinite" nuclear matter, analytical conclusions based on quantum mechanical effects have been obtained for a variety of lattice types with varying spin- and isospin-layering topologies. Those studies have examined the binding properties of the unit structures of the most common of the cube-based lattices known from crystallography, as shown in Fig. 6 and summarized Table 3.

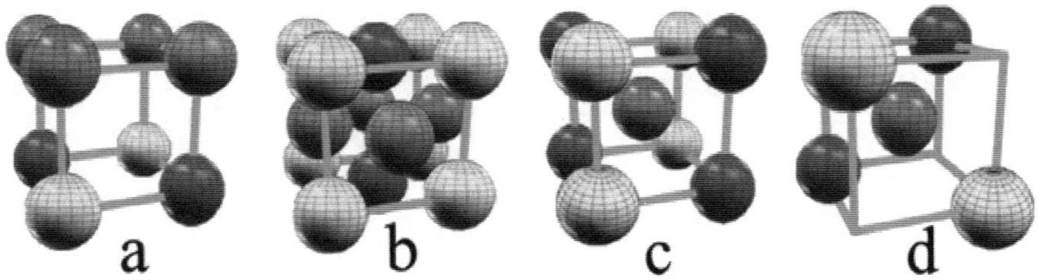


**Figure 6.** The geometry of four of the most common crystal lattices found in nature: (a) simple cubic packing (scp), (b) face-centered cubic (fcc) close-packing, (c) body-centered cubic (bcc) packing, and (d) diamond packing. The two varieties of nucleon, protons and neutrons, are depicted in red and blue, and the spin-up and spin-down properties are distinguished by the different shades of red and blue.

Most importantly, differences among the lattice substructures imply between 4 and 12 nearest-neighbors (i.e., the number of strong nucleon-nucleon interactions per nucleon). Insofar as nuclear force effects are already negligible at 3.0 fermi, the different numbers of nearest and second-nearest neighbors will be crucial in determining nuclear binding effects, while more distant neighbors can be ignored in first approximation. Depending upon assumptions concerning spin and isospin symmetries within the lattice, several candidate models can be enumerated. Comprehensive comparisons among all theoretical possibilities would be useful, but neutron star research has already provided rough guidelines. That is, consideration of both the Coulomb repulsion among protons and the magnetic dipole interactions between nucleons indicates that lattices with (i) isospin layering, and (ii) within each isospin layer, antiferromagnetic alignment of magnetic dipoles will produce greater nuclear binding [26]. Configurations with randomized spin and isospin (etc.) configurations may also be of interest, but whatever tentative answers are obtained from theoretical analysis, they can then be rigorously tested within the framework of "experimental theory", i.e., molecular dynamics simulations.

In conclusion, although various nuclear force parametrizations can be implemented for simulating nuclear dynamics, the remaining unresolved issue at the nuclear level is the purely spatiotemporal question of lattice geometry (lattice type and intranuclear spin/isospin symmetries). Answers are not yet in hand, but theoretical considerations dating from 1937 [27], rediscovered in 1958 [28] and developed in the 1970s–1990s [7,29–33] indicate the importance of the antiferromagnetic fcc lattice with isospin layering (Fig. 7).

**Table 3.** The characteristics of five of the most common crystal lattices [25].

| Lattice type | fcc/bcc | bcc | scp | Diamond |
|---|---|---|---|---|
| Number of nearest neighbours | 12 | 8 | 6 | 4 |
| Nearest-neighbour distance (fm)[a] | 2.0262 | 1.9697 | 1.8052 | 1.7896 |
| Number of second-nearest neighbour | 6 | 6 | 8 | 8 |
| Second-nearest neighbor distance (fm) | 2.8655 | 2.2744 | 2.5529 | 2.9223 |
| Coulomb force between neighbouring protons (MeV)[b] | 0.7107 | 0.6331 | 0.7977 | 0.4927 |

[a]Nearest-neighbor distance that implies a density equivalent to that of the nuclear core (0.17 nucleons $fm^{-3}$).
[b]Assuming structures with alternating isospin layers, the Coulomb force is that at the nearest-neighbor distance in the fcc/hcp and scp lattices, with a larger Coulomb effect in the scp lattice because of the shorter nearest-neighbor distance. In the bcc and diamond lattices, the Coulomb force is that between second-nearest neighbors (which is as close as protons come to one another in these lattices), with a stronger effect in the more compact bcc lattice.

**Figure 7.** Illustration of the *realistic* spatial dimensions of the nuclear core. While the colors and the spherical structure of nucleons are fanciful "artistic renditions", the size of the nucleons and the density of their packing in the nuclear core are drawn using the empirical values. In the antiferromagnetic fcc nuclear lattice model, the spin and isospin layering implies attractive nearest-neighbor magnetic dipole interactions within each layer and a means for reducing Coulomb repulsion between the proton layers, respectively. The inherent tetrahedral (alpha particle) and octahedral regions in the close-packed lattice are apparent on the right. Note that the fcc "unit cube" contains eight tetrahedral spaces and a total of four octahedral spaces (one at the center of the cube and 12 octahedral quadrants on the 12 edges of the cube). This geometry gives the unit cube an overall density approximately one half that of the tetrahedron. A nearest-neighbor distance (d) of 2.026 fm implies a cube edge distance of 2.865 fm. Since each unit cube contains a total of four nucleons (six hemispheres on the cube faces and eight octants at the corners), the density of the nuclear interior (shown left and center) is $4/(2.865)^3$ nucleons per $fm^3$ or 0.170 n/$fm^3$, which is the known core density [6].

## 5. Nuclear Structure Theory

The foundations of the independent-particle model (IPM) were laid by Eugene Wigner [27] and later exploited in the development of the shell model by Mayer and Jensen – all three receiving Nobel Prizes for nuclear structure theory in 1963. Although Wigner clearly stated that the symmetries of the quantum numbers in the nuclear Hamiltonian had an inherent fcc structure, the spatiotemporal geometrical implications were generally neglected in the development of "non-spatiotemporal" nuclear structure theory through the 1950s and 1960s. In the recent development of computational nuclear lattice theory, spatiotemporal considerations have become inevitable. Unfortunately, most researchers have resorted to the simple cubic packing scheme (for its obvious algorithmic simplicity) and not made comparisons among different lattice configurations. Here I show that, despite the slightly more complex lattice geometry of the close-packing schemes (see the Appendix), the fcc lattice exhibits a remarkably easy-to-understand isomorphism with the conventional IPM – a geometrical fact first illustrated with a pen-and-ink freehand drawing by Wigner in *Physical Review* [27] 80 years ago!

Inevitably, the first impression of such visual depictions of the nuclear texture is one of classical mechanics, but Wigner's main argument concerned the quantal symmetries inherent to the nuclear Hamiltonian. Whatever may be the correct physical interpretation of those symmetries, they are fundamentally a consequence of the well- established quantum texture (noted by subscripts *n, l, j, m, s, i* and parity for each nucleon) that is the essence of the Schrödinger wave equation used in nuclear quantum mechanics. In many publications (e.g., [7,25,29–35]), we have shown that all of the nucleon quantum numbers have unambiguous geometrical definitions within the framework of the fcc lattice (Eqs. (1)–(7)):

$$\text{principal}, n = (|x| + |y| + |z| - 3)/2, \tag{1}$$

$$\text{orbital angular momentum}, \ l = (|x| + |y|)/2, \tag{2}$$

$$\text{total angular momentum}, \ j = (|x| + |y| - 1)/2, \tag{3}$$

$$\text{azimuthal}, m = (|x| \ (-1)^{(x-1)/2})/2, \tag{4}$$

$$\text{spin}, s = ((-1)^{(x-1)/2}/2, \quad (5)$$

$$\text{isospin}, i = ((-1)^{(z-1)/2}/2, \quad (6)$$

$$\text{parity}, \ \pi = \text{sign}(xyz), \quad (7)$$

where all quantum numbers are defined in terms of each nucleon's unique set of $x, y, z$ coordinates in Cartesian space. *All* of the nucleon shells/subshells and their occupancies are thereby reproduced. That fact indicates that the IPM and the lattice model are fundamentally isomorphic, but they clearly differ in implying: a diffuse, gaseous nuclear interior, on the one hand, or a high-density nuclear interior where nucleon–nucleon interactions are local, on the other.

In fact, the physical interpretation of the mathematical correspondence between the symmetries of the nuclear Hamiltonian and the fcc lattice is far from obvious. Both the "orbital angular momentum", $l$, and the "total angular momentum, $j$, cannot be realistically interpreted as indicating the "orbiting" of nucleons within the extremely dense nuclear interior around a central potential-well of the nuclear force. Both angular momentum values for nucleons must therefore be understood as indicative of the orbital motion of charge within the individual nucleon itself. In fact, that assumption leads to realistic magnetic dipole moments that are dependent on nucleon $j$-values, as summarized by the so-called Schmidt lines of conventional nuclear structure theory, but leaves questions unanswered concerning the internal structure of the nucleons. The non-orbiting of nucleons within the nuclear volume is of course the reason why all of the details of nuclear structure physics cannot be explained simply in terms of a highly-miniaturized version of the $s$-, $p$, $d$-, $f$-orbitals for electrons in the theory of atomic structure.

Suffice it to say that, based on Eqs. (1)–(7), there is a deep correspondence between the fcc lattice and the known symmetries of the nucleus [7]. Its physical interpretation still needs the attention of nuclear structure theorists, but it is worth reiterating that nothing similar is found with other nucleon lattices. The known symmetries of nucleon quantum states – as successfully summarized in the shell model from 1950 – is reproduced in its entirety uniquely in the fcc lattice. It is for this reason that the rather sophisticated computational work currently being done in "nuclear lattice effective field theory" needs to be expanded to include a comparison of, at the very least, both scp and fcc lattices, in order to make a fair-minded, objective evaluation of candidate lattice configurations.

The dangers inherent to proceeding on the basis of unanalyzed starting assumptions is fully apparent in the nuclear lattice simulations already published by Epelbaum, Meißner and colleagues [10, 13–24]. On the one hand, they have implemented a modern version of the nuclear potential that is far more sophisticated than the nuclear force effects in earlier lattice model simulations of multi-fragmentation effects or nuclear binding energies [1,12]. On the other hand, they have examined nuclear binding effects only in the scp lattice and, moreover, using an unusual parametrization of the nuclear potential that leads to counterintuitive results (Fig. 8).

The nuclear potential used in obtaining the results shown in Fig. 8 allows multiple nucleons to reside at any given lattice site [10,13–24]. Indeed, experiments at much higher energies (>300 MeV) than are likely to occur in stable nuclei suggest that there is some degree of nucleon "transparency", but, at the low energies of nucleon–nucleon binding in nuclear structure physics (<10 MeV), nucleons typically exhibit an impenetrable "hard core" ($Q > 300$ MeV) of 0.6 fm (Fig. 5). Clearly, simulation results very different from those illustrated in Fig. 8 would be obtained from a parametrization that forbids overlap of nucleons. As Meißner and colleagues themselves have commented, "A concerted effort should be made to improve the current computational algorithms to handle interactions with more short-range repulsion" [21].

The assumption of an scp lattice is also deeply suspect. In the alpha-clustering studies by the Meißner group, they report alpha clusters (four nucleons on a given lattice site) for $^{16}O$, with again counter-intuitive geometries. The lowest-energy 4-cluster configuration was reported as "tetrahedral" (Fig. 8), but is actually a quarter-pyramid, not a regular tetrahedron. The difference between these two geometries is that the quarter -pyramid has three alpha clusters

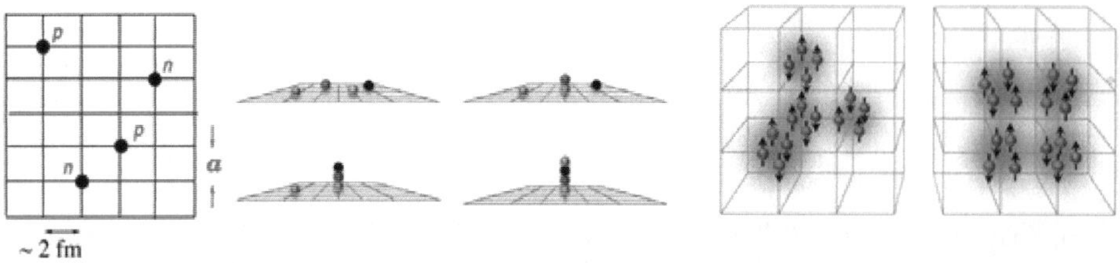


**Figure 8.** The "nuclear lattice effective field theory" advocated by Meißner and colleagues. (*Left and Center*) The basic dimensions of the 3D scp lattice and the occupancy of lattice sites with up to four nucleons (spin up/down protons and neutrons). A huge number of configurations of 16 nucleons distributed at random within a $6 \times 6 \times 6$ lattice volume (216 grid sites) was generated and total nuclear binding energies calculated [18]. (*Right*) The lowest-energy alpha particle structure obtained for $^{16}O$ was a quarter-pyramid, but was described as "tetrahedral" and the second-lowest-energy structure was square planar. Neither result requires supercomputer calculations, and both are a simple consequence of the high binding energy of four nucleons located on the same lattice site (due to the a priori assumption of a "soft" nuclear potential with no short-range repulsion)..

bound to a common vertex, but without nearest-neighbor bonds among the three remaining alphas, whereas a regular tetrahedron of alphas would have fully six nearest-neighbor alpha-alpha bonds. A regular tetrahedral configuration was not obtained in the Meißner simulation for the simple reason that an scp lattice constrains all nucleons to sites on the cubic coordinates of the lattice; there are no tetrahedral substructures. As such, the reported "reproduction" of the traditional tetrahedral alpha structure of the conventional alpha-cluster model, is utterly false. Their alpha structure is "tetrahedral" only in the sense that any non-planar configuration of four points gives a 4-vertex, 6-edge, 4-sided structure can be described as tetrahedral, but the claim for a *regular* tetrahedron requires that all six edges are of the same length.

For anyone interested in the possibility of publishing in the mainstream nuclear physics literature, it is relevant to point out that the Meißner paper [18] reporting a "tetrahedral" alpha structure for $^{16}O$ was published in one of the world's leading physics journals, *Physical Review Letters*. The computational result indicating a "tetrahedral" geometry is clearly misleading, but in fact most of the methodological detail has been published elsewhere – notably, in 76 (!) papers by Meißner in the *European Physical Journal G: Hadrons and Nuclei*. By chance, this is the same journal where Meißner has been the Editor responsible for "theoretical reviews" for the past 13 years [36].

Again, the counterintuitive results are a direct result of the choice of (i) an scp lattice, and (ii) a parametrization of the nuclear force that does *not* reproduce the known nucleon hard-core. Such assumptions about the lattice type may fall within the realm of plausible hypotheses, but the results obtained through computer simulations then need to be compared between (i) an fcc lattice, and (ii) an scp lattice, and most importantly (iii) utilizing parametrizations of the nuclear force that realistically reproduce the nucleon hard-core radius of 0.6 fm. Without the realism of a nuclear force that is deduced from low-energy experimental results, it is questionable why one should bother with the lattice simulations in the first place.

## 6. LENR

Finally, let us consider the conundrum of LENR research. The intellectual challenge of low-energy nuclear reactions has many aspects, but the clearest indications that LENR is truly nuclear come from the published works on nuclear transmutations. Notable are the papers by Tadahiko Mizuno [37], Yasuhiro Iwamura [38], George Egely [39], and the group that examined the ash obtained from Andrea Rossi's E-Cat [40]. Many other, unreplicated reports of transmutation are to be found in the literature, but they are, without exception, highly suspect from the perspective of conventional nuclear physics. The same, of course, could be said about the early reports of the fission of Uranium isotopes.

Only once the phenomenon was shown to be replicable – with reliable, repeated detection of neutron radioactivity, heat, and fission transmutation products – did there emerge a consensus among scientists that fission was real. Technological applications followed immediately. The much lower energies involved in LENR and the apparently much more delicate chemistry that produces nuclear active environments have made the phenomena harder to reproduce. When experimental LENR systems have become established, it will then be possible to focus on the low-energy dynamics of the nucleus – i.e., traditional nuclear structure physics – rather than the extremely high-energy phenomena that continue to fascinate most theorists.

Many constructive suggestions about how to demonstrate LENR phenomena and proceed to applications have been made. The most promising theoretical developments concern the production of neutrons from low-lying electron states [41,42] – leading ultimately to a diversity of low-energy nuclear reactions. Unfortunately, most proposals for experimental research imply massive funding that is unlikely to be realized until the fundamental phenomena themselves are shown to be replicable and open to detailed analysis. In contrast, a rather modest theoretical research project could be undertaken strictly within the framework of molecular dynamics-like simulations at the nuclear level. While experimentalists deal with the issues of the mechanisms for inducing LENR, quantitative simulations could, in principle, predict the likely nuclear reaction products for comparison against experimental findings.

## 7. Conclusions

A "return" to Wigner's ideas from 1937 [27] may appear to be "regressive" to some theorists, but it is relevant to note that theoretical progress in clarifying the so-called nuclear force has been remarkably slow since the 1930s. Authors of modern textbooks on nuclear physics are still obliged to note that admittedly imperfect, contradictory nuclear models are employed specifically *because* fundamental questions concerning the nuclear force remain unanswered. The theoretical stalemate concerning the nuclear models contrasts sharply with the remarkable advances in both experiment and practical applications of all things nuclear. Although theoretical nuclear physics is *not* totally unhinged pseudoscience, its lack of spatiotemporal grounding and the absence of a unified understanding of the nuclear force are noteworthy deficiencies. Nevertheless, despite those shortcomings, computational power has grown significantly in the past two decades and, indeed, the empirical landscape of nuclear stability, instability and excitability has been mapped out in impressive detail. As a consequence, prospects for achieving a self-consistent understanding of nuclear structure – even while using a make-shift, imperfect model of the nuclear force – have improved remarkably.

Contrary to first impressions, recasting nuclear structure theory within the lattice representation of quantal symmetries is a less radical renovation than is currently appreciated by theorists still wedded to the weak analogies and unrealistic models of the 1950s and 1960s. Principally because the fcc lattice symmetries map directly onto the known quantal symmetries of the nucleus, there is reason to believe that the present generation of computationally intensive nuclear structure simulations will produce coherent answers that finally put to rest the "many mutually contradictory models" of traditional nuclear structure theory. At the very least, it can be said today that the properties of the nuclear force *known* from nucleon-nucleon scattering experiments are consistent with the lattice perspective on nuclear structure. A return to spatiotemporal nuclear theory would be welcomed.

Finally, we can look forward to applications of lattice simulations to the problems in LENR. The transmutation data from many groups are some of the most tantalizing new findings in nuclear physics since the discovery of fission in 1938. In parallel with continuing efforts by the experimentalists, attempts at sorting out the wheat from the chaff using fermi-level molecular dynamics simulations would be useful.

## Appendix. An Algorithm in C for Generating Sequential Shells of fcc Coordinates

Computational "nuclear lattice effective field theory" is arguably on track for providing numerically satisfying answers to many of the old puzzles in nuclear structure theory. The chore of comparing various lattice configurations has just begun, but, in whatever way that such work may proceed, it will be necessary to use parametrizations of the nuclear force that reproduce the known hard-core repulsion of the nuclear force. From there, it will be of interest to make comparisons between, at the very least, various spin-, and isospin-configurations within scp and fcc lattices. Because of the algorithmic simplicity of simple-cubic packing, most previous studies have examined exclusively the scp lattice, but it is, on the contrary, uniquely the fcc lattice that exhibits interesting symmetries related to the symmetries known in traditional nuclear structure theory [7]. For that reason, the algorithm below may be useful for producing isospin-layered, antiferromagnetic fcc lattice coordinates in sequential shells containing 4,16,40,80,140,224, and 336 nucleons ($Z = N$). An Excel spreadsheet with similar results [34] and a Fortran program which includes bcc and diamond lattices can be found elsewhere [35]. See the following program.

```
#include <stdio.h>
#include "math.h"
int x, y, z, N, L, JJ, M, i, K=1, f, parity, nuc[241][3][2];
int Nvalue[241][2], Lvalue[241][2], Svalue[241][2], Jvalue[241][2], Mvalue[241][2], Pvalue[241][2];

int main(int argc, char** argv){
   printf(" N(x   y   z) N  L   J    M    S   P   P(x   y   z) N  L   J    M    S   P ¥n");
   for(N=0;N<=6;N=N+1){               /* N-shells up to N=6 to give 168 protons and 168 neutrons */
      for(L=N;L>=0;L=L-1){            /* L-subshells, higher L-values come first: L = N,..,2,1,0 */
         JJ=(2.0*((float)L+.5));      /* JJ is 2*j-value, J = |L+spin| = 1, 3, 5, ...            */
         for(M=-JJ;M<=JJ;M=M+2){      /* M is 2*m-value.  M = -JJ, ..., -3, -1, 1, 3, ..., JJ    */
            for(i=-1;i<=1;i=i+2){     /* isospin: neutrons = -1, protons = 1                     */
               f=((i+1)/2);           /* 0 and 1 for neutrons and protons                        */
               x=(abs)(M)*(pow(-1.0,(float)M/2.0+.5));
               y=(JJ+1.0-abs((int)x))*pow(-1.0,-i/2.0+JJ/2.0+M/2.0+.5);
               z=(2.0*N+3.0-abs((int)x)-abs((int)y))*pow(-1.0,-i/2.0+N-JJ/2.0-1.0);
               nuc[K][0][f]=(int)x;
               nuc[K][1][f]=(int)y;
               nuc[K][2][f]=(int)z;
               Nvalue[K][f]=(abs((int)x)+abs((int)y)+abs((int)z)-3)/2;       /* N is an integer */
               Lvalue[K][f]=L;                                                /* L is an integer */
               Jvalue[K][f]=(abs(nuc[K][0][f])+abs(nuc[K][1][f]))-1;          /* J is 2*j        */
               if(fmod((float)nuc[K][0][f]+1.0,4.0)==0.0) Svalue[K][f]=-1;    /* -1 "spin down"  */
               else Svalue[K][f]=1;                                           /* 1 "spin up"     */
               Mvalue[K][f]=(abs(nuc[K][0][f]))*Svalue[K][f];                 /* M is 2*m        */
               parity=nuc[K][0][f]*nuc[K][1][f]*nuc[K][2][f];                 /* P is +1 or -1   */
               if(parity>=0) Pvalue[K][f]=1; else Pvalue[K][f]=-1;
            } /* end of isospin i-loop */

printf("%d%3d %3d %3d  %2.f %2.f %2d/2 %2d/2 %2d/2 %2d   %3d %3d %3d  %2.f %2.f %2d/2 %2d/2 %2d/2 %2d ¥n",
K, nuc[K][0][0], nuc[K][1][0], nuc[K][2][0], (float)Nvalue[K][0], (float)Lvalue[K][0], Jvalue[K][0],
Mvalue[K][0], Svalue[K][0], Pvalue[K][0], nuc[K][0][1], nuc[K][1][1], nuc[K][2][1], (float)Nvalue[K][1],
(float)Lvalue[K][1], Jvalue[K][1], Mvalue[K][1], Svalue[K][1], Pvalue[K][1]);
            K=K+1;
         } /* end of M-loop */
      } /* end of L-loop */
   } /* end of N-loop */
} /* end of main function */
```

Sample output where fcc lattice coordinates [$n(xyz)$ and $p(xyz)$] are followed by their quantum numbers.

| | $N(x,y,z)$ | | | $N$ | $L$ | $J$ | $M$ | $S$ | $P$ | $p(xyz)$ | | | $N$ | $L$ | $J$ | $M$ | $S$ | $P$ |
|---|---|---|---|---|---|---|---|---|---|---|---|---|---|---|---|---|---|---|
| 1 | 1 | −1 | −1 | 0 | 0 | 1/2 | 1/2 | 1/2 | 1 | 1 | 1 | 1 | 0 | 0 | 1/2 | 1/2 | 1/2 | 1 |
| 2 | −1 | 1 | −1 | 0 | 0 | 1/2 | −1/2 | −1/2 | 1 | −1 | −1 | 1 | 0 | 0 | 1/2 | −1/2 | −1/2 | 1 |
| 3 | −3 | −1 | −1 | 1 | 1 | 3/2 | 3/2 | 1/2 | −1 | −3 | 1 | 1 | 1 | 1 | 3/2 | 3/2 | 1/2 1/2 | −1 |
| 4 | 1 | 3 | −1 | 1 | 1 | 3/2 | 1/2 | 1/2 | −1 | 1 | −3 | 1 | 1 | 1 | 3/2 | 1/2 | 1/2 | −1 |
| 5 | −1 | −3 | −1 | 1 | 1 | 3/2 | −1/2 | −1/2 | −1 | −1 | 3 | 1 | 1 | 1 | 3/2 | −1/2 | −1/2 | −1 |
| 6 | 3 | 1 | −1 | 1 | 1 | 3/2 | −3/2 | −1/2 | −1 | 3 | −1 | 1 | 1 | 1 | 3/2 | −3/2 | −1/2 | −1 |
| 7 | 1 | −1 | 3 | 1 | 0 | 1/2 | 1/2 | 1/2 | −1 | 1 | 1 | −3 | 1 | 0 | 1/2 | 1/2 | 1/2 | −1 |
| 8 | −1 | 1 | 3 | 1 | 0 | 1/2 | −1/2 | −1/2 | −1 | −1 | −1 | −3 | 1 | 0 | 1/2 | −1/2 | −1/2 | −1 |

## References


[1] N. Bohr, Atomic theory and mechanics, In *Philosophical Writings of Niels Bohr,* Vol. 1, 1925, p. 48, reprinted by Woodbridge, CN: Ox Bow Press, 1987.

[2] A. Plotnitsky, *Niels Bohr and Complementary*, Chapter 3, Springer, New York, 2013.

[3] R.M. Littauer, H.F. Schopper and R.R. Wilson, *Phy. Rev. Lett.* **7** (1961) 144.

[4] I. Sick and D. Trautmann, Proton root-mean-square radii and electron scattering, *Phy. Rev. C* **89** (2014) 012201.

[5] I. Sick, *Prog. Particle Nucl. Phys.* **55** (2005) 440.

[6] R. Hofstadter, *Rev. Modern Phys.* **28** (1956) 214.

[7] N.D. Cook, *Models of the Atomic Nucleus*, 2nd Edn., Springer, New York, 2010, freely available at: www.res.kutc.kansai-u.ac.jp/~cook/PDFs/ man2.pdf.

[8] I. Angeli and K.P. Marinova, Table of experimental nuclear ground state charge radii: An update, *Atomic Data and Nuclear Data Tables* **99** (2013) 1

[9] R.B. Wiringa, *Rev. Modern Phys.* **65** (1993) 231.

[10] U.-G. Meißner, A new tool in nuclear physics: Nuclear lattice simulations, *Nucl. Phys. News* **24** (2014) 11.

[11] W. Bauer, D.R. Dean, U. Mosel and U. Post, New approach to fragmentation reactions – The nuclear lattice model, *Phys Lett B* **150** (1985) 53

[12] W. Bauer, U. Post, D.R. Dean and U. Mosel, The nuclear lattice model of proton-induced multifragmentation, *Nucl. Phys A* **452** (1986) 699

[13] D. Lee, Lattice simulations for few- and many-body systems, arXiv:0804.3501v2 (2008).

[14] B. Borasoy, E. Epelbaum, H. Krebs, D. Lee and U.-G. Meißner, *Euro. Phys. J. A* **31** (2007) 105.

[15] E. Epelbaum, H.-W. Hammer and U.-G. Meißner, *Rev. Modern Phys.* **81** (2009) 1773.

[16] E. Epelbaum, H. Krebs, T.A. Lahde, D. Lee,U.-G. Meißner and G. Rupak, *Phy. Rev. Lett.* **112** (2014) 10.

[17] E. Epelbaum, H. Krebs, A. Lahde, D. Lee and U.-G. Meißner, *Euro. Phys. J. A* **49** (2013) 82.

[18] S. Elhatisari, N. Li, A. Rokash, J. M. Alarcon, D. Du, N. Klein, B.-N. Lu, U.-G. Meißner, E. Epelbaum, H. Krebs, T. A. Lahde, D. Lee and G. Rupak, Nuclear binding near a quantum phase transition, *Phy. Rev. Lett.* **117** (2016) 132501.

[19] E. Epelbaum H. Krebs, T.A. Lähde, D. Lee, U.-G. Meißner and G. Rupak, *Phy. Rev. Lett.* **112** (2014) 102501.

[20] U.-G. Meißner, *EPJ Web of Conferences* **66** (2014) 01012.

[21] T.A. Lähde, E. Epelbaum, H. Krebs, D. Lee, U.-G. Meißner and G. Rupak, Lattice effective field theory for medium-mass nuclei, *Phys. Lett. B* **732** (2014) 110.

[22] T.A. Lähde, T. Luu, D. Lee, U.-G. Meißner, E. Epelbaum, H. Krebs and G. Rupak, *Euro. Phys. J. A* **51** (2015) 92.

[23] S. Elhatisari, D. Lee, G. Rupak, E. Epelbaum, H. Krebs, T.A. Lähde, T. Luu and U.-G. Meißner, *Nature* **528** (2015) 111.

[24] U.-G. Meißner, T.A. Lähde and T. Luu, Pushing the boundaries of nuclear physics with lattice simulations, in, *NIC Symposium 2016*, K. Binder, M. Mueller, M. Kremer and A. Schnurpfeil (Eds.), NIC Series, Vol. 48, 2016.

[25] N.D. Cook, *J. Phys. G: Nucl. Particle Phys.* **20** (1994) 1907.
[26] V. Canuto and S.M. Chitre, *Nature Phys. Sci.* **243** (1973) 63.
[27] E. Wigner, *Phy. Rev.* **51** (1937) 106.
[28] F. Everling, Notarized document (1958) and many references at www.everling-nuclear-lattice-model.com/ (2016).
[29] K.J. Lezuo, *Atomkernenergie* **23** (1974) 285.
[30] N.D. Cook, *Atomkernenergie* **28** (1976) 195.
[31] V. Dallacasa, *Atomkernenergie* **31** (1981) 143.
[32] N.D. Cook and V. Dallacasa, *Phy. Rev. C* **36** (1987) 1883.
[33] N.D. Cook, Is the lattice gas model a unified model of nuclear structure? *J. Phys. G* **25** (1999) 1213.
[34] N.D. Cook, available at: http://www.res.kutc.kansai-u.ac.jp/~cook/QuantumNumbers.xls.
[35] N.D. Cook, Computing nuclear properties in the FCC model, *Computers Phys.* **3** (1989) 73, available at: http://www.res.kutc.kansai-u.ac.jp/~cook/PDFs/CompPhys.pdf.
[36] U-.G. Meißner, https://www.itkp.uni-bonn.de/~meissner/cvr.pdf.
[37] T. Ohmori and T. Mizuno, Nuclear transmutation reaction caused by light water electrolysis on tungsten cathode under incandescent conditions, *Infinite Energy* **5** (1999) 27.
[38] Y. Iwamura, T. Itoh and S. Tsuruga, Transmutation reactions induced by deuterium permeation through nano-structured Pd multilayer thin film, *Current Sci.* **108** (2015) 628.
[39] G. Egely, Nanodust fusion, *Infinite Energy* **11** (2012) 102.
[40] G. Levi, E. Foschi, T. Hartman, B. Höistad, R. Pettersson, L. Tegnér and H. Essén, Indication of anomalous heat energy production in a reactor device, arXiv:1305.3913v3 (2013).
[41] J.L.Paillet and A. Meulenberg, Special relativity: the source of electron deep orbits, *Found. Phys.,* online 09/02/2017.
[42] R. Godes, http://brillouinenergy.com/science/the-quantum-reaction-hypothesis/.

J. Condensed Matter Nucl. Sci. 24 (2017) 75–86

Research Article

# Catalytic Mechanism of LENR in Quasicrystals based on Localized Anharmonic Vibrations and Phasons

V. Dubinko*

*NSC "Kharkov Institute of Physics and Technology," Kharkov, Ukraine*

D. Laptev

*B. Verkin Institute for Low Temperature Physics and Engineering, Kharkov, Ukraine*

K. Irwin

*Quantum Gravity Research, Los Angeles, USA*

---

**Abstract**

We propose a mechanism explaining high catalytic activity of quasicrystals (QCs), which is based on unusual dynamics of atoms at special sites in QCs, namely, localized anharmonic vibrations (LAVs) and phasons. With the vibrations, one deals with a large amplitude (fractions of an angstrom) time-periodic oscillations of a small group of atoms around their stable positions in the lattice, known also as discrete breathers, which can be excited in regular crystals as well as in QCs. On the other hand, phasons are a specific property of QCs, which are represented by very large amplitude (angstrom) oscillations of atoms between two quasi-stable positions determined by the geometry of a QC. Large amplitude atomic motion in LAVs and phasons may result in time-periodic driving of adjacent potential wells occupied by hydrogen ions (protons or deuterons). The rate of tunneling of the particle through the potential barrier separating the wells is shown to be drastically enhanced by the driving. These results support the concept of nuclear catalysis in QCs that can take place at special sites provided by their inherent topology.


*Keywords:* Quasicrystals, Localized anharmonic vibrations, Low energy nuclear reactions, Nuclear active sites, Phasons, Tunneling

---

## 1. Introduction

The tunneling through the Coulomb potential barrier during the interaction of charged particles presents a major problem for the explanation of low energy nuclear reactions (LENR) observed in solids [1–3]. Corrections to the cross section of the fusion due to the screening effect of atomic electrons result in the so-called "screening potential," which

*E-mail: vdubinko@hotmail.com

© 2017 ISCMNS. All rights reserved. ISSN 2227-3123

is far too weak to explain LENR observed at temperatures below the melting point of solids. Nobel laureate Julian Schwinger proposed that a substantial suppression of the Coulomb barrier may be possible at the expense of *lattice vibrations* [4,5]. The fusion rate of deuteron–deuteron or proton–deuteron oscillating in adjacent lattice sites of a metal hydride, according to the Schwinger model, is about $10^{-30}$ $s^{-1}$ [6], which is huge as compared to the conventional evaluation by the Gamow tunnel factor ($\sim 10^{-2760}$). However, even this is too low to explain the observed excess heat generated, e.g. in Pd cathode under $D_2O$ electrolysis. The fusion rate by Schwinger is extremely sensitive to the amplitude of *zero-point vibrations* (ZPV) of the interacting ions, which has been shown to increase under the action of time-periodic driving of the harmonic potential well width [6]. Such a driving can be realized in the vicinity of *localized anharmonic vibrations* (LAVs) defined as large amplitude ($\sim$ fractions of an angstrom) time-periodic vibrations of a small group of atoms around their stable positions in the lattice. A sub-class of LAV, known as *discrete breathers*, can be excited in regular crystals by heating [1–3,7] or irradiation by fast particles [8]. Based on that, a drastic increase of the D–D or D–H fusion rate with increasing number of driving periods has been demonstrated in the framework of the modified Schwinger model [6,8].

One of the most important practical recommendations of the new LENR concept is to look for the *nuclear active environment* (NAE), which is enriched with nuclear active sites, such as the LAV sites. In this context, a striking *site selectiveness* of LAV formation in disordered structures [9] allows one to suggest that their concentration in quasicrystals (QCs) may be very high as compared to regular crystals where discrete breathers arise homogeneously, and their activation energy is relatively high. Direct experimental observations [10] have shown that in the decagonal quasicrystal $Al_{72}Ni_{20}Co_8$, *mean-square thermal vibration amplitude* of the atoms at special sites substantially exceeds the mean value, and the difference increases with temperature. This might be the first experimental observation of LAV, which has shown that they are arranged in just a few nanometers from each other, so that their average concentration was about $10^{20}$ per cubic cm that is many orders of magnitude higher than one could expect to find in periodic crystals [1–3,7]. Therefore, in this case, one deals with a kind of *'organized disorder'* that stimulates formation of LAV, which may explain a strong catalytic activity of quasicrystals [11].

In addition to the enhanced susceptibility to the LAV generation, QCs exhibit unique dynamic patterns called *phasons*, which are represented by *very* large amplitude ($\sim$ angstrom) quasi time-periodic oscillations of atoms between two quasi-stable positions determined by the geometry of a QC. It is natural to expect that the driving effect of phasons can exceed that of LAVs due to the larger oscillation amplitude in phasons. The main goal of the present paper is to develop this concept to the level of a quantitative comparison between the driving/catalytic action of LAVs and phasons, which could be used to suggest some practical ways of catalyzing LENR.

The paper is organized as follows. In the next section, the Schwinger model [4,5] and its extension [6] are shortly reviewed to demonstrate an importance of time-periodic driving of potential wells in the LENR triggering.

In Section 3, we extend our analysis beyond the model case of infinite harmonic potential (the tunneling from which is impossible) and obtain numerical solution of Schrödinger equation for a particle in a *non-stationary* double well potential, which is driven time-periodically imitating the action of a LAV or phason. We show that the rate of tunneling of the particle through the potential barrier separating the wells is enhanced drastically by the driving, and it increases strongly with increasing amplitude of the driving. In Section 4, we present some examples of dynamic patterns in QCs and their clusters and discuss the ways of experimental verification of the proposed concept. The summary and outlook is given in Section 5.

## 2. Schwinger Model of LENR in an Atomic Lattice Modified with Account of Time-periodic Driving

According to Schwinger [4], the effective potential of the deuteron–deuteron (D–D) or proton–deuteron (P–D) interactions is modified due to averaging ${}_0\langle\ \rangle_0$ related to their *zero-point vibrations* (ZPV) in adjacent harmonic potential wells, where ${}_0\langle\ \rangle_0$ symbolizes the phonon vacuum state. This means that nuclei in the lattice act not like point-like charges, but rather (similar to electrons) they are "smeared out" due to quantum oscillations in the harmonic potential wells near the equilibrium positions. The resulting effective Coulomb interaction potential$_0$ $\langle V_c(r)\rangle_0$ between a proton and a neighboring ion at a distance $r$ can be written, according to [4] as

$$_0\langle V_c(r)\rangle_0 = \frac{Ze^2}{r}\sqrt{\frac{2}{\pi}}\int_0^{r/\Lambda_0} dx \exp\left(-\tfrac{1}{2}x^2\right) \approx \begin{cases} r \gg \Lambda_0 : \frac{Ze^2}{r} \\ r \ll \Lambda_0 : \left(\frac{2}{\pi}\right)^{1/2}\frac{Ze^2}{\Lambda_0}, \end{cases} \tag{1}$$

where $Z$ is the atomic number of the ion, $e$ is the electron charge, $\Lambda_0 = (\hbar/2m\omega_0)^{1/2}$ is the ZPV amplitude, $\hbar$ is the Plank constant, $m$ is the proton mass, and $\omega_0$ is the angular frequency of the harmonic potential. A typical value of $\Lambda_0 \sim 0.1$ Å, which means that the effective repulsion potential is saturated at approximately *several hundred eV* as compared to *several hundred keV* for the unscreened Coulomb interaction. Schwinger estimated the rate of fusion as the rate of transition out of the phonon vacuum state, which is reciprocal of the mean lifetime $T_0$ of the vacuum state, which can be expressed via the main nuclear and atomic parameters of the system [5,6]:

$$\frac{1}{T_0} \approx 2\pi\omega_0 \left(\frac{2\pi\hbar\omega_0}{E_{\rm nucl}}\right)^{1/2}\left(\frac{r_{\rm nucl}}{\Lambda_0}\right)^3 \exp\left[-\frac{1}{2}\left(\frac{R_0}{\Lambda_0}\right)^2\right], \tag{2}$$

where $E_{\rm nucl}$ is the nuclear energy released in the fusion, which is transferred to the lattice producing phonons (*that explains the absence of harmful radiation in LENR*), $r_{\rm nucl}$ is the nuclear radius, $R_0$ is the equilibrium distance between the nuclei in the lattice.

For D–D $\Rightarrow$ $He^4$ fusion in PdD lattice, the mass difference $E_{\rm nucl} = 23.8$ MeV. Assuming $r_{\rm nucl} = 3\times10^{-5}$Å, $\Lambda_0 = 0.1$ Å (corresponding to $\omega_0 = 320$ THz) and $R_0 = 0.94$ Å as the equilibrium spacing of two deuterons placed in *one site* in a hypothetical $PdD_2$ lattice, Schwinger estimated the fusion rate to be $\sim10^{-19}$ s$^{-1}$ [5]. For a more realistic situation, with two deuterons in *two adjacent sites* of the PdD lattice, one has $R_0 = 2.9$ Å. Even assuming a lower value of $\omega_0 = 50$ THz corresponding to larger $\Lambda_0 = 0.25$ Å [6], Eq. (2) will results in the fusion rate of $\sim10^{-30}$ s$^{-1}$, which is too low to explain the observed excess heat generated in Pd cathode under $D_2O$ electrolysis.

The above estimate is valid for the fusion rate between D–D or D–H ions in regular lattice sites. However, the ZPV amplitude can be increased locally under time-periodic modulation of the potential well width (that determines its eigenfrequency) at a frequency that exceeds the eigenfrequency by a factor of $\sim$2 (*the parametric regime*). Such regime can be realized for a hydrogen or deuterium atom in metal hydrides/deuterides, such as NiH or PdD, in the vicinity of LAV [2,3]. Under parametric modulation, ZPV amplitude increases exponentially fast (Fig. 1a) with increasing number of oscillation periods $N = \omega_0 t/2\pi$ [6]:

$$\Lambda_N = \Lambda_0\sqrt{\cosh(g_\omega \pi N)}, \quad \Lambda_0 = \sqrt{\frac{\hbar}{2m\omega_0}}, \tag{3}$$

where $g_\omega \ll 1$ is the amplitude of parametric modulation, which is determined by the amplitude of LAV. For example, $g_\omega = 0.1$ corresponds to the LAV amplitude of $\sim$0.3 Å in the PdD lattice with $R_0 = 2.9$ Å, which is confirmed by molecular dynamic simulations of gap discrete breathers in NaCl type crystals [7]. Substituting Eq. (3) into the Schwinger Eq. (2) one obtains a drastic enhancement of the fusion rate with increasing number of oscillation periods $N$ (Fig. 1b):

$$\frac{1}{T_N} \approx 2\pi\omega_0 \left(\frac{2\pi\hbar\omega_0}{E_{\rm nucl}}\right)^{1/2}\left(\frac{r_{\rm nucl}}{\Lambda_N}\right)^3 \exp\left[-\frac{1}{2}\left(\frac{R_0}{\Lambda_N}\right)^2\right], \tag{4}$$

The parametric driving considered above requires rather special conditions similar to those in gap breathers in diatomic crystals [7], while in many other systems, e.g. in metals [12], oscillations of atoms in a discrete breather have different amplitudes but the same frequency. This case is closer to the driving of the potential well *positions* with the frequency equal to the potential eigenfrequency. Such driving does not increase the ZPV amplitude since the wave packet dispersion remains constant, however, the mean oscillation energy grows with time as [13]:

$$\langle E\rangle = \frac{\hbar\omega_0}{2} + \frac{g_x^2\hbar\omega_0}{16}\left[\omega_0^2t^2 + \omega_0 t\sin 2\omega_0 t + \sin^2\omega_0 t\right], \tag{5}$$

where $g_x$ is the relative amplitude of the position driving. Accordingly, one could expect an acceleration of the escape from a potential well of a finite depth similar to the parametric driving.

In reality, one is interested in the effect of potential well driving on the *tunneling* through the barrier of finite height between the wells as a function of the driving frequency and strength (amplitude). An analytical solution of the *non-stationary* Schrödinger equation even for the simplest case of a double well potential cannot be obtained. In the following section, we will analyze a *numerical solution* of Schrödinger equation for a particle in a double well potential, which is driven time-periodically imitating the action of a LAV or phason.

## 3. Tunneling in a Periodically Driven Double Well Potential

Consider the Schrödinger equation for a wave function $\psi(x,t)$of a particle with a mass $m$in the non-stationary double-well potential $V(x,t)$:

$$i\hbar\frac{\partial}{\partial t}\psi(x,t) = -\frac{\hbar^2}{2m}\frac{\partial^2}{\partial x^2}\psi(x,t) + V(x,t)\psi(x,t), \tag{6}$$

$$V(x,t) = \frac{m\omega_0^2}{2}\left[\frac{a(t)}{x_0^2}x^4 - b(t)x^2\right], \quad x_0 = \sqrt{\frac{\hbar}{m\omega_0}}, \tag{7}$$

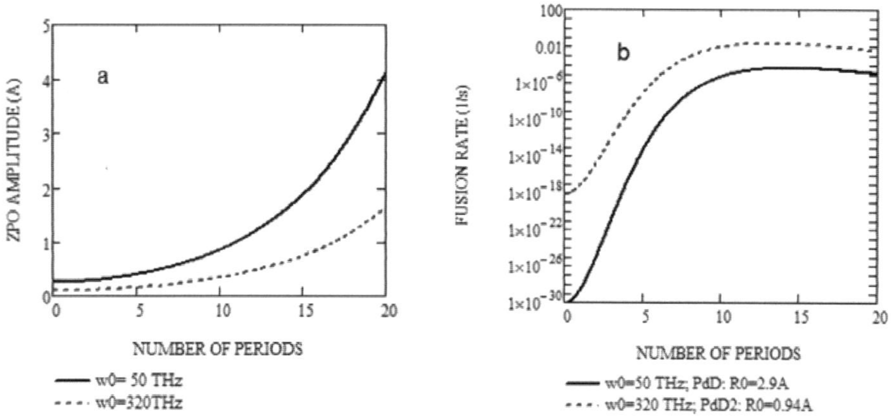


**Figure 1.** (a) Zero-point vibration amplitude of deuterium ions vs. $N$ in the *parametric regime* [6] for different $\omega_0$ according to Eq. (3) at $g = 0.1$. (b) D–D fusion rate D–D ⇒ $He^4$ + 23.8 MeV in PdD lattice according to Eq. (4) for deuterium ions in PdD lattice oscillating near equilibrium positions in one site ($R_0 = 0.94$ Å) or in two neighboring lattice sites ($R_0 = 2.9$ Å).

where $a\,(t)$ and $b\,(t)$ are the dimensionless parameters that determine the form and the driving mode of the potential shown in Fig. 2.

$$a\,(t) = \frac{1}{2\sqrt{\alpha}}\left[\alpha - \beta\cos\left(\Omega t\right)\right], \quad b\,(t) = \frac{1}{2\sqrt{\alpha}}\sqrt{\alpha - \beta\cos\left(\Omega t\right)}, \tag{8}$$

$\Omega = 2\omega_0$ is the driving frequency of the eigenfrequencies $\omega_{\text{eigen}}$ and positions $x_{\min}$ of the potential wells in the vicinity of the minima given by

$$\frac{\omega_{\text{eigen}}}{\omega_0} = \sqrt{2b} = \sqrt[4]{1 - \frac{\beta}{\alpha}\cos\left(2\omega_0 t\right)} \approx \left[1 - \frac{\beta}{4\alpha}\cos\left(2\omega_0 t\right)\right], \quad g_\omega = \frac{\beta}{2\alpha} \ll 1, \tag{9}$$

$$\begin{aligned} \frac{x_{\min}}{x_0} &= \pm\sqrt{\frac{b}{2a}} = \pm\frac{1}{\sqrt{2}}\frac{1}{\sqrt[4]{\alpha - \beta\cos(2\omega_0 t)}} = \frac{1}{\sqrt{2}}\frac{1}{\alpha^{1/4}\sqrt[4]{1 - \frac{\beta}{\alpha}\cos(2\omega_0 t)}} \\ &\approx \frac{x_{\min}(0)}{x_0}\left(1 + \frac{\beta}{4\alpha}\cos\left(2\omega_0 t\right)\right), \quad g_x \equiv \frac{\beta}{4\alpha} \ll 1 \end{aligned}, \tag{10}$$

From Eqs. (9) and (10) it follows that the driving under consideration results in a *simultaneous* time-periodic modulation of the potential well *positions* and *eigen frequencies* with amplitudes $g_x$ and $g_\omega$, respectively. Therefore, we are dealing here with a synergetic effect of the two mechanisms considered separately for a harmonic oscillator in the previous section and in [13].

Initial state of the system is described by a wave function of the Gaussian form placed near the first energy minimum (Fig. 2a):

$$\psi\left(x, t_0 = 0\right) = \frac{1}{\sqrt[4]{\pi x_0^2}}\exp\left(-\frac{\left(x - x_{\min}\right)_0^2}{2x_0^2}\right). \tag{11}$$

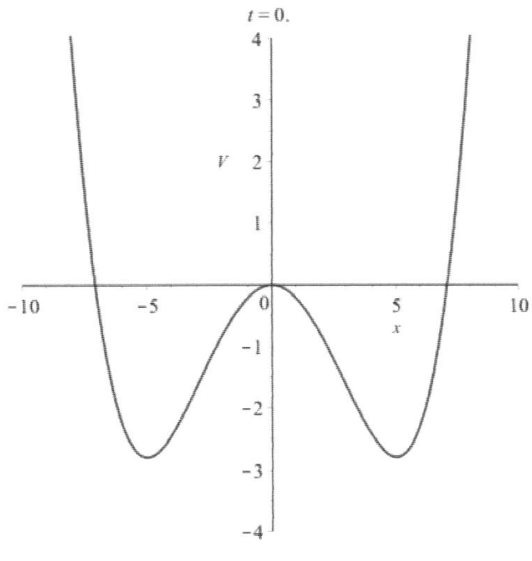


**Figure 2.** Double-well potential given by Eq. (7) at $\alpha = 0.0005$, $\beta = 0.0001$, which corresponds to the ratio of the potential depth to ZPV energy given by $1/8\sqrt{\alpha} \approx 5.6$.

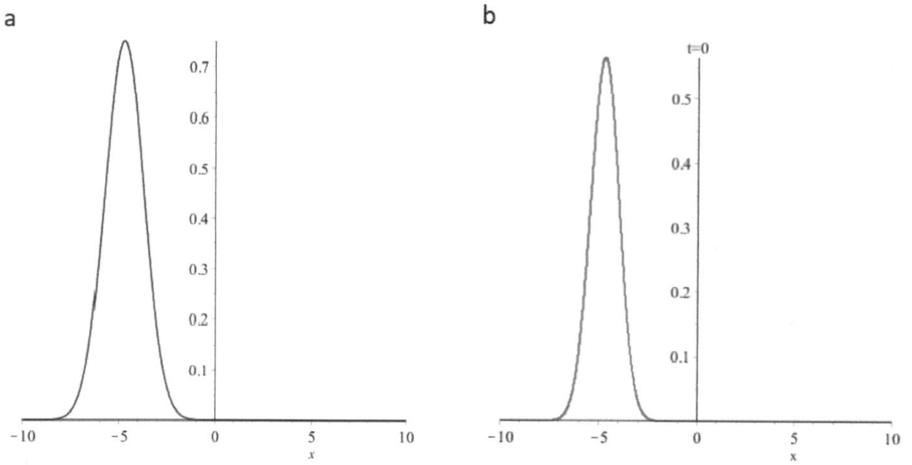


**Figure 3.** (a) Initial wave function $\psi\,(x, t_0 = 0)$ and (b) the probability distribution to find the particle at the point $x\;:\;\rho\,(x, t_0 = 0) = |\psi\,(x, t_0 = 0)|^2$ in the left potential well shown in Fig. 1.

The probability distribution of finding the particle at the point $x$ is given by $\rho\,(x, t_0 = 0) = |\psi\,(x, t_0 = 0)|^2$, which is shown in Fig. 3b. It can be seen that the probability density is concentrated at $x_{\min} \approx 4.73$, which means the particle spends most of its time at the bottom of the potential well.

At the selected parameters, the potential depth to ZPV energy ratio is given by $1/8\sqrt{\alpha} \approx 5.6$, which is a typical ratio for solid state chemical reactions. This means that the particle energy is 5.6 times lower than the energy required to 'jump' over the barrier into another well. The mean time of tunneling through the barrier from a *stationary* potential well is very large, as can be seen from Fig. 4 showing the probability distribution of the particle at different moments of time $t = 2\pi/\omega_{\text{eigen}}$, measured in the oscillator periods. For example, $t = 1000$ corresponds to 1000 "attempts" to escape from the left well. However, one can see that the probability to find the particle in the right well is still

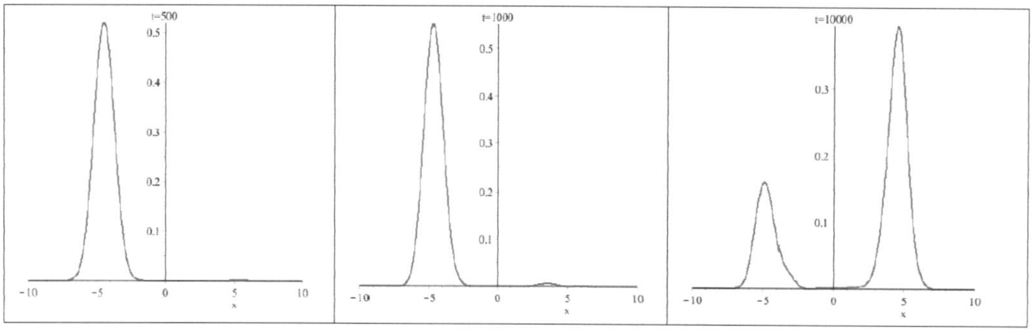


**Figure 4.** The probability distribution of the particle at different moments of time $t = 2\pi/\omega_{\text{eigen}}$ in *stationary potential wells*: $\alpha = 0.0005$; $\beta = 0$.

negligibly small. Only at $t = 10\,000$, does it become higher than the probability to find the particle in the left well.

The situation becomes dramatically different in the case of time-periodically driven wells, as demonstrated in Fig. 5 for the two driving frequencies $\Omega = \omega_{\rm eigen}$; $2\omega_{\rm eigen}$. In both cases, already at $t = 100$, the probability of finding the particle in the right well becomes comparable with the probability of finding the particle in the left well. This means that the mean escape (tunneling) time has decreased by $\sim$2 *orders of magnitude* due to the driving with a comparatively small driving amplitude $g_\omega = 2g_x = 0.1 \ll 1$.

The driving frequency effect is different from that obtained for a harmonic oscillator [13], where two sharp peaks were observed at resonant frequencies $\Omega = \omega_{\rm eigen}$ and $\Omega = 2\omega_{\rm eigen}$. Due to a *simultaneous* time-periodic modulation of the potential well *positions* and *eigenfrequencies*, the accelerating effect of driving depends non-monotonously on the driving frequency with a several maximums lying between $\omega_{\rm eigen}$ and $2\omega_{\rm eigen}$.

Finally, dependence of the tunneling time on the driving amplitude is shown in Fig. 6. It appears that increasing the amplitude by a factor of two results in decreasing the mean tunneling time by an order of magnitude. This example demonstrates the importance of the time-periodic driving of the potential wells in the vicinity of LAVs and phasons in the reactions involving quantum tunneling.

In the following section, we consider some characteristic examples of LAVs and phasons in quasicrystals.

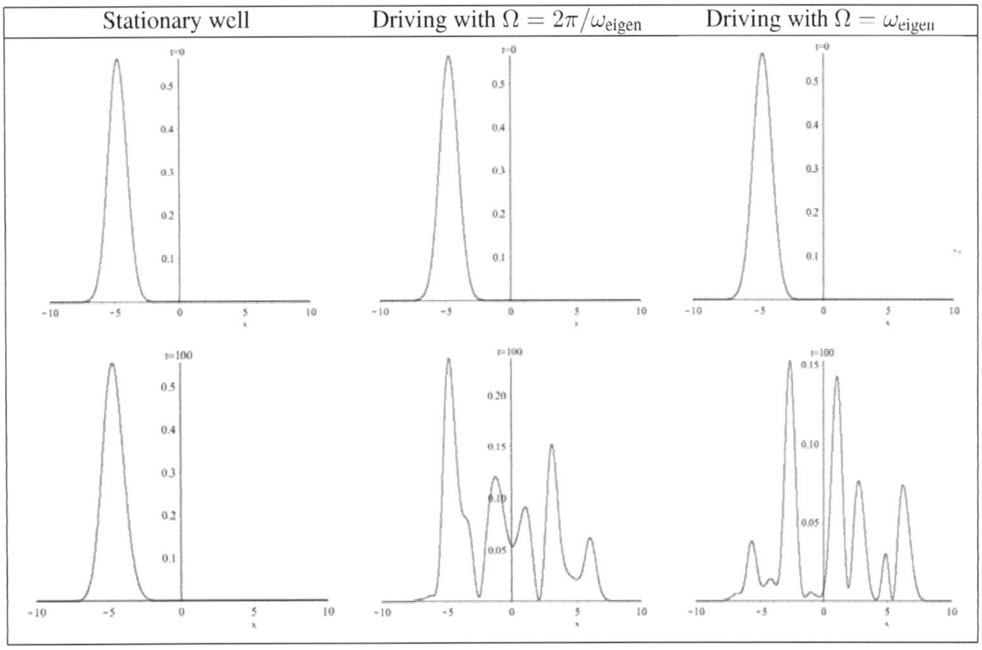


**Figure 5.** The probability distribution of the particle at different moments of time $t = 2\pi/\omega_{\rm eigen}$ in stationary potential wells ($\alpha = 0.0005$; $\beta = 0$) and under the potential driving ($\alpha = 0.0005$; $\beta = 0.0001$) corresponding to $g_\omega = \beta/2\alpha = 0.1$; $g_x = \beta4\alpha = 0.05$. The driving frequency $\Omega$ is indicated in the figure.

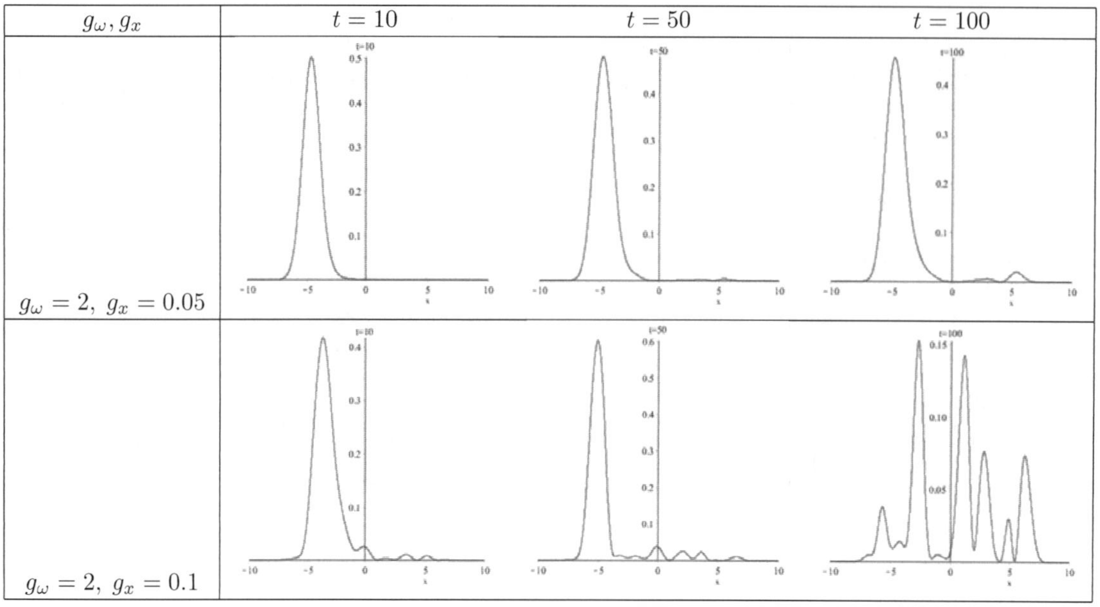


**Figure 6.** The probability distribution of the particle at different moments of time under the potential driving at $\Omega = \omega_{\text{eigen}}$, $\alpha = 0.0005$; $\beta = 0.00005 \div 0.0002$, corresponding to different driving amplitudes $g_\omega$, $g_x$ as indicated in the figure.

## 4. LAVs and Phasons in Nanocrystals and Quasicrystals

The fact that the energy localization manifested by LAV does not require long-range order was first realized as early as in 1969 by Ovchinnikov, who discovered that localized long-lived molecular vibrational states may exist already in simple molecular crystals ($H_2$, $0_2$, $N_2$, NO, CO) [14]. He realized also that stabilization of such excitations was connected with the *anharmonicity* of the intramolecular vibrations. Two coupled anharmonic oscillators described by a simple set of dynamic equations demonstrate this idea:

$$\begin{aligned} \ddot{x}_1 + \omega_0^2 x_1 + \varepsilon\lambda x_1^3 &= \varepsilon\beta x_2, \\ \ddot{x}_2 + \omega_0^2 x_2 + \varepsilon\lambda x_2^3 &= \varepsilon\beta x_1, \end{aligned} \tag{12}$$

where $x_1$ and $x_2$ are the coordinates of the first and second oscillator, $\omega_0$ are their zero-point vibrational frequencies, $\varepsilon$ is a small parameter, and $\lambda$ and $\beta$ are parameters characterizing the *anharmonicity* and the *coupling* force of the two oscillators, respectively. If one oscillator is displaced from the equilibrium and starts oscillating with an initial amplitude, $A$, then the time needed for its energy to transfer to another oscillator is given by the integral:

$$T = \frac{\omega_0}{\varepsilon\beta} \int\limits_0^{\pi/2} \frac{\mathrm{d}\varphi}{\sqrt{1 - \left(A^2\gamma/4\right)^2 \sin^2\varphi}}, \qquad \gamma = \frac{3\lambda}{\beta}, \tag{13}$$

from which it follows that the full exchange of energy between the two oscillators is possible only at sufficiently small initial amplitude: $A^2\gamma/4 < 1$. In the opposite case, $A^2\gamma/4 > 1$, the energy of the first oscillator will *always be larger* than that of the second one. And for sufficiently large initial amplitude, $A \gg \sqrt{4/\gamma}$, there will be practically no sharing of energy, which will be localized exclusively on the first oscillator.

Thus, Ovchinnikov has proposed the idea of LAV for molecular crystals, which was developed further for any nonlinear systems possessing *translational symmetry*; in the latter case, LAVs have been named *discrete breathers* (DBs) or *intrinsic localized modes* (ILMs). Now, we are coming back to the idea of LAV arising at *"active sites"* in defected crystals, quasicrystals and nanoclusters. As noted by Storms, "Cracks and small particles are the Yin and Yang of the cold fusion environment". A physical reason behind this phenomenology is that in topologically disordered systems, sites are not equivalent and band-edge phonon modes are intrinsically localized in space. Hence, different families of LAV may exist, localized at different sites and approaching different edge normal modes for vanishing amplitudes [9]. Thus, in contrast to perfect crystals, which produce DBs homogeneously, there is a *site selectiveness* of energy localization in the presence of spatial disorder, which has been demonstrated by means of atomistic simulations in biopolymers [9], metal nanoparticles [15] and, *experimentally*, in a decagonal *quasicrystal* $Al_{72}Ni_{20}Co_8$ [11].

The crystal shape of the nanoparticles (cuboctahedral or icosahedral) is known to affect their catalytic strength [16], and the possibility to control the shape of the nanoparticles using the amount of hydrogen gas has been demonstrated both experimentally by Pundt et al. [17], and by means of atomistic simulations by Calvo et al. [18]. They demonstrated that above room temperature the *icosahedral phase* should remain stable due to its higher entropy with respect to cuboctahedron. And icosahedral structure is one of the forms quasicrystals take, therefore one is tempted to explore further the *link between nanoclusters and quasicrystals*.

Figure 7 shows the structure of $Pd_{147}H_{138}$ cluster containing 147 Pd and 138 H atoms having minimum free energy configuration, replicated using the method and parameters by Calvo et al. [18]. In particular, Fig. 9(b) reveals the presence of H–H–H chains aligned along the $I$-axis of the cluster. This ab initio simulation points out at the possibility of excitation of LAVs in these chains, with a central atom performing large-amplitude anharmonic oscillations and its neighbors oscillating in quasi-harmonic regime [19], which is similar to that considered in [7] for regular diatomic lattice of NaCl type. Such oscillations have been argued to facilitate LENR [2,3], and in the present paper we develop this concept further.

Let us consider phasons observed in a decagonal *quasicrystal* $Al_{72}Ni_{20}Co_8$ [11] and a possible link between LAVs and phasons.

Abe et al. [11] have measured by means of high resolution scanning transmission microscope (STEM) temperature dependence of the so-called Debye–Waller (DW) factor in decagonal *quasicrystal* $Al_{72}Ni_{20}Co_8$. DW factor is

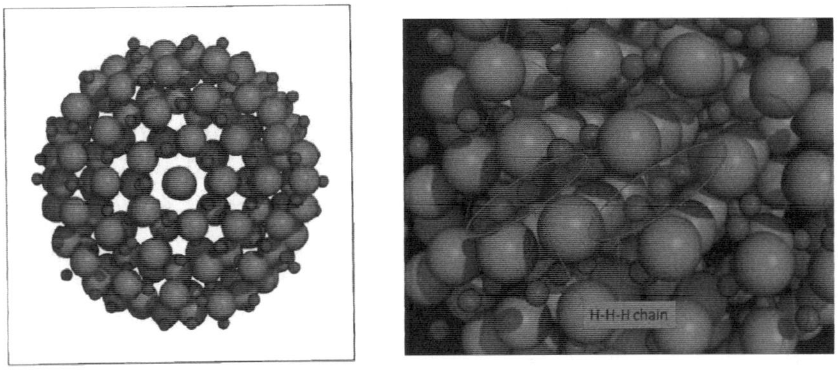


**Figure 7.** (*Left*) Structure of PdH cluster containing 147 Pd and 138 H atoms having minimum free energy configuration, replicated using the method and parameters by Calvo et al. [18]; (*Right*) H–H–H chains in the nanocluster, which are viable sites for LAV excitation [19].

determined by the mean-square vibration amplitude of the atoms. The vibrations can be of thermal or quantum nature depending on the temperature. The authors demonstrated that the anharmonic contribution to Debye–Waller factor increased with temperature much stronger than the harmonic (phonon) one. This was the first *direct observation* of a "local thermal vibration anomaly", i.e. LAVs, in our terms. The experimentally measured separation between LAVs was about 2 nm, which meant that their mean concentration was about $10^{20}$ per $cm^3$ that is many orders of magnitude higher than one could expect to find in periodic crystals [7].

The LAV amplitude dependence on temperature fitted by two points at 300 and 1100 K has shown that the maximum LAV amplitude at 1100 K = 0.018 nm (Fig. 8a). What is more, it appears that LAVs give rise to phasons at T >990 K, where a phase transition occurs, and additional quasi-stable sites $\beta$ arise near the sites $\alpha$. The phason amplitude of 0.095 nm (Fig. 8b) is an *order of magnitude larger* than that of LAVs. Thus, on the one hand, the driving amplitude induced by phasons is larger than that by LAVs, but on the other hand, phason oscillations may be less time-periodic (more stochastic), which requires more detailed investigations of the driving stochasticity effect on tunneling, as discussed in the following section.

## 5. Discussion and Outlook

In the present paper, we presented a numerical solution to the Schrödinger equation for a particle in a *non-stationary* double well potential, which is driven *time-periodically* imitating the action of a LAV or a phason on the reaction cite in their vicinity. We have shown that the rate of tunneling of the particle through the potential barrier separating the wells can be enhanced by *orders of magnitude* with increasing number of driving periods. This effect is novel, since it differs qualitatively from a well-studied effect of resonance tunneling [20–22], a.k.a. Euclidean resonance (an easy penetration through a classical nonstationary barrier due to an under-barrier interference). In the latter case, the tunneling rate has a sharp peak as a function of the particle energy when it is close to the certain *resonant value* defined by the non-stationary field. Therefore, it requires a very specific parametrization of the tunneling conditions. In contrast to that, the time-periodic driving of the potential wells considered above, results, first of all, in a sharp and continuous (not quantum) increase of the ZPV amplitude and energy [6,13], which in its turn increases the tunneling rate. This result is closely related to the *correlation effects*, proposed by Dodonov et al. [23] and analyzed in details by Vysotskii

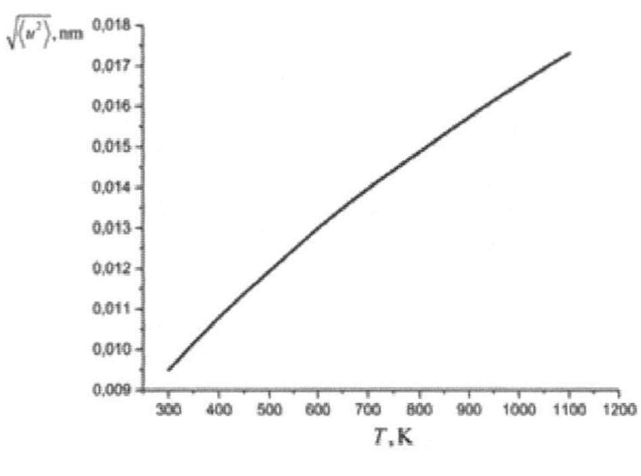


**(a)**

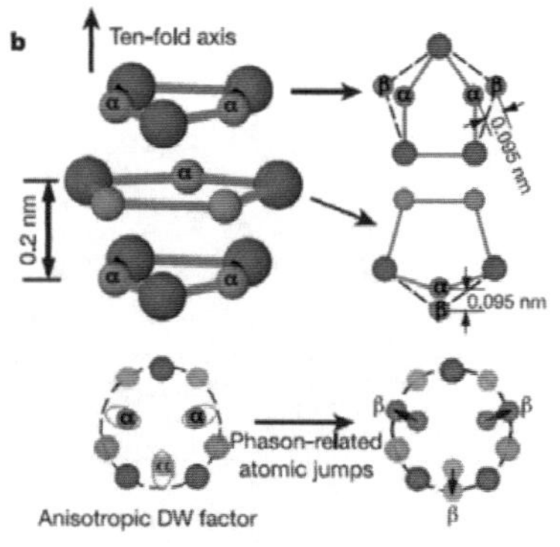


**(b)**

**Figure 8.** (a) LAV amplitude dependence on temperature in $Al_{72}Ni_{20}Co_8$, fitted by two points at 300 and 1100 K, according to Abe et al. [11]. The maximum LAV amplitude at 1100 K is 0.018 nm. (b) LAVs give rise to phasons at $T > 990$ K, where a phase transition occurs, and additional quasi-stable sites $\beta$ arise near the sites $\alpha$. The phason amplitude of 0.095 nm is an order of magnitude larger than that of LAVs.

et al. [24] who predicted a giant increase of sub-barrier transparency (up to hundreds of orders of magnitude) during the increase of the so-called *correlation coefficient* at special periodic action on a quantum system. This prediction was based on the numerical calculation of the time dependence of the correlation coefficient $r(t)$ for the case of a periodically driven *harmonic potential*, in which case it was shown that

$$|r(t)| \underset{t\to\infty}{\longrightarrow} 1$$

at any *initial energy* of the oscillator, which was argued to result in a tunneling through the potential barrier of *any height* after sufficient number of driving periods. However, as shown more recently by Dubinko and Laptev [6], the oscillator energy also increases exponentially with time in the parametric regime considered in [24], which poses a limit to the correlation coefficient increase in a *double well* system (which is more relevant for the tunneling analysis than the infinite parabolic well considered in [6,24]).

As shown in the present paper, the tunneling probability increases strongly with increasing strength of the driving, which is related to the amplitude of the non-linear dynamic phenomenon that causes the driving. As we have demonstrated in the previous section, the driving amplitude induced by phasons may be larger than that induced by LAVs by an order of magnitude, which implies that phasons may be stronger catalysts than LAVs. However, further research is needed in order to make more definite conclusions, since the phason dynamics itself is an activated process driven by thermal or quantum fluctuations. Therefore, phasons can hardly induce a strictly time-periodic driving considered in the present paper. The tunneling rate through a fluctuating barrier in the presence of a periodically driving field has been shown to decrease with increasing fluctuation strength [25]. One may expect similar effects due to fluctuations in the cases of LAV and phason driven tunneling, which requires further investigation.

In conclusion, the present results support the concept *of nuclear catalysis* in QCs taking place at special sites caused by their inherent topology. This makes QCs a logical model to explain the structure of the microscopic nuclear active environments, or *hot-spots* observed by experimentalists in TEM analysis.

**Acknowledgements**

The authors would like to thank Dmitry Terentyev for designing Fig. 7 and Dan Woolridge – LAV animation [19]. VD and DL gratefully acknowledge financial support from Quantum Gravity Research.

**References**

[1] V. I. Dubinko, *J. Condensed Matter Nucl. Sci.* **14** (2014) 87.
[2] V. I. Dubinko, Quantum tunneling in gap discrete breathers, *Lett. Materials* **5** (2015) 97.
[3] V. I. Dubinko, *J. Condensed Matter Nucl. Sci.* **19** (2016) 1–12.
[4] J. Schwinger, *Z. Phys.* D **15** (1990) 221.
[5] J. Schwinger, Nuclear energy in an atomic lattice, *The First Annual Conference on Cold Fusion*, University of Utah Research Park, Salt Lake City, Utah, National Cold Fusion Institute, 1990.
[6] V. I. Dubinko and D. V. Laptev, *Lett. Materials* **6** (2016) 16.
[7] L.Z. Khadeeva and S.V. Dmitriev, *Phys. Rev.* B **84** (2011)144304.
[8] V. I. Dubinko, *J. Micromech. Mol. Phys.* **1** (2016) 165006.
[9] V.I. Dubinko and F. Piazza, *Lett. Materials* **4** (2014) 273.
[10] M. Yoshimura and A.P. Tsai, *J. Alloys Compound* **342** (2002) 451.
[11] E. Abe, S.J. Pennycook and A.P.Tsai, *Nature* (London) **421** (2003) 347.
[12] D. Terentyev, A. Dubinko, V. Dubinko, V. Dmitriev and E. Zhurkin, *Modelling Simul. Mater. Sci. Eng.* **23** (2015) 085007.
[13] V.I. Dubinko, A.S. Mazmanishvili , D.V. Laptev and J. F. R.Archilla, *J. Micromech. Mol. Phys.* **1** (2016) 1650010.

[14] A.A. Ovchinnikov, *Zh. Eksp. Teor. Fiz.* **57** (1969) 2137.
[15] H. Zhang and J. F. Duglas, *Soft Matter* **9** (2013) 1266.
[16] B. Corain, G. Schmid and N. Toshima (Eds.), *Metal Nanoclusters in Catalysis and Materials Science, The Issue of Size Control*, Elsevier , Amsterdam, 2008, p. 470.
[17] A. Pundt, A. Dornheim, M. Guerdane, H. Teichler, H. Ehrenberg, M.T. Reetz and N.M. Jisrawi, *Eur. Phys. J.* D **19** (2002) 333.
[18] F. Calvo and A. Carre, *Nanotechnol.* **17** (2006) 1292.
[19] D. Woolridge, *Quantum Gravity Research*, Los Angeles, CA 90290, USA.
https //www.dropbox.com/s/557akac5zoiynqs/LAV%20animation%20in%20PdH%20cluster.mp4?dl=0.
[20] B.I. Ivlev and V. I. Melnikov, in *Quantum Tunneling in Condensed Media*, A. Leggett and Yu. Kagan (Eds.), NorthHolland, Amsterdam, 1992.
[21] J.P. Palomares-Báez, B. Ivlev and J.L. Rodríguez-López, *Phys. Rev.* **A 76** (2007) 052103.
[22] B.I. Ivlev, *Phys. Rev.* **A 66** (2002) 012102.
[23] V.V. Dodonov, V.I. Man'ko, *Phys. Lett.* **A 79** (2/3) (1980) 150.
[24] V.I. Vysotskii and M.V. Vysotskyy, *Eur. Phys. J.* A (2013) . DOI 10.1140/epja/i2013-13099-2.
[25] A. Saha and P. Sarkar, *Int. J. Quantum Chem.* **97** (2004) 914.

J. Condensed Matter Nucl. Sci. 24 (2017) 87–97

Research Article

# Statistical Mechanics Models for $PdH_x$ and $PdD_x$

Peter L. Hagelstein*
*Massachusetts Institute of Technology, Cambridge, MA, USA*

**Abstract**

In recent years we have developed simple empirical mean-field statistical mechanics models for $PdH_x$ and $PdD_x$ that include both O-site and T-site occupation. A brief account of this work is given here. At low loading it is impossible to fit solubility data over a wide range of temperature assuming only O-site occupation; however, with T-site occupation a good fit is obtained and an O-site to T-site excitation energy near 105 meV is found. Results from neutron diffraction studies at modest loading, and resistance and solubility measurements at high pressure and high loading, are used to infer O-site to T-site excitation energies and develop a smooth fit. The resulting statistical mechanics models have been optimized to provide systematic results at high loading for PdH and PdD, and also to fit the isotherms of the phase diagram of PdH. Estimates for the O-site energy as a function of loading and temperature have been developed in the latter case.


*Keywords:* Mean field model, Palladium hydride, Phase diagram, Solubility, Tetrahedral occupation

## 1. Introduction

Since the announcement by Fleischmann and Pons of excess heat in the PdD system there has been interest among researchers interested in the effect to understand how it works. The problem is highly multi-disciplinary. A variety of issues that might be considered conventional are involved: electrochemical reactions that impact deuterium loading; diffusion of the deuterium within the cathode; the deuterium chemical potential and octahedral site occupation; physical chemistry and materials science issues at very high loading; and vacancy formation and occupation. Then there are issues that require new physics to understand, specifically concerning whatever physical processes are involved that enable nuclear energy to be expressed without commensurate energetic particles.

Over the last few years we have taken an interest in developing simple statistical mechanics models that describe the loading of hydrogen and deuterium in bulk palladium [1–3]. Such models can be useful for simulating cathode loading in the Fleischmann–Pons experiment; they can help in understanding what sites are occupied by interstitial H and D; the models can help interpret loading isotherms in terms of the energies of interstitial H/D states; and it may be possible to model what happens at high loading where the D/Pd ratio approaches or exceeds unity. In this short review

*E-mail: plh@mit.edu.

we provide a brief overview of some of the results; a much more complete technical presentation of the models and how they were derived from statistical mechanics and experiment is provided in [1–3].

## 2. Modeling Solubility in the $\alpha$-phase

Palladium hydride may be the most studied of the metal hydrides, both theoretically and experimentally. Near room temperature hydrogen and deuterium are known from experiment to occupy octahedral sites, a conclusion strongly supported by density functional and quantum chemistry calculations. Experimental solubility studies at low loading have been carried out over a wide range of temperatures, and theoretical models for hydrogen occupation of octahedral sites in alpha phase $PdH_x$ have been available since the 1930s.

We focused on a data set published by Clewley et al. [4], which is of interest because it provides consistent data over a wide temperature range, and also because the results were provided in a form that is convenient to connect with theoretically. The data in this work is plotted as a function of

$$\frac{p(1-\theta)^2}{\theta^2}, \tag{1}$$

which is useful since according to Fowler and Smithells [5] one would expect in the $\alpha$-phase an isotherm of the form

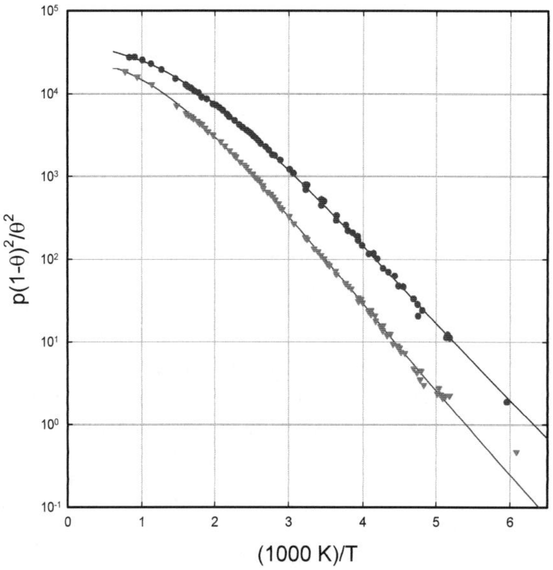


**Figure 1.** Plot of $p(1-\theta)^2/\theta^2$ for $\alpha$-phase $PdD_x$ as a function of (1000 K)/$T$ from [1]; data set of Clewley et al. [4] for $PdH_x$ (*red circles*); optimized empirical model for $PdH_x$ (*red line*); data set of Clewley et al. [4] for $PdD_x$ (*blue circles*); optimized empirical model for $PdD_x$ (*blue line*).

$$\frac{\theta}{1-\theta} = f(T)\sqrt{p}e^{\Delta E/k_B T} \tag{2}$$

if only octahedral occupation occurs in a model with constant $\Delta E$. When plotted on a log-lin plot as a function of $1/T$ one would expect only minor deviations from a straight line due to the weak dependence of $f(T)$ on temperature; however, the data (see Fig. 1) shows a marked deviation from linearity at elevated temperature, which can be interpreted as resulting from the availability of (and occupation of) additional sites beyond the octahedral sites.

We found it to be impossible to develop acceptable fits to the data based on models for alpha phase $PdH_x$ and $PdD_x$ with O-site occupation alone; however, when we allowed for a combination of O-site and T-site occupation then good agreement could be obtained [1] as shown in Fig. 1. The O-site to T-site excitation energy derived from fitting the model with the data set was found to be 106.5 meV for $PdH_x$ and 105.3 meV for $PdD_x$. Such a low energy excitation energy for T-site occupation was unexpected based on most of the relevant literature on solubility; however, recent modeling with density functional theory codes and quantum chemistry codes is qualitatively consistent with these results.

## 3. O-site to T-site Excitation Energy

The low O-site to T-site excitation energy at low loading would suggest that T-site occupation might be important at higher loading. Although there are many computations of the excitation energy from density functional calculations and from quantum chemistry calculations, there has not been any previous systematic effort to connect the results with experimental solubility studies. There have been arguments made in the literature where it was suggested that information about the excitation energy could be obtained from estimates for the barrier energy associated with H or D diffusion in Pd; for $\alpha$-phase $PdH_x$ at low temperature the barrier energy is listed as 0.13 eV [6], which is a bit higher than our estimate derived from the $\alpha$-phase solubility. The barrier energy near room temperature for the $\alpha$-phase is higher, for PdD the barrier energy is listed as 0.206 meV [6], perhaps indicative of the participation of higher energy states (see [7]). For $\beta$-phase PdH Baranowski provides a fit dependent on temperature with a corresponding barrier energy of 281 meV [8].

We would conclude from this that the O-site to T-site excitation energy is probably less than 280 meV near an H/Pd loading of 0.6. However, we require a parameterization of the O-site to T-site excitation energy in order to implement a statistical mechanics model for H/D solubility in Pd above the $\alpha$-phase. We made use of solubility data at high loading and quasi-elastic neutron scattering studies at lower loading to construct an empirical estimate for the O-site to T-site excitation energy as a function loading shown in Fig. 2 (as discussed in detail in [2]). Here we give only a brief outline of the approach. The analysis of the $\alpha$-phase region outlined above provides an estimate at zero loading. We applied the same statistical mechanics model to all available solubility data at high loading [2], including solubility data indicative of above unity D/Pd loading published by Baranowski [9]. We found that it was possible to estimate the excitation energy by matching the model to the change of slope in the resistance ratio data at high pressure, and that the resulting model was consistent with other high pressure data as well as with the resistance ratio calibrations of Crouch-Baker et al. [10]. The resulting excitation energies for both PdH and PdD at a loading of unity was estimated to be about 225 meV (shown in Fig. 2). There are a small number of neutron diffraction experiments, such as the study of Pitt and Gray [11], which can be interpreted in terms of T-site occupation. In Ref. [2] the data was analyzed to give the (inferred) data points shown in green between about $\theta = 0.08$ and $\theta = 0.42$.

A parameterization for $PdD_x$ was developed based on a least squares fit of the data points shown to

$$\Delta E(\theta) = \frac{\alpha_0 + \alpha_1\theta}{1+\beta\theta} \tag{3}$$

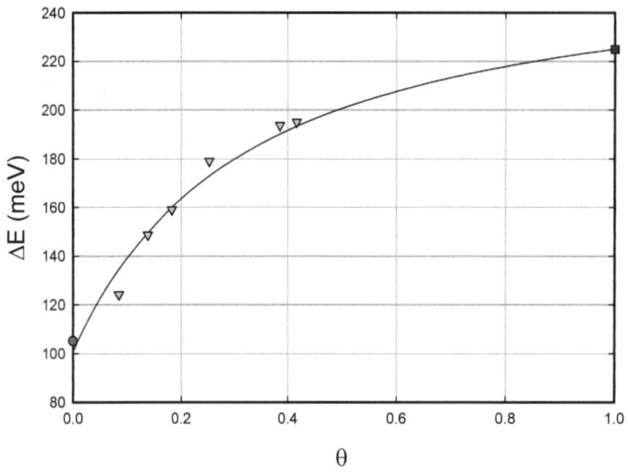


**Figure 2.** O-site to T-site excitation energy as a function of loading [2]; inferred from data of Pitt and Gray [11] (*green triangles*); estimated from high pressure loading data of Baranowski et al. [9] (*blue square*); excitation energy from $\alpha$-phase $PdD_x$ analysis of [1] (*red circle*); empirical fit (*dark blue line*).

resulting in the fitting parameters

$$\alpha_0 = 100.676\,\text{meV}, \quad \alpha_1 = 824.259\,\text{meV K}^{-1},$$

$$\beta = 3.1108\,\text{K}^{-1}. \tag{4}$$

The resulting fit can be seen in Fig. 2. We had not anticipated the apparent consistency between the O-site to T-site excitation energy from the analysis of data from such disparate sources. Also, we were expecting to find the excitation energies for PdH and PdD to be different by on the order of 10–20 meV; it came as a surprise that very similar results were obtained near $\theta = 0$ and $\theta = 1$. It is probable that the O-site to T-site excitation energy is also a (weak) function of temperature; but at this point there is insufficient data available to develop a parameterization that takes into account both loading and temperature.

## 4. Solubility at High Loading

Models in the literature based on O-site loading predict an upper limit for the H/Pd or D/Pd ratio of unity, as would be expected if there are no other sites that could be occupied. As mentioned above Baranowski [9] reported high pressure experiments with PdD in which it was concluded that a D/Pd loading higher than unity had been developed. In electrochemical studies with PdD in our field there have been claims that a D/Pd ratio greater than unity have been obtained. This presents no particular difficulty if T-site occupation occurs (or if a different phase is present), but is problematic in models with only O-site occupation.

It is possible to make use of the Sarkur–Tetrode chemical potential of the ideal gas law for the development of statistical mechanics model at low pressure and loading; however at high loading the equivalent pressure can be as large as 1 GPa, where hydrogen deviates significantly from an ideal gas. This provided us with motivation to examine

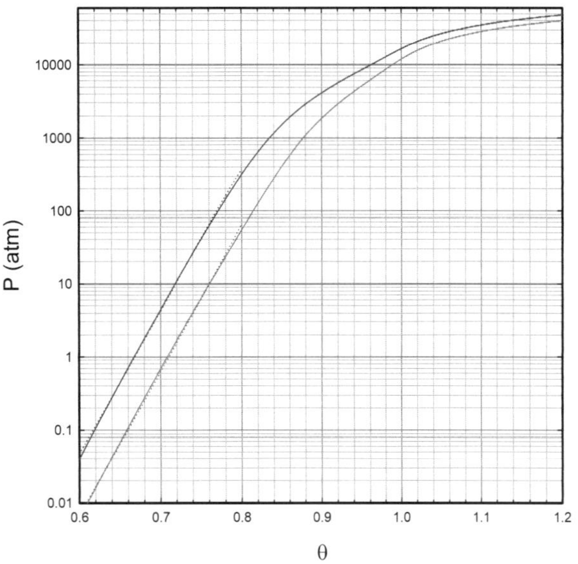


**Figure 3.** Model pressure as a function of loading $\theta$ at 300 K [2]; PdH (*red line*); low-pressure empirical model of Baranowski et al. [9] for PdH (*dark red dotted line*); PdD (*blue line*); low-pressure empirical model of Baranowski et al. [9] for PdD (*dark blue dotted line*).

models for the fugacity of $H_2$ and of $D_2$ gas [12]. We concluded that the difference between the best of the recent fugacity models for $H_2$ and $D_2$ was much greater at high loading than would be expected from the equation of state (even so the differences are small), so that it would be appropriate to make use of the $H_2$ fugacity to model $D_2$. We are able to get reasonably good agreement with the high pressure loading data using a model that includes both O-site and T-site occupation, taking advantage of the O-site to T-site excitation models discussed above; the model pressure composition isotherms that result for $PdH_x$ and $PdD_x$ are shown in Fig. 3 [2].

## 5. Phase Boundary Near a Loading of Unity

Reviewers for this paper and previous papers have raised the issue of the absence of an obvious phase boundary between the $\beta$-phase below $\theta = 1$ and a tetrahedral phase above $\theta = 1$. We were expecting to see a discontinuity around $\theta = 1$ in the isotherm, perhaps with a factor of two or three increase in pressure required to proceed past $\theta = 1$. Some of the early models that we explored in which a large O-site to T-site excitation energy was assumed showed exactly such an effect. However, when we made use of all available high pressure solubility data, it became clear that the excitation energy was relatively low (225 meV is not a particularly large excitation energy in this context), and that the transition from the $\beta$-phase to the T-site phase in the model was gentle. The O-site and T-site fractional occupation from the models are shown in Fig. 4 [2].

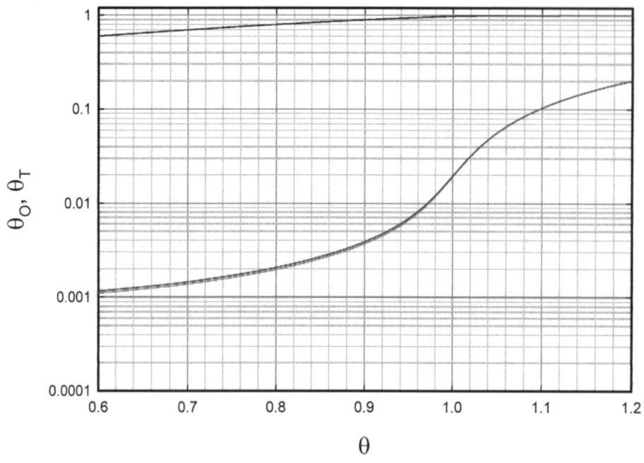


**Figure 4.** O-site fraction for PdH (*upper dark red line, under upper dark blue line*) as a function of $\theta$; T-site fraction for PdH (*lower red line*); O-site fraction for PdD (*upper dark blue line*); T-site fraction for PdD (*lower blue line*).

## 6. Modeling the Phase Diagram

We were interested in the possibility of applying the simple statistical mechanics models with both O-site and T-site occupation to model the isotherms of the phase diagram. Since these isotherms involve different data sets and also different temperatures, we thought that it might be possible to check for consistency with our model for the O-site to T-site excitation energy. In addition, we were interested in developing estimates for the O-site energy as a function of loading and temperature, which would only be possible if fitting were done starting at low loading. We found that the a good fit was possible if we made use of the O-site to T-site excitation energy from the earlier study (shown in Fig. 2) [3]; the resulting fitted isotherms are shown in Fig. 5. The O-site energies consistent with this phase diagram are illustrated in Fig. 6. The optimization of the phase diagram fit in the case of no T-site occupation leads to unphysical results for the O-site energy curves (the mean O-site energy decreases with temperature at some loadings, which is inconsistent with a simple configuration splitting of the O-site energies). If we fit with a model that optimizes the O-site to T-site excitation energy, then we end up with a very good mathematical fit to the phase diagram but once again the resulting O-site energies are unphysical.

Note that according to this model, the energy of H in $PdH_x$ is lower than in $H_2$ at all loadings. This will also be true for the energy of D in $PdD_x$, which is not in agreement from the conclusion derived from the enthalpy curves widely used in the literature.

## 7. Summary and Discussion

It has long been taken as settled that the solubility of $PdH_x$ and $PdD_x$ can be accounted for with interstitial hydrogen/deuterium in octahedral sites. This is consistent with early theoretical models, and also with early neutron scattering studies of PdD. However, we found it impossible to account for the solubility data of Clewley et al. with such models. It was only when we allowed for T-site occupation that good agreement could be obtained, with a relatively low O-site to T-site excitation energy near 105 meV. Quasi-elastic neutron diffraction studies of $PdD_x$ at elevated temperature showed peaks consistent with T-site occupation which is supportive of this approach, and which provides

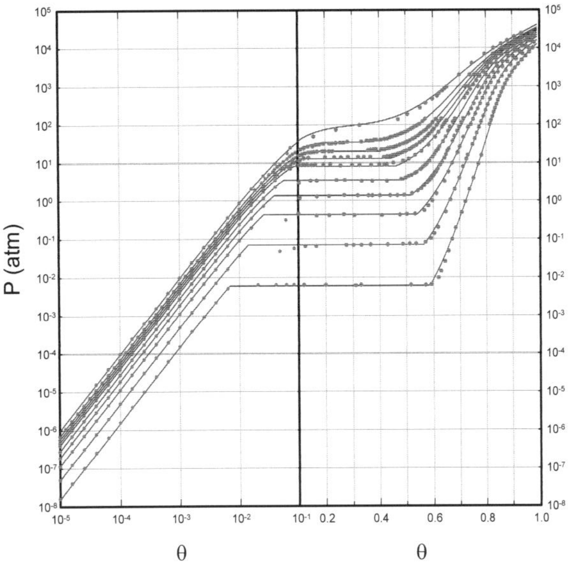


**Figure 5.** Isotherms and data points for the model with temperature-dependent 6th-order polynomial fit of $E_O$ and an empirical $\Delta E$ model, from [3]. The isotherms from low to high are for temperatures of 20, 70, 120, 160, 200, 243, 270, 300, 340, and 433°C.

estimates for the O-site to T-site excitation energy at intermediate loading. At high pressure there are some solubility studies which allow for an estimate of the excitation energy, and we found it possible to obtain agreement between the models and experimental data with an excitation energy near 225 meV. We made use of the statistical models for O-site and T-site occupation to model the phase diagram of $PdH_x$, and found good agreement when we made use of our empirical model for the O-site to T-site excitation energy with temperature-dependent O-site energies that are physically plausible.

In view of the DFT calculations of Yoshinari [7], a mean field model based on O-site and T-site occupation constitutes a significant simplification of a complicated situation. For example, one might imagine some time in the future where DFT computations for $PdH_x$ become accurate at the meV level; in this case a better statistical mechanics model could be developed based on the occupation of O-sites on the same footing with the different excited states systematically. Perhaps such a model would be able to predict solubility as well as diffusion coefficients from the same more general formulation. Such a project seems well beyond our capability today, which is why the much simpler O-site and T-site mean field model we have developed may be useful at the present time.

In the future we would like to make use of the approach outlined in this paper to develop a mean-field statistical mechanics model for palladium deuteride. Our hope is that these models might stimulate new experiments that can address the issue of T-site occupation, so that we might have more confidence in our understanding of these systems. We are also interested in pursuing modeling of the NiH system, due to its importance in Condensed Matter Nuclear Science.

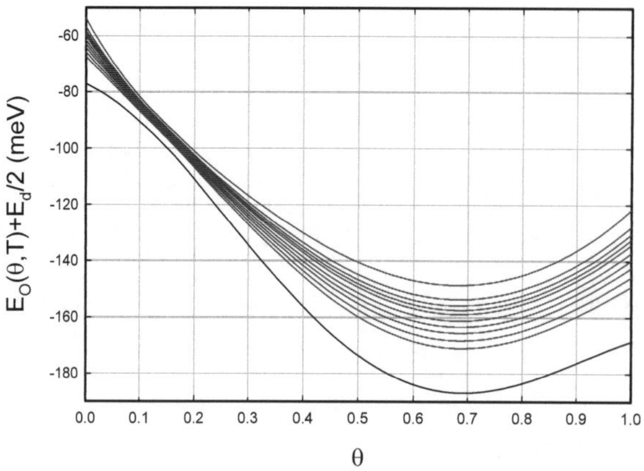


**Figure 6.** O-site energies as a function of loading from Ref. [3] for the different temperatures (20, 70, 120, 160, 200, 243, 270, 300, 340, and 433°C) of the isotherms of Fig. 4 (*blue lines*); extrapolation to $T = 0$ (*black line*).

## Appendix A. Discussion on Points Raised by the Reviewer

The reviewer has raised a number of issues which we consider in this Appendix.

### Appendix A.1. Possible phases at high loading

The reviewer has written:

> Apparently the reviewer and the author have a basic difference in how they interpret the behavior of crystal structures such as that of PdD. Generally, structures of atoms tend to change from one form to another rather than have the atoms move into tetrahedral sites, which requires high energy. For example, in the Zr–H system, the cubic form of ZrH converts to a tetrahedral form designated $ZrH_2$ when sufficient $H_2$ pressure is applied. Presumably the same transformation might occur in the Pd–H system, which the author notes.
>
> Evidence summarized by Storms in [14] shows a limit to the cubic phase near but slightly less than $PdH_1$. A H/Pd ratio greater than this value would presumably be a mixture of cubic PdH and tetragonal $PdH_2$. Although this conclusion has not been proven, it at least needs to be acknowledged when discussing the possibility of tetrahedral occupancy and its influence on the limit of the cubic phase.

To understand what crystal structures are present at a D/Pd loading above unity it will be necessary to do X-ray diffraction in order to determine the Pd substructure, and neutron diffraction to determine the D substructure. Until unambiguous measurements are done that settle the issue, it is clear that people will continue to make spirited cases for one structure or another. Certainly the modeling the the author has carried out is not intended as a substitute for such much needed experiments.

The author is of the opinion that the Fukai structure is thermodynamically preferred at a loading above about $PdD_{0.95}$ near room temperature. Near room temperature atomic self-diffusion of Pd atoms is slow, so that it might take a large number of years for superabundant vacancies to form in large amounts. If so, then it may be that the Pd sublattice remains FCC, and interstitial H or D fills O-sites first and then more highly excited states subsequently. This is presumed in the model under discussion in this paper.

In most DFT and in other simulations H is found to occupy O-sites as the lowest energy available sites. Certainly there are excited orbitals available [7], which the T-site is an example. The model discussed in the paper adopts a simplistic picture of an O-site and a T-site as anharmonic wells, which is useful in the sense that there is additional freedom within the model to connect with experimental data quantitatively. It would certainly be possible to develop better statistical models that take advantage of more realistic excited state energies once they are known, and also to take account of more accurate many-body energies associated with the various configurations once reliable estimates become available. If and when the crystal structure in the vicinity of unity loading and above is clarified, there will be no problem developing statistical models that take into account a different crystal structure and interstitial state energies.

### Appendix A.2. Role of tetrahedral occupation

The reviewer has written:

> Based on previous papers, the author apparently is intent on showing how D can occupy tetrahedral sites as the location of the LENR fusion reaction. Although this is a worthy goal, the conclusions must also be consistent with the conventional behavior of PdD. This reviewer is only asking that this consistency, or lack thereof, be acknowledged. The author concludes that inclusion of tetrahedral occupancy provides a better fit to the available data. While this might be true, the reviewer would like to have this conclusion made more obvious.

The initial goal associated with the various studies summarized in this paper was to see whether a statistical model could be developed that might be relevant to D/Pd loading near and above unity. The large majority of statistical models in the literature are simply not applicable to the case of interstitial loading above unity, primarily because they are restricted to O-site occupation. In this sense the research was exploratory.

However, once work started in earnest on this kind of model it became possible to make a connection with many experimental measurements. As discussed above, it is impossible to account for alpha-phase solubility data with a mean-field model based only on O-site occupation (with either harmonic or anharmonic potentials). The neutron diffraction data at modest loading implicates T-site occupation (which the author was not aware of during the initial development of the model). And it was possible within the model to develop fits which connected with experimental data near and above unity loading. Of course analogous statistical models can be developed based on other schemes and crystal structures that would similarly be able to connect with experimental data sets at high loading – there is nothing unique in this sense about the model under discussion in this paper.

The hope was to develop a statistical model that might be used systematically over the different regions, including low and high loading, as well as low and high temperatures. If we would like to develop a reasonably general simulation model for the Fleischmann Pons and related experiments, such a model would be extremely helpful.

It was never the case, as the reviewer suggests, that the tetrahedral site in bulk PdD was considered by the author as the site for any LENR fusion reaction. As has been discussed a great many times [13], this author considers monovacancies (or their equivalent in the Fukai structure) to be where deuteron-deuteron reactions occur in the F–P experiment.

### Appendix A.3. Isotherm of Santandra and Behrens

The reviewer has written:

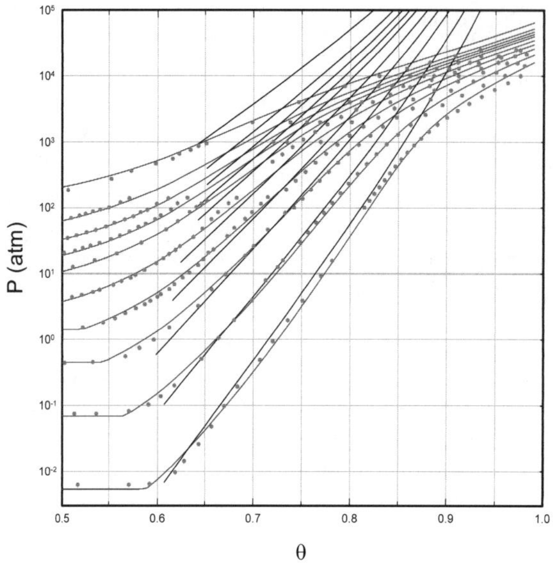


**Figure 7.** Close-up of data points (red circles) and model isotherms (*blue lines*) from Fig. 5 displayed with isotherms from the Santandrea and Behrens parameterization (*black lines*).

Santandrea and Behrens [15] give the following equations as their description of the $H_2$ pressure in equilibrium with $\beta$-PdH based on the data available at the time.

$$P(\mathrm{H}_2) = \exp\left(-\frac{4693}{T} + 10.97\right), \quad \text{miscibility gap},$$

$$P(\mathrm{H}_2) = \exp\left(12.9 + 2\ln\frac{\theta}{1-\theta} - \frac{12070 - 10830\,\theta}{T}\right), \quad \beta\text{–PdH}.$$

How do these equations compare to the equations the author obtains using his treatment? Can the author provide a better fit to all available data based on his approach? If so, please provide equations showing the effect of pressure in a form that can be used by an experimenter.

Perhaps the place to start might be to compare the isotherm of Santandrea and Behrens in the beta phase with the set of PCT points that we used, and with our model isotherms. The results are shown in Fig. 7. One sees that the Santander and Behrens fit works best near room temperature and at modest pressures (we found a similar restriction in applicable range in Baranowski's parameterization). Our model is discussed in detail in [3].

## References

[1] P.L. Hagelstein, O-site and T-site occupation of $\alpha$-phase $PdH_x$ and $PdD_x$, *J. Condensed Matter Nucl. Sci.* **17** (2015) 67–90.

[2] P.L. Hagelstein, Empirical models for octahedral and tetrahedral occupation in PdH and in PdD at high loading, *J. Condensed Matter Nucl. Sci.* **17** (2015) 35–66.

[3] P.L. Hagelstein, Models for the phase diagram of palladium hydride including O-site and T-site occupation, *J. Condensed Matter Nucl. Sci.* **20** (2016) 54–80.

[4] J.D. Clewley, T. Curran, T. B. Flanagan and W. A. Oates, Thermodynamic properties of hydrogen and deuterium dissolved in palladium at low concentrations over a wide temperature range, *Journal of the Chemical Society, Faraday Transactions 1: Physical Chemistry in Condensed Phases* **69** 449–458 (1973)

[5] R. H. Fowler and C.J. Smithells, A theoretical formula for the solubility of hydrogen in metals, *Proc. Roy. Soc. London, Series A, Math. Phy. Sci.* **160** 37–47 (1937).

[6] Y. Fukai, *The Metal–Hydrogen System*, Springer, New York, 1993.

[7] O. Yoshinari, Origin of 'Inverse Isotope Effect' of hydrogen diffusion in palladium, In *Defect and Diffusion Forum* **312** (2011) 295–300.

[8] S. Majorowski and B. Baranowski, Anomalous behavior of diffusion coefficients of hydrogen and deuterium in high content palladium hydride and deuteride, *Electronic Structure and Properties of Hydrogen in Metals*, Plenum Press, New York, 1983, pp. 519–524.

[9] B. Baranowski, S. M. Filipek, M. Szustakowski, J. Farny and W. Woryna, Search for "cold-fusion" in some Me–D systems at high pressures of gaseous deuterium, *J. Less Common Metals* **158** (1990) 347–357.

[10] S. Crouch-Baker, M.C.H. McKubre and F.L. Tanzella, Variation of resistance with composition in the $\beta$-phase of the H–Pd system at 298 K, *Zeitschrift für Physikalische Chemie* **204** (1998) 247–254.

[11] M.P. Pitt and E.M. Gray, Tetrahedral occupancy in the Pd–D system observed by in situ neutron powder diffraction, *Europhys. Lett.* **64** (2003) 344.

[12] P.L. Hagelstein, Equation of state and fugacity models for $H_2$ and for $D_2$ *J. Condensed Matter Nucl. Sci.* **16** (2015) 23–45.

[13] P.L. Hagelstein, Current status of the theory and modeling effort based on fractionation, *J. Condensed Matter Nucl. Sci.* **19** (2016) 98–109.

[14] E.K. Storms, The explanation of low energy nuclear reaction, Infinite Energy Press, Concord, NH, 2014, pp. 127–128.

[15] R.P. Santandrea and R.G. Behrens, A review of the thermodynamics and phase relationships in the palladium–hydrogen, palladium–deuterium and palladium–tritium systems, *High Temperature Materials and Processes* **7** (1986) 149.

J. Condensed Matter Nucl. Sci. 24 (2017) 98–113

Research Article

# Developing Phonon–Nuclear Coupling Experiments with Vibrating Plates and Radiation Detectors

Florian Metzler, Peter L. Hagelstein* and Siyuan Lu

*Massachusetts Institute of Technology, Cambridge, MA, USA*

**Abstract**

Excess heat has been reported in cold fusion experiments since 1989; however, there is at present no accepted explanation for what mechanisms are involved. Over the past decades a general theory has been developed which seems applicable to excess heat and other anomalies systematically; but in this case we do not yet have unambiguous experimental support for the phonon–nuclear coupling and enhanced up-conversion and down-conversion mechanism. This has motivated experimental studies with which we hope to develop relevant experimental results from which clear tests of theory can be made. A facility has been developed with which we are able to induce vibrations in metal plates from about 10 kHz up to about 10 MHz and then measure the relative displacement. With a high-power piezo transducer we have driven a steel plate at 2.23 MHz to produce a vibrational power of 100 W. We are able to detect X-rays with film, scintillator and camera, with low-cost sensitive scintillator/PMT detectors, and with an Amptek X-123 detector. We also have detectors that can see gamma and neutron emission.


*Keywords:* Karabut experiment, kHz and MHz transducers, Phonon–nuclear coupling, Radiation detection, Up-conversion

## 1. Introduction

Following the announcement of excess heat in the Fleischmann–Pons experiment [1,2] there have been a large number of studies performed within our community seeking excess heat in PdD in electrochemical and in gas loading experiments [3]. Correlations between excess heat and current density, loading, temperature, $^4$He, RF emission, neutron emission and gamma emission have all been pursued; as well as tests for excess heat with Pd alloys and with other metals; and also with hydrogen and mixed H/D loading. A presumption implicit in this research is that by understanding more about the system we might come closer to understanding how it works microscopically. Due to the large amount of energy produced, and absence of commensurate chemical products, the origin of the energy must be nuclear. When nuclear energy is created in conventional nuclear fission or fusion reactions, energetic particles are observed in amounts commensurate with the energy; in this case we are able to understand how the microscopic reactions work by detecting the energetic particles and comparing with theoretical predictions from nuclear physics. In

*E-mail: plh@mit.edu

the Fleischmann–Pons experiment the absence of commensurate energetic particles then prevents us from making use of standard nuclear methods to study microscopic mechanisms, with the result that after almost 28 years there remains no agreement as to what mechanism is responsible for the excess heat effect. This has hindered scientific progress, and has made it difficult to interact with skeptical colleagues who are not convinced that there is a real effect worthy of study.

Basic physics issues associated with the excess heat effect have been considered in our group for many years. The absence of commensurate energetic particles in connection with excess heat production has been identified as the most fundamental of the theoretical issues, which for us implies the existence of a mechanism capable of down-conversion from the MeV regime down to an energy scale relevant to condensed matter energy scales at the eV level or below [4]. In 2002 a relatively simple mechanism was proposed that could lead to a strongly enhanced down-conversion effect that appears to be sufficiently strong to be relevant to the problem. Subsequently an interaction mechanism was identified in which internal degrees of freedom of a nucleus is coupled to the center of mass motion [5]. This coupling, combined with the down-conversion mechanism, can provide a foundation for a theoretical description of the excess heat effect, as well as other anomalies. Continued progress on the models provides encouragement that an acceptable theoretical solution will emerge; however, a connection between the emerging theory and experiment is needed. For example, if there really is a coupling between internal nuclear degrees of freedom and vibrations, then there should be observable consequences independent of the Fleischmann–Pons experiment. If a large quantum can be down-converted, then this or the inverse up-conversion effect should be observable, again independent of the Fleischmann–Pons experiment.

These considerations have in part provided motivation for an experimental effort focused on providing experimental feedback in response to the theoretical ideas that have been put forth. There are a number of experiments which have been reported previously in which interpretations involving up-conversion are possible; these include collimated X-ray emission in the Karabut experiment [6]; collimated X-ray emission in the waterjet experiments of Kornilova and coworkers [7]; neutron emission from a steel sample that is vibrated in the experiments of Cardone and coworkers [8]; and fracture experiments of Carpinteri and coworkers [9] where neutron emission and other effects are reported when large granite samples containing iron are fractured. We are interested in whether unambiguous experimental evidence for phonon–nuclear coupling can be developed, and whether a case can be made for up-conversion or down-conversion in the phonon–nuclear coupled system.

## 2. Input from Theory

Models based on the lossy spin–boson system describe up-conversion and down-conversion as being essentially the same theoretically, so that we expect to understand the down-conversion mechanism from experiments involving up-conversion. In this model a highly excited oscillator is linearly coupled with many identical two-level systems in the presence of loss. In order for the up-conversion of many oscillator quanta to create a single two-level excitation to occur within the model with a significant rate, the coupling must be sufficiently strong so that [10]

$$(\hbar\omega_0)^3 V_0 \left\langle \sqrt{S^2 - m^2}\sqrt{n} \right\rangle > 2 \times 10^{-4} \Delta E^4, \tag{1}$$

where $\hbar\omega_0$ is the phonon energy, $V_0$ the coupling strength, $n$ the number of phonons in the mode, $\Delta E$ the excitation energy, and $\sqrt{S^2 - m^2}$ is the Dicke number. This constraint has the best chance to be satisfied when the (nuclear) excitation energy $\Delta E$ is minimized, when the (vibrational phonon) oscillator energy $\hbar\omega_0$ is maximized, when $n$ is large so that the (phonon) mode is highly excited, and when the Dicke number is large which means that many nuclei are involved.

These considerations suggest that our focus should be on low energy nuclear transitions (to have a small $\Delta E$), some of which are listed in Table 1, high frequency vibrational modes, strong vibrational excitation, and large samples.

**Table 1.** Some important low-energy nuclear transitions [7].

| Isotope | Natural abundance (%) | Energy (keV) | Half-life | Multipolarity |
|---|---|---|---|---|
| Hg-201 | 13.2 | 1.565 | 81 ns | M1+E2 |
| Ta-181 | 100 | 6.237 | 6.05 $\mu$s | E1 |
| Fe-57 | 2.1 | 14.4129 | 98.3 ns | M1+E2 |

The phonon–nuclear interaction that arises from a relativistic model has E1 multipolarity, which suggests that the 6.237 keV transition in $^{181}$Ta is particularly interesting to us. M1 transitions are possible with two-phonon exchange, but we would expect them to be weaker. The 1.565 keV transition is also very interesting since it is the lowest energy transition from the ground state of all the stable nuclei. We note that collimated X-ray emission in the Karabut experiment is reported near 1.5 keV [6], which is compatible with an interpretation of this experiment as demonstrating up-conversion of vibrations to produce phase-coherent excitation of this transition in $^{201}$Hg nuclei as an impurity near the cathode surface. The cathode is mechanically coupled to a cathode support and large stainless steel vacuum chamber, which is of interest here in that inadvertently a large number of $^{57}$Fe nuclei can share common vibrations in the system. Perhaps up-conversion of vibrations occurs due to the 14.4 keV transition in the vacuum chamber, with the excitation transferred to produce excitation in the $^{201}$Hg nuclei below the cathode surface resulting ultimately in collimated X-ray emission. In order for a highly-excited oscillator strongly coupled to many two-level systems at energy $\Delta E_1$ in the presence of loss to up-convert to two-level systems at a lower energy $\Delta E_2$, we expect conversion to be hindered unless [12]

$$\frac{(\hbar\omega_0)^3 V_0 \left\langle \sqrt{S^2 - m^2}\sqrt{n} \right\rangle}{(\Delta E_1)^3} > 2 \times 10^{-4} \Delta E_2. \tag{2}$$

Earlier efforts to apply the up-conversion model to calculate up-conversion of vibrations in the MHz range to produce excitation in the keV regime were basically unsuccessful. However, we have recently understood that the models are compatible with up-conversion in stages, where instead of up-converting from MHz frequencies to produce excitation in the keV regime in a single step, the numbers work much better if the up-conversion involves intermediate steps. In this case up-conversion of a strongly excited oscillator at $\omega_0$ coupled with many two-level transitions at $\Delta E$ in the presence of loss would be expected to produce excitation in a vibrational mode at higher frequency $\omega_1$ when

$$\frac{(\hbar\omega_0)^3 V_0 \left\langle \sqrt{S^2 - m^2}\sqrt{n} \right\rangle}{\Delta E^3} > 2 \times 10^{-4} \hbar\omega_1. \tag{3}$$

In this case we might imagine up-conversion from MHz vibrations to GHz vibrations; then a second step of up-conversion of GHz vibrations to THz vibrations; and finishing up with up-conversion to the keV regime. There are issues associated with momentum conservation in the coupling matrix elements since disparate wavelength are involved; however, this may still work out since $^{57}$Fe is present at low levels (and hence is not present at every site).

Excitation transfer experiments seem possible as a way to test the proposed phonon–nuclear coupling, where excited state $^{181}$Ta (or $^{57}$Fe) nuclei are produced as a result of radioactive decay of $^{181}$W (or $^{57}$Co); where gamma emission from nuclei away from where the initial excitation is produced would signal that excitation transfer has occurred. If the proposed interpretation of up-conversion in the Karabut experiment is right, then perhaps it is possible to develop new experiments where (large) steel samples are vibrated and X-ray emission sought from a small amount of mercury added near the surface. As noted above we have conjectured that neutron emission and transmutation claimed in some of Cardone's experiments [8], where steel samples are vibrated very strongly at 20 kHz, may be a result of up-conversion resulting in neutron emission [13]. We note reports from Carpinteri of the observation of neutron emission and elemental anomalies in fractured granite containing iron [9]. In recent experiments reported by Cardone and coworkers liquid mercury is vibrated very strongly resulting in apparent elemental and isotopic anomalies [14], which also permits an interpretation of up-conversion. If so, then perhaps we might be able to see anomalies in samples with low energy nuclear transitions that are vibrated strongly, and test whether the conjectures from theory are correct or not.

## 3. Experimental, Generation and Observation of Vibrations

In earlier experiments carried out at SRI and at MIT, capacitive coupling through air or PVDF (polyvinylidene fluoride) was used to stimulate vibrations in the range of 10–50 MHz, where the relatively high frequency range was chosen to maximize the phonon energy following the arguments above [15]. Due to the various technical difficulties encountered in this work, and in the initial MIT efforts, we decided to work at lower frequencies so that the experiments might be easier generally, and also to reduce issues with noise that plagued the earlier work.

### 3.1. Wide-band transducers and laser amplitude modulation

In our system sinusoidal signals originate in a computer controlled tunable signal generator; they can be amplified by a linear amp; and then the signal is coupled through a dual directional coupler to a transducer coupled to a metal plate (Fig. 1).

To detect the vibrational response we initially attempted to make use of FM modulation using a laser in a homodyne detection system; however, we found that the signals detected were due instead to amplitude modulation. When we first collected spectra in the MHz regime, a large number of resonances combined to produce a complicated spectrum.

For a rectangular plate the (unloaded) compressional resonances are at

$$f_{lmn} = \sqrt{\left(\frac{c_L l}{2L_x}\right)^2 + \left(\frac{c_L m}{2L_y}\right)^2 + \left(\frac{c_L n}{2L_z}\right)^2} \tag{4}$$

**Figure 1.** Steel plate 3 mm thick, $12'' \times 12''$ transverse dimensions with broadband transducer.

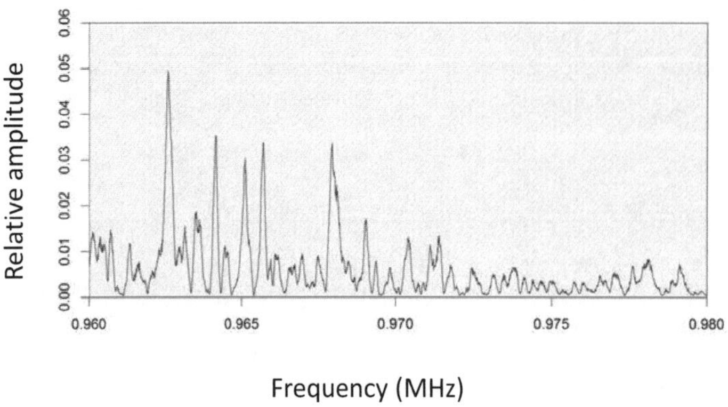


**Figure 2.** Compressional spectrum for the steel plate of Fig. 1 near the $n = 1$ fundamental.

where $c_L$ is the compressional sound speed, where $L_x$, $L_y$ and $L_z$ are the plate dimensions in the three directions. We are most interested in the fundamental (where $l = m = 0, n = 1$) and the low harmonics ($n = 2, 3, \ldots$) since these are transversely uniform in the ideal case. In order to find the fundamental we tried working with a shallow incident laser beam, hoping for a stronger response from the fundamental and harmonics than from modes with transverse structure. This was helpful, but we still observed many lines in the vibrational spectrum as shown in Fig. 2. In this case we worked with a steel plate 3 mm thick, motivated by the 3 mm thick steel plates used by Kornilova and coworkers in their collimated X-ray experiments with a waterjet [7]. An example of a spectrum in the vicinity of the 3rd and 4th harmonic of the compressional fundamental of a 3 mm Ta plate (Fig. 3) is shown in Fig. 4.

The sound speed for Ta was measured to be $4230 \pm 40$ m/s, where the uncertainty in this case is associated with

**Figure 3.** Ta plate 3 mm thick, 6"×6" transverse dimensions with a broadband transducer.

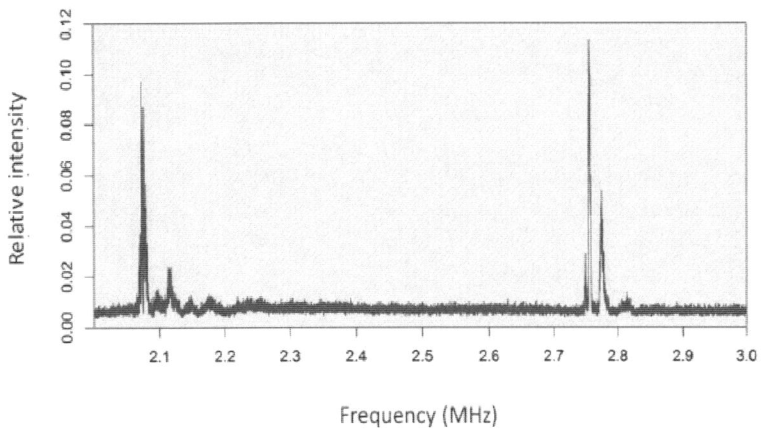


**Figure 4.** Compressional spectrum for the Ta plate of Fig. 3 near the $n = 3$ and $n = 4$ harmonics.

the plate thickness; this value is close to the 4100 m/s that can be found on tabulations. Note that loading by the relatively massive transducer causes mode splitting, corresponding to stronger vibrations at different locations on the plate surface.

### 3.2. Gel and water vibration

It was noticed that the ultrasound gel used to make contact between the transducer and the plate would seem to move during the frequency scans. The eye is not sufficiently fast to detect MHz motion, which means the gel would undergo a low frequency modification when the plate was driven near the fundamental. Subsequent experimentation with water (which was quickly lost) and with oil (which remained for a longer time) led to an ability to determine precise frequencies of maximum deformation which are produced at the fundamental.

**Figure 5.** Piezoelectric transducer on Styrofoam.

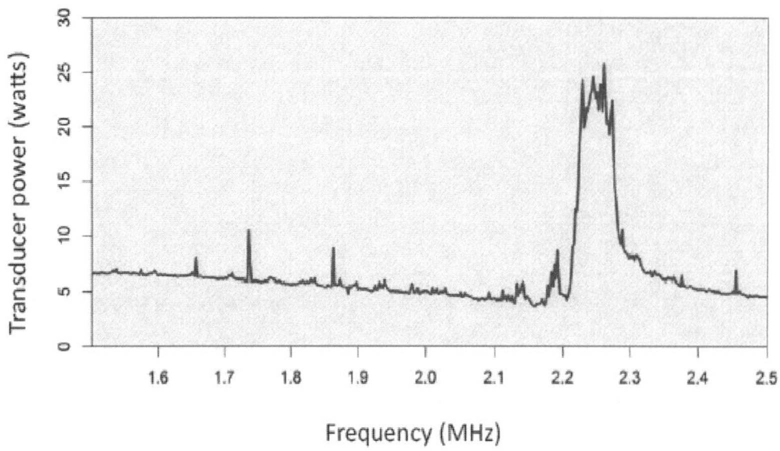


**Figure 6.** Transducer power spectrum (*black*) in the vicinity of its resonance on Styrofoam.

### 3.3. High-power piezo transducer

We obtained two high-power piezoelectric transducers with resonances near 2.23 MHz. One of the transducers on Styrofoam is shown in Fig. 5, with the transducer power spectrum (in isolation) shown in Fig. 6. The resonance of the piezo transducer is reasonably narrow, so in order to couple power between this transducer and a plate we need to work

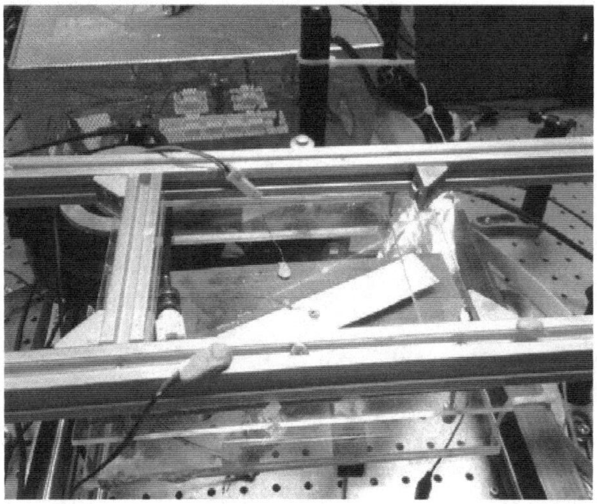

**Figure 7.** Piezoelectric transducer mounted on a matched steel plate.

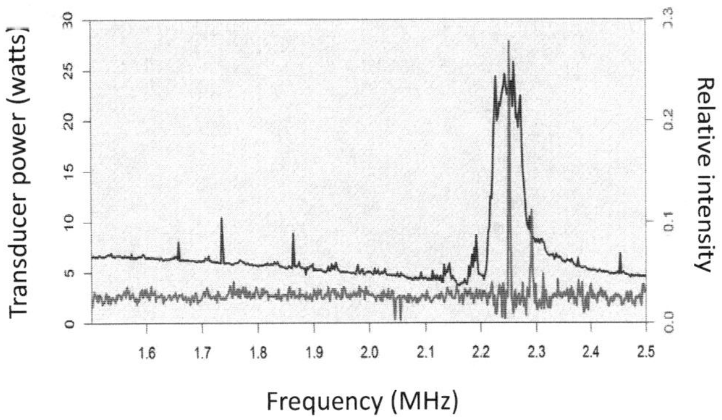


**Figure 8.** Transducer power spectrum (*black*) in the vicinity of its resonance on Styrofoam, and plate resonance (*red*).

with a plate with a matched resonance. This has been done by machining a thicker plate to the required thickness, and also by using plates ordered from a supplier that are made with thickness corresponding to a resonance. The resulting steel plate resonance is shown along with the transducer resonance in Fig. 7. We are able to drive the matched steel plate by placing the piezo transducer on the steel plate (Fig. 8) with an ultrasonic coupling gel between them.

### 3.4. Temperature measurements

We have been interested in how much vibrational energy is present (since up-conversion in the model depends on the level of vibrational excitation), and also in how much vibrational power is dissipated (since loss is important for up-conversion). The first thought was to use a Doppler measurement to determine the surface displacement, but we have as yet not managed to measure suitable Doppler signals for this purpose in the MHz regime. Additionally, the

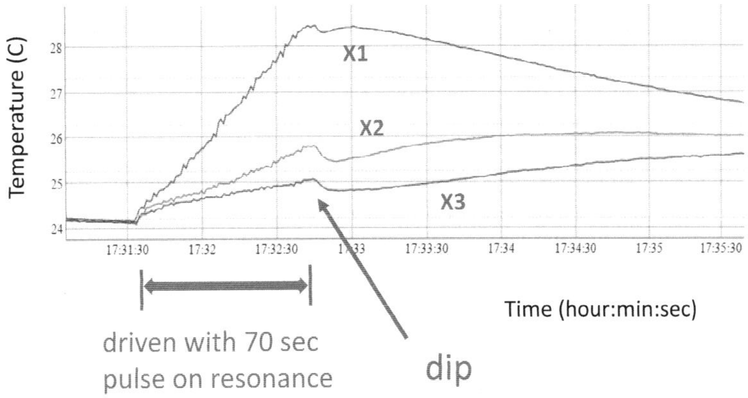


**Figure 9.** Transducer power spectrum on Styrofoam (*blue*) in the vicinity of the steel plate resonance (*red*).

**Figure 10.** Plate driven on resonance at 2.1 MHz with thermal sensors.

displacement would have to be on the order of an optical wavelength to see sidebands, and it was not obvious at the time that we should expect such a large displacement based on the applied power.

Subsequently we noticed that the broad band transducers heated significantly, which we wanted to monitor to prevent overheating, and that the plates heated enough to be warm to the touch. Measurements of the plate temperature at different locations (Fig. 9) showed a slow temperature component that could be attributed to thermal conduction of heat from the transducer, and a fast component due to the dissipation of the vibrational energy. This effect was seen both when the transducer was turned on, and when turned off; a dip correlated with transducer turn off is illustrated in Fig. 10.

It was possible to develop an approximate calibration of the thermal response of the plate by putting in known thermal power from a resistive heater and measuring the temperatures. The thermal mass is substantial as can be seen in the linear temperature rise prior to the dip in Fig. 10. The incremental temperature in subsequent tests was fit to a single pole model

$$\left(\tau\frac{\mathrm{d}}{\mathrm{d}t}+1\right)\Delta T(t) = \left(\frac{\Delta T}{\Delta P}\right)P(t) \tag{5}$$

with a calibration constant $\Delta P/\Delta T$ of about 2.85 K/W and a time constant $\tau$ (due to the heat capacity) near 500 s. When the plate was driven on resonance with the high-power piezo transducer we observed a fast (delocalized) temperature increase of about 2.5°C with a 5-second long burst; according to the model this would correspond to about 100 W dissipated vibrational power in the harmonic. While we have found a simple single pole model adequate to allow an independent estimate of the vibrational power, a more sophisticated model would be required to account accurately for the position-dependent transport effects evident in the data shown in Fig. 10.

### 3.5. Ring-down measurement

Some attempts were made to determine the relaxation time of the vibrational energy making use of a ring-down measurement based on the amplitude modulation signal. Based on preliminary results the vibrations at 2.23 MHz probably decay with a time constant at or below 1.5 ms.

## 4. Experimental, Radiation Detection

In this section we describe some of the radiation detectors that we have worked with so far.

### 4.1. X-ray film, optical film, and light tight covers

Before this effort began, there was an exploratory attempt to measure X-rays from a steel plate in a Kornilova type of water-jet experiment, where X-ray film was used. In support of this experiment JET Energy Inc acquired a Kodak M35 X-OMAT processor to develop the X-ray film. Film from this exploratory experiment was processed and digitized using a Vidar VXR-16 Dosimetry Pro running RIT 113 Version 3.14 software.

X-ray film is exposed by visible light, and initially most films used were found to be exposed due to light leaks. Some following effort was devoted to debugging this and other issues, where optical film was used with light tight covering based on aluminum foil and with thin graphite sheets. Aluminum foil developed pinholes immediately, but the graphite was found to be somewhat less susceptible to pinholes. We were able to demonstrate direct X-ray exposure of optical film at an X-ray facility down to about 10 keV, with no obvious advantage from scintillators placed over the film. Preliminary experiments in the lab and at a waterjet were done with ISO 400 optical film (which we could develop in house), but no exposure that could be attributed to X-rays was seen.

### 4.2. Weak Fe-55 source and film

This motivated us to pursue X-ray diagnostics further in several directions. Subsequently we worked with a weak (15 $\mu$Ci) Fe-55 source which allowed us to quantify the sensitivity of different detectors near 5.9 and 6.5 keV. With 1 and 2 day exposures of the source close to the film we could see a clear image, but only the faintest remnant was apparent on a light table after 1 h exposure (Fig. 11). It was found that optical film with higher ISO was much more sensitive. With ISO 1600 film we are able to see weak images from exposures as short as 10 min. As we have not yet succeeded in observing anomalous X-ray emission in any of our experiments so far, we have not yet returned to the use of X-ray film due to its lower sensitivity compared to other X-ray detection methods.

### 4.3. Scintillator with optical film and camera

Harder X-rays are imaged on film using planar scintillators, which is important since direct X-ray absorption by the film is reduced, and since the light produced increases with X-ray energy. We tried a number of different scintillators

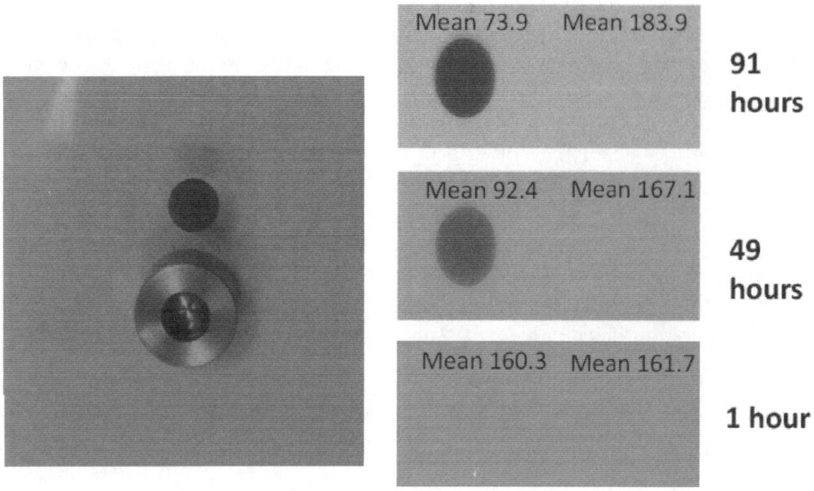


**Figure 11.** 15 $\mu$Ci Fe-55 source and image produced on film (*left*); images on optical film for different exposures (*right*).

(Al-covered YAG:Ce, BC-400, BC-404, and ZnS), with some improvement above 10 keV, but no enhancement with the Fe-55 source compared with direct optical film exposure. We also tried experiments with scintillator and a sensitive camera normally used for astronomy in the hope of having real-time imaging, but the signals from the scintillator were weak compared to noise (Fig. 12).

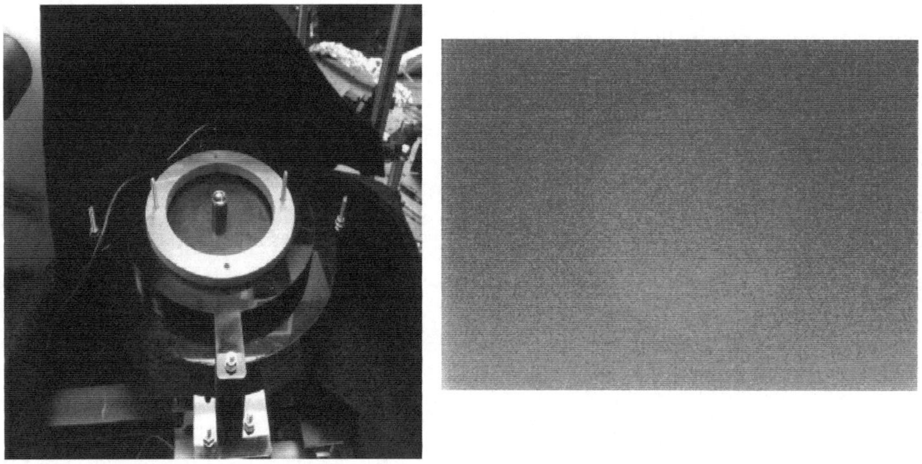

**Figure 12.** Scintillator and camera (*left*); image from camera CCD for a strong Co-57 source above the scintillator (*right*).

**Figure 13.** ZnS scintillator/PMT X-ray detector.

### 4.4. Scintillator and PMT detectors

Since the CCD array of the camera was insensitive and noisy, and since the sensitivity of the film is marginal relative to more sensitive detectors, we were motivated to pursue X-ray detection based on a scintillator/PMT combination. It is possible to work with scintillators with high photon yield, even though there is some loss in the yield at low energy, and make use of the very good sensitivity of low cost photomultipliers to end up with a very sensitive X-ray detector. The X-ray is converted to many optical photons in the scintillator, some of which are detected by the PMT. An X-ray spectrum can be constructed following pulse shape analysis of the PMT output, which is normally done using sophisticated and moderately expensive pulse shape analyzer electronics. In recent years it has been recognized that much cheaper sound cards are capable of being used for this when the count rate is low, which has led to a new field of low cost gamma and X-ray detectors for students and hobbyists.

An X-ray detector was built by Florida-based iRad Inc, based on a 250 $\mu$ ZnS scintillator (normally used for alpha detection) and Hamamatsu 3" PMT (model R6233), with a thin radfilm cover (Fig. 13), with the analog signals read by a Focusrite 192 kHz low-noise USB audio interface and analyzed with open source Theremino MCA software. It is very much more sensitive than film: X-ray fluorescence from the Mn K-alpha ionized by Fe-55 sources is apparent in seconds. A minor drawback is that the resolution is poor; as can be seen in the Ti K-alpha spectrum at 4.5 keV shown in Fig. 14, which has a line width close to a keV. The absorption of the few microns of aluminum on the radfilm prevents us from seeing XRF K-alpha lines of silicon through calcium, but we have seen X-ray fluorescence from the Mg K-alpha at 1.25 keV.

A modified version of this detector with a Be window is being built, and we hope to test it in coming months. The scintillator is thin so that the detector is not sensitive to X-rays of more than about 20 keV. We have also acquired a similar detector with a 2" thick Saint Gobain BC-408 scintillator and Photonis 3" PMT (model XP5312) for gamma detection.

In addition, we have available an Amptek X-123 X-ray detector which has better energy resolution but poorer

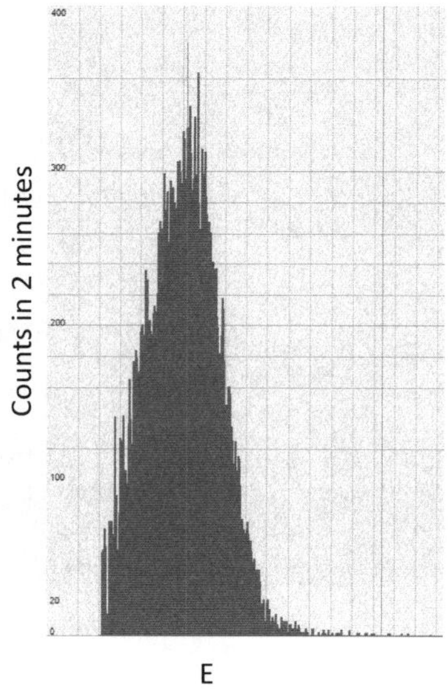


**Figure 14.** Two minute XRF spectrum of Ti K-alpha (4.5 keV) with background subtraction, from an X-ray fluorescence experiment with Fe-55 sources.

sensitivity than the scintillator/PMT detectors discussed above.

### 4.5. Neutron detection

We are also interested in the possibility of seeing whether neutron emission can be seen from vibrated steel samples, as has been claimed by Cardone and coworkers [10]. We have recently obtained a Wendi-2 He-3 neutron detector from Thermo Fisher, which has a listed sensitivity of 0.01 $\mu$Sv/h. Data taken with this detector are shown in Fig. 15; we see a background count rate on the order of 0.2 cpm, which based on the 0.84 cps/$\mu$Sv/h listed conversion is about 4 nSv/h. This is lower than a 0.45 cpm value measured outside, indicating some shielding by the building our lab is located in; and is also lower than the neutron background value of 14.5 nSv/h quoted in Ref. [16]. Motivated by the observations of Cardone and by Carpinteri, we have attempted a quick exploratory measurement looking for neutron emission from a steel plate vibrated with the piezo transducer at 2.23 MHz above 100 W (conditions very different than in the experiments mentioned above), with no counts obvious above background.

## 5. Experimental, Sample Preparation

We work with steel and copper plates from McMaster-Carr, and tantalum plates from Baoji Hanz Material Technology Co. Ltd. Some discussion is appropriate concerning the further processing of the samples.

### 5.1. Mercury on copper

We have conjectured that the 1565 eV transition in $^{201}$Hg is responsible for collimated X-ray emission near 1.5 keV in the Karabut experiment [11], which due to the absence of absorption edges in the spectrum must be on or just below

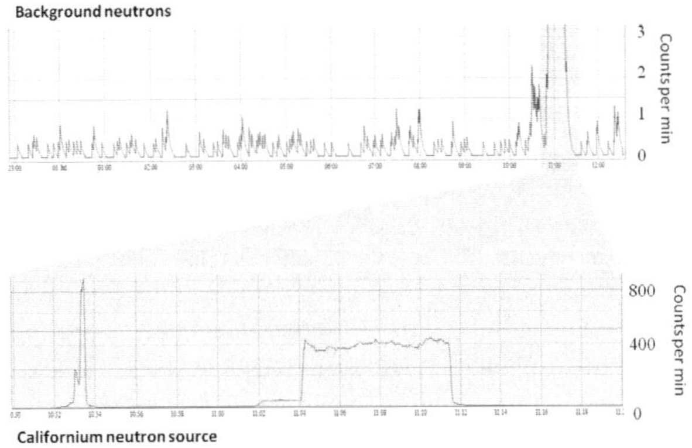


**Figure 15.** Laboratory background time history (*top*); elevated count rate with a Cf-252 neutron source (*bottom*).

the surface; the proposal is that it is present as a low-level contamination from the discharge. We might expect Hg to be present in our plates at low level as an uncontrolled impurity; however, since we have as yet no signals an obvious question can be raised as to whether we have sufficient Hg present to given us sufficient emission to detect. Hg binds strongly to copper, a fact exploited by Tanzella to plate a small amount of Hg onto the surface of copper foils for the tests at SRI [15]. We plan to make use of this approach to make available Cu foils with surface Hg that we can attach to a driven steel resonator, in the hope of clarifying whether this is how collimated X-rays are produced in the Karabut experiment.

### 5.2. Cobalt-57

In an excitation transfer experiment [17], a radioactive Co-57 source can produce excited state Fe-57 in the vicinity of the source; we will monitor for evidence that the excitation is transferred elsewhere in the plate. It is possible to acquire $^{57}CoCl_2$ in solution (from Eckert & Ziegler) and evaporate it on steel or other metal surfaces.

### 5.3. Tungsten-181

In a second version of the experiment radioactive W-181 can be generated in a localized region through $^{181}Ta(p,n)^{181}W$ reactions [18], and in this case we will monitor for signs that the 6.237 keV excited state has been induced away from the localized bombarded region.

## 6. Conclusion and Discussion

As discussed in the first two sections of the paper, a theoretical model with the potential for accounting systematically for many of the anomalies reported in our field has been developed over the past decades. According to this model there is a coupling between vibrations and internal nuclear degrees of freedom, and there is a mechanism that involves a highly excited oscillator coupled to many identical two-level systems that in the presence of loss predicts a strongly enhanced up-conversion and down-conversion effect. We would like to carry out experiments that would test these ideas. Excitation transfer requires the least in the way of new physics according to the models, and we are interested in excitation transfer tests with Co-57 to produce excited states at 14.4 keV in Fe-57, and in W-181 to produce excited states at 6.2 keV in Ta-181. Up-conversion to produce excited states at 1.56 keV in Hg-201 is perhaps the next "easiest" goal. Up-conversion to produce gammas at higher energy is expected to be harder, and neutron emission from steel would probably be the "hardest" effect to produce in that it would require the most up-conversion of the experiment under consideration.

Up-conversion or down-conversion in the lossy spin–boson model is critically dependent on the presence of a strong loss effect. To date the majority of our experiments in the lab and with the waterjet have been done under conditions where the loss has been minimized; this can be understood in the context of the laboratory experiments in that we have been interested in maximizing the vibrational displacements in order to be able to see that we have succeeded in inducing vibrations and to be able to understand where the resonances are. An important issue in the experiments currently under consideration is to add loss, and to quantify the amount of loss present. For example, if we imagine that the nuclear energy in the two-laser experiment goes into THz vibrations, we know that such high frequency vibrational modes are very lossy, with a Q value on the general order of 10. If we interpret the collimated X-ray bursts in the Karabut experiment [6] as informing us as to the vibrational frequencies involved, then we might infer a frequency on the order of 10 kHz [19] and an associated time constant on the order of a few milliseconds. If so, then the Karabut experiment can be viewed as involving a low-Q and hence very lossy resonator.

### Acknowledgements

We are grateful for equipment and advice provided by Dr. Mitchell Swartz and Dr. Alex Frank in the realm of X-ray diagnostics as well as for general feedback from Dr. Mitchell Swartz, Gayle Verner, and Jeff Tolleson during seminars and workshops. We would like to express appreciation to Industrial Heat for support in the form of unrestricted gift that we used for this work.

### References

[1] M. Fleischmann and S. Pons, Electrochemically induced nuclear fusion of deuterium, *J. Electroanal. Chem. Interfacial Electrochem.* **261** (1989) 301–308.

[2] M. Fleischmann, S. Pons, M. W. Anderson, L. J. Li and M. Hawkins, Calorimetry of the palladium-deuterium-heavy water system, *J. Electroanal. Chem. Interfacial Electrochem.* **287** (1990) 293–348.

[3] E Storms, *The Science of Low Energy Nuclear Reaction: A Comprehensive Compilation of Evidence and Explanations about Cold Fusion*, World Scientific, Singapore, 2007.

[4] P.L. Hagelstein, Current status of the theory and modeling effort based on fractionation, *J. Condensed Matter Nucl. Sci.* **19** (2016) 98–109.

[5] P.L. Hagelstein, Quantum composites: A review, and new results for models for Condensed Matter Nuclear Science, *J. Condensed Matter Nucl. Sci.* **20** (2016) 139–225.

[6] A.B. Karabut, E.A. Karabut and P.L. Hagelstein, Spectral and temporal characteristics of X-ray emission from metal electrodes in a high-current glow discharge, *J. Condensed Matter Nucl. Sci.* **6** (2012) 217–240.

[7] A.A. Kornilova, V.I. Vysotskii, N.N. Sysoev, N.K. Litvin, V.I. Tomak and A.A. Barzov, Generation of intense X-rays during ejection of a fast water jet from a metal channel to atmosphere, *J. Surface Investigation: X-ray, Synchrotron and Neutron Techniques* **4** (2010) 1008–1017.

[8] F. Cardone, G. Cherubini, M. Lammardo, R. Mignani, A. Petrucci, A. Rosada, V. Sala, and E. Santoro, Violation of local Lorentz invariance for deformed space–time neutron emission, *Eur. Phys. J. Plus* **130** (2015) 55.

[9] A. Carpinteri, G. Lacidogna, A. Manuello and O. Boria, Piezonuclear fission reactions from earthquakes and brittle rocks failure: evidence of neutron emission and non-radioactive product elements, *Exp. Mech.* **53** (2013) 345–365.

[10] P.L. Hagelstein and I.U. Chaudhary, Coherent energy exchange in the strong coupling limit of the lossy spin–boson model, *J. Condensed Matter Nucl. Sci.* **5** (2011) 116–139.

[11] P.L. Hagelstein, Bird's eye view of phonon models for excess heat in the Fleischmann–Pons experiment, *J. Condensed Matter Nucl. Sci.* **6** (2012) 169–180.

[12] P.L. Hagelstein and I.U. Chaudhary, Models for phonon–nuclear interactions and collimated X-ray emission in the Karabut experiment, *J. Condensed Matter Nucl. Sci.* **13** (2014) 177–222.

[13] P.L. Hagelstein and I.U. Chaudhary, Anomalies in fracture experiments, and energy exchange between vibrations and nuclei, *Meccanica* **50** (2015) 1189–1203.

[14] F. Cardone, G. Albertini, D. Bassani, G. Cherubini, E. Guerriero, R. Mignani, M. Monti, A. Petrucci, F. Ridolfi, A. Rosada and F. Rosetto, Nuclear metamorphosis in mercury, *Int. J. Modern Phys. B* **30** (2016) 1550239.

[15] F. Tanzella, J. Bao, M.C.H. McKubre and P.L. Hagelstein, Seeking X-rays and charge emission from a copper foil driven at MHz frequencies, *J. Condensed Matter Nucl. Sci.* **19** (2016) 110–118.

[16] M. Brugger, P. Carbonez, F. Pozzi, M. Silari and H. Vincke, New radiation protection calibration facility at CERN, *Radiation Protection Dosimetry* **161** (2013) 181–184.

[17] P.L. Hagelstein and I.U. Chaudhary, Coupling between the center of mass and relative degrees of freedom in a relativistic quantum composite and applications, *J. Condensed Matter Nucl. Sci.* **24** (2017).

[18] V.A. Dornow, J. Binder, A. Heidemann, G.M. Kalvius and G. Wortmann, Preparation of narrow-line sources for the 6.2 keV Mössbauer resonance of $^{181}$Ta, *Nucl. Instr. Methods* **163** (1979) 491–497.

[19] P.L. Hagelstein, Probabilistic models for beam, spot, and line emission for collimated X-ray emission in the Karabut experiment, *J. Condensed Matter Nucl. Sci.* **22** (2017) 53–73.

J. Condensed Matter Nucl. Sci. 24 (2017) 114–122

Research Article

# Coupling between the Center of Mass and Relative Degrees of Freedom in a Relativistic Quantum Composite and Applications

Peter L. Hagelstein*
*Massachusetts Institute of Technology, Cambridge, MA, USA*

Irfan U. Chaudhary
*Lahore University of Engineering and Technology, Pakistan*

**Abstract**

If we consider the nucleus as a relativistic composite, then we are able to derive from a many-particle Dirac model a coupling between the center of mass motion and internal nuclear degrees of freedom. This interaction can be rotated out in free space, but has the potential to give rise to new physics when two or more nuclei exchange phonons with a common vibrational mode. The simplest example of such a system is a homonuclear diatomic molecule in a frozen matrix, for which we are able to develop an expression for the second-order phonon–nuclear interaction that can result in a splitting of the nuclear energy levels as a result of excitation transfer between the nuclei. The phonon-nuclear coupling is an E1 interaction, so the low energy 6.237 keV E1 transition in $^{181}$Ta is special; this motivates an interest in molecular $^{181}Ta_2$ as a candidate for a Mössbauer experiment where the splitting might be observable. We also consider excitation transfer in the case of a macroscopic a Ta plate.


*Keywords:* Down-conversion, Excitation transfer, Mössbauer effect, Phonon–nuclear coupling, Theory

## 1. Introduction

In most quantum mechanical treatments of interacting quantum systems, it is possible to specify Hamiltonians for the systems independently in the absence of an interaction, and then to identify what kind of interaction to use when they are coupled together. For excess heat in the Fleischmann–Pons experiment, a theoretical explanation would likely involve an interaction between the condensed matter environment and internal nuclear states. Consequently, we should be able to start with models for nuclear states in isolation, and models for condensed matter systems in isolation, and identify the coupling between them.

*E-mail: plh@mit.edu.

© 2017 ISCMNS. All rights reserved. ISSN 2227-3123

It was precisely this issue that contributed in an important way to the general dismissal of the field back in 1989. The internal nuclear states are small and localized, with an energy scale not particularly well matched to the electronic or vibrational degrees of freedom of the lattice. The only coupling possible as was argued are the weak second-order electrostatic coupling (argued to be analogous to tidal forces in gravitation); or magnetic interactions which we might expect also to be weak. Mismatched energy scales combined with a weak coupling can never produce the dramatic effects claimed by Fleischmann and Pons, so it was argued.

Over the past decades we have been considering a theoretical explanation for the Fleischmann–Pons effect which is based on the down-conversion of the large nuclear energy quantum into a great many smaller condensed matter quanta, with a focus on vibrational quanta [1]. In the models that have emerged, a relatively weak coupling can result in much larger cooperative effects when many nuclei interact together. A down-conversion of the large nuclear quantum would be able to account for the absence of energetic particles commensurate with the energy produced, and also account for the very large acceleration of the associated reaction rates. In an incoherent model we would expect the reaction rate to be very slow due to the (very small) tunneling factor being squared; but in a coherent model the rate can be much faster, since the maximum rate is linear in the tunneling factor.

Nuclei are composite particles, with internal degrees of freedom associated with the nuclear states, and with center of mass coordinates that have to be considered lattice degrees of freedom [2]. In a nonrelativistic model there is a clean separation between internal and center of mass degrees of freedom, so that the only coupling is through external field interactions that depend on both. The situation is more complicated in the relativistic version of the problem, where there is in addition a relativistic coupling that can be attributed to the change in the nuclear potential when the nucleus is in motion. In free space we would expect there to be no consequence of this interaction, since it is possible to transform to a frame in which the nucleus is at rest. However, the same is not true for a nucleus in a lattice. The thought is that this relativistic coupling can mediate interactions between internal nuclear states and lattice vibrations, and it would do so in ways that we as yet have little intuition about.

In what follows we will consider the separation of internal and center of mass degrees of freedom in a composite, the associated coupling terms that might mediate interactions, and then consider new experiments where we might be able to study this basic interaction.

## 2. The Nucleus as a Relativistic Quantum Mechanical Composite

A relativistic description for a nucleus in an external field can be developed modeling the nucleons as Dirac spin 1/2 particles through the Hamiltonian

$$\hat{H} = \sum_j \beta_j mc^2 + \sum_j \boldsymbol{\alpha}_j \cdot c[\hat{\mathbf{p}}_j - q_j \mathbf{A}(\mathbf{r}_j)] + \sum_{j<k} \hat{V}_{jk}(\mathbf{r}_j - \mathbf{r}_k) + \sum_j q_j \Phi(\mathbf{r}_j), \tag{1}$$

where $\boldsymbol{\alpha}_j$ and $\beta_j$ are Dirac matrices. We have used the notation $\hat{V}_{jk}(\mathbf{r}_j - \mathbf{r}_k)$ to indicate a two-body interaction including the strong force and Coulomb interaction; however, modern nucleon–nucleon interaction models are also momentum dependent and often include three-body interactions [3] (for which the arguments that follow can be generalized). We presume that the interaction also includes projection operators needed to prevent continuum dissolution connected with Brown-Ravenhall disease [4].

We can introduce center of mass ($\mathbf{R}$ and $\hat{\mathbf{P}}$) and relative ($\boldsymbol{\xi}_j$ and $\hat{\boldsymbol{\pi}}_j$) operators according to

$$\mathbf{R} = \frac{1}{N}\sum_j \mathbf{r}_j, \qquad \hat{\mathbf{P}} = \sum_j \hat{\mathbf{p}}_j \tag{2}$$

$$\boldsymbol{\xi}_j = \mathbf{r}_j - \mathbf{R}, \qquad \hat{\boldsymbol{\pi}}_j = \hat{\mathbf{p}}_j - \frac{\hat{\mathbf{P}}}{N} \tag{3}$$

assuming $N$ nucleons in the nucleus. The external field Hamiltonian written in terms of center of mass and relative coordinates is

$$\begin{aligned}\hat{H} =& \frac{1}{N}\sum_j \boldsymbol{\alpha}_j \cdot c[\hat{\mathbf{P}} - Q\mathbf{A}(\mathbf{R})] + \sum_j \beta_j mc^2 + \sum_j \boldsymbol{\alpha}_j \cdot c\hat{\boldsymbol{\pi}}_j + \sum_{j<k} \hat{V}_{jk}(\boldsymbol{\xi}_j - \boldsymbol{\xi}_k) \\ &+ \sum_j q_j \Phi(\mathbf{R} + \boldsymbol{\xi}_j) - \sum_j \boldsymbol{\alpha}_j \cdot c\left[q_j \mathbf{A}(\mathbf{R} + \boldsymbol{\xi}_j) - \frac{Q}{N}\mathbf{A}(\mathbf{R})\right].\end{aligned} \tag{4}$$

We can make use of a Taylor series expansion of the scalar and vector potentials to write [2]

$$\begin{aligned}\hat{H} =& \sum_j \beta_j mc^2 + \frac{1}{N}\sum_j \boldsymbol{\alpha}_j \cdot c(\hat{\mathbf{P}} - Q\mathbf{A}) + Q\Phi + \sum_j \boldsymbol{\alpha}_j \cdot c\hat{\boldsymbol{\pi}}_j + \sum_{j<k} \hat{V}_{jk} \\ &+ \left(\sum_j q_j \boldsymbol{\xi}_j\right) \cdot \nabla\Phi - \sum_j \boldsymbol{\alpha}_j \cdot c\left(q_j - \frac{Q}{N}\right)\mathbf{A} - \sum_j \boldsymbol{\alpha}_j \cdot cq_j(\boldsymbol{\xi}_j \cdot \nabla)\mathbf{A} + \cdots\end{aligned} \tag{5}$$

In this expression the scalar and vector fields ($\Phi$ and $\mathbf{A}$) and their derivatives are evaluated at $\mathbf{R}$. In the first line we see a relativistic description of the composite (as a particle) interacting with external fields, and also a relativistic description of the internal problem; however, both the Hamiltonian for the center of mass and for the relative degrees of freedom share the same mass terms, and are hence coupled together in this description. In the second line there are dipole interactions with the scalar and vector potentials; and higher-order interactions are included in the terms that follow.

Nuclei in a condensed matter environment act as nonrelativistic composite particles, while we may remain interested in a relativistic description for the internal proton and neutron dynamics. We have worked with a partial Foldy-Wouthuysen transformation which results in a rotated Hamiltonian in which the center of mass degrees of freedom are nonrelativistic while the relative degrees of freedom are modeled relativistically. The transformation can be written as

$$\hat{H}' = \mathrm{e}^{i\hat{S}}\left(\hat{H} - i\hbar\frac{\partial}{\partial t}\right)\mathrm{e}^{-i\hat{S}} \tag{6}$$

with

$$\hat{S} = -i\frac{1}{2Mc^2}\sum_j \beta_j \boldsymbol{\alpha}_j \cdot c(\hat{\mathbf{P}} - Q\mathbf{A}). \tag{7}$$

The low-order terms of the rotated Hamiltonian are [2]

$$
\begin{aligned}
\hat{H}' \to\ & \frac{|\hat{\mathbf{P}} - Q\mathbf{A}|^2}{2M} \frac{1}{N} \sum_j \beta_j + Q\Phi - \frac{\hbar Q}{2M} \frac{1}{N} \sum_j \beta_j \mathbf{\Sigma}_j \cdot \mathbf{B} - \frac{\hbar^2 Q}{8M^2c^2} \nabla \cdot \mathbf{E} \\
& + \frac{\hbar Q}{8M^2c^2} \sum_j \mathbf{\Sigma}_j \cdot \Big[ (\hat{\mathbf{P}} - Q\mathbf{A}) \times \mathbf{E} - \mathbf{E} \times (\hat{\mathbf{P}} - Q\mathbf{A}) \Big] \\
& + \sum_j \beta_j mc^2 + \sum_j \boldsymbol{\alpha}_j \cdot c\hat{\boldsymbol{\pi}}_j + \sum_{j<k} \hat{V}_{jk} \\
& + \sum_j \Big[ q_j \Phi(\mathbf{R} + \boldsymbol{\xi}_j) - \frac{Q}{N}\Phi(\mathbf{R}) \Big] - \sum_j \boldsymbol{\alpha}_j \cdot c \Big[ q_j \mathbf{A}(\mathbf{R} + \boldsymbol{\xi}_j) - \frac{Q}{N}\mathbf{A}(\mathbf{R}) \Big] \\
& + \frac{1}{M} \sum_j \beta_j (\hat{\mathbf{P}} - Q\mathbf{A}) \cdot \hat{\boldsymbol{\pi}}_j + \frac{1}{2Mc} \sum_{j<k} \Big[ (\beta_j \boldsymbol{\alpha}_j + \beta_k \boldsymbol{\alpha}_k) \cdot (\hat{\mathbf{P}} - Q\mathbf{A}), \hat{V}_{jk} \Big]. \qquad (8)
\end{aligned}
$$

This rotated Hamiltonian is very interesting. The first two lines describe the center of mass degrees of freedom of the nucleus as a nonrelativistic composite interacting with the external field; the third line includes a relativistic description of the relative degrees of freedom appropriate for a nuclear structure calculation; the fourth line includes the lowest-order coupling between the two degrees of freedom due to external field interactions; and in the last line are relativistic terms that couple the center of mass and relative degrees of freedom. We are interested in these terms (in the last line) for the possibility that they provide the coupling between lattice vibrations and internal nuclear degrees of freedom responsible for anomalies in Condensed Matter Nuclear Science.

## 3. Finite Basis Approximation

In our models we have focused in earlier work on simple two-level system models to describe the nuclear states, in part because the mathematics becomes much more complicated when more general models are used. Here it will be convenient to consider the more general finite basis approximation, which is easily adapted for describing two-level systems. We consider the finite basis approximation written as

$$
\Psi(\{\boldsymbol{\xi}\}, \mathbf{R}, t) = \sum_n \Psi_n(\mathbf{R}, t) \phi_n(\{\boldsymbol{\xi}\}), \qquad (9)
$$

where $\{\boldsymbol{\xi}\}$ denotes the set of relative position vectors, where $\Psi(\{\boldsymbol{\xi}\}, \mathbf{R}, t)$ is the dynamic wave function for the nucleus including center of mass and relative degrees of freedom, and the summation is over terms that depend only on the center of mass ($\Psi_n(\mathbf{R}, t)$) or relative ($\phi_n(\{\boldsymbol{\xi}\})$) degrees of freedom. The individual $\phi_n(\{\boldsymbol{\xi}\})$ describe nuclear states, and associated with each nuclear state is a center of mass wave function $\Psi_n(\mathbf{R}, t)$.

Of interest is to develop coupled-channel equations for the different center of mass wave functions, projecting out the internal degrees of freedom. To proceed it will be convenient to make use of a vector and matrix formalism, with $\boldsymbol{\Psi}(\mathbf{R}, t)$ a vector made up of all of the different $\Psi_n(\mathbf{R}, t)$ associated with each of the channels. We can write the coupled-channel equations in vector and matrix form as [2]

$$
\begin{aligned}
i\hbar\frac{\partial}{\partial t}\boldsymbol{\Psi}(\mathbf{R},t) = {} & \mathbf{M}c^2\boldsymbol{\Psi}(\mathbf{R},t) + \left(\frac{|\hat{\mathbf{P}}-Q\mathbf{A}|^2}{2M} + Q\Phi - \frac{Q}{NM}\mathbf{S}\cdot\mathbf{B} - \frac{\hbar^2 Q}{8M^2c^2}\nabla\cdot\mathbf{E}\right)\boldsymbol{\Psi}(\mathbf{R},t) \\
& + \frac{Q}{4M^2c^2}\mathbf{S}\cdot\left[(\hat{\mathbf{P}}-Q\mathbf{A})\times\mathbf{E} - \mathbf{E}\times(\hat{\mathbf{P}}-Q\mathbf{A})\right]\boldsymbol{\Psi}(\mathbf{R},t) - \mathbf{d}\cdot\mathbf{E}\boldsymbol{\Psi}(\mathbf{R},t) - \mathbf{j}\cdot\mathbf{A}\boldsymbol{\Psi}(\mathbf{R},t) \\
& + \mathbf{a}\cdot c(\hat{\mathbf{P}}-Q\mathbf{A})\boldsymbol{\Psi}(\mathbf{R},t) + \cdots
\end{aligned}
\tag{10}
$$

In this case elements of the $\mathbf{a}$ matrix are

$$
\mathbf{a}_{n,n'} = \left\langle \phi_n(\{\boldsymbol{\xi}\}) \left| \frac{1}{Mc}\sum_j \beta_j \hat{\boldsymbol{\pi}}_j + \frac{1}{2Mc}\sum_{j<k}\left[\beta_j\boldsymbol{\alpha}_j + \beta_k\boldsymbol{\alpha}_k, \hat{V}_{jk}\right] \right| \phi_{n'}(\{\boldsymbol{\xi}\}) \right\rangle . \tag{11}
$$

The selection rule associated with these matrix elements are consistent with E1 electric dipole transitions, with the possibility of M2, E3, M4 coupling at higher order.

## 4. Phonon–nuclear Coupling for a Diatomic Molecule in a Matrix

The problem of phonon–nuclear coupling in a homonuclear diatomic molecule is of special interest to us, since it is the simplest system in which there may be a possibility of observing the relativistic phonon–nuclear interaction discussed above. The basic idea is that phonon–nuclear coupling has the potential to introduce a splitting in the nuclear states of the molecule, which might be seen in a Mössbauer experiment. At low temperature rotational motion in a gas can be eliminated; however, in a frozen matrix it is possible for the molecule to have a fixed orientation in a nonspherical confining potential, which eliminates averaging over angles and potentially allows for more information to be obtained from a Mössbauer spectrum.

We can include the relativistic phonon-nuclear interaction in a model for the nuclear states and vibrational states for a diatomic molecule in a matrix with the Hamiltonian [5]

$$
\hat{H} = \mathbf{M}_1c^2 + \mathbf{M}_2c^2 + \frac{|\hat{\mathbf{P}}_1|^2}{2M} + \frac{|\hat{\mathbf{P}}_2|^2}{2M} + V(\mathbf{R}_1,\mathbf{R}_2) + \mathbf{a}_1\cdot c\hat{\mathbf{P}}_1 + \mathbf{a}_2\cdot c\hat{\mathbf{P}}_2. \tag{12}
$$

We expect the phonon–nuclear coupling to be weak, so that we can make use of perturbation theory to develop a second-order interaction; the resulting model can be written as

$$
\hat{H} \to \mathbf{M}_1c^2 + \mathbf{M}_2c^2 + \frac{|\hat{\mathbf{P}}_1|^2}{2M} + \frac{|\hat{\mathbf{P}}_2|^2}{2M} + V(\mathbf{R}_1,\mathbf{R}_2) + \hat{U}_{12}, \tag{13}
$$

where the second-order indirect interaction can be written (assuming that the molecule is in the vibrational ground state) as [5]

$$
\hat{U}_{12} = -\frac{Mc^2(\hbar\omega_0)^2}{2(\Delta E)^2}\frac{(\mathbf{a}_1\cdot\mathbf{R}_{21})(\mathbf{a}_2\cdot\mathbf{R}_{21})}{|\mathbf{R}_{21}|^2}. \tag{14}
$$

This interaction is qualitatively different than what would be expected from the quadrupole interaction associated with an electric field gradient. The form seems similar algebraically to the nuclear spin–spin interaction

$$\hat{H}_{ss} = -\frac{\mu_0}{4\pi}\mu_N^2 \frac{g_1 g_2}{|\mathbf{R}|^3}\left[\mathbf{I}_1 \cdot \mathbf{I}_2 - 3\frac{(\hat{\mathbf{I}}_1 \cdot \mathbf{R}_{21})(\hat{\mathbf{I}}_2 \cdot \mathbf{R}_{21})}{|\mathbf{R}_{21}|^2}\right] \tag{15}$$

but we would expect the splitting from phonon–nuclear coupling to be distinguishable if the interaction is sufficiently strong.

There remains the issue of which nuclear transition we might be most interested in. We see that the interaction is inversely proportional to the square of the transition energy, which favors low energy nuclear transitions. Since the $\mathbf{a} \cdot c\hat{\mathbf{P}}$ interaction is an E1 interaction at lowest order we seek a low energy nuclear transition with E1 symmetry. Of the stable nuclei, the 6.237 keV E1 transition in $^{181}$Ta is favored in this regard, with the 25.651 keV transition in $^{161}$Dy a distant second.

Unfortunately, the electronic ground state of molecular $Ta_2$ is predicted to be an electronic $^3\Sigma_g^-$ state [6] (we would have preferred an electronic singlet state), which is unfortunate because of the additional complications of the coupling between the electronic and nuclear spins. Although Mössbauer experiments have been done on $Fe_2$ in a frozen argon matrix [7], and optical spectra have been taken with $Ta_2$ in a matrix [8], a Mössbauer experiment with $Ta_2$ in a matrix to detect the splitting due to phonon–nuclear coupling looks to be a difficult one. At present we do not yet have a reliable estimate for the $\mathbf{a}$-matrix elements, so that it is at present unknown from theory whether the splitting is expected to be observable.

## 5. Excitation Transfer in a Plate

Given the headaches involved associated with Mössbauer experiments involving homonuclear diatomic $^{181}Ta_2$ in a frozen argon matrix, we seek an alternate experiment in which phonon-nuclear coupling might lead to effects more easily observed. For example, suppose that a $^{181}$Ta nucleus were excited in a plate made up of $^{181}$Ta ground state nuclei, then perhaps it might be possible to observe an excitation transfer effect in which the excited state is transferred elsewhere in the plate.

In order to describe this kind of system, we work with a model that keeps track of the nuclear internal energy through a diagonal matrix $\mathbf{M}c^2$, the lattice vibrational energy in the different modes, and the relativistic coupling between the internal nuclear degrees of freedom and center of mass motion. We can write

$$\hat{H} = \sum_j \mathbf{M}_j c^2 + \sum_j \mathbf{a}_j \cdot c\hat{\mathbf{P}}_j + \sum_{\mathbf{k},\sigma} \hbar\omega_{\mathbf{k},\sigma}\hat{a}^\dagger_{\mathbf{k},\sigma}\hat{a}_{\mathbf{k},\sigma}. \tag{16}$$

The momentum operator for a monatomic crystal lattice is

$$\hat{\mathbf{P}}_j = \sum_{\mathbf{k},\sigma} \mathbf{u}_{\mathbf{k},\sigma}\sqrt{\frac{M\hbar\omega_{\mathbf{k},\sigma}}{2N_L}}\left(\frac{\hat{a}_{\mathbf{k},\sigma}e^{i\mathbf{k}\cdot\mathbf{R}_j^{(0)}} \quad \hat{a}^\dagger_{\mathbf{k},\sigma}e^{-i\mathbf{k}\cdot\mathbf{R}_j^{(0)}}}{i}\right), \tag{17}$$

where $\mathbf{u}_{\mathbf{k},\sigma}$ are unit vectors that specify the direction of displacement; where $\hbar\omega_{\mathbf{k},\sigma}$ is the energy of a phonon with wave vector $\mathbf{k}$ and polarization $\sigma$; where $\mathbf{R}_j^{(0)}$ is the equilibrium position of nucleus $j$; and there are $N_L$ atoms in the plate.

We can think of the nuclear and vibrational system in the absence of coupling as the unperturbed system $\hat{H}_0$, and the coupling term as an interaction $\hat{V}$

$$\hat{H} = \hat{H}_0 + \hat{V}. \tag{18}$$

Making use of these identifications we can develop a second-order model based on infinite-order Brillouin–Wigner theory to write

$$\hat{H} \rightarrow \hat{H}_0 + \hat{V}\left[E - \hat{H}_0\right]^{-1}\hat{V} \tag{19}$$

from which second-order interactions (such as excitation transfer) can be analyzed.

We consider second-order interactions where whatever phonons are created are also destroyed, and vice versa, so that the overall phonon part of the interaction is resonant. In this case we can write for the second-order coupling between two sites

$$\begin{aligned}
&\hat{V}\left[E - \hat{H}_0\right]^{-1}\hat{V}\Big|_{\text{resonant}}(1,2)\\
&= \frac{Mc^2}{2N}\sum_{\mathbf{k},\sigma}(\mathbf{a}_1\cdot\mathbf{u}_{\mathbf{k},\sigma})\hbar\omega_{\mathbf{k},\sigma}\hat{a}_{\mathbf{k},\sigma}\mathrm{e}^{i\mathbf{k}\cdot\mathbf{R}_1^{(0)}}\left[E - \hat{H}_0\right]^{-1}(\mathbf{a}_2\cdot\mathbf{u}_{\mathbf{k},\sigma})\hat{a}^{\dagger}_{\mathbf{k},\sigma}\mathrm{e}^{-i\mathbf{k}\cdot\mathbf{R}_2^{(0)}}\\
&+ \frac{Mc^2}{2N}\sum_{\mathbf{k},\sigma}(\mathbf{a}_1\cdot\mathbf{u}_{\mathbf{k},\sigma})\hbar\omega_{\mathbf{k},\sigma}\hat{a}^{\dagger}_{\mathbf{k},\sigma}\mathrm{e}^{-i\mathbf{k}\cdot\mathbf{R}_1^{(0)}}\left[E - \hat{H}_0\right]^{-1}(\mathbf{a}_2\cdot\mathbf{u}_{\mathbf{k},\sigma})\hat{a}_{\mathbf{k},\sigma}\mathrm{e}^{i\mathbf{k}\cdot\mathbf{R}_2^{(0)}}\\
&+ \frac{Mc^2}{2N}\sum_{\mathbf{k},\sigma}(\mathbf{a}_2\cdot\mathbf{u}_{\mathbf{k},\sigma})\hbar\omega_{\mathbf{k},\sigma}\hat{a}_{\mathbf{k},\sigma}\,\mathrm{e}^{i\mathbf{k}\cdot\mathbf{R}_2^{(0)}}\left[E - \hat{H}_0\right]^{-1}(\mathbf{a}_1\cdot\mathbf{u}_{\mathbf{k},\sigma})\hat{a}^{\dagger}_{\mathbf{k},\sigma}\mathrm{e}^{-i\mathbf{k}\cdot\mathbf{R}_1^{(0)}}\\
&+ \frac{Mc^2}{2N}\sum_{\mathbf{k},\sigma}(\mathbf{a}_2\cdot\mathbf{u}_{\mathbf{k},\sigma})\hbar\omega_{\mathbf{k},\sigma}\hat{a}^{\dagger}_{\mathbf{k},\sigma}\mathrm{e}^{-i\mathbf{k}\cdot\mathbf{R}_2^{(0)}}\left[E - \hat{H}_0\right]^{-1}(\mathbf{a}_1\cdot\mathbf{u}_{\mathbf{k},\sigma})\hat{a}_{\mathbf{k},\sigma}\mathrm{e}^{i\mathbf{k}\cdot\mathbf{R}_1^{(0)}}.
\end{aligned} \tag{20}$$

In order to reduce this we require a separation of the $\mathbf{a}$-matrix elements into components which excite ($\mathbf{a}^{+}$) and which de-excite ($\mathbf{a}^{-}$) according to

$$\mathbf{a}_j = \mathbf{a}_j^{+} + \mathbf{a}_j^{-}. \tag{21}$$

With these we obtain an approximate phonon-averaged interaction

$$\begin{aligned}
&\left\langle \hat{V}\left[E - \hat{H}_0\right]^{-1}\hat{V}\Big|_{\text{resonant}}(1,2)\right\rangle \rightarrow \frac{Mc^2}{(\Delta E)^2}\frac{1}{N}\times\\
&\sum_{\mathbf{k},\sigma}(\hbar\omega_{\mathbf{k},\sigma})^2\left[(\mathbf{a}_1^{+}\cdot\mathbf{u}_{\mathbf{k},\sigma})(\mathbf{a}_2^{-}\cdot\mathbf{u}_{\mathbf{k},\sigma})e^{-i\mathbf{k}\cdot[\mathbf{R}_2^{(0)}-\mathbf{R}_1^{(0)}]} + (\mathbf{a}_2^{+}\cdot\mathbf{u}_{\mathbf{k},\sigma})(\mathbf{a}_1^{-}\cdot\mathbf{u}_{\mathbf{k},\sigma})e^{i\mathbf{k}\cdot[\mathbf{R}_2^{(0)}-\mathbf{R}_1^{(0)}]}\right],
\end{aligned} \tag{22}$$

where we have kept terms that preserve total nuclear excitation. The expectation value in this expression is taken over the vibrational degrees of freedom. Individual intermediate terms in the calculation depend on the degree of excitation, but due to the strong destructive interference that occurs in the end there is no dependence on lattice excitation.

An approximate evaluation is possible in the limit that we model the dispersion relation as linear up to a cut-off momentum $K$. We may write for the long range limit of the interaction

$$\left\langle \hat{V}\left[E-\hat{H}_0\right]^{-1}\hat{V}\Big|_{\text{resonant}}(1,2)\right\rangle$$
$$\rightarrow -\frac{Mc^2}{(\Delta E)^2}\frac{V}{N}\frac{1}{2\pi^2}\frac{\hbar^2}{R_{21}^5}\int_0^{KR_{21}}\xi^3\sin\xi\,\mathrm{d}\xi\left\{c_{\mathrm{L}}^2\frac{(\mathbf{a}_1^+\cdot\mathbf{R}_{21})(\mathbf{a}_2^-\cdot\mathbf{R}_{21})+(\mathbf{a}_2^+\cdot\mathbf{R}_{21})(\mathbf{a}_1^-\cdot\mathbf{R}_{21})}{|\mathbf{R}_{21}|^2}\right.$$
$$\left.+c_{\mathrm{T}}^2\left[\mathbf{a}_1^+\cdot\mathbf{a}_2^-+\mathbf{a}_2^+\cdot\mathbf{a}_1^- -\frac{(\mathbf{a}_1^+\cdot\mathbf{R}_{21})(\mathbf{a}_2^-\cdot\mathbf{R}_{21})+(\mathbf{a}_2^+\cdot\mathbf{R}_{21})(\mathbf{a}_1^-\cdot\mathbf{R}_{21})}{|\mathbf{R}_{21}|^2}\right]\right\}, \quad (23)$$

where $c_{\mathrm{L}}$ is the longitudinal sound speed, and where $c_{\mathrm{T}}$ is the transverse sound speed. The integral can be done analytically, leading to

$$\int_0^{KR}\xi^3\sin\xi\,\mathrm{d}\xi = -(KR)[(KR)^2-6]\cos(KR)+3[(KR)^2-2]\sin(KR). \quad (24)$$

For large separation the resonant part of the second-order interaction Hamiltonian falls off as $1/|\mathbf{R}_{21}|^2$, a result that suggests the possibility of observing delocalization of excitation in a Ta plate (although this interaction is very weak).

If we extend the argument to higher order

$$\hat{H} \rightarrow \hat{H}_0+\hat{V}\left[E-\hat{H}_0\right]^{-1}\hat{V}+\hat{V}\left[E-\hat{H}_0\right]^{-1}\hat{V}\left[E-\hat{H}_0\right]^{-1}\hat{V}+\cdots \quad (25)$$

then we find that the analogous resonant fourth-order interaction depends explicitly on the number of phonons present. This is interesting because if a highly excited vibrational mode is present it may be possible to observe excitation transfer from an initial location, where the excitation is created, to another location which might be distant, as long as both locations interact with the (same) highly excited mode.

## 6. Conclusions and Discussion

A nucleus is a compound particle made up of protons and neutrons. If we make use of a relativistic description based on a Dirac spin 1/2 model for the neutrons and protons, we find a coupling between the center of mass motion and the internal nuclear degrees of freedom. This coupling can be rotated out to lowest order in free space (and completely eliminated in a Poincaré invariant model), but the same rotation that works in free space generates a large number of coupling terms in the case of many interacting nuclei in a lattice [2].

The conceptually simplest system in which observable effects might be detected is in homonuclear diatomic $^{181}\mathrm{Ta}_2$, where this relativistic phonon-nuclear coupling would produce additional level splittings beyond what is expected from conventional hyperfine splitting in a Mössbauer experiment. Unfortunately, the resulting experiment suffers from a number of practical headaches which makes it a difficult experiment to do; and we currently lack an estimate from theory for the strength of the interaction, so it is not clear whether the splitting is sufficiently large to be observed. A theoretical effort is currently under way to develop an estimate for the phonon-nuclear coupling matrix element.

The splitting in the $^{181}\mathrm{Ta}_2$ molecule can be thought of as due to phonon-mediated excitation transfer between the two nuclei, which motivates us to consider the possibility of an excitation transfer experiment involving a much larger number of Ta nuclei. If an excited $^{181}\mathrm{Ta}$ nucleus is created in a Ta plate, then it may be possible to detect excitation transfer through a delocalization of the emitted radiation. We discussed the resonant second-order interaction in this case, which as a result of destructive interference is independent of the level of vibrational excitation. This interaction is expected to fall off in proportion to the inverse square of the distance between the nuclei. At fourth-order the resonant contribution depends explicitly on lattice excitation, which suggests that substantially greater delocalization might occur in a plate with a strongly excited vibrational mode. This motivates us to consider an experiment where a Ta plate is irradiated to produce localized $^{181}$W (which decays 66% of the time to populate the 6.237 keV state in $^{181}$Ta [9] and has been used as a Mössbauer source for the 6.237 keV transition [10]), and then vibrated in order to stimulate excitation transfer.

In previous work we described enhanced up-conversion and down-conversion effects with the lossy spin–boson model [1]; we would expect that excitation transfer may also be enhanced by loss.

## References

[1] P.L. Hagelstein, Current status of the theory and modeling effort based on fractionation, *J. Condensed Matter Nucl. Sci.* **19** (2016) 98–109.

[2] P.L. Hagelstein, Quantum composites: A review, and new results for models for Condensed Matter Nuclear Science, *J. Condensed Matter Nucl. Sci.* **20** (2016) 139–225.

[3] R. Machleidt and D.R. Entem, Chiral effective field theory and nuclear forces, *Phys. Reports* **503** (2011) 1–75.

[4] G.E. Brown and D.G. Ravenhall, On the interaction of two electrons, *Proc. Roy. Soc. London A: Math., Phy. Eng. Sci.* **208** (1951) 552–559.

[5] P.L. Hagelstein and I.U. Chaudhary, Possibility of observing a center of mass interaction in a Mössbauer experiment, J. Condensed Matter Nucl. Sci., to appear. submitted for publication.

[6] A.C. Borin and J.P. Gobbo, Electronic structure and chemical bonding in the ground and low-lying electronic states of $Ta_2$, *Int. J. Quantum Chem.* **111** (2009) 1306.

[7] P.H. Barrett, H. Paul and T.K. McNab, Hyperfine interactions of $Fe^{57}$ in a frozen argon matrix, *Phys. Rev. Lett.* **25** (1970) 1601.

[8] Z. Hu, B. Shen, J.R. Lombardi and D.M. Lindsay, Spectroscopy of mass-selected tantalum dimers in argon matrices, *J. Chem. Phys.* **96** (1992) 8757.

[9] J.L. Campbell and B. Martin, Internal conversion of the 6.2 keV transition in $^{181}$Ta, *Zeitschrift für Physik A Atoms and Nuclei* **277** (1976) 59–64.

[10] G. Kaindl, D. Salomon and G. Wortmann, Quadrupole splitting of the 6.2-keV $\gamma$-rays of $^{181}$Ta in rhenium metal, *Phys. Rev. Lett.* **28** (1972) 952.

J. Condensed Matter Nucl. Sci. 24 (2017) 123–129

Research Article

# Stabilization of Nano-sized Pd Particles under Hydrogen Atmosphere

T. Hioki*, A. Ichiki and T. Motohiro

*Materials Science and Energy Engineering Division, Green Mobility Research Institute, Institutes of Innovation for Future Society, Nagoya University, Furo-cho, Chikusa-ku, Nagoya 464-8603, Japan*

**Abstract**

To enhance the reproducibility of anomalous heat effects reported in gas loaded experiments using nano-sized metal particle systems, stabilization of the nano-sized metal particles under hydrogen atmosphere and at elevated temperatures seems to be important. In this paper, the synthesis of nano-Pd particles embedded in the meso-pores of a mesoporous-silica is reported. We have succeeded in synthesizing nano-Pd particle systems with an average Pd particle size of 3.60 nm and a standard deviation of 0.64 nm. It was found that the synthesized nano-Pd particle system is very stable, i.e., the average size of the Pd particle is unchanged when exposed to a hydrogen atmosphere at temperatures ranging from room temperature up to 773 K. The hydrogen absorption capacity of the nano-Pd particles in the mesoporous silica host at room temperature was smaller than that of the Pd bulk.


*Keywords:* Elevated temperature, Hydrogen atmosphere, Hydrogen absorption capacity, Mesoporous silica, Nano-size, Pd nanoparticles, Stabilization, Synthesis

## 1. Introduction

Nano-sized particles of Pd, Ni, and their alloys have attracted attention because these particle systems have been reported to produce anomalous heat when they interact with hydrogen isotope gases at room temperature [1–6] and elevated temperatures [4,7–9]. However, the anomalous heat production phenomenon has not been widely recognized in the scientific communities. One of the reasons seems to be the lack of reproducibility of the phenomenon.

It is well known that nano-sized metallic particles are easy to coalesce and grow under reductive environments at elevated temperatures. Nano-sized Pd or PdO particles have been observed to grow once they are exposed to hydrogen even at room temperature [10]. The heat of hydrogen absorption for Pd and/or the heat of reduction of PdO are enough to cause the growth. Therefore, if the anomalous heat production phenomenon is closely related to the existence of nano-sized metallic particles, the growth of the particles under hydrogen atmosphere has to be suppressed to improve the reproducibility of the anomalous heat production and to clarify the mechanism of the phenomenon.

In this paper, a trial to stabilize nano-sized Pd particles under hydrogen atmosphere is reported: we have tried to precipitate Pd particles only inside the pores of mesoporous silica's (MPS) using the conventional incipient wetness

*Corresponding author. E-mail: hioki@gvm.nagoya-u.ac.jp.

© 2017 ISCMNS. All rights reserved. ISSN 2227-3123

impregnation method followed by a process of washing out the mixture of MPS powder and Pd precursor with n-octane [11]. With the washing out process, the Pd precursor adhered to the outer wall of the mesoporous silica could be removed completely. The Pd/MPS system thus prepared contained only Pd particles with sizes ranging from 2 to 5 nm and showed excellent stability under hydrogen atmosphere, i.e., X-ray diffraction patterns and transmission electron microscopy confirmed that the sizes of the Pd particles in these systems were unchanged when the systems were exposed to hydrogen even at an elevated temperature of 773 K.

## 2. Experiment

### 2.1. Synthesis method

The MPS used in the present study was TMPS-4R obtained from Taiyo Kagaku Co. Ltd. The average pore size of the MPS was 4.2 nm. To precipitate Pd particles inside the pores of the MPS, $Pd(NO_3)_2$ aqueous solution was used as a Pd precursor. TMPS-4R powder was added to a mixed solution of the Pd precursor and citric acid. Then, the mixture was stirred with a magnetic stirrer for about 24 h at room temperature. After the impregnation process, the Pd precursor was separated from TMPS-4R powder with a centrifuge. Then, the TMPS-4R powder was replenished with n-octane and subjected to ultrasonic treatment to remove the Pd precursor adhered to the outer surface of the TMPS-4R particle. Subsequently, the resultant mixture was dried at 423 K for about 10 h and calcined at 673 K for 2 h in air.

### 2.2. Characterization method

X-ray diffraction (XRD) analyses using Cu$K\alpha$ radiation were performed to characterize the crystallographic changes of Pd particles and host MPS. The XRD measurements were performed before and after being exposed to deuterium of 0.2 MPa at elevated temperatures.

The concentration of the Pd in the TMPS-4R was determined to be 6.7 wt% with ICP-analysis operated at Aichi Center for Industry and Science Technology.

The morphology and sizes of Pd particles were observed with transmission electron microscopy (TEM) operated at 200 kV.

A standard volumetric method (Sievert's method) was used to evaluate the hydrogen absorption properties of Pd nanoparticles in the samples. The measurements were carried out at room temperature up to hydrogen equilibrium pressure of 0.02 MPa. The measurements were performed three times for each sample. The first run was employed as a preliminary treatment to reduce PdO in the as-prepared sample to metallic Pd. Thereby, in the first run more hydrogen was consumed than in the following second run and third runs. The data obtained in the second run was almost the same as that of the third run. Therefore, the data of the second run was used to evaluate the hydrogen absorption capacity of the Pd nanoparticles embedded in the MPS.

## 3. Results and Discussion

### 3.1. Characterization of Pd particles embedded in TMPS-4R

In Fig. 1, XRD profiles for the Pd/TMPS-4R are shown for the samples as-synthesized and after exposure to deuterium of 0.2 MPa for ~1 h at room temperature. It is seen that in the as-synthesized sample there exists PdO with tetragonal crystal structure, and the PdO crystals are reduced to metallic Pd with a face centered cubic crystal structure, once the sample is exposed to hydrogen. From the width of the XRD profiles, the average grain sizes of the PdO and Pd are estimated using the Scherrer equation [12],

$$D = 0.9\lambda/\beta\cos\theta, \tag{1}$$

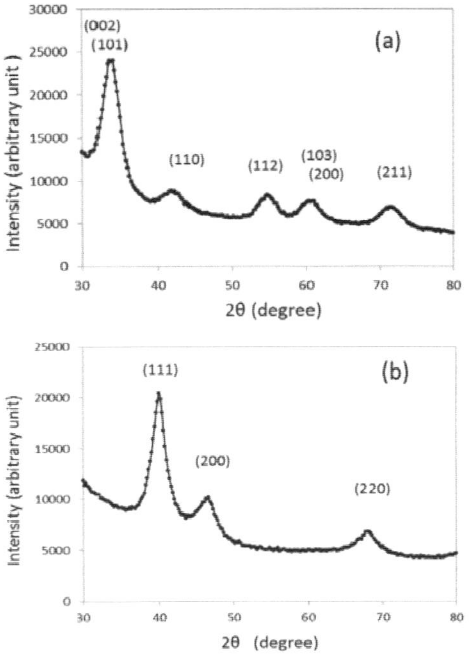


**Figure 1.** XRD profiles for (a) as-synthesized sample, indices are for PdO and (b) the sample after exposure to deuterium at room temperature, indices are for Pd.

where $D$ is the grain size (nm), $\lambda$ the wavelength of the incident X-ray (nm), $\beta$ the full width at half maximum (FWHM) of the peak (radian) and $\theta$ the center angle of the peak.

For the as-synthesized sample, the average grain size of PdO calculated with a (101) peak was 3.1 nm. After exposure to deuterium, the PdO crystals was reduced to metallic Pd and the average grain size of the metallic Pd particles calculated with a (111) peak was 3.9 nm. The grain size of PdO crystals are considerably smaller than the size of the pores of the MPS host, which strongly suggests that almost all the PdO particles are inside the pores of the MPS. Upon exposure to deuterium, the PdO crystals were reduced to metallic Pd and the size of the resultant Pd particle was still smaller than 4.2 nm, the pore size of TMPS-4R. This fact suggests that almost all the Pd crystals remain inside the pores of the MPS host. The metallic Pd crystals were a little larger than the initial PdO crystals. Because the heat of the reduction reaction is rather large, the sample is heated when it is exposed to deuterium. This heat is assumed to result in the growth of Pd crystals. However, it is noted that the size of the Pd crystals is still a little smaller than the size of the pores of the MPS host. The growth of the Pd crystals is considered to be limited by the size of the pores. This fact also strongly suggests that the Pd crystals are inside the channels of the MPS host.

Figure 2 shows the TEM images for the sample after exposure to deuterium. As seen in Fig. 2 (a) at low magnification, no coarse Pd particles exist on the outside surface of the MPS host. As suggested in Fig. 2 (b), the Pd particles seem to be embedded inside the pores of the TMPS-4R. This is consistent with the XRD results.

To determine the size of the Pd particles with TEM observation, 150 Pd particles in the high-magnification images were randomly chosen and the size of the particle image was measured. The result is shown in Fig. 3.

The average size of the Pd particles was 3.60 nm, which was close to the value estimated with the XRD profile.

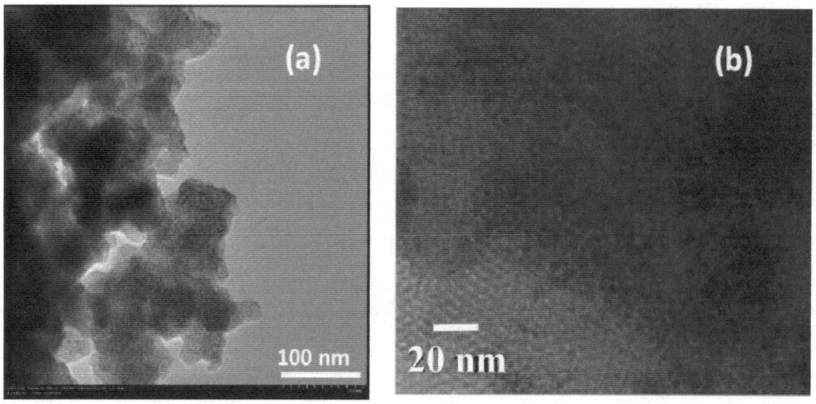


**Figure 2.** TEM observation of Pd/TMPS-4R after deuterium exposure at room temperature. (a) Low-magnification image and (b) high-magnification image.

The standard deviation was 0.64 nm, which indicated that the distribution for the size of the Pd particles is very narrow. Figure 3 indicates that 99% of the Pd crystals are smaller than 5.0 nm. Furthermore, 89% are smaller than 4.2 nm, the pore size of the TMPS-4R, indicating that the Pd crystals are largely inside the pores of the host MPS.

From the XRD and TEM analyses, it has been clearly demonstrated that almost all the Pd particles are inside the pores of the TMPS-4R and no coarse Pd particles larger than 5 nm exist in the sample. Furthermore, upon exposure to deuterium, the size of the PdO particles was largely unchanged. This is contrary to what happened with the samples previously used in flow calorimetry experiments showing heat evolution upon pressurization with hydrogen isotope gases [6,10], e.g., for Pd nanopowder AY-4030$^{\mathrm{TM}}$ used in the previous experiments, the average size of the Pd particles

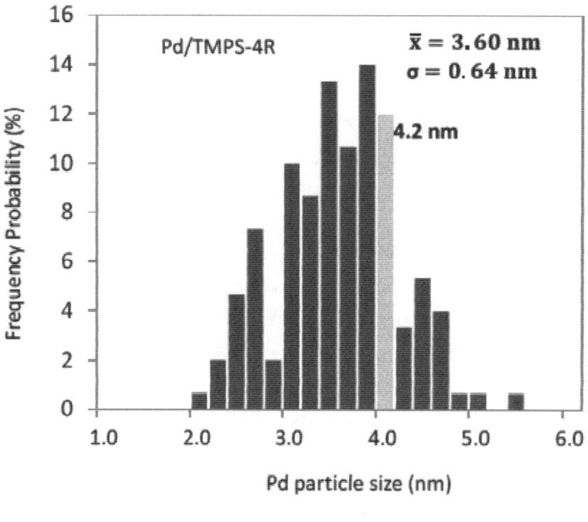


**Figure 3.** Size distribution for Pd particles in Pd/TMPS-4R.

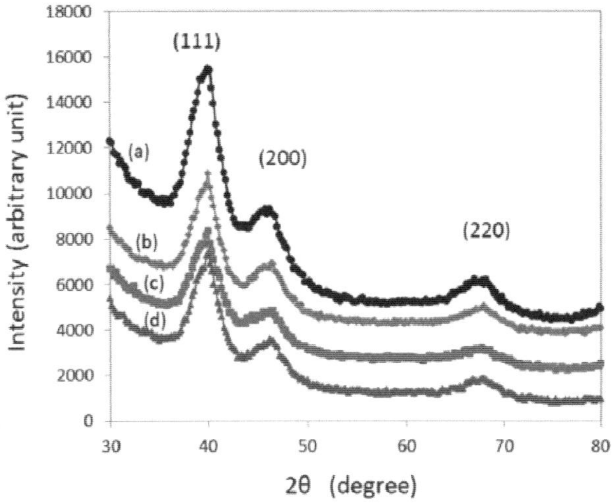


**Figure 4.** XRD profiles after heat treatment under $D_2$ atmosphere at (a) 300 K, (b) 613 K, (c) 673 and (d) 773 K.

was initially 10–20 nm, whereas that of the Pd particles after hydrogen absorption measurements at room temperature was increased to about 50 nm [10].

### 3.2. Thermal stability of Pd particles embedded in TMPS-4R

To investigate the stability of the Pd particles embedded in the pores of TMPS-4R, a sample of Pd/TMPS-4R was heat treated under deuterium atmosphere of 0.2 MPa for 2 h at 300, 613, 673 and 773 K. The XRD profiles after each

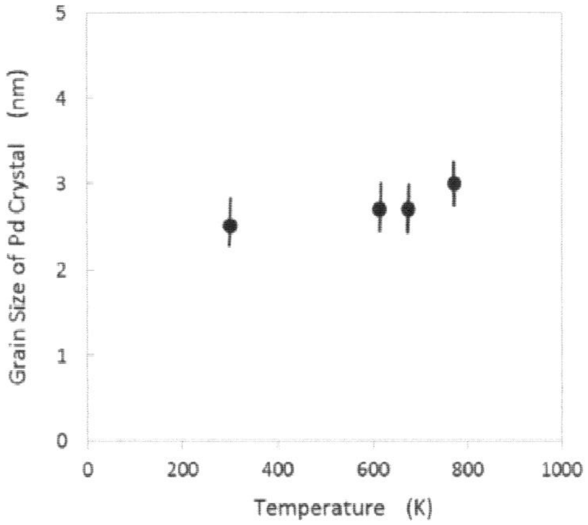


**Figure 5.** The size of Pd crystals VS. temperature at which the sample was heat-treated under 0.2 MPa deuterium.

treatment are shown in Fig. 4.

It is seen that the peaks are broad even at the elevated temperature of 773 K. The size of the Pd particles was calculated using Eq. (1) for the (111) peak, and the result is shown in Fig. 5.

The grain size is almost unchanged up to 773 K within the error bars. This fact indicates that the Pd crystals inside the pores of the MPS host are stable under hydrogen atmosphere even at 773 K.

### 3.3. Hydrogen absorption property

To investigate the effect of grain size on the hydrogen absorption property of Pd, the hydrogen absorption capacity was measured at room temperature using the standard volumetric method for a powder sample of Pd/TMPS-4R. In Fig. 6, the results are compared with those for a reference sample of bulk Pd.

The reference sample was prepared by mixing Pd powder (AY4030$^{TM}$: Tanaka Kikinzoku Kogyo Co. Ltd.) and pure TMPS-4R powder.

As seen in Fig. 6, the ratio of absorbed hydrogen atoms to Pd atoms, H/Pd, at 0.02 MPa is 0.65 for the reference sample, which is in good agreement with literature values for Pd bulk. For Pd/TMPS-4R, the value is 0.50, which is smaller than the bulk value. It is also noted that the concentration limit of the solid solution phase ($\alpha$ phase) is increased and the plateau region (($\alpha + \alpha'$) phase) is narrowed in the Pd/TMPS-4R sample. These results are consistent with those for polymer-coated Pd nanoparticles with sizes 2.6± 0.4 and 7.0± 0.9 nm as reported by Yamauchi et al. [13].

## 4. Conclusion

We have succeeded in preparing a Pd-nanoparticle/mesoporous silica system, where a large percent of the Pd particles are confined in the pores of the MPS host, using TMPS-4R with a pore size of 4.2 nm: the average size of the Pd crystals was less than the pore size of the host MPS. The Pd nanocrystals confined in the pores of TMPS-4R were thermally

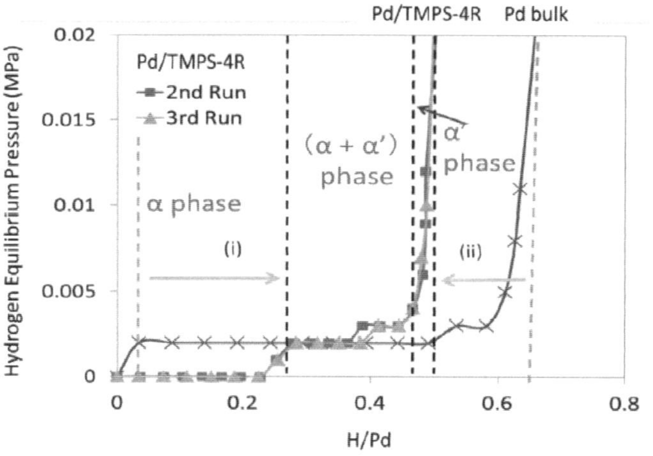


**Figure 6.** Hydrogen absorption capacity VS. equilibrium pressure at room temperature for Pd/TMPS-4R and Pd bulk.

very stable: their sizes were nearly unchanged when they were heated up to 773 K under hydrogen atmosphere. The hydrogen absorption capacity per Pd atom, H/Pd, for a Pd/TMPS-4R sample was smaller than the value for Pd bulk.

## Acknowledgements

We would like to thank Y. Ichikawa and K. Kawai for experimental support. A part of this work was supported by Grant-in-Aid for challenging Exploratory Research 16K13718 from Japan Society for the Promotion of Science.

## References

[1] Y. Arata and Y. Zhang, *Proc. ICCF10*, Cambridge, August 24–29, 2003, p.139.
[2] Y. Arata and Y. Zhang, *J. High Temp. Soc.* **34** (2008) 85.
[3] A. Kitamura, T. Nohmi, Y. Sasaki, A. Taniike, A. Takahashi, R. Seto and Y. Fujita, *Phys. Lett.* **A 373** (2009) 3109.
[4] F. Celani, A. Spallone, E. Righi, G. Trenta, V. Andreassi, A. Marmigi, P. Quercia, G. Cappuccio, D. Hampai, P. Marini, V. DiStrfano, M. Nakamura, F. Todarello, E. Purchi, A. Mancini, U. Mastromatteo, F. Falcioni, M. Marchesini, P. DiBiagio, P.G. Sona, F. Fontana, L.Gamberale and D. Garbelli, *Proc. ICCF13*, Dagomys, Russia, June25–July1, 2007, p. 181.
[5] J. Biberian and N. Armanet, *Proc. ICCF13*, Dagomys, Russia, June25–July1, 2007, p. 170.
[6] T. Hioki, N. Sugimoto, T. Nishi, A. Itoh and T. Motohiro, *J. Condensed Matter Nucl. Sci.* **13** (2014) 223.
[7] F. Celani, E. F. Marano, B. Ortenzi, S. Pella, S. Bartalucci, F. Micciulla, S. Bellucci, A. Spallone, A. Nuvoli, E. Purchi, M. Nakamura, E. Righi, G. Trenta, G. L. Zangari and A. Ovidi, *J. Condensed Matter Nucl. Sci.* **13** (2014) 56.
[8] A. Kitamura, A. Takahashi, R. Seto, Y. Fujita, A. Taniike and Y. Furuyama, *Current Sci.* **108** (4) (2015) 589.
[9] A. Kitamura, E. F. Marano, A. Takahashi1, R. Seto, T. Yokose, A. Taniike and Y. Furuyama, Heat evolution from zirconia-supported Ni-based nano-composite samples under exposure to hydrogen isotope gas, *Proc. JCF16*, 2016, p. 1.
[10] T. Hioki, H. Azuma, T. Nishi, A. Itoh, S. Hibi, J. Gao and T. Motohiro, *J. Condensed Matter Nucl. Sci.* **4** (2011) 69.
[11] Y. Ichikawa, T. Hioki, N. Sugimoto, A. Ichiki and T. Motohiro, *J. Nanosci. Nanotechnol.* **16** (2016) 12947.
[12] A. L. Patterson, *Phys. Rev.* **56** (1939) 978.
[13] M. Yamauchi, R. Ikeda, H. Kitagawa and M. Tasaka, *J. Phys. Chem.* C **112** (2008) 3294.

J. Condensed Matter Nucl. Sci. 24 (2017) 130–145

Research Article

# Increased PdD anti-Stokes Peaks are Correlated with Excess Heat Mode

Mitchell R. Swartz*

*JET Energy Inc., Wellesley Hills, MA 02481, USA*

Peter L. Hagelstein†

*Massachusetts Institute of Technology, Cambridge, MA, USA*

**Abstract**

Volume-enhanced Coherent Multi-wavelength Optical Reflection Electric-driven (CMORE) spectroscopy successfully differentiates active states in LANR nanomaterials. All the anti-Stokes peaks are relatively missing in the undriven mode for all of the nanostructured materials. Weak anti-Stokes peaks are elicited from Pd (and Ni and their alloy) nanostructured material in $ZrO_2$. But when NANOR®-type components are electrically driven, there is diversity in outcome. When driven in the avalanche mode, the anti-Stokes peaks differ considerably from those which appear during the excess heat (XSH)-producing or desired mode. The anti-Stokes peak(s) differ in energy, amount, and in what stimulates their appearance. However, normal anti-Stokes peaks return when the electrical drive creates "avalanche mode" characterized by higher electrical current at decreasing voltage. The avalanche anti-Stokes peaks are many, and they are lower energy than the XSH mode produced anti-Stokes peak (described main text). By contrast, successful cold fusion is heralded by a large increase in the anti-Stokes to Stokes (aS/S) ratio, and the generated anti-Stokes peak for the desired and XSH-producing state is very different from the avalanche-generated multiple anti-Stokes peaks. That XSH-related peak is singular and at higher energy. This distinguishing, higher energy, single, anti-Stokes peak is also not seen in the "off" state or the avalanche (undesirable) mode. Our analysis finds that the excess-heat produced anti-Stokes peak is matched to the Stokes line of PdD. We also find that the several lower energy anti-Stokes in avalanche mode (unsuccessful regarding CF/LANR) are matched to the many Stokes peaks for zirconia. In the desired electric-driven XSH-producing mode, the two-terminal deuterided NANOR®-type CF/LANR component has a measured Boltzmann Stokes ratio ∼1.3. Analysis of the phonon gain heralds $\sim 7^{\pm 0\ 15}$ acoustic phonons assisting nuclear reactions and a core peak calculated Stokes temperature of circa 1645 K. Therefore, these findings confirm a role for PdD acoustic phonons in successful CF/LANR.



*Keywords:* Avalanche mode, CMORE spectroscopy, Excess heat mode, NANOR®-type LANR component, Phonon gain

*Dr. Mitchell R. Swartz ScD, MD, EE, E-mail: mica@theworld.com.

†E-mail: plh@mit.edu

## 1. Background

### 1.1. Dry NANOR®-type CF/LANR preloaded components

Dry preloaded NANOR®-type LANR (lattice assisted nuclear reaction) components have been described in the literature [1–9]. NANOR®-type components have been used to investigate material science [2–9] and radiation physics [5,6] of their active LANR systems. These reports were expanded to demonstrate that several electrical transconduction states exist, but that only one is active, desired, and capable of producing excess heat (XSH) [7,8]. Later, it was demonstrated that CMORE spectroscopy [8], based on Raman spectroscopy, would be quite useful in distinguishing active LANR states because they are electrically driving them at their peak optimal operating point. This report expands that and shows (cf. Fig. 1) that for successful LANR, the anti-Stokes component matches the Stokes component for PdD. By contrast, non-active, unsuccessful, electrical avalanches have never produced XSH, and have anti-Stokes components that match the $ZrO_2$ Stokes components and not the PdD.

### 1.2. Successful LANR requires considerable engineering

LANR success is rewarded by XSH, which means that the energy producing reactions have generated *de novo* helium into the lattice ($\sim 10^{12}$ for every watt-second [10]), and there were adequate conditions to enable energy transfer to the lattice and then to appear as XSH [1,12]. There may also be other reactions.

However, the LANR method which Fleischmann and Pons first taught in March 1989 (aqueous, low impedance, $Pd/D_2O/Pt$) had problems, including inefficiency and non-reproducibility.

This created havoc for those inexperienced in metallurgy, electrochemistry, and physics [13]. One major problem to achieve successful cold fusion has been the difficulty in achieving high D/Pd loadings above $\sim$0.70 near room

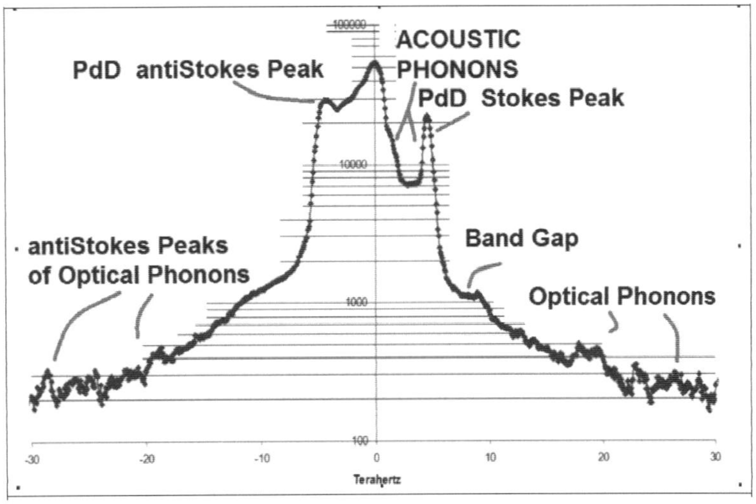


**Figure 1.** Log intensity Raman spectrum obtained during XSH mode (desired, non-avalanche). This is a logarithmic presentation of coherent dual wavelength electric-driven volume-enhanced reflection spectroscopy of a preloaded, correctly driven, $ZrO_2$PdD NANOR®-type component. The $x$-axis here presents the frequency difference from the main laser frequency. Identified are the Stokes and anti-Stokes peaks of PdD, and the band gap between acoustic and optical phonons; with the appearance of the optical phonon modes at their reported locations. The $ZrO_2$ Stokes peaks and anti-Stokes peaks are not labeled, but are in Fig. 6.

temperature, and then maintaining that, sometimes for weeks. Simply put, the rapid increase in deuterium chemical potential acts to limit further loading, but success requires high loading of >85% for $PdD_x$ hydrides. In most initial efforts, loading was not even considered. Other problems have included the control of vacancies, adequate incubation time, concomitant flux, inadvertent quenching conditions, and lack of critical control of input power. Many "negative" results are due to a failure to operate the system at the optimal operating point, which is an optimum peak in the XSH and power gain curves as a function of input electrical power [13–16]. The problems with loading, and later with optimal operating point manifolds (OOPs) are why initial efforts to replicate successful LANR were so difficult and failed to show XSH [13].

### 1.3. Raman spectroscopy

Raman spectroscopy is a powerful tool to study vibrations within molecules, materials, and nanostructured materials to obtain knowledge of the momenta and energies from the returned optical spectrum. When used to examine materials on metal surfaces, there is a large enhancement of the electric field from the illuminating irradiation, which results in surface enhanced Raman spectroscopy (SERS). When used to examine vibrations of a lattice, the term Brillouin scattering spectroscopy is used. Raman spectroscopy was previously used to examine PdH [17–19].

## 2. Experimental

### 2.1. Materials

A NANOR®-type component is a hermetically sealed CF/LANR (cold fusion/lattice assisted nuclear reactions) nanomaterial, preloaded with D and arranged as a two-terminal electrical component. They are designed to avoid leakage, enabling stabilization and activation of these materials. The central core generating the XSH in the desired state involves $ZrO_2PdNiD$, $ZrO_2PdD$, and $ZrO_2NiD$ and similar materials [3,20,21].

### 2.2. Methods

#### 2.2.1. *Signal pickup by CMORE spectroscopy*

The black granular nanomaterials fortunately give very large signals. The volume enhancement is probably from the black nanostructured CF/LANR materials which are black, electrically insulating, and therefore will accept light deeply into the material, unlike a conductive metal electrode. A previous paper has given the experimental details of CMORE-spectroscopy [8], which uses two wavelengths of illumination from two lasers.

This was to co-illuminate the target, located below the irradiated surface, and into the volume of a sample of interest. The high intensity coherent illumination was used to elicit Rayleigh, Brillouin, and Raman bands, and then to identify the possible roles of acoustic and optical phonons during CF/LANR, and possibly distinguish the desired active, XSH-producing, state from the "off"-state, and the undesired inactive avalanche state (Fig. 2). A second lower power intensity irradiation at a second wavelength enables wavelength calibration and thus semiquantitative measurements. In this study, a 532 nm laser was used to elicit the Raman spectra, and a weaker 635 nm laser was used for calibration. The green laser (532 nm peak) had a power output level of about 150 mW. The red orange laser (635 nm peak) had a power output level of about ~2 mW for energy calibration. Laser polarization is usually important to the degree that the nanomaterial is electrically conductive. However, LANR nanomaterials which are electrically insulating (~megohms to gigohms or more), and are porous, black like bituminous coal, are different. There the light enters and has a volume interaction.

The sample illuminated was the core of a NANOR®-type LANR component capable of being electrically driven. Optics for collimation, control of beam direction, band pass filters, beam splitters, were used to obtain optical beam

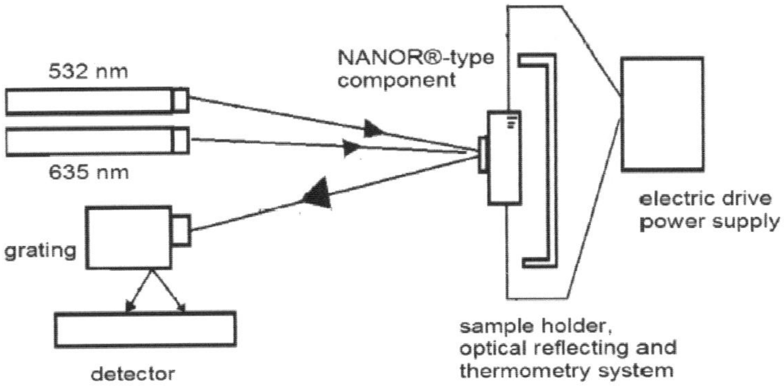


**Figure 2.** Schematic of coherent multi-wavelength optical reflection electric-driven spectroscopy (CMORE) experiments.

overlap on the sample's surface, facilitating interaction with the sample while being electrically driven. The important result is that illumination yields many photons which enter the black material. The multifaceted craterous surface leads to a very significant volumetric photon–sample interaction [8]. As a result of the interaction, what was initially just two initial frequencies coming from the two lasers, finally becomes a more complicated spectrum, decorated with the addition of photons of energy both above and below the two frequencies of said lasers.

### 2.2.2. *Determination of Stokes states*

The Boltzmann anti-Stokes to Stokes (aS/S) ratio was measured by a method enabling the simultaneous electrical driving of a $ZrO_2PdD$ CF/LANR system in several states while simultaneously visualizing both acoustic and optical phonons using dual wavelength coherent stimulated volume-enhanced Rayleigh–Brillouin–Raman spectroscopy. When the emitted photons have less energy than the initial absorbed photon, then this energy difference is the Stokes shift. The energy goes into the lattice, and the difference is in the "red-shift" direction. On the other hand, if the emitted photon has more energy, the energy difference is called an anti-Stokes shift. That extra energy is derived from energetic phonons in a crystal lattice that give up their energy to create the "blue-shift" from the initial two frequencies. To resolve the spectra [8], the reflected (exit) beam leaves the illuminated sample through an optical slit and then passes through a diffraction grating where it is resolved into different energies (or in equivalent systems into wavelengths or frequencies). The intensity is measured by a detector such as a CCD or other sensitive photodetector. Analysis is made by integrating the incoming detected optical information which has been sorted by the grating. The final plots show the total energy of the reflected beams in a distribution of amplitude (intensity) as a function of wavelength.

### 2.2.3. *Determination of sample activity by calorimetry*

The LANR preloaded, stabilized NANOR®-type components are driven by a high voltage circuit up to a peak of 3000 V. This voltage is delivered in any run to either the NANOR or the ohmic control and used to thermally calibrate the calorimeter. Input power is defined as $V \times I$. There is no thermoneutral correction in denominator. Therefore, the observed power is a lower limit. The instantaneous power gain (power amplification factor (non-dimensional)) is defined as $P_{\text{output}}/P_{\text{input}}$. The energy is calibrated by at least one electrical joule control (ohmic resistor) used frequently, and with time integration for additional validation. The excess energy, when present, is defined as ($P_{\text{output}} - P_{\text{input}}$) × time. Data acquisition is taken from voltage, current, and temperature sensors at multiple sites. Data acquisition sampling is at data rates of 0.20–1 Hz, with 16–24$^{+}$ bit resolution; voltage accuracy $0.015^{\pm 0\ 005}$ V, temperature accuracy <0.6°C). The noise power of the Keithley current sources driving the reactions is generally ∼10 nW or less [15].

### 2.2.4. *Electrical driving components*

The sample or component was electrically driven, or not, while it was irradiated by the two lasers while physically maintained in position. The NANOR®-type component was electrically activated and controlled, and was designed to include and use several controls: several metals and material controls such as $ZrO_2Pd$ and $ZrO_2Ni$, pn-junctions (as an energy conversion control), a thermal ohmic control, a non-functional unloaded NANOR®-type component control, an inactive undriven component as a control, an inactive driven avalanche-mode component as a control, and a component driven in the desired active mode.

The electrical power supply is capable of delivering an electric current (Norton equivalent) or electric voltage (Thevenin equivalent) through the two wires which connect to the sample. The system has several electrical states, the simplest of which are "off", meaning that the sample is not electrically driven by the electrical power supply, and "on" where there may be a range of electrical driving levels of electrical power (watts). Within the holding container there are temperature detectors, and a heat flow detector, linked to a temperature measurement and analytic unit. This enables thermometry to measure any possible incremental temperature change, and therefore analysis of possible heat output from the sample.

After driving the component and the control in each run, the power and energy gain for both the component and the ohmic control were separately determined both by approximations such as input-power-normalized delta-$T$ ($\mathrm{d}T/P_{\text{in}}$), and input-power-normalized heat flow (HF/$P_{\text{in}}$), and also directly by semiquantitative calorimetry. In the latter, the amount of output energy is determined from the released heat produced during the temperature rise, and then comparing that to the input energy.

### 2.2.5. *Control – Rhodamine B deposited upon filter paper*

In order to test the system, and here to show the optical purity of the two lasers as well, dual wavelength reflection spectroscopy was performed on rhodamine B, as a control. The generated spectrum acts as a control to demonstrate the impact of the system on a rhodamine B aqueous solution of which the Stokes, anti-Stokes, and fluorescence bands are well known. Figure 3 presents the chemical structure, what a solution of the rhodamine B does to two incoming laser beams, and the resultant spectrum by this system. This control experiment used a 7.6 mmol rhodamine B solution irradiated by the two laser wavelengths (532 and 635 nm). They are labeled in the figure, and the Stokes, anti-Stokes and fluorescence bands are clearly seen, demonstrating the effectiveness of the CMORE-spectroscopic system. The second laser peak permits semiquantitative wavelength (and thus energy) calibration.

## 3. Results

### 3.1. Stokes peaks of $ZrO_2$

$ZrO_2$ has several Raman peaks between 5.7 and 19 THz. These can be seen in Fig. 4a which is a Raman spectrum of monoclinic $ZrO_2$ [22]. Pure zirconia's CMORE signature ($ZrO_2$ without any other added material) is consistent with this and is shown in Fig. 4b where the two laser initiation wavelengths of 532 and 635 nm are obvious. However, note that the first peak has an extended array of Stokes peaks observed falling to its right, in an intensity-decreasing sequence. This optical signature of pure zirconia is unusual among all the materials so far examined. $ZrO_2$ has a thick

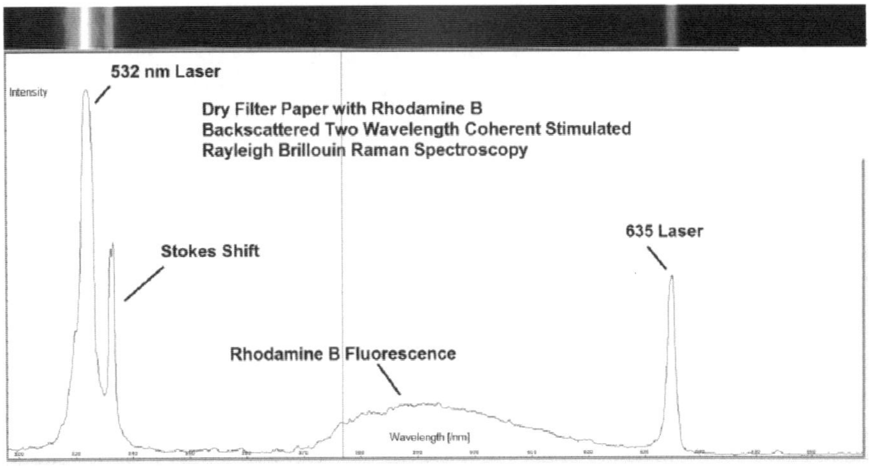


**Figure 3.** Spectrum from rhodamine B on filter paper. Dual wavelength reflection spectroscopy of a rhodamine B control. This CMORE-spectroscopic spectrum is a control experiment of rhodamine B (7.6 mmol aqueous solution). The vertical axis is non-linear and represents the intensity of the returned backscatter along with the reflected optical information, plotted as intensity as a function of wavelength. The horizontal axis plots the decreasing frequency to the right. The Stokes, anti-Stokes and fluorescence bands are seen in their well-known locations, and the two laser initiation wavelengths (532 and 635 nm) are seen and labeled. The actual image is located above the graph.

richness of Stokes peaks, and the anti-Stokes peaks are missing in $ZrO_2$ nanostructured materials observed by CMORE spectroscopy, as has been reported [8]. Given that Stokes peaks herald loss of energy of incident surface irradiation, it certainly seems possible that $ZrO_2$ is a molecular system able to dissipate energy through phonons (Fig. 4b). Why is there a significant loss of anti-Stokes peaks in the CF/LANR nanomaterials by this spectroscopy? Is it the same reason it is coal black?

### 3.2. Stokes peaks of $ZrO_2$ PdD – loaded and undriven (null, "off state")

Cryogenic Raman spectroscopy of PdD shows a band gap between acoustic and optical phonons, near 9 THz in those studies [23]. The PdD Phonon density of states [23] is shown in Fig. 5a. It reveals a peak before 6 THz and a band gap between 7 and 8 THz.

As a further control, the dual wavelength CMORE reflection spectroscopy was used to examine an undriven Nanor®-type component. Figure 5b is the CMORE spectrum from a loaded, electrically undriven Nanor®-type Component ($ZrO_2$–PdD) no driving electric current - electrically "Off". Functionally, that makes the observed optical output to be that of the raw material, $ZrO_2$PdD with no additional electrical drive or activation. It can be seen that the Stokes peak derived by the CMORE spectrum of $ZrO_2$–PdD shows the peak before 5.4 THz, consistent with the Rowe data, as are the other two peaks in the optical phonon region.

### 3.3. Stokes peaks of driven $ZrO_2$PdD in the electrical avalanche state. All CF/LANR activities are quenched by electrical avalanche breakdown

To simplify, in addition to the undriven "off" state, there are two electrically driven states for a NANOR®-type LANR component. Therefore, three (3) electrical driving states exist [7]. First, there is the undesirable "avalanche" state, associated with increasing electrical currents at falling electrical potentials. Second, by contrast, the desired "XSH"

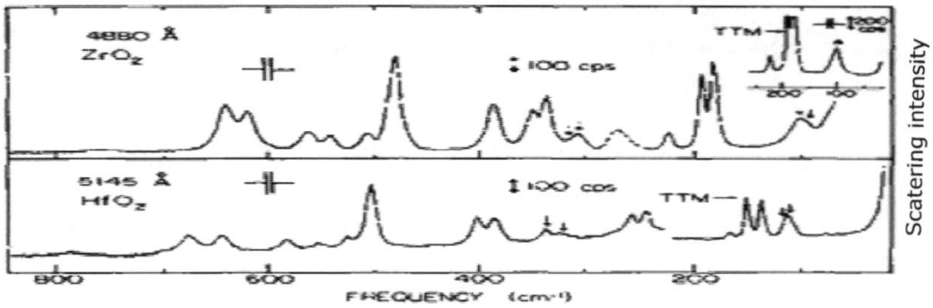


**(a)**

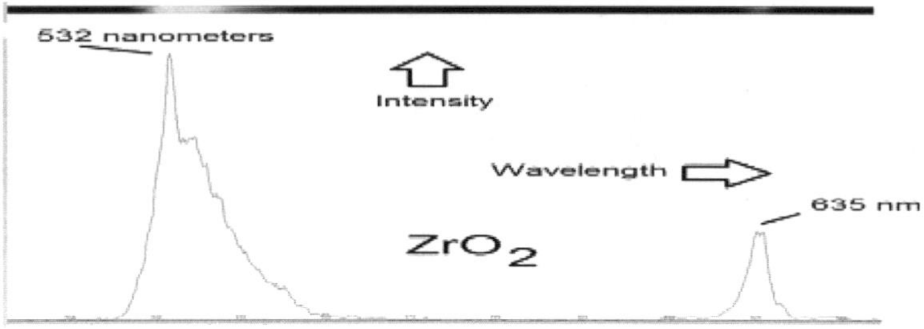


**(b)**

**Figure 4.** Raman and dual wavelength reflection CMORE spectroscopy of $ZrO_2$. (a) (upper) Raman spectrum from monoclinic $ZrO_2$ [22]. This is plotted in reverse along the $x$-axis compared to other figures shown. (b) (lower) CMORE-spectroscopic spectrum of $ZrO_2$ (zirconia without any other added material). Here, as in the rest of the figures, the horizontal axis plots the increasing wavelength to the right. The extended, and unique, multi-peak Stokes sequence of zirconia is seen along with the two laser initiation wavelengths (532 and 635 nm).

state produces very large amounts of heat beyond the normal electrical dissipation expected - XSH. In that case, high electrical impedance is maintained.

Figure 5c shows the avalanche behavior by presenting the dual wavelength electric-driven volume-enhanced reflection spectroscopic signature of a loaded (over)-driven NANOR®-type component ($ZrO_2PdD$) in its avalanche mode. The loaded nanomaterial, $ZrO_2PdD$, in a modified two terminal, dry, NANOR®-type LANR electrical component was electrically driven at 1000 V initially (direct current) which enabled a current of about 1.35 mA. Thereafter, the voltage decreased with time to about 500 V. This behavior is what is termed "avalanche mode" and usually results from excessive applied voltage. The ambient temperature was 26.7°C.

### 3.4. Stokes peaks of driven $ZrO_2PdD$ in the desired XSH state

CMORE (Dual Wavelength Electric-Driven Volume-Enhanced Reflection) spectroscopy is capable of rapidly differentiating the different, and active, states in LANR nanomaterials. The functional, desired XSH state of a correctly driven NANOR®-type component ($ZrO_2PdD$) in its XSH mode is indicated by a unique signature using dual wavelength electric-driven volume-enhanced reflection spectroscopy. In Figs. 1, 5c and 6, the XSH mode can be seen by a unique reflected optical backscatter along with the reflected optical beams from the component in its desired XSH state.

Figure 5c shows the response of the loaded, deuterated material, $ZrO_2PdD$, in a NANOR®-type component to the dual wavelength reflection spectroscopy driven in all three electrical states. In this figure, the curves are superimposed.

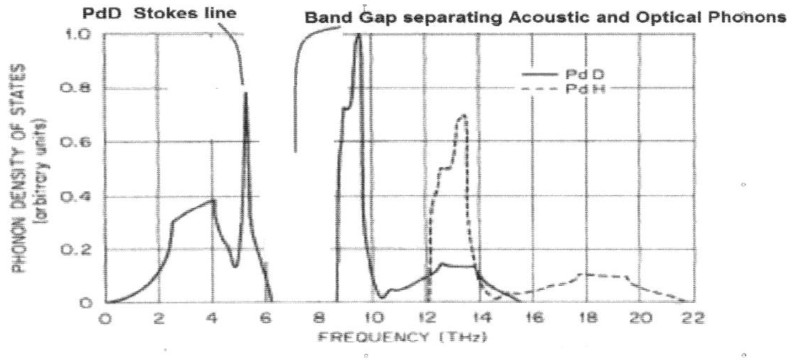


(a)

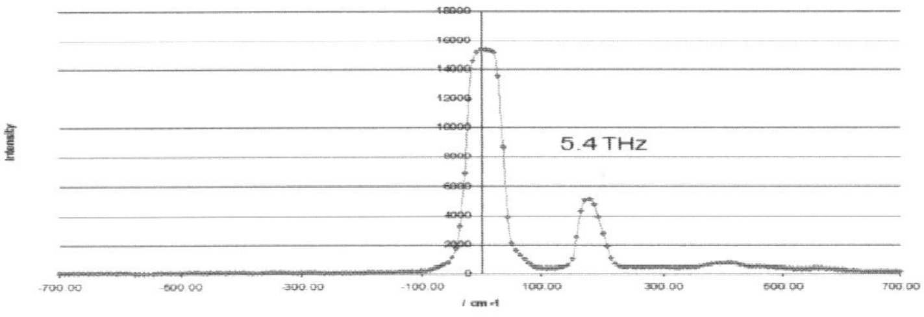


(b)

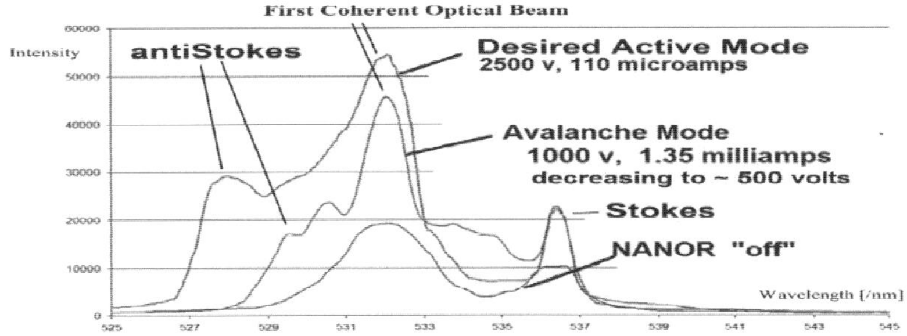


(c)

**Figure 5.** Comparison of PdD phonon density of states to CMORE spectra of $ZrO_2$–Pd and PdD. (a) (top) The PdD Phonon density of states [23] shows a peak before 6 THz and at 7–8 THz a gap. The band gap (hole) can be seen between the acoustic and optical phonons. The PdD Stokes line and band gap are marked. (b) (middle) CMORE spectrum of $ZrO_2$–PdD shows the peak before 5.4 THz, consistent with the Rowe data. The vertical axis represents the intensity of the returned backscatter plotted as intensity as a function of wavelength. The horizontal axis plots the decreasing frequency to the right. The output of the first laser is shown as the peak on the left side, located at 532 nm. (c) (bottom) Dual Wavelength electric-driven volume-enhanced reflection spectroscopy of a preloaded $ZrO_2$PdD NANOR®-type component in three electrical states (drive modes). Shown are spectra of the same NANOR®-type CF/LANR component in three different electronic states, resolved by dual wavelength coherent electric-driven volume-enhanced reflection spectroscopy. The three modes (responses) are the undriven "off"-state, the unwanted electrical avalanche mode [7], and the optimal operational state, the "Desired Active Mode" where excess energy is being released.

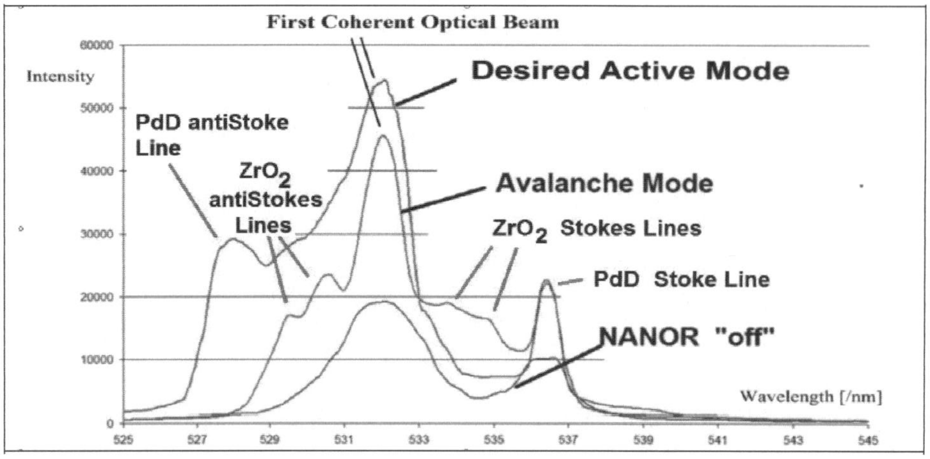


**Figure 6.** Assignment of anti-Stokes peaks to zirconia and PdD. These are the overlaid spectra of the three Different Electronic States Optical Signatures for the same preloaded $ZrO_2$PdD NANOR®-type CF/LANR component in three different electrical drive modes. Shown are spectra of the same NANOR®-type CF/LANR component (NANOR® 7-6) in three different electronic states, resolved by dual wavelength coherent electric-driven volume-enhanced reflection spectroscopy. The three modes (responses) are the undriven off-state, the unwanted electrical avalanche mode [7], and the optimal operational state, the desired active mode, where excess energy is being released. The plots show reflected optical intensity as a function of wavelength, initially from two incident coherent optical beams but then reflected by backscatter along with the initial optical beams from the core.

The nanomaterial NANOR®-type CF/LANR component was properly, correctly electrically driven at 2500 V which produced an electrical current of about 0.11 mA. The sample maintained its high impedance (compared to that seen during avalanche mode) during the very short run, and there was no electrical avalanche quenching the desired reactions. Shown is the output as intensity as a function of wavelength, as returned by backscatter along with the reflected optical beams from the volume-enhanced interactions. Note closely that in the best, most preferred, heat producing mode, the so-called "desired state", there is a much larger-than-expected anti-Stokes peak. This is the unique reflected optical output from the desired "XSH" state where large amounts of energy are being released. Intensity is shown as a function of wavelength, as returned by backscatter along with the reflected optical beams. This is the observed optical output in the desired correctly driven active "state" as observed by the diagnostic.

Figures 1, 5c, and 6 show the dramatic increase of the anti-Stokes component for an activated NANOR®-type CF/LANR component. For the activated NANOR®-type CF/LANR component, there was a major, dramatic increase of the anti-Stokes component and the aS/S ratio both greater than expected. It has been discovered that there exist increased levels of anti-Stokes-type peaks for all nanostructured materials undergoing electrical drive, however, the aS/S ratio goes above one only in active, desired mode. In fact, not only is the anti-Stokes peak of the "desired state" much higher in amplitude than expected for an initial Boltzmann statistic calculation, it is beyond any known calculated temperature, and beyond what could accrue due to curve shift to a very slight amount. These measurements demonstrate that the best XSH results occur (with the CF/LANR nanomaterial in the active, desirable mode) with an aS/S peak ratio greater than 1 (Figs. 1, 5c, and 6).

There are many critically new findings: Importantly, the unusual aS/S ratio exists ONLY with active NANOR®-type LANR component in XSH producing mode, characterized by a larger electrical impedance which precedes electrical breakdown. Thus, the CMORE spectroscopic signature heralds the "desired mode" or XSH production mode and only results from proper drive voltage, maintaining high impedance, and avoiding quenching materials and quenching states, as discussed elsewhere [1,3,7,13,15] and demonstrated conclusively again in this report.

Furthermore, the avalanche mode-induced anti-Stokes peaks differ considerably from those observed in the XSH mode. All the anti-Stokes peaks are relatively missing in the undriven mode. But when electrically driven, there is diversity. When nanostructured NANOR®-type components are driven in avalanche mode, the anti-Stokes peaks differ considerably from those which appear during the XSH producing "desired mode". They differ in energy, and amount, and what stimulates their appearance.

The avalanche anti-Stokes peaks are many, but are lower energy than the XSH mode produced anti-Stokes peak. These match the many Stokes peaks of zirconia. By contrast, the XSH-generated anti-Stokes peak is an alternative to the avalanche-generated many peaks. That peak is singular and at one higher energy which matched the Stokes line of PdD. Thus, there appear to be two electrical heat-generating mechanisms which both dissipate energy (both applied and possibly generated internally), and they are distinguished by two entirely different end products, amounts of output heat production, and CMORE signatures.

In the unwanted mode, the anti-Stokes peaks of the electrical avalanche indicate conventional dissipation including undesirable electronic and lattice vibrations. In contrast, in the desired XSH-producing mode, the phonon gain heralded by the large anti-Stokes peaks link to unlocking the excited $^4$He* and coupling the energy released to the lattice as whole.

### 3.5. $ZrO_2$PdD – Optical phonons

In a solid lattice, alloyed with deuterons, there are multiple modes of vibration. The acoustical phonons are those where the palladium and deuterons move in phase in the long wavelength limit, located in reciprocal space at the center of the Brillouin zone. By contrast, optical phonons have out-of-phase vibrations between neighboring atoms. They are called "optical" because in ionic crystals they can be excited by the electromagnetic radiation, with the positive ions moving one way while their negatively charged neighbors move the other way. In $k$ (momentum)-space, the phonons appear as manifolds in two branches. The lower manifold contains the acoustical branch, and the higher manifold contains the optical branch. There is an energy gap between them when a single metal is alloyed with hydrogen or deuterium.

To better observe the optical phonons, a logarithmic vertical axis is used to enable other peaks of less intensity, like the optical phonons, to stand out. CMORE-spectroscopy resolves and measures all phonon bands, which can be seen at optical and acoustic frequencies.

In Figs. 1, 5c, and 6, the electrical driving mode of the two-terminal NANOR®-type CF/LANR component is the desired XSH mode and only results from proper drive voltage, maintaining high impedance, avoiding quenching materials and quenching states. Importantly, the phonon peaks visualized and measured in the present diagnostic, are both direct Evidence of possible optical phonons being involved, and also corroborating the indirect Evidence of Cravens, Letts, and Hagelstein [24]. The optical phonon bands are not yet well resolved in these first CMORE systems, but the need to do so clearly remains, and explorations should be fruitful.

### 3.6. Acoustic phonons from Pd enable XSH mode

CMORE spectroscopy has revealed a greater-than-normal intensity of anti-Stokes peaks (Figs. 1, 5c, and 6), and therefore acoustic phonon density, only during XSH mode compared to both the "off" state and compared to the avalanche mode. CF/LANR activity is now absolutely linked with acoustic phonons – and only the PdD lattice enables them. There are some other important findings.

Volume-enhanced electrically driven multiwavelength optical (CMORE) spectroscopy images acoustic and optical phonon prevalence and their diversity. Although optical phonons had previously been considered key to the energy

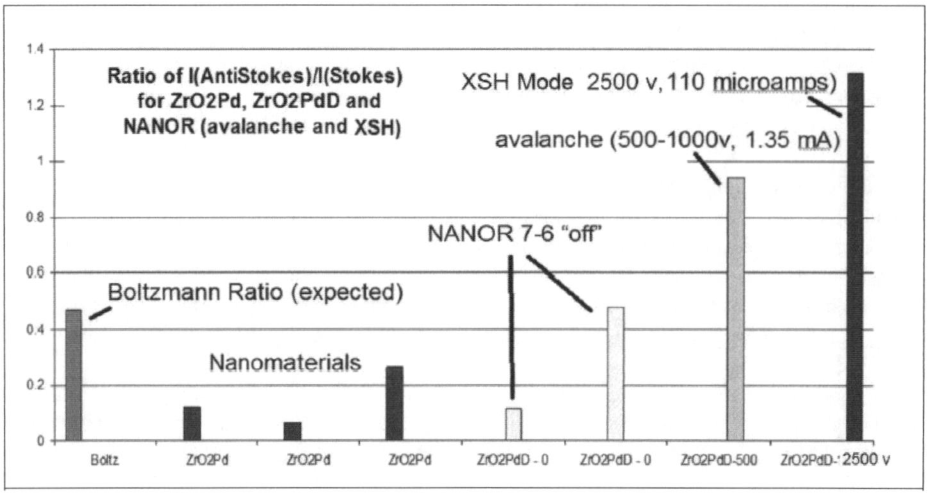


**Figure 7.** Results: Boltzmann ratios exceed expectations for XSH mode compared to a series of materials, and a NANOR®-type LANR component in non-XSH modes. Histogram of measured Boltzmann ratios and what is expected shown are the experimental Boltzmann ratios measured for a series of materials, and a NANOR®-type LANR component in its three modes.

transfer of $^4$He* formed *de novo*, the present diagnostic system has determined a role for acoustic phonons in active CF/LANR systems.

The present diagnostic has revealed new direct evidence of acoustic PdD phonons in successful CF/LANR, using a NANOR®-type CF/LANR component driven correctly in its XSH mode.

Previously, cold fusion investigators have not been able to directly see phonons in their hydrogen loaded systems, although they have been inferred indirectly through beat frequency two wavelength systems [24], and thought to be mainly optical phonons [11]. Our findings confirm an important role for phonons in successful CF/LANR. They go further because this new diagnostic enables direct visualization of the actual phonon states of the hydrogen loaded material, even while it is being electrically driven. These findings confirm the role for zirconia anti-Stokes in avalanche mode (unsuccessful CF/LANR), and also herald the presence of PdD anti-Stokes peaks during successful CF/LANR. Most importantly and specifically, we have now discovered that PdD acoustic phonons result from, or are required for, successful CF/LANR.

This diagnostic has detected the existence and visualized a large single anti-Stokes peak, heralding phonon gain, which is ONLY associated with active CF/LANR XSH-production. Also, attention is directed to the fact that in the desired electric-driven XSH-producing mode, characterized by very high electrical impedance, the two-terminal deuterided NANOR®-type CF/LANR component has a measured Boltzmann–Stokes ratio $\sim$1.3.

There are several important implications. First, acoustic phonon gain either results from, or is required for, a cold fusion process which produces energy gain in its XSH mode. Successful cold fusion creates an aS/S ratio $>1$ during XSH.

Second, in retrospect, this discovery may also be consistent with the fact that only acoustic phonons are used to calculate the thermal conductivity of a material. Third, how do the XSH-generated acoustic phonons enable the creation of, or de-excitation loss of, $^4$He* (which is the precursor to *de novo* $^4$He seen with the XSH in aqueous systems [10,13,14])?

## 4. Interpretation

### 4.1. Summary of Boltzmann ratio results

In physics, the calculated Boltzmann statistic ratio (also known as the Boltzmann factor) is derived from the frequency and the temperature. Numerically, it is calculated from the exponential of the ratio of two energies. Those energies are the energy of the difference in energies and the thermal energy (which is the Boltzmann constant times the thermodynamic temperature in Kelvin). This is done because the ratio thus reflects the magnitude of the expected populations of the two states.

The results of measuring $R_{\mathrm{aS/S}}$ in several types of CF/LANR nanomaterials and during different types of electrical activation can be simply summarized in Fig. 7. Figure 7 is a histogram that presents both the Boltzmann statistic ratio and the Boltzmann aS/S Intensity ratios which were experimentally measured by the CMORE-spectroscopy for a series of materials, and a NANOR®-type LANR component in its three electrical modes (states). In Fig 7, the horizontal axis shows and identifies eight different categories of said $ZrO_2Pd$, $ZrO_2PdD$ nanomaterial, and three electrical drive states, and what is expected normally. The vertical axis shows the amplitude of the theoretical Boltzmann statistic ratio and the experimental Boltzmann aS/S Intensity ratios.

Figure 7 shows by, histogram of the experimentally measured Boltzmann ratios, and the first column in the histogram shows exactly what is expected by calculation. To the right of that "expectation" column (expected), are shown the experimental Boltzmann ratios actually measured for a series of nanomaterials, and then a NANOR®-type LANR component in its three electrically driven modes.

Then, the next columns involve electrical activation: first electrical avalanche, and then the XSH producing (desired) mode. When values are inserted into the conventional formulae, the expected Boltzmann statistic ratio is a number between 0.4 and 0.6, as can be seen in the first column of Fig. 7. The other columns were derived experimentally for this report. The Boltzmann aS/S Intensity Ratio is derived from the actual measured relative intensities of the anti-Stokes peak divided by the intensity of the Stokes peak. Here, those values were determined by CMORE-spectroscopy. Then, the ratio is derived from the intensity of the anti-Stokes peak divided by the intensity of the Stokes peak. The histogram shows the collected information. It was obtained from the data obtained using measured Boltzmann ratios for a variety of related samples; including the relevant $ZrO_2Pd$ nanomaterial, before heavy hydrogen is added, and after the loading deuterons to a very high level [3], and then with the two-terminal component in several electrical drive states.

### 4.2. Implications of the aS/S ratio >1 during XSH

Why is the anti-Stokes/Stokes ratio so high with XSH mode in $ZrO_2PdD$? There are important implications of an XSH-related aS/S ratio greater than one, from/during the desired cold fusion/LANR state. First, it can indicate, and so far is ONLY associated with, the desired electronic state of cold fusion (lattice assisted nuclear reaction; CF/LANR) activity in a NANOR-type CF/LANR component.

Second, that ratio is significant evidence of acoustic phonon gain, which itself is present only while the preloaded NANOR-type component is driven electrically. The highest levels indicate that the active CF/LANR system, confirmed by the CMORE-spectroscopy, is driven properly. Third, that ratio is far outside of the range expected normally by temperature alone. It passes through a region which is beyond any known calculated temperature [8]. It is a level greater than could be accounted for by a prosaic curve shift secondary to temperature.

### 4.3. Calculation of effective temperature during XSH

The fact that the Boltzmann Stokes ratio is $\sim$1.3 reveals more about successful cold fusion. Because the Stokes and anti Stokes energies are derivable, the effective Stokes temperature can be computed from the experimental data to reveal the core temperature during the XSH mode. The next equation derives the calculated Boltzmann Stokes Ratio and the calculated Boltzmann Statistic ratio.

The equation involves temperature, and is

$$R_{\mathrm{aS/S(calculated)}} = \text{Intensity(anti-Stokes)/Intensity(Stokes)} = \exp(-E_{\mathrm{diff}}/k_{\mathrm{B}}T), \tag{1}$$

where $E_{\mathrm{diff}}$ is the energy difference between the two states is $5.96\times10^{-21}$ J and $k_{\mathrm{B}}$ is Boltzmann 's constant.

The calculated temperature at core is, therefore, $\sim$1645 K ($\sim$1372°C). Temperature results from the number of states available to a system and it will be important to map this finding by depth to determine its relevance as a real, physical value [8].

**Table 1.** Variables used in calculation

| | |
|---|---|
| $R_{\mathrm{s}}$ | the Stokes ratio = Intensity(anti-Stokes)/Intensity(Stokes) |
| $k_{\mathrm{B}}$ | Boltzmann's constant = $1.38066\times10^{-23}$ J/K |
| $T$ | temperature in degrees, Kelvin |
| $E_{\mathrm{diff}}$ | = $\Delta E$ an energy difference |

## 4.4. Calculation of phonon number during XSH

How does the desired XSH mode actually create the anti-Stokes components at levels higher than observed for all other material precursor and avalanche mode? The Stokes and anti-Stokes energies are derivable, and therefore the number of phonons involved during XSH can be computed. As discussed in solid state texts [25,26], with an acoustic frequency of 4.9 $^{\pm 0\ 1}$ THz, there are $\sim$7 $^{\pm 0\ 15}$ phonons involved. One important point is that this is much lower than has ever been expected before in most theoretic analyses of energy transfer. Another possible point is that this number may be related to previously considerations of the palladium lattice Pd vacancy which is surrounded by six deuterons [8,27].

## 4.5. Spin Bose model might include acoustic phonons

Hagelstein incorporates the optical phonons in his theory. It explains the massive energy cooperative transfer from the megavoltage energy of the $^4$He* to the lattice in a coherent de-excitation process which proceeds by way of lossy spin bosons involving phonons [11], and possibly magnons [4,9]. This study heralds that acoustic phonons have an indelible role.

## 4.6. Phuson theory corroborated

The proof that phonons are involved as a cohort for the XSH in active cold fusion (LANR) systems supports the PHUSON theory [12] of coherent energy transfer to the lattice. The PHUSON is the quantum of that energy transfer, in a process which is consistent with conventional physics, and where the energy then appears as XSH [1,13]. The PHUSON theory explains why there is a relative absence of strong neutron and gamma ray emissions in CF (LANR). The gamma emission branch from the excited state of $^4$He* is actually spin-forbidden for both hot and cold fusion [12]. However, at higher hot fusion temperatures the restriction is slightly lifted. This is consistent to what is seen for both hot and cold fusion. Thus, the PHUSON theory correctly describes the relative absence of neutron emissions in CF (LANR). The only nuclear branches available are those whose band gaps are surmountable by the available activation energy (limited by the ambient temperature and incident radiation). The neutron emission branch is more than 1 MeV above the first excited state ($^4$He*). Hot fusion has large activation energies available (it is "hot" LANR/CF is not. In LANR, given the actual much smaller amount of thermal energy, $k_B T$, available for cold fusion (~1/25 eV), absence of adequate activation energy decisively means that that branch is NOT available, as it is for hot fusion.

### 4.7. Explanation for the large signals

The major problem of the Raman effect is that it produces very weak signals because the photon conversion efficiencies are less than $10^{-18}$. As Sir Chandrasekhara Raman stated, the effect has "excessive feebleness" [28]. These very small signals result because the proportionality constant $\chi(3)$, known as the "third-order susceptibility", is only linearly proportional to the local oscillator density. However, that is not the case when irradiation is made using coherent lasers which yield a much larger signal because with coherent illumination there results phase-matching conditions and quadratic dependence on the number of local oscillators. Thus, even with the same selection rules, the coherent irradiation Raman effect yields a much greater signal intensity (circa $10^6$ times greater) [29,30].

## 5. Conclusions

### 5.1. Implications for improvement of LANR systems

CMORE spectroscopy differentiates active states in LANR nanomaterials. The present diagnostic generates spectra in real time which are able to illuminate, visualize, and help identify at least three states for a CF/LANR material or component which is potentially active. The system can determine which state the material or component is in, in real time, even as it is electrically driven, and may be able to determine time constants for changes between levels. Most importantly, it can saliently detect the desired reactions just as products begin to appear and accumulate (e.g. heat). Therefore, it has great use because it reveals both the desired, and undesired, reactions and states and may enable better control, and possible detection of other states of operation – and therefore reactions and products.

Future studies should attempt to resolve loading of the PdD by determining the displacement of the anti-Stokes from the main peak and relating that to the Pd–Pd bond energies which decrease as loading proceeds and the lattice separation expands [31]. Other investigators may elect to use CMORE spectroscopy to resolve other peaks such as D–D.

Another advantage includes the immediate telegraphing of the state, rather than having impatient skeptics await the appearance of XSH. The use of CMORE-spectroscopy has implications for examining substates in material science and metallurgy, cold fusion physics, and in the electrical engineering of CF/LANR systems. Other spectroscopies only identify materials, while the present diagnostic reveals the electrical state of the component while continuing the means to electrically drive, control, and monitor that component and state as desired.

Volume-enhanced CMORE spectroscopy successfully differentiates active states in LANR nanomaterials. All the anti-Stokes peaks are relatively missing in the undriven mode for all of the nanostructured materials. Weak anti-Stokes peaks are elicited from Pd (and Ni and their alloy) nanostructured material in $ZrO_2$. But when NANOR®-type components are electrically driven, there is diversity in outcome. When driven in the avalanche mode, the anti-Stokes peaks differ considerably from those which appear during the XSH-producing, or desired, mode. The anti-Stokes peak(s) differ in energy, and amount, and differ in what stimulates their appearance. Normal anti-Stokes peaks return when the electrical drive creates "avalanche mode", characterized by higher electrical current at decreasing voltage. The avalanche anti-Stokes peaks are many, and they are lower energy than the XSH mode produced anti-Stokes peak. By contrast, successful cold fusion is heralded by a large increase in the aS/S ratio, and the generated anti-Stokes peak for the desired and XSH-producing state is very different from the avalanche-generated many anti-Stokes peaks. That XSH-related peak is singular and at higher energy. This distinguishing, higher energy, single, anti-Stokes peak (which also heralds phonon gain) is also not seen in the "off" state or the avalanche (undesirable) mode. Our analysis finds that the excess-heat produced anti-Stokes peak is matched to the Stokes line of PdD. We also find that the several lower energy anti-Stokes in avalanche mode (unsuccessful regarding CF/LANR) are matched to the many Stokes peaks for zirconia. In the desired electric-driven XSH-producing mode, the two-terminal deuterided NANOR®-type CF/LANR component has a measured Boltzmann–Stokes ratio $\sim$1.3.

Analysis of the phonon gain heralds $\sim 7^{\pm 0\ 15}$ acoustic phonons assisting nuclear reactions and a core peak calculated Stokes temperature of circa 1645 K. Therefore, these findings confirm a role for PdD acoustic phonons, in the loaded lattice, during successful CF/LANR which produces XSH.

## Acknowledgments

The authors acknowledge and thank for their very helpful comments, editorial assistance, and support: Gayle Verner, Jeffery Tolleson, Alex Frank, Joshua Gyllinsky, Florian Metzler, Charles Entenmann, Christy Frazier, Dennis Cravens, Dennis Letts, Frank Gordon, Pamela Mosier-Boss, Lawrence Forsley, Robert Smith, Brian Ahern, Jeff Driscoll, Louis DeChiaro, for their help, ideas and suggestions. This effort was supported by JET Energy Inc. and the New Energy Foundation. NANOR® is a registered trademark of JET Energy Incorporated. NANOR®-technology and other technology described here are protected by patents pending.

## References

[1] M.R. Swartz, G. Verner, J. Tolleson and P. Hagelstein, Dry, preloaded NANOR®-type CF/LANR components, *Current Sci.* **108** (4) (2015) 595.

[2] M.R. Swartz and P.I. Hagelstein, Demonstration of energy gain from a preloaded $ZrO_2$–PdD nanostructured CF/LANR quantum electronic device at MIT, *J. Condensed Matter Nucl. Sci.* **13** (2014) 516. www.iscmns.org/CMNS/JCMNS-Vol13.pdf.

[3] M.R. Swartz, G. Verner and J. Tolleson, Energy gain from preloaded $ZrO_2$–PdNi–D nanostructured CF/LANR quantum electronic components, *J. Condensed Matter Nucl. Sci.* **13** (2014) 528. www.iscmns.org/CMNS/JCMNS-Vol13.pdf.

[4] M.R. Swartz, G. Verner, J. Tolleson, L. Wright, R. Goldbaum and P.L. Hagelstein, Amplification and restoration of energy gain using fractionated magnetic fields on $ZrO_2$–PdD nanostructured components, *J. Condensed Matter Nucl. Sci.* **15** (2015) 66. www.iscmns.org/CMNS/JCMNS-Vol15.pdf.

[5] M.R. Swartz, Incremental high energy emission from a $ZrO_2$–PdD nanostructured quantum electronic component CF/LANR, *J. Condensed Matter Nucl. Sci.* **15** (2015) 92. www.iscmns.org/CMNS/JCMNS-Vol15.pdf.

[6] M.R. Swartz, G. Verner, J. Tolleson, L. Wright, R. Goldbaum and P.L. Hagelstein, Imaging of an active NANOR®-type LANR component using CR-39, *J. Condensed Matter Nucl. Sci.* **15** (2015) 81. www.iscmns.org/CMNS/JCMNS-Vol15.pdf.

[7] M.R. Swartz, P. Hagelstein and G. Verner, Impact of electrical avalanche through a $ZrO_2$-NiD nanostructured CF/LANR component on its incremental excess power gain, *ICCF-19*, Padua, Italy, April 16, 2015.

[8] M.R. Swartz, Optical detection of phonon gain distinguishes an active cold fusion/LANR component, *J. Condensed Matter Nucl. Sci.* **20** (2016) 29. www.iscmns.org/CMNS/JCMNS-Vol20.pdf.

[9] M.R. Swartz, Oscillating excess power gain and magnetic domains in NANOR®-type CF/LANR component, in preparation.

[10] M.H. Miles, R.A. Hollins, B.F. Bush, J.J. Lagowski and R.E. Miles, Correlation of excess power and helium production during $D_2O$ and $H_2O$ electrolysis using palladium cathodes, *J. Electroanal. Chem.* **346** (1993) 99–117.

[11] P.L. Hagelstein, Current status of the theory and modeling effort on fractionation, *J. Condensed Matter Nucl. Sci.* **19** (2016)98–109.

[12] M.R. Swartz, Phusons in nuclear reactions in solids, *Fusion Technol.* **31** (1997) 228–236.

[13] M.R. Swartz, Survey of the observed excess energy and emissions in lattice assisted nuclear reactions, *J. Scientific Exploration* **23** ( 4) (2009) 419–436.

[14] M.R. Swartz, Excess power gain using high impedance and codepositional LANR devices monitored by calorimetry, heat flow, and paired stirling engines, *Proc. ICCF14* **1** (2008) 123. ISBN: 978-0-578-06694-3, 123 (2010). www.iscmns.org/iccf14/ProcICCF14a.pdf.

[15] M.R. Swartz and G. Verner, Excess heat from low electrical conductivity heavy water spiral-wound Pd/$D_2O$/Pt and Pd/$D_2O$–$PdCl_2$/Pt devices, *Condensed Matter Nuclear Science, Proc.ICCF-10*, World Scientific, Singapore, ISBN 981-256-564-6, 29-44 (2006) 45–54.

[16] M.R. Swartz, Consistency of the biphasic nature of excess enthalpy in solid state anomalous phenomena with the quasi-1-dimensional model of isotope loading into a material, *Fusion Technol.* **31** (1997) 63–74.

[17] R. Sherman, H.K. Birnbaum, J.A. Holy and M.V. Kleinet, *Phys. Lett.* **62A** (1977) 353.

[18] M. Fleischmann, P.R. Graves, LR. Hill and J. Robinson, *Chem. Phys. Lett.* **95** (4,5) (1983) 322–324.

[19] K. Tsuchiya, Al. Watanabe, M. Ozaki and S. Sasabe, Observation of optical phonon in palladium hydrides using Raman spectroscopy, *Proc. ICCF-14*, D.J. Nagel and M.E. Melich (Eds.), pp. 338–342. ISBN: 978-0-578-06694-3, 343, (2010).

[20] Y. Arata and Y.C. Zhang, Observation of anomalous heat release and helium-4 production from highly deuterated palladium fine particles, *Jpn. J. Appl. Phys.* **38** Part 2, No. 7A (1999) L774–L776.

[21] M.R. Swartz, Deuterium production and light water excess enthalpy experiments using nickel cathodes, *J. New Energy* **1** (3) (1996) 219. www.iscmns.org/FIC/J/JNE1N3.pdf.

[22] E. Anastassakis, B. Papanicolaou and I.M. Asher, *J. Phys. Chem. Solids* **36** (1975) 667.

[23] J. M. Rowe et al., *Phys. Rev. Lett.* **33** (1974) 1297.

[24] D. Cravens, D. Letts and P.L. Hagelstein, Progress on two-laser experiments, *Proc. ICCF15* (2009). http://lenr-canr.org/acrobat/Hagelsteinprogresson.pdf).

[25] C. Kittel, *Introduction to Solid State Physics*, Wiley, New York, 1976.

[26] N.W. Ashcroft and and N.D. Mermin, *Solid State Physics*, Holt, Rinehart and Winston, NY, 1976.

[27] D. Letts, A method to calculated excess power, *Infinite Energy* **112** (2013) 63.

[28] C.V. Raman, *Scientific Papers of C.V. Raman*, S. Ramaseshan (Ed.), Vol. 1, *The Scattering of Light*, Indian Academy of Sciences, Bangalore, India, 1988.

[29] Ji-xin Cheng, E.O. Potma and X.S. Xie, Coherent anti-Stokes Raman scattering correlation spectroscopy: probing dynamical processes with chemical selectivity, *J. Phys. Chem.* A, **106** (2002) 8561–8568.

[30] L. Evans Conor and X.S. Xie, Coherent anti-Stokes Raman scattering microscopy: chemical imaging for biology and medicine, *Ann. Rev. Anal. Chem.* **1** (2008) 883–909.

[31] P.W. Atkins and J. de Paula, *Physical Chemistry, Thermodynamics, Structure, and Change*, 10th Edition, W.H. Freeman, VA, 2014. ISBN-13: 978-1429290197.

J. Condensed Matter Nucl. Sci. 24 (2017) 146–167

Research Article

# Fusion of Light Atomic Nuclei in Vacuum and in Solids and Two Ways of Mastering Nuclear Fusion Energy

V.F. Zelensky*,†

*National Science Center Kharkov Institute of Physics and Technology, Kharkov, Ukraine*

**Abstract**

Two ways of mastering nuclear fusion energy, viz., controlled thermonuclear fusion and cold fusion, are considered. This paper deals with the intensive cold fusion model, hereinafter referred to as the *chemo*nuclear fusion hypothesis. The determining role of virtual photons and electrons of anomalous internal $\gamma$-conversion in cold fusion is shown. The *chemo*nuclear fusion hypothesis for the 2D-, HD- and $(^7\text{Li}–^1\text{H})_{\text{gas}}$-transition metal systems provides an explanation of cold fusion. It is demonstrated that this hypothesis can serve as a base for the development of cold fusion reactors.


*Keywords:* Anomalous $\gamma$-conversion, *Chemo*nuclear fusion, Cold fusion

## 1. Introduction

Nowadays, the radical solution of ecological and energy problems of the Earth is associated by scientists with the development of nuclear fusion, which present an inexhaustible source of energy. In recent years, consideration has been given to two ways of achieving nuclear fusion: creation of controlled nuclear fusion reactors ("thermonuclear fusion" [1]), and creation of cold fusion reactors (HTE-Cat, "Rossi reactor") [3,4].

It is well known that the non-recognition of any cold fusion research results by the scientific community is based on the universal approval of three theoretical restrictions on the occurrence of low-temperature transmutation of chemical elements [5]: (a) the impossibility of overcoming the Coulomb barrier, (b) extremely small cross-section values of weak processes, and (c) the low probability of multiparticle collisions.

The analysis of numerous cold fusion results (including our results) has led us to the conclusion that the three restrictions listed above are valid only for reactions occurring in vacuum [11,12]. In cold fusion conditions, when the reactions take place in solid-state matter with the participation of virtual photons and electrons of anomalous internal $\gamma-$conversion, the restrictions may be removed, and the process of fusion would be intensified (Section 3.2). This

*E-mail: vgamov@kipt.kharkov.ua.
†Deceased.

phenomenon has been investigated on 2D-, HD- and $(^{7}Li–^{1}H)_{gas}$-transition metal cold fusion systems. We begin our consideration with the 2D-palladium system, which has received the most study.

## 2. Three Channels of 2D-fusion Reactions in Vacuum and the Controlled Thermonuclear Fusion Reactor Development

It is known that 2D-fusion reactions in vacuum can be carried out with various probabilities at three channels (see Table 1).

As follows from Table 1, the current practical interest is based on the cross-section values of reactions 1 and 2 in Table 1, because the reaction probability in channel 3 is seven orders lower. For this reason, the thermonuclear scientists of the whole world focus their research efforts on the realization of reactions reactions 1 and 2 in Table 1, and in the first place, on investigating the easier processes, viz. the d–T reactions. Historically, this was the first path to nuclear fusion development [1].

As is known, the nuclear reaction cross-section for fusion

$$\sigma = (S(E)/E) \cdot P(E) \tag{1}$$

is given by the product of "internal nuclear cross-section"

$$\sigma_0 = S(E)/E \tag{2}$$

by the quasi-classical probability $P(E)$ of charged particle penetration through the Coulomb barrier $V(r)$:

$$P(E) = \exp\left[-2W(E)\right],$$

$$W(E) = \int_{r_1}^{r_2} \sqrt{2\mu(V(r) - E)}\,\mathrm{d}r/h = \sqrt{2\mu\langle V(E)\rangle}\,|r_2 - r_1|\,/h, \tag{3}$$

where $S(E)$, being "the astrophysical factor," is a slowly varying function of particle energy, which at a low relative energy of interacting particles in case of non-resonant nuclear reactions is constant, $S(E) = S_0$ (for non-resonant. reactions of 2D-fusion $S_0 \approx 0.11$ MeV barn); $r_2 - r_1$ – classical "rotational displacement points" at moving of one of particles in the field of another; $\mu$ – the reduced mass of interacting particles; $\langle V(E)\rangle$ – medial height of the potential barrier lying above level $E$.

From (3) follows that the barrier width $(r_2 - r_1)$ which is a linear function of coordinates, influences penetrability of a barrier more essentially, than its medial height $\langle V(E)\rangle$ does entering in (3) in the form of square root.

**Table 1.** Reaction channels of 2D-fusion in vacuum

| Ch. No. | Reaction (within brackets particle energy in MeV) | Penetrable barrier, MeV [7] | | Energy yield (MeV) | No. of reactions per 1 W in $10^{12}$ $s^{-1}$ of energy yield | Reaction cross-section at $E = 1$ 1 keV, $cm^2$[8] | Reaction yield on channels (%) |
|---|---|---|---|---|---|---|---|
| | | Coul. | Nucl. | | | | |
| 1 | d+d →$^3$He (0.82) +n (2.45) | 0 | 2.2 | 3.27 | 1.9 | $\sim 6 \times 10^{-33}$ | ~ 50 |
| 2 | d + d→T(1.01) + p(3.02) | 0.4 | 2.2 | 4.03 | 1.6 | $\sim 6^{-33}$ | ~ 50 |
| 3 | d + d →$^4$He +γ(23.8) | 0.4 | 0 | 23.8 | 0.26 | $\sim 1 \times 10^{-40}$ | $\sim 1 \times 10^{-5}$ |

From (2) and (3) it follows that the particle's energy, the Coulomb barrier width and the astrophysical factor value could be the factors defining the reaction rate of deuterium fusion in vacuum. However, for double reactions of traditional hot fusion of deuterium only the first of the named factors has a practical value.

The most widespread method of overcoming Coulomb forces in thermonuclear research is to heat the medium containing atoms of deuterium (or deuterium and tritium) to a temperature high enough that the kinetic energy of these atoms ensures their collision energy is sufficient to trigger fusion reactions by approaching the nuclei. The scientific basis of thermonuclear fusion reactors is well studied [1]. High-temperature deuterium–tritium or deuterium plasma serve as an actuation medium for a controlled thermonuclear reactor. For a reactor producing DT-reactions or 2D-fusion reactions, channels 1 and 2 of Table 1 will be active. A fusion reaction in the thermonuclear reactor occurs with collisions of non-screened ("bare") nuclei that have randomly oriented spins. The low cross sections of these reactions are a main source of difficulty to making a controlled thermonuclear reactor. The temperature of the deuterium-tritium mixture in such a reactor equals approximately $2\times 10^8$ K (10 keV), and for deuterium plasmas this temperature is even higher. The problem of maintaining this temperature over time, sufficiently high enough for a positive energy yield, presents colossal engineering difficulties. To overcome these difficulties major projects with thousands of researchers worldwide have been working for many decades.

**Table 2.** Some hypothetical nuclear processes involving bineutron and neutron in systems: deuterium–palladium and deuterium–hydrogen–palladium [6,11,12].

$$\begin{aligned}(\mathrm{D}^+ + \mathrm{e}^- + \mathrm{D}^+ + \mathrm{e}^- + \mathrm{D}^+) &\rightarrow (\mathrm{D}^+ + \mathrm{e}^- + \mathrm{D}^+) + \mathrm{D}^+ \rightarrow (^4\mathrm{He}^* + \mathrm{e}^-) + \mathrm{D}^+ \\ &\rightarrow {}^4\mathrm{He}_{\mathrm{g\ s}} + \mathrm{e}^-_{\mathrm{conv}}(\Sigma E_{\mathrm{conv}}) + \mathrm{D}^+ \rightarrow {}^4\mathrm{He}_{\mathrm{g\ s}} + {}^2\mathrm{n} + \nu_{\mathrm{e}},\end{aligned} \tag{4}$$

$$\begin{aligned}(\mathrm{D}^+ + \mathrm{e}^- + \mathrm{D}^+ + \mathrm{e}^- + \mathrm{H}^+) &\rightarrow (\mathrm{D}^+ + \mathrm{e}^- + \mathrm{D}^+) + \mathrm{H}^+ \rightarrow (^4\mathrm{He}^* + \mathrm{e}^-) + \mathrm{H}^+ \\ &\rightarrow {}^4\mathrm{He}_{\mathrm{g\ s}} + \mathrm{e}^-_{\mathrm{conv}}(\Sigma E_{\mathrm{conv}}) + \mathrm{H} \rightarrow {}^4\mathrm{He}_{\mathrm{g\ s}} + \mathrm{n} + \nu_{\mathrm{e}},\end{aligned} \tag{5}$$

$$46^{104}\mathrm{Pd}^* + {}^2\mathrm{n} \rightarrow {}^{106}_{46}\mathrm{Pd}^* \rightarrow \begin{cases} \rightarrow f \left\langle \begin{array}{l} \mathrm{Nu}_1(^{58}_{26}\mathrm{Fe}), \\ \mathrm{Nu}_2(^{49}_{20}\mathrm{Ca}), \end{array} \right. & \text{(5a)} \\ \rightarrow {}^{105}_{45}\mathrm{Pd} + \mathrm{n}(8.48\,\mathrm{eV}), & \text{(5b)} \\ \rightarrow {}^{105}_{45}\mathrm{Rh}(\beta,\ 35.4\,\mathrm{h}) + \mathrm{p}(7.2\,\mathrm{eV}), & \text{(5c)} \\ \rightarrow {}^{102}_{44}\mathrm{Ru} + \alpha(11.6\,\mathrm{eV}), & \text{(5d)} \\ \rightarrow {}^{106}_{46}\mathrm{Pd} + \mathrm{e}^-_{\mathrm{conv}}(\Sigma E_{\mathrm{conv}}), & \text{(5e)} \end{cases}$$

$$\mathrm{d} + {}^2\mathrm{n} \rightarrow \mathrm{T}(1.56\,\mathrm{eV}) + \mathrm{n}(4.64\,\mathrm{eV}) \rightarrow \mathrm{d} + \mathrm{T}(1.56\,\mathrm{eV}) \rightarrow \alpha(6.9 - 1.44\,\mathrm{eV}) + \mathrm{n}(12.5 - 17.7\,\mathrm{eV}), \tag{6}$$

$${}^{104}_{46}\mathrm{Pd} + \mathrm{n} \rightarrow {}^{105}_{46}\mathrm{Pd}^* \rightarrow {}^{105}_{46}\mathrm{Pd}_{\mathrm{g\ s}} + \mathrm{e}^-_{\mathrm{conv}}(\Sigma E_{\mathrm{conv}}). \tag{7}$$

The numeral $\Sigma E_{\mathrm{conv}}$ designates the total energy of nuclear reaction which is carried away by conversion electrons ($e^-_{\mathrm{conv}}$).

At the same time, as was noted for the first time in [7], we can expect that under certain conditions the fusion reaction 3 in Table 1, as an energy source, will be favored. Authors of this work pay attention to the fact that reaction 3 in Table 1 for its realization demands only overcoming a rather low Coulomb barrier ∼0.4 MeV, while reaction (1) is interfaced to a rupture of nucleus of deuterium 2.2 MeV, and a reaction 2 with that and another: ∼0.4 and 2.2 MeV. Because the tunneling reactions to an extremely strong degree depend on barrier height and barrier width, there are reasons to expect that under conforming conditions reaction 3 can become a favorable source of pollution-free nuclear energy. However, implementing this concept presents huge difficulties.

The problem is that the low cross section of a reaction on channel 3 is due to the fact that only one particle figures in the outcome. As is well known, for such reactions, the laws of conservation of energy and impulse can be fulfilled simultaneously only under conditions when the lifetime of the excited nucleus exceeds the time necessary for a discharge of the nucleus excitation energy.

The energy discharge of the excited nucleus of helium by $\gamma$-quanta radiation is impossible. According to the selection rules, the birth of a photon is forbidden for transitions when initial and final states of the nucleus both have spins equal to zero as it takes place for the ground state and for the first two excited states of $^4He^*$. In the case of 2D-fusion, an excited virtual $^4He^*$ compound-nucleus exist for $\sim 10^{-22}$ s [9]. Until now, no process has been proposed which would make it possible to remove the excitation energy of the nucleus in such a short time, and consequently it is considered that a fundamental prohibition of a reaction passing on channel 3 (Table 1) must exist.

## 3. Cold Fusion – 2D-Fusion Reaction in Solids ("Chemonuclear" Fusion)

### 3.1. Introduction

The second way to master nuclear fusion energy is by exploring cold fusion. This was disclosed in 1989 when Fleischman and Pons published sensational results of their research of this phenomenon [2][a]. Unlike the enormous technical difficulties developers of thermonuclear fusion reactors have met, practical development of cold fusion, in particular, development of $(^7Li–^1H)_{gas}$–nickel system was much easier and has far outstripped the creation of a theoretical basis of this phenomenon. The E-Cat reactor presentation was held by S. Focardi and A. Rossi in March 2011 in Italy [3]. Since then work of "Rossi reactor" "cell" has already been successfully reproduced in a number of countries (Russia, China, Kazakhstan, etc.) [10].

In 2012, our claims to explain the intensive (self-sustaining) cold fusion [11][b] were published. The fusion model offered in this work is the *chemo*nuclear fusion hypothesis is a complex, two-stage, cluster fusion process of light nuclei in deuterium–metal and natural hydrogen–metal systems (see below) which has passed from the moment of the publication of works [11] four years has undergone experimental testing and has gained further development in [6], in this paper and confirmation in our verification experiment [13]. *Chemo*nuclear fusion hypothesis is consistent also with the results carried out in this period by a group of independent experts in Lugano (Switzerland) the HTE-Cat reactor study [14]. Today this gives reason to consider a *chemo*nuclear fusion hypothesis [6] a consistent qualitative intensive cold fusion model in 2D-, HD- and $(^7Li–^1H)_{gas}$-transition metal systems.

The main positions of the *chemo*nuclear fusion hypothesis are set forth in Section 3.2 for the example of the Fleischmann and Pons experiment analysis.

Let us focus on the history of this model development.

[a] Fleischman–Pons electrolysis experiment and its results interpretation from the perspective of the *chemo*nuclear fusion hypothesis is described in Section 3.2.

[b] This work was published in journal in April 2013 [12].

### 3.2. Intensive cold fusion – 2D-fusion reactions in solids with virtual photons and electrons of anomalous $\gamma$-conversion participation

In our approach to studying cold fusion, we proceeded from the assumption that the probability of tunneling through the Coulomb barrier at temperatures of cold fusion experiments is extremely small and virtually eliminates the occurrence of nuclear reactions. With this in mind, in one of our earlier works [15] (1991), we already assumed that a two-stage cold fusion natural process and formation in cold fusion conditions of the accelerator mechanisms is able to accelerate charged particles to energies up to ~1 keV. We proposed to call this phenomenon "nuclear fusion in matter" (NFM). We used this phenomenon name in [6,11–13] and here.

Since we first published studies on this method, anomalously high heat generation and low radioactivity from cold fusion processes have given us reasons to intuitively expect that cold fusion is a result of nuclear reactions occurring in 2D-fusion on "light," not interfaced with the rupture of nuclear bonds, neutronless channel 3 (in Table 1). The confidence level in this type of fusion process increased after the Bush and Lagowski experiments [16] obtained correlations between the energy generation and $^4$He yield. This work has been confirmed as a reaction of cold fusion [15]. We expected and stated that in such conditions the interdiction of the reaction on channel 3 had disappeared. This result did not contradict the foundations of modern nuclear physics because, as mentioned above, the prohibition of photon birth relates to events in vacuum. Removal of excitation energy of the nucleus in solid-state matter can be carried out by the "anomalous" internal $\gamma$-conversion virtual photon on which birth the specified above interdiction does not apply. In solid-state matter a virtual photon transmits energy to the "anomalous" internal $\gamma$-conversion electron and process of energy discharge of the excited nucleus is terminated by energy absorption of a conversion to an electron by the matrix.

The anomalous process of conversion by "penetration" electrons is the process in which the electron penetrates into the nucleus more deeply than the nucleon with which it interacts in the course of conversion at the moment of nuclear transition (the "penetration effect"). One essential role of this effect in the case of strongly slowed $\gamma$-transitions was specified for the first time in Church and Wenser's work [17] in 1956. In that same work, the experimental results relating to this effect were analyzed. Many authors have conducted further similar analysis of experiments, the phenomenon has gained wide recognition, and tables of known anomalous cases of the coefficient of electronic $\gamma$-conversion [18] have been compounded.

Let us estimate the parameters of an internal $\gamma$-conversion, which could ensure – in the conditions of cold 2D-fusion – a timely energy discharge of the excited $^4\mathrm{He}^*$ nucleus and, thereby, would remove the prohibition for fusion passing in the "light" channel 3, Table 1. For this purpose, let us use the scheme of 2D-fusion reaction decay channels (Fig. 1) from Ref. [19].

According to this scheme $^4\mathrm{He}^*$ nucleus excitation energy decreases from 23.8 to 19.8 MeV ($\Delta E = 4$ MeV) on channels 1 and 2 in Table 1 forbids the decay of the nucleus. The further energy discharge of $^4\mathrm{He}^*$ excited nucleus is carried out already by standard for these conditions $\gamma$-conversion $\Delta E \leq 9$ keV [11,12] ("standard" $\gamma$-conversion in this case is energy transfer from nucleus to electrons owned by an electronic shell of a quasimolecule (the cluster–heavy superficial atom of matrix). Let us estimate, proceeding from an uncertainty principle [50]

$$\Delta E\,\Delta t \approx h, \tag{8}$$

the greatest possible distance $\Delta R$ of an electron from the excited $^4\mathrm{He}^*$ nucleus on which the virtual photon still can carry out transmission to an electron energy $\Delta E = 4$ MeV and time necessary for transmission $\Delta t$.

$$\Delta t = h/\Delta E \approx 1.6 \times 10^{-22}\mathrm{s}, \tag{9}$$

$$\Delta R = \Delta t\, c \approx h/\Delta E\, 3 \times 10^{10}\mathrm{s} \approx 4.9 \times 10^{-12}\mathrm{cm}. \tag{10}$$

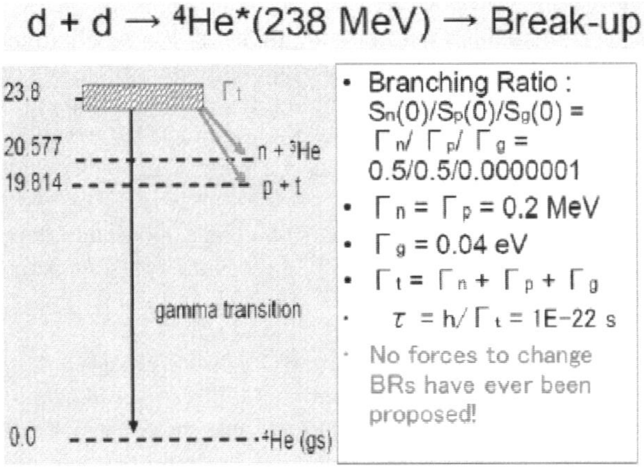


**Figure 1.** 2D-fusion reaction decay channels.

The results obtained can be treated so that in the conditions of cold fusion in a deuterium-loaded metal matrix are the transitive formations, the "deuterium microclusters" in which distance "trapped electron"– merge of deuterons zone does not exceed the specified above value of $\Delta R \approx 4.9\times 10^{-12}$ cm. It is important to note that the close values for $\Delta t$ and $\Delta R$ can be gained with another approach. According to A.S. Davidov's theoretical calculations, the maximum admissible radius of a deuteron in the conditions of cold fusion can reach $r \sim 4\times 10^{-12}$ cm [20]. Then time $\Delta t$, necessary for transmission by virtual $\gamma$-quantum to an electron captured in "microcluster" of excitation energy portion $\Delta E = 4$ MeV will be $\Delta t = r/c \approx 1.33\times 10^{-22}$ s, and the distance $\Delta R = r \approx 4\times 10^{-12}$ cm.

Note that the obtained value of $\Delta R$ is five times smaller than of muon orbits radius ($4.9\times 10^{-12}$ and $25\times 10^{-12}$ cm, respectively). Therefore, the process of fusion in conditions of intense cold fusion should proceed at velocities significantly higher compared to the muon catalysis.

The conversion electron yield in cold fusion experiments is poorly understood.

However, as has been indicated in our study [11], the radiation detected by Karabut in cold fusion experiments outside the discharge chamber (see [21]), may give evidence to the yield of 5–6 MeV electrons, and this is in qualitative agreement with the *chemo*nuclear 2D-fusion hypothesis. The yield of conversion electrons of the $\sim 10^6$ eV energy range in the course of intense cold 2D-fusion is attested by the results of work [22]. However, the authors of this work and the author of [21] do not consider the participation the phenomena of anomalous $\gamma$-conversion in the process of fusion [11,12].

These data and the results of theoretical studies by Gryzinski, Barut and Vigier (see below) have encouraged us to start work on the development of the two-stage, cluster cold fusion model, i.e., the *chemo*nuclear fusion hypothesis. The main requirement of the model was that it had to explain the functioning of the anomalous internal $\gamma$-conversion in cold fusion conditions with parameters $\Delta R$ and $\Delta t$ obtained above.

The first person who has predicted the existence of "microclusters" ($H^{+}+e^{-}+H^{+}$ quasimolecules) in the hydrogen-filled transition metal matrix was Gryzinski.

The idea that electrons are responsible for tunneling of protons through the Coulomb barrier was developed and advanced for the first time by Gryzinski in his research on chemical bonds in research from 1967 to 1987 [23–26]. Studying within the limits of the classical physics the bound states of the system consisting of two protons and an electron, Gryzinski discovered that there is a whole class of solutions when the attraction between protons and negatively charged electron predominates over pushing apart of protons, and two originally free nuclei can come nearer to each other over very small distances and organize a quasimolecule ($H^+ + e^- + H^+$). Gryzinski concludes that overcoming of the Coulomb barrier during atomic collisions by tunneling of particles is a process of three bodies, one of which is an electron. Addressing the phenomenon of cold fusion Gryzinski in 1989 in work [26] in "Nature" journal concludes: "... basically, two deuterons in the presence of an electron can merge at any temperature," and in 1990 he published his quasimolecular cold fusion model [27].

In 1990 Barut [28], using the Bohr–Zommerfeld quantum theory of metals, came to a similar conclusion about the existence in a three bodies system (two deuterons and one electron) of the superbonded metastable quasimolecule -$D_2^+$, which gained for quasimolecule energy bond a value of 50 keV. In 1992 Vigier [29] published an analysis very similar to Barut's analysis.

For various reasons, the cold fusion model proposed by Gryzinski [27] has not received recognition [11,12]. One of the main reasons for this is the fact that, as noted in review [30], in real cold fusion conditions in condensed matter, the existence in the proposed model of strict symmetry of the quasimolecule, and consequently, it's long existence – is an extremely unlikely event. The second reason is that in research specially carried out to search for a superbonded quasimolecule, the particle has not been confirmed [31]. An essential drawback in the model [27] is that the timely energy discharge mechanism of a fused nucleus is not offered [6].

In the *chemo*nuclear fusion hypothesis [11,12] the "Gryzinski quasimolecule" exists already only as a short-lived transition state of the interacting particles system. In work [6] the *chemo*nuclear fusion hypothesis gained further development. In this work the necessity of participation of a "Gryzinski quasimolecule" cluster in a successful cold fusion experiment is shown. We will demonstrate how the *chemo*nuclear fusion hypothesis explains the sensational Fleischman and Pons electrolysis experimental results [2].

The saturation of a palladium matrix by deuterium serves as the basic process of Fleischman and Pons experiments. Saturation of a transition metal matrix by deuterium in electrolysis experiments is followed by a growth in matrix of internal tensions and accumulating of a large number of "voids" in it with sizes ranging from macrocracks to subatomic size pores [11,12]. In the process of fusion, the voids in the matrix are filled by deuterium. It is important to note that in voids deuterium is in the form $D_2$ molecules. Irradiation of molecular deuterium by a flow of energetic conversion electrons is followed by generation in a deuterium of $D_2^+$ diatomic ions. According to Gryzinski, Refs. [23–27] representations developed in works, a certain part of diatomic ions of a deuterium will be transformed at the same time to a quasimolecule $D^+ + e^- + D^+$ ("Gryzinskis quasimolecule" [11,12]) in which an attraction between deuterons and a negatively charged electron prevail over pushing apart of deuterons. In these conditions, the two initially free nuclei may approach each other to very small distances, when the processes of anomalous internal $\gamma$-conversion is brought into action. As appears from the assessment given above, with a decrease of a quasimolecular cluster to a size $R$ = 4–4.9× $10^{-12}$ cm becomes possible an energy discharge of a fused $^4$He* nucleus by a virtual photon of anomalous internal $\gamma$-conversion from 23.8 MeV to 19.8 MeV (see Fig. 1). As a result, a $\Delta E = 4$ MeV energy portion on chain (11) is transferred to the matrix:

$$\begin{aligned} \mathrm{D}^+ + \mathrm{e}^- + \mathrm{D}^+ &\rightarrow {}^4_2\,\mathrm{He}^*(23.8\,\mathrm{MeV}) \\ &\rightarrow {}^4_2\mathrm{He}^*(19.8\,\mathrm{MeV}) + \mathrm{e}^-_{\mathrm{conv}}(E = 4\,\mathrm{MeV}) \\ &\rightarrow {}^4_2\mathrm{He}_{\mathrm{g.s}} + \mathrm{e}^-_{\mathrm{conv}}(E = 4\,\mathrm{MeV},\ \Sigma E = 4\,\mathrm{MeV}) + \mathrm{e}^-_{\mathrm{conv}}(E \le 9\,\mathrm{keV},\ \Sigma E = 19.8\,\mathrm{MeV}). \end{aligned} \quad (11)$$

where $E$-conversion electron energy, $\Sigma E$- total energy carried away by the conversion electrons. A quasimolecule of such size limits will be called a "Gryzinski microcluster."

As it has been noted earlier (see p. 5), the decrease in the excitation energy of $^4$He* below 19.8 MeV makes it impossible for the $^4$He* to decay with the yield of neutrons and charged particles. The time of 1.3–1.6× $10^{-23}$ s of the transfer of the energy portion $\Delta E = 4$ MeV to the matrix is comparable with the lifetime of the $^4He^*$ excited nuclei ($\sim 10^{-22}$ s [9]). This gives reason to assume that at least some $^4He^*$ excited nuclei involved in the cold fusion process have time to lose excitation energy to a level where decay is impossible. A further energy discharge of this part of fused $^4He^*$ nuclei will be realized with conversion electrons of the "standard" internal $\gamma$-conversion, $\Delta E \leq 9$ keV (see above)[c]. The qualitative picture of the process is represented by the chain (11). The fusion process by the chain is not connected with the nuclear bond rupture, it requires overcoming of a rather low Coulomb barrier (0.4 MeV), and therefore has an intense character.

From the foregoing, it follows that cold fusion intensity is directly connected with the concentration of voids in the matrix filled with deuterium. In Sections 4, 7 and 8 it will be shown that deuterium-hydrogen and $(^7Li)_{gas}$-hydrogen mixtures can also be used as working gases, while the working gas-filled nickel or other transition metals can serve as matrix. It is just in this sense that we further use the terms "working gas" and "matrix."

Previously, in our paper [6], we called the transition metal matrix containing a high concentration of cracks, pores or other voids filled with this gas, a "pseudo-composite." The mechanism considered above of cold fusion occurrence in this matrix is in good agreement with the well-known observation that: "The matrix, containing a large number of cracks, gaps and other voids is the only common condition in all successful cold fusion experiments medium" [32].

Along with the described process, another possible variant of the cluster mechanism of intensive cold fusion has been offered (see [11,12]), where "the Gryzinski quasimolecule," preaccelerated in the electric field is responsible for the fusion process. The matter is that, as shown in [11,12], formation in a matrix of cracks and gaps, and also spatial separation of electric charges by a flow of conversion electrons create in system in intensive cold fusion conditions of an inner electric fields capable to accelerate charged particles to energy $\sim 1$ keV.

Then part of "Gryzinski quasimolecules" formed in the conditions of intensive cold fusion can be exposed to a fusion [11,12] variant in a cluster mechanism. According to this mechanism, the "Gryzinski quasimolecule," getting in electric field, accelerated and built along a direction of movement. Colliding with heavy atom built quasimolecular cluster immerses into an electronic cloud of heavy atoms and forms, in turn, a quasimolecule (quasimolecular cluster-heavy atom) with it. In this case an energy discharge of a fused $^4He^*$ nucleus considered above occurs in extreme conditions – in the "Gryzinski microcluster," immersed in an electronic cloud formed by a cluster and a heavy atom of a quasimolecule, in the conditions of electronic supershielding in immediate proximity from a nucleus of a heavy atom (~0,1 Å [11,12]).

The matrix microvolume where there is an act of fusion, acquires on a short period ($\sim 10^{-7}$–$10^{-18}$ s) anomalously high local values of nucleus and electron density and extremely high level of fusion energy, unattainable for macrolaboratories. We can expect, that in these conditions intensity of such phenomena as Compton scattering of $\gamma$-radiation, weak interaction reactions (see also Sections 5, 7 and 8) and others, can be extremely change. Fusion energy by a standard and anomalous internal $\gamma$-conversion is transmitted to matrix through electrons and the cycle of fusion is repeated [6].

In both the considered variants, the cold fusion process requires for it implementation involvement of the phenomena of two – gas (the first stage) and intercluster – "condensed" (the second stage) mediums. Experimental conditions determine contribution to the fusion process of each described above processes.

From considered follows that the cold fusion is a complex process, which in the characteristic distance $\Delta R$ between the interacting particles can be conditionally divided into two stages. The first stage – the characteristic distance $\Delta R \geq 10^{-8}$ cm, occurs in the gas medium and "Gryzinski quazimolecule" formation ends. The second stage occurs into the quazimolecule at continuous convergence ("condensation") of interacting particles from $10^{-8}$ to $5\times 10^{-12}$ cm. The second stage ends by the merge of nuclei. In this sense, it should be understood the stages names, used here.

[c]In [11,12] the attention that measuring devices in experiences of cold fusion were always closed by protective films, impenetrable for electrons of such energies is paid. Such circumstance has led to impossibility of electrons of such energies registration.

Thus, main "secret" of high intensity cold 2D-fusion that the transition state of the interacting particles which is created in the course of cold fusion – "Gryzinski microcluster," provides a timely energy discharge of a fused $^4He^*$ nucleus through participation in process of a discharge of virtual photons and electrons of anomalous internal $\gamma$-conversion. The last opens reactions normally forbidden in the vacuum "light" channel 3 of Table 1, which is not interfaced with rupture of nuclear bonds. As it was noted reaction of fusion on this channel requires overcoming of rather low Coulomb barrier (0.4 MeV) and therefore has high intensity.

As follows from the considered, the important role in the *chemo*nuclear fusion belongs to conversion electrons. In addition to the function considered above of providing a timely discharge of the $^4He^*$ excited state, with direct participation of conversion electrons in the *chemon*uclear fusion hypothesis, there are also two more processes listed below.

(1) Conversion electrons ionize working gas, viz. molecular deuterium, and, thereby, for the account of energy of nuclear fusion carry out reproduction of ionic clusters. It is a primary factor defining efficiency of conversion electrons in the course of cold fusion. Efficiency of conversion electrons "using" is defined by thickness of an "active" "pseudo-composite" layer and depth of conversion electrons run. Obviously, the efficiency of conversion electrons will reach high rates depending on the diffusion saturation of an active layer by deuterium and the growth of concentration of voids in this layer. After reaching the limits of admissible concentration, the rate of increase of intensity of the process will be slowed down and reach saturation._
(2) Intensive cold fusion is a two-stage process. The melting of a matrix and the elimination of "voids" forbids the first stage and thereby prohibits the fusion process. That is why melting of a matrix should be considered as the "passive" protection method to prevent runaway reactions in cold fusion reactors.

From these considerations it follows that conversion electrons provide an effective feedback and high intensity of cold fusion process due to its participation [6,11,12].

## 4. Cold Fusion in 2D-, HD- and $(^7Li–^1H)_{gas}$-Transition Metal Systems

Consideration of the *chemo*nuclear 2D-fusion hypothesis causes us to anticipate that qualitatively similar processes of fusion will take place in other systems of light nuclei: HD- and $(^7Li–^1H)_{gas}$-transition metal. At the same time for each system the "Gryzinski quasimolecule" and "Gryzinski microcluster" will have its own element composition (see the left-hand side of reactions (12)–(14)).

In all these cases, cold fusion is caused by what in the system pass the "light," not interfaced with rupture of nuclear bonds, fusion reactions. It is testified by the inert gases ($^3He$, $^4He$) yield as reaction product and absence of essential radioactivity in the course of fusion. For the passing of such reactions overcoming only rather low Coulomb barrier, inherent to light nuclei, is required. It can be one of (or a few of) the reactions listed below:

$$D^+ + e^- + D^+ \rightarrow {}^4_2He + e^-_{conv} (\Sigma E = 23.8\,MeV), \tag{12}$$

$$D^+ + e^- + H^+ \rightarrow {}^3_2He + e^-_{conv} (\Sigma E = 5.5\,MeV), \tag{13}$$

$$({}^7_3Li^+ + e^- + H^+)_{gas} \rightarrow {}^8_4Be + e^-_{conv} (\Sigma E = 17.25\,MeV), \tag{14}$$

$${}^8_4Be\,(\tau \approx 10^{-15}\,s) \rightarrow {}^4_2He + {}^4_2He + E(0.095\,MeV). \tag{15}$$

In a vacuum, these reactions are accompanied by the emission of high-energy $\gamma$-radiation, and have extremely low intensity. The latter caused by the smallness of $\sigma_{\rm nuc}$, determined by $\gamma$-quantum radiation. Therefore, in a fuel cycle of a traditional thermonuclear reactor, the reaction on third neutronless 2D-fusion channel is not considered.

In matter, as noted in Section 3, a fusion reaction occurs inside the "Gryzinski microcluster" and removal of nucleus excitation energy is carried out by a virtual photon of "anomalous" internal $\gamma$-conversion. In such a case, excitation energy of a fused nucleus on a chain: the excited nucleus–a virtual photon–electron anomalous internal $\gamma$-conversion-matrix transfers to matrix. The interdictions connected with $\gamma$-quantum radiation are eliminated, and the process of fusion on reactions (12)–(14) becomes intense. This is the main reason for the high intensity of the fusion processes in the "Rossi reactor" and in the experimental source of *chemo*nuclear HD-fusion created in the present work. The *chemo*nuclear fusion hypothesis well explains the fuel cycle and nuclear processes accompanying "Rossi reactor" operation and our *chemo*nuclear fusion power source operation (see Sections 7 and 8).

## 5. Weak Interaction Reactions Amplification and Bineutrons Generation under the Intensive Cold Fusion Conditions

The above-described extreme conditions, under which the reactions of *chemo*nuclear fusion occur, create preconditions for a dramatic amplification of weak interaction reactions due to intensive 2D-, HD- and $(^7\text{Li}\text{–}^1\text{H})_{\text{gas}}$-fusion reactions. Direct evidence of dramatic amplification of weak interaction reactions in the conditions of intense cold fusion is observed in the expanded deuterium reproduction in "co-deposition" experiments [33]. In these experiments deuterium concentration in the light-water electrolyte increased up to 70–80% during the experiment. The high generation level of heat in the HTE-Cat reactor evidences the high rate of deuterium reproduction in active zone, since the deuterium reproduction – a link in the chain of reactor fuel cycle processes (see Section 8)

Another example of the weak interaction reaction amplification under conditions of intense cold fusion is the existence in these conditions of a bineutron bound state. In works [11,12] the generation of a bineutron in these conditions is shown in the example of fusion reaction in a $\text{D}_3^+$-cluster. If in the process of *chemo*nuclear fusion one of three deuterons, which have got to the collision zone of a $\text{D}_3^+$-cluster with a heavy superficial atom, owing to any reasons, do not participate in process of a merge of deuterons, that deuteron having appeared in extreme conditions merging with two other cluster deuterons zone, such deuteron can be involved in a reaction responsible for a birth of a bineutron with the Pokropivny mechanism [34]:

$$\text{D} + \text{e}^-_{\text{conv}} \to {}^2\text{n} + \nu_\text{e}, \tag{16}$$

where $^2$n is a bineutron.

As noted in Zeldovich and Novikov's book [35], "... excess-neutron nuclei are stabilized by the presence of electrons." More certain, this is indicated by Pokropivny [34]: "... a necessary condition for obtaining a quasi-stable bineutron on an endothermic $e$-capture reaction is a high energy and density of electrons ..." In cold fusion conditions (the second variant of fusion, see Section 3.2) a bineutron is born in the vicinity ($\sim$ 0,1 Å [11,12]) of the nucleus of a heavy atom of a matrix, and from birth until its absorption by the nucleus of a heavy atom may not go beyond the quasimolecule electronic cloud. The high density of an electronic cloud contributes to the existence of a bineutron bound state, which closely agrees with the results of many cold fusion experiments (see [6,11–13] and the present work).

## 6. Experimental Confirmation of Chemonuclear Fusion Hypothesis

### 6.1. Chemonuclear fusion hypothesis confirmation in cold fusion experiments

As specified in Section 3.2, as the most important direct experimental confirmation of the *chemo*nuclear fusion hypothesis, the well-established observation made by many researchers can serve: "The matrix, containing a large number of cracks, gaps and other voids is the only common condition in all successful cold fusion experiment medium" [32]. Such description in [32] of the medium responsible for cold fusion, matches the description of a "pseudo-composite" matrix in the *chemo*nuclear fusion hypothesis (see Section 3.2). In [6,11–13] and in this paper, we consider the numerous other experimental results that can be interpreted as evidence of the validity of the *chemo*nuclear fusion hypothesis.

These results are the reproduction of deuterium, nuclei fission and anomalous isotopic composition impurity generation, energetic particles generation (neutrons, protons, tritons and $\alpha$-particles), the low radioactivity of cold fusion processes, etc. Some of these results are considered below.

### 6.2. Bineutronic mechanism of energetic $\alpha$-particles, protons and neutron generation

Emission of energetic (10–14 MeV) $\alpha$-particles is one of the well-established and yet not explained phenomena of cold fusion [36,37]. Other phenomena that have not been explained include a proton yield with energies in the range of 5.8–7.8 MeV [36] and a neutron yield with energies ranging from 3 to 10 MeV in [37,38]. Let us show that the emission of such particles may be due to the occurrence of a reaction involving a bineutron in these experiments.

At absorption of a bineutron, the nucleus of heavy atom is excited to a level when excitation energy of the nucleus $E_{\mathrm{exc}}$ (excitation energy) essentially exceeds the energy of nucleon separating – $E_{\mathrm{nucl\ sep}}$ (nucleon separation). In that case the removal of excitation energy of the nucleus, if it has not had time to be divided [11,12], will occur due to an emission of nucleons n, p or their clusters d, T, $^3$He, $^4$He. $\gamma$-Quanta radiation by the nucleus in an excitation high-energy range ($E_{\mathrm{exc}} > E_{\mathrm{nucl\ sep}}$) occurs with a probability, essentially lower, than the probability of nucleons emission, because the constant of an electromagnetic interaction is two orders below a strong interaction constant (see Table 2).

In Table 2 a process of a nucleus discharge by emission of fast particles by reactions (5b)–(5d) is shown.

In Table 3, compiled by us from reactions (5b)–(5d) (Table 2), the calculated values of energies of these particles with ones observed in experiments are compared. From Table 3, it follows that all calculated values of energies are in close agreement with the experiment.

### 6.3. Anomalous isotopic composition impurities generation

The process of fast particle generation in Section 6.2 and the fission process [11,12] is accompanied by an accumulation process of the anomalous isotopic composition of impurities in the matrix, which is also observed in many works on cold fusion [32].

It is especially necessary to pay attention to the extreme changes of nickel isotopic composition (see Table 4 and Section 8) found in the Lugano experiment [14] and in our verification experiment [13]. Concentration decrease of all nickel isotopes, except the $^{62}$Ni isotope, where there is a concentration raise is easily explained by the assumption that

**Table 3**. High-energy $\alpha$-particles, protons and neutrons in the *chemo*nuclear fusion hypothesis of deuterium [6,11,12].

| Palladium isotopes | 102 | 104 | 105 | 106 | 107 | 108 | 110 |
|---|---|---|---|---|---|---|---|
| $E_\alpha$ (Rated) (MeV) | 14.44 | 11.63 | 12.05 | 11.46 | 12.0 | 10.14 | 8.76 |
| $E_\alpha$ (Experimental) (MeV) [36,37] | 14.0–9.2 | | | | | | |
| $E_{\mathrm{p}}$ (Rated) (MeV) | 8.86 | 7.2 | 6.74 | 5.26 | 5.45 | – | – |
| $E_{\mathrm{p}}$ (Experimental) (MeV) [36,37] | 7.8 | – | – | 5.6 | – | – | – |
| $E_{\mathrm{n}}$ (Rated) (MeV) | 7.54 | 8.84 | 9.48 | 6.47 | 9.47 | 6.09 | 5.7 |
| $E_{\mathrm{n}}$ (Experimental) (MeV) [37,38] | 10–3 | | | | | | |

**Table 4.** Nickel isotopic structure in nature, in "ashes" of HTE-Cat reactor [6] and in present work.

| No. | Isotope | Natural composition (%) | Isotope composition in "ashes," (%) [6] | Composition on a cathode surface (the present work) (%) | Proton energy in reaction $^x$Ni $+^2$n $\to$ $^{x+1}$Co $+$p $(E)$) (MeV) | Columbs barrier $E_c$ (MeV) | $(E_p - E_c)$ (MeV) |
|---|---|---|---|---|---|---|---|
| 1 | $^{58}$Ni | 67.76 | 0.8 | 66.52 | 10.7 | $\sim 7 \pm 1$ | +3.7 |
| 2 | $^{60}$Ni | 26.16 | 0.5 | 26.72 | 10.24 | $\sim 7 \pm 1$ | +3.2 |
| 3 | $^{61}$Ni | 1.25 | 0 | 1.24 | 6.06 | $\sim 7 \pm 1$ | -1 |
| 4 | $^{62}$Ni | 3.66 | 98.7 | 4.42 | 3.9 | $\sim 7 \pm 1$ | -3.1 |
| 5 | $^{64}$Ni | 1.16 | 0 | 1.1 | – | – | – |

the observed changes of the isotopic composition is the result of bineutron absorption by the nickel nucleus in reaction (17) (see [11,12], Table 11):

$$ {}^{x}_{28}\mathrm{Ni} + {}^{2}\mathrm{n} \to {}^{x+2}_{28}\mathrm{Ni}^* \to {}^{x+1}_{27}\mathrm{Co} + \mathrm{p}(E). \tag{17} $$

The reaction of bineutron absorption by $^{62}$Ni nucleus cannot be accompanied by the yield of a proton (Coulomb barrier interdiction, proton energy is lower than a Coulomb barrier, see Table 4) and, therefore, there is found in [14] and in our verification experiment [13] the accumulation of the $^{62}$Ni isotope takes place.

### 6.4. Tritium generation

Tritium generation is one of the first reliably established and yet not explained cold fusion phenomena [32,39,40]. The intensity of this process is 5–9 orders above intensity of a neutrons yield. This indicates that the tritium yield in these experiments is not related directly with occurrence of traditional hot fusion.

The results of our experiments (Table 5) on research of tritium generation in the impulse electric discharge process in heavy water electrolyte and an electrolyte representing a solution 0.2 M $K_2CO_3$ [41] in normal water are presented.

Two features of the results of these experiments should be noted: very high level of accumulation of tritium in heavy water electrolyte, and tritium accumulation in an electrolyte of normal water (although at essential lower rates). In both experiments the accumulation level of tritium far exceeds a possible error that leaves no doubts in the reality of tritium generation process in these experiments.

As shown in Table 2, in *chemo*nuclear fusion hypothesis of deuterium generation of tritium is a result of deuteron reactions with a bineutron on the channel (18) this table:

$$ \begin{aligned} \mathrm{d} + {}^{2}\mathrm{n} &\to \mathrm{T}(1.56\,\mathrm{MeV}) + \mathrm{n}(4.64\,\mathrm{MeV}) \to \mathrm{d} + \mathrm{T}(1.56\,\mathrm{MeV}) \\ &\to \alpha(6.9 - 1.44\,\mathrm{MeV}) + \mathrm{n}(12.5 - 17.7\,\mathrm{MeV}). \end{aligned} \tag{18} $$

**Table 5.** Tritium generation from passing an impulse electric discharge in electrolytes with deuteroxide and water of natural composition [41]

| Experiment | Electrolyte composition | Material of the Cathode–anode | Content of tritium (Bk/kg*) | Background excess |
|---|---|---|---|---|
| 1 | $D_2O$+1M Li | Nb–W | 1340 | 450 times |
| 2 | 0.2 M $K_2CO_3$+$H_2O$ | Ta–W | 160 | 53 times |

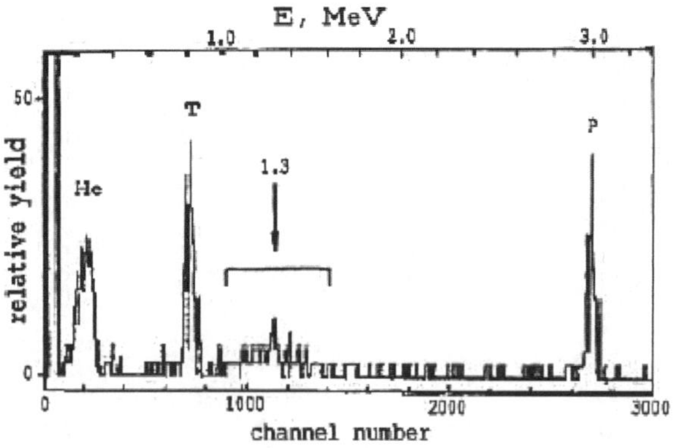


**Figure 2.** The energy distribution of charged particles from a sample Ti irradiated with ions $D_n^+$ ($E = 25$ keV) to a dose of $1\times 10^{19}cm^{-2}$. The peak in the middle of the spectrum belongs to the neutrons having energy ~1.0–1.6 MeV [43].

Figure 2 shows the energy spectrum of the particles emitted by a titanium target during the cryogenic implantation process of deuterons (with energy 25 keV) in our experiments in 1989 [42] and 1998 [43]. The detector in this experiment was covered by a nickel film with 0.57 $\mu$m thickness. The analysis of spectrums (Fig. 2) with the count of energy losses and the conforming spectrum peak shift (caused by a proper traditional 2D-fusion of energetic $\alpha$-particles, tritons and protons output), has allowed us to determine that in spectrum Fig. 2 making along with process of traditional hot fusion of deuterium the processes responsible for an of tritons generation with energy 1.5–1.6 MeV (peak $\sim$ 1.3 MeV on a spectrum, Fig. 2) participates.

As shown in Table 2, a source of tritons with such energy is reaction (18), initiated by a bineutron. This is additional convincing evidence of the reliability the *chemo*nuclear fusion hypothesis of deuterium, which is described in the present work.

### 6.5. Three $\alpha$-particles yield

Let us focus on one phenomena well established in cold fusion experiments – the three $\alpha$-particles yield from a very small volume of the detector Fig. 3. In [44–46] this phenomenon is interpreted as evidence of a fission reaction passing of carbon nucleus under the influence of a fast neutron:

$$C + n \rightarrow 3\alpha + n + Q.$$

At emission of $\alpha$-particles, made in work [46] energy estimations of an initiating (26) reaction neutron have gained values $E_n \approx 13.25$–$13.47$ MeV. The generally accepted generation mechanism of such neutrons does not exist today.

Considered as a source of neutrons the secondary DT-reaction of hot fusion [46] demands for the realization of passing of primary 2D-reactions at the level never watched in experiments on cold fusion. The attention was paid to this circumstance in [43] (1998), and also in work of [48] (2000). As appears from Table 2, in *chemo*nuclear fusion conditions responsible for fission of carbon nucleus are neutrons initiated by bineutrons in the chain of reactions (18). As it was already noted, tritons having energy 1.5–1.6 MeV, generated in this chain of reactions are well watched on a spectrum (see Fig. 2) that is the additional testimony in favors for this statement.

### 6.6. Low radioactivity of cold fusion processes

The low radioactivity of the cold fusion processes is one of the best known and attractive features of this phenomena. In the *chemo*nuclear fusion hypothesis low radioactivity is explained by a series of reasons. Let us list the main ones.

(1) The reactions responsible for the fusion process (12)–(13) are not due to a rupture of nuclear bonds.
(2) The energy generated in the course of nuclear fusion is carried away by conversion electrons. This eliminates a second potential source of $\gamma$-radiations in cold fusion experiments.
(3) The process of nucleus fission – the third potential source of a radioactivity in the cold fusion conditions, occurs with intensity 8–10 orders lower in comparison with the first group phenomena, and as it was noted in [11,12], is not accompanied by formation of radioactive products, $\gamma$-quanta emission and energetic heavy particles.
(4) An energy discharge of an excited nucleus occurs due to the emission of nucleons and their clusters [6,11,12].

Thus, *chemo*nuclear fusion hypothesis adequately explains reasons where there is low radioactivity in cold fusion experiments.

To conclude this section, we note that Refs. [6,11,12] demonstrate the close accordance expected of Table 2 reaction results with the well-established results from cold fusion experiments, and can be considered experimental confirmation of the participation of weak interaction reactions and bineutron reactions in cold fusion processes, and the *chemo*nuclear hypothesis as a qualitative model of cold fusion process in the 2D-, HD- and $(^{7}Li–^{1}H)_{gas}$-transition metal systems.

Of course, any of such consideration without a serious study of anomalous $\gamma$-conversion mechanism in relation to such "exotic" object, as "Gryzinski quazymolecule," and no comprehensive study of the other features of the fusion model, can only have the hypothetical character. However, as shown in [6,11,12] a compliance with the conclusions of

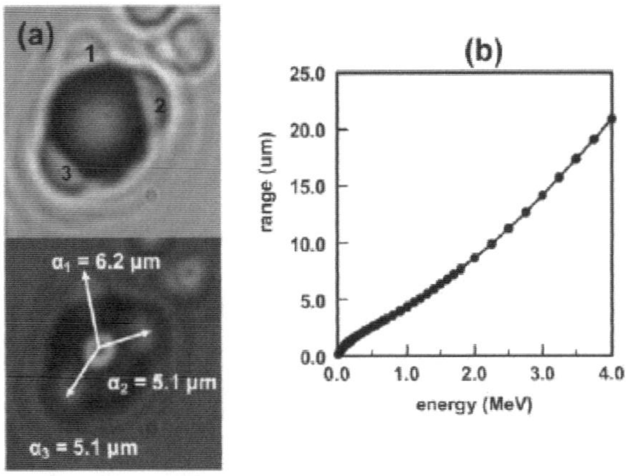


**Figure 3.** Photos of the triple track [46]: (a) Photo of the triple track shown. The top image was obtained with the optics focused on the surface of the CR-39 detector. The bottom image is an overlay of two photos taken at different focusing depths (surface of the detector and the bottom of the pits). On the bottom image, the bright circle observed in the track indicates where the neutron impacted the carbon atom. Arrows are drawn to show the directions the alpha particles traveled after the carbon atom shattered. The distances traveled by the alpha particles are indicated. (b) Linear energy transfer (LET) curve for alpha particles in CR-39 calculated using the SRIM-2003.26 code of Ziegler and Biersack [47].

**Table 6.** The radioactivity composition of lithium in nature and reactor HTE-Cat "ash".

| Isotope | Natural composition (%) | The radioactivity composition in leach, method SIMS (%) | An isotopic composition in "ash," method JCP-MS (%) |
|---|---|---|---|
| $^6$Li | 7 | 92.1 | 57.5 |
| $^7$Li | 93 | 7.9 | 42.5 |

the review with well-established numerous experimental results is, in our opinion, a serious reason for broad deployment of operations on the practical use of cold fusion and for a further deep study of this phenomenon.

## 7. Pilot Chemonuclear Fusion Energy Generator Development and Testing (Verification Experiment)

The model of intensive (self-sustained) cold 2D-fusion developed in works [11,12] – the *chemo*nuclear fusion hypothesis – explains this phenomenon at a qualitative level. Since the publication of Ref. [11] four years ago, the model has received an experimental verification in [14] and in our papers of 2013–2015. The model modernized to account for the experimental verification results is shown in [6] and here. In work [13] results of our 2013–2015 work on the creation and testing of a *chemo*nuclear fusion energy source (verification experiment) are presented. In this section, the main results of this work are shown.

The source of *chemo*nuclear energy is created on the basis of gas-discharge installation (KD-installation) [41], equipped with a tool-kits set for studying of the physical phenomena accompanying cold fusion.

The verification experiment details include:

(1) Studied system – HD-nickel system.
(2) Working gas – a mix of $D_2$ (85%) and $H_2$ (15%).
(3) Heat generating materials – "pseudo-composite" – the nickel containing high concentration of filled by gas voids: micro cracks, ruptures, pores, etc. (see [6,13] and Section 3 of the present work). "The pseudo-composite" layer is created on a surface of nickel electrodes by special processing of electrodes in the gas discharge in the energy source chamber [6,13].

In verification experiment it has been established:

(1) Thermal power of a source ∼35 W, efficiency – 1.8–1.9.
(2) Responsible for process of cold fusion in HD-nickel system is an intercluster reaction (13).

This confirms:

(1) generation of rare $^3$He gas atoms in the experiment,
(2) heat generation is not accompanied by emission of $\gamma$-radiation,
(3) emission of conversion electrons.

A sharp drop in resistance of an order of magnitude (from 8–7 to 2 k$\Omega$) confirms the last claim when the energy source reaches a heat generation mode. This effect is well reproduced.

(1) Work by an energy source in a mode of intensive cold fusion is accompanied by change of isotopic structure of nickel in a superficial "pseudo-composite" layer of electrodes. In Section 6.2, it is shown that changes of nickel isotopes concentrations are a result of bineutron absorption of nickel nucleus on reaction (17).

Based on the bineutron absorption dramatically change the nickel isotopic composition in the course of experiments in Lugano [14], on our verification experiment [13], and on the emission of fast particles in cold fusion experiments

[11,12] – the *chemo*nuclear fusion hypothesis is justified both for the cold fusion "HTE-Cat" reactor and for our pilot *chemo*nuclear energy source.

Neither in the course of experiment, nor in the subsequent "flashing" was any $\gamma$-radiation above background revealed. This absence of the induced activity in experiments with nickel can be explained by a deep reorganization of a nickel nucleus in the conditions of intensive cold fusion. This question demands the further study (see also Sections 3, 5 and 8).

## 8. Cold Fusion HTE-Cat Reactors Creation – A New Stage in Nuclear Power Engineering Development

### 8.1. HTE-Cat reactor development

Today the history of the evolution of a new scientific direction – cold fusion, transfers in a responsible phase. The process of practical exploration of this phenomenon begins.

As it was noted, the beginning of exploration of cold fusion it is possible to consider development of the first thermal reactor of cold fusion E-Cat with a power of 1 MW [3]. Presentation of this reactor was made by Focardi and Rossi in March 2011 in Italy. In subsequent in the USA, where Rossi has transferred the research, on the same principle a more effective generator HTE-Cat with operating temperature 1000–1200°C has been created that gives a possibility of its usage in superheated steam production (600°C) for turbines of a modern power plant.

Rossi in an interview with "The Huffington Post" in October, 2015, declared that Leonardo Technologies Inc. already has orders to manufacture of 600,000 generators of a different power levels for the total of three billion US-dollars, from Europe, the USA and other parts of the world. This year he planned to begin commercial production of the generators (including major units of 1 MW capacity).

Unfortunately, very little data about the construction of these reactors and the composition of the fuel has been published. As declared in Ref. [4], the fuel in the HTE-Cat reactor is a mixture of powders of lithium aluminum hydride ($LiAlH_4$) and a nickel powder. When heated, lithium aluminum hydride undergoes three stages of decay:

$$3LiAlH_4 \rightarrow Li_3AlH_6 + 2Al + 3H_2 \quad \text{(first stage)}, \tag{19}$$

$$2Li_3AlH_6 \rightarrow 6LiH + 2Al + 3H_2 \quad \text{(second stage)}, \tag{20}$$

$$2LiH + 2Al \rightarrow 2LiAl + H_2 \quad \text{(third stage)}. \tag{21}$$

In work [4] it is noted that the third stage of decay becomes reversible and that LiH pressure equilibrium at 500°C has a value of 0.25 bar.

According to developers of HTE-Cat reactor, nickel in composite fuel plays role of the catalytic agent and in the reactor operation process is not spent. For initiation of reactor operation the fuel is necessary to heat up to temperature when lithium and hydrogen intensively diffuse into nickel. In crystal lattice of nickel heated to 600–1200°C the lithium and hydrogen nuclei merge at collision ("Rossi effect"), forming two $^4$He nuclei on a reaction

$$^7_3\mathrm{Li} + ^1_1\mathrm{H} \rightarrow ^8_4\mathrm{Be}^* \rightarrow ^4_2\mathrm{He} + ^4_2\mathrm{He} + 17.35\,\mathrm{eV}. \tag{22}$$

The authors of work [4] suppose that in this reaction the energy of fusion is released in the form of a kinetic energy of "hot" helium nuclei, which are thermalized in a crystal lattice of nickel and the material of the walls of the reactor. Such a result, according to the authors, makes the "Rossi effect" an ideal phenomenon for nuclear energy utilization with a total absence of radioactive materials and radiation.

### 8.2. Experiment in Lugano

In October, 2014, results of studying the HTE-Cat [14] reactor operation which was carried out in Lugano (Switzerland) by a group of independent experts was published. According to this group, reactor fuel – a mixture of nickel powders and hydrogen containing $LiAlH_4$ solution – in the amount of 1 g was placed in the reactor before its start. Reactor start was carried out by heat of an active core of the reactor by a resistive heater. The total of the excess energy produced over 32 days in the reactor was $\sim$1.5 MW h. Experts note that this amount of energy is much more than can be produced by the small volume of fuel in the reactor from any of known chemical energy sources.

The analysis of the used fuel ("ash") has shown that except nickel there are lithium and aluminum impurities with amounts, allowing to conclude that it was in the nickel powder before it was filled in the reactor with $LiAlH_4$ powder added. In the "ash" iron, oxygen and carbon are also found. Deuterium presence in "ash" is not revealed. In work [14] it is revealed that in the course of experiment significant modifications of the isotopic composition of nickel (Table 4) and lithium (Table 6) take place.

"These results are difficult to correspond with the fact that radioactivity at experiment carrying out has not been discovered," authors of expertise [14] state that the process of fast nucleus $^4He^*$ on reaction (22) thermalization declared by Rossi should be accompanied by radiation which could not be missed in their experiments.

Thus, the results gained by a group of experts in Lugano do not confirm the model of the HTE-Cat reactors declared by the developer.

As was already noted (see Section 3.2), the experiment in Lugano [14] (work of Rossi "cell") has already been successfully reproduced in a number of the countries [10].

### 8.3. HTE-Cat reactor fuel cycle in chemonuclear fusion hypothesis

The "Rossi reactor" operation is successfully explained by the *chemo*nuclear fusion hypothesis.

As noted above, developers of the HTE-Cat reactor believe the reaction is fusion of lithium and hydrogen nuclei in a fuel cycle of the reactor as shown on reaction (22) and in which the reaction with nickel is a catalyzer and is not consumed in the operation process. However, this was not confirmed in [14] as explianed in Section 8.2.

At the same time lithium with hydrogen reaction (23):

$$^7_3\mathrm{Li} + ^1_1\mathrm{H} \rightarrow ^8_4\mathrm{Be}^* \rightarrow ^8_4\mathrm{Be} + \gamma(17.25\,\mathrm{eV}), \tag{23}$$

Has a strongly pronounced resonance character and reaches a maximum level at energy of 440 keV of a proton [49]. In vacuum this reaction is accompanied by high-energy ($E$ = 17.25 MeV) $\gamma$-quanta emission and has low intensity caused by the small scale of $\sigma_{\mathrm{nuc}}$, defined by $\gamma$-quanta radiation. However, as it is shown many times (see Section 3.2), the intercluster character of cold fusion reaction provides a timely removal of energy of the excited fused nucleus (in this case this nucleus of beryllium $^8Be^*$) due to a virtual photon and electron of anomalous $\gamma$-conversion participation. In such a case, the fusion process in "Rossi reactor" becomes intense.

Immediately after producing a reaction (14) ($\tau \approx 10^{-15}$ s) the $^8Be$ nucleus decays on two low-energy $\alpha$-particles (reaction (15)). Decay from a ground state is possible [49] as the mass of $^8Be$ at 0.095 MeV exceeds the mass of two $\alpha$-particles. This explains the observations in the Lugano experiments [14] of the burning of $^7Li$ isotope in the fuel, and the absence of $\gamma$-radiation in this experiment.

Thus, in the *chemo*nuclear fusion hypothesis the fuel cycle of "Rossi reactor" is accompanied by a following set of nuclear reactions:

(a) The fusion processes are not related to a break-up of nuclear bonds considerably enhanced with medium *chemo*factors intercluster reaction (14) and, as a result, after a very short delay there is reaction (15) (see Section 4).

The lower "gas" index says that fuel has to be heated to temperature when the lithium component of fuel (LiH or Li) is in a gaseous state.

$$\begin{aligned}({}^7_3\mathrm{Li}^+ + \mathrm{e}^- + \mathrm{H}^+ + \mathrm{e}^- + \mathrm{H}^+ + \mathrm{e}^- + \mathrm{H}^+)_{\mathrm{gas}} &\to \left[{}^8_4\mathrm{Be}^* + (\mathrm{H}^+ + \mathrm{e}^- + \mathrm{H}^+)\right] \\ &\to \left[{}^8_4\mathrm{Be} + \mathrm{e}^-_{\mathrm{conv}}(17.25\,\mathrm{eV}) + (\mathrm{H}^+ + \mathrm{e}^- + \mathrm{H}^+)\right] \\ &\to {}^4_2\mathrm{He} + {}^4_2\mathrm{He} + \mathrm{D} + \nu_\mathrm{e} + \mathrm{e}^-. \end{aligned} \tag{24}$$

(b) Reactions of weak interaction. First of all, these the intercluster [11,12] reactions of a deuteron generation which are dramatically strengthened in the conditions of intensive cold fusion:

and a bineutron [34]:

$$\mathrm{D} + \mathrm{e}^-_{\mathrm{conv}} \to {}^2\mathrm{n} + \nu_\mathrm{e}. \tag{25}$$

(c) Reactions of bineutron interaction with a nickel, working gas and others elements of reactor active zone nuclei.

Reaction (17) of bineutron absorption by a nickel nucleus (Sections 6.3 and 7) which is already considered above is important for a fuel cycle of "Rossi reactor."

(d) Secondary reactions.

Interactions reactions of fast protons from reaction (17) with elements of reactor active zone nuclei. First of all it is reaction (26):

$${}^{\mathrm{x}}_{28}\mathrm{Ni} + {}^1\mathrm{p} \to {}^{\mathrm{x}+1}_{29}\mathrm{Cu} \tag{26}$$

and other element of active zone of reactor nuclei.

Generating on reaction (26) copper nuclei with the exception of stable $^{63}$Cu and $^{65}$Cu isotopes decays by one of channels:

$${}^{x+1}_{29}\mathrm{Cu} \to {}^{x+1}_{28}\mathrm{Ni} + \mathrm{e}^+ + \nu_\mathrm{e} \quad \text{(positron decay)}, \tag{27}$$

$${}^{x+1}_{29}\mathrm{Cu} + \mathrm{e}^- \to {}^{x+1}_{28}\mathrm{Ni} + \nu_\mathrm{e} \quad \text{(electron capture)}. \tag{28}$$

Nuclear reactions "products" in "Rossi reactor":

- On Item (a) $^4$He, conversion electrons and excessive energy will be observed.
- "Products" of interaction of a bineutron with a matrix and working gas must follow reactions according to Items (b) and (c): fast nucleons yield of p, n and their clusters of d, T, $^3$He, $^4$He, generations of the anomalous isotopic composition impurities, $\gamma$-radiation of radioactive impurity nuclides and others. We can expect that in verification experiment first of all will be the following: (1) an anomalous isotopic composition of lithium ($^7$Li burning) and nickel (the concentration of $^{62}$Ni growth and remaining isotopes burning out), (2) $\gamma$-rays of 0.067 MeV line $T_{1/2} = 1.65$ h [6,13]. Reactions according to Item (b) (reaction (24)) can be registered on growth of the deuterium content in a work medium.
- Reactions product yield on Item (d) (secondary reactions) studied a little now. However, as it was noted in [51], in long-term experiment in a cell of "Rossi reactor" accumulation of copper atoms with anomalous isotopic composition is observed. In work [14] in similar experiment revealed a presence of iron, oxygen and carbon. These questions demand further studying.

All listed above the expected accordingly the *chemo*nuclear fusion hypothesis results, except $\gamma$-activity, observed at "Rossi reactor" cell operation.

Thus, the *chemo*nuclear fusion hypothesis well describes a fuel cycle and nuclear processes of "Rossi reactor" provided that as the nucleus initiating fusion process in this case the $^7$Li-nucleus serves. The reaction of $^7$Li with hydrogen is responsible for the process of fusion. The expected outcomes of the reactions such as the excess heat, the variations in the isotopic composition of lithium and nickel, have been reliably confirmed by the Lugano experiment [14]. As well as in our verification experiment, at studying operation of "Rossi reactor" radioactivity had not been found. The absence of $\gamma$-radiation in the nickel experiments may be caused in this case by nuclear transformation. As it was noted above, the problem of physical phenomena taking place in the matrix microvolume, where the event of nuclear fusion takes place, demands further studies (see also Sections 3,5 and 7). The presence of deuterium in the spent fuel, which may be due to its high temperature, has not been observed, too. The experimental data on the deuterium content in the working gas are unavailable.

## 9. Conclusion

In recent years two ways of mastering nuclear fusion energy have been defined.

(1) Historically the first way of solving this problem was the heating of medium comprising atoms of deuterium (or deuterium and tritium) up to the temperature when the kinetic energy of the atoms at their impacts would be sufficient to cause the mutual approach of nuclei necessary for realization of fusion reactions ("thermonuclear fusion") [1]. The scientific base of the controlled nuclear fusion reactor has been studied reasonably well. The working matter in this reactor is a high-temperature deuterium–tritium or deuterium plasma. In the first case, the DT-fusion reactions are responsible for reactor operation, and in the second case, these are the 2D-fusion reactions on channels 1 and 2 (see Table 1). In the controlled thermonuclear reactor, the occurrence of reactions in the "light" channel (3), which exhibits no nuclear bond rupture (Table 1), is fundamentally prohibited (see Section 2) since in vacuum there are no conditions for the synthesized $^4$He* nucleus to discharge energy.

Fusion reactions in the thermonuclear reactor occur at collisions of unscreened (bare) nuclei having randomly oriented spins. Low nuclear reaction cross-sections in these conditions cause the main difficulties on the way of creating the controlled thermonuclear reactor. The deuterium–tritium mix temperature in this reactor equals $\sim 2\times10^8$ K ($\sim$10 keV), for the deuterium plasma this temperature is still higher. The problem of maintaining this temperature for the time sufficient to gain positive energy yield presents great technical difficulties, which have been attacked for many decades by numerous teams of scientists from different countries [1].

(2) The second way of fusion energy mastering is the development of cold fusion by Fleischman and Pons, who reported in 1989 the sensational results of their investigations into this phenomenon [2]. Unlike the technical difficulties encountered by the developers of the controlled thermonuclear fusion reactor, the practical development of the cold fusion reactor, in particular, the creation of the $(^7\text{Li-}^1\text{H})_{\text{gas}}$-nickel system was a much easier task and is far ahead in the formulation of theoretical foundations of the phenomenon. The first cold fusion reactor ("E-Cat reactor") based on these reactions was presented by Focardi and Rossi in March 2011, in Italy. Since then, the "Rossi reactor" "cell" operation has been successfully reproduced in a number of countries (Russia, China, Kazakhstan, etc.) [10]. However, there is no generally accepted model of this phenomenon today.

Our first paper claiming to explain intensive (self-sustaining) cold fusion was published in 2012 [11]. During the four years since publication, the fusion model presented in [11], – the *chemo*nuclear fusion hypothesis, has gained further development in the present work and in [6]. The model was checked and confirmed by the results of our verification experiment [13]. The *chemo*nuclear fusion hypothesis is in good agreement with the results of comprehensive study of the HTE-Cat reactor [14], made during that period in Lugano (Switzerland) by a group of independent experts. So, it gives us reason to now treat the *chemo*nuclear fusion hypothesis as a self-consistent, qualitative model of intensive cold fusion in 2D-, HD- and $(^7Li–^1H)_{gas}$-transition metal systems [6].

In the *chemo*nuclear fusion hypothesis, cold fusion, being a complex, two-stage cluster process, is determined by nuclear reactions not involving nuclear bond rupture, viz., the 2D-reactions on the neutron-free "light" channel of 2D fusion, and also, HD- and $(^7Li-^1H)_{gas}$-reactions (see Section 4). These reactions are accompanied by high-energy gamma emission and are of extremely low intensity. This is due to the smallness of $\sigma_{nuc}$ governed by $\gamma$-quantum radiation. In cold fusion conditions, the cluster nature of fusion provides an anomalous $\gamma$-conversion function. As a result, the excitation energy of the synthesized nucleus is transferred to the matrix by the following chain: excited nucleus–virtual $\gamma$-quantum–anomalous $\gamma$-conversion electron–matrix. This eliminates the prohibitions associated with the $\gamma$-quantum radiation, and the process of intense cold fusion starts. It is one of the most important causes of cold fusion occurrence in media.

The extremely high intensity of cold fusion is determined by the joint action of the following factors ("*chemo*factors").

(a) The transition state formed during fusion in the interacting particle system – the "Gryzinski quasimolecule" – provides a timely energy discharge of the synthesized nucleus due to participation of virtual photons and electrons of anomalous internal $\gamma$-conversion in the discharge process. This removes restrictions specified by $\gamma$-quantum radiation. When this occurs, the fusion process is intense in character, as it proceeds without rupture of nuclear bonds, and it requires overcoming a low Coulomb barrier peculiar to light nuclei (see Sections 3 and 4).

(b) According to the second variant of the *chemo*nuclear fusion hypothesis (see Section 4), the processes of nuclear fusion and synthesized nucleus formation as parts of the synthesis event take place under extreme conditions, namely, in the "Gryzinski microcluster" immersed in the electron cloud of quasimolecule (Gryzinski microcluster–heavy matrix atom), under electronic superscreening conditions in the immediate vicinity of the heavy atomic nucleus ($\sim$0.1 Å), in the matrix microvolume, where the density of nuclei and electrons as well as the released energy of nuclear fusion reach for a short time ($10^{-7}$–$10^{-18}$ s) anomalously high values, unattainable in most laboratories.

(c) In the extreme conditions of intense cold fusion, the cluster nature of fusion promotes the dramatic enhancement of weak interaction reactions owing to nuclear fusion reactions that result in the generation of deuterons and bineutrons (see Section 5).

(d) A "pseudo-composite" is assumed to be an optimum matrix for the initiation of intense cold fusion. The transition metal matrix has a high concentration of cracks, pores and other voids filled with a working gas. This choice is proved by the following final statement of numerous cold fusion papers: "The matrix, containing a large number of cracks, gaps and other voids is the only thing common to all successful cold fusion experiment media." [32].

(e) The *chemo*nuclear fusion hypothesis is a two-stage process. The matrix melting forbids the first stage (processes in the gas medium) and, thereby, terminates the fusion process. Therefore, melting of a metal matrix should be considered as a technique of "passive" protection in cold fusion reactors.

(f) The cold fusion process is provided with an effective feedback mechanism. The energy of nuclear fusion is transferred through the conversion electron emission channel to the matrix, and is spent (partially) for reproduction (and under certain conditions, expanded reproduction) of the fusion process, for working gas ionization and cluster ion generation, and also, for the formation of internal electric fields, acceleration of ion clusters and other processes.

(3) In view of the above, the HTE-Cat generator can be considered a cold fusion reactor, the operation of which is based on the *chemo*nuclear fusion reaction in the $(^7Li–^1H)_{gas}$-nickel system, while the KD installation represents a prototype reactor using the HD-nickel system. As regards to *chemo*nuclear fusion, it is hoped to be a source of clean and safe energy.

(4) The *chemo*nuclear fusion hypothesis is a hypothetical, qualitative model of intense cold fusion. Even now, the investigation level of the model attained so far opens the way for practical mastering of the phenomenon. However, the development of the quantitative model calls for further deep investigation of the cold fusion mechanism and the model optimization with a careful choice of the "pseudo-composite" matrix structure, the temperatures and pressure of the working gas, etc. The use of the magnetic field can be a promising direction in improving the model efficiency. The tasks of primary importance in the development of the research program include the meeting of the ecological and safety requirements under operating conditions of cold fusion reactors.

Mastering of *chemo*nuclear fusion energy as the direction opening the way to clean, nuclear-safe and practically inexhaustible source of energy should become the priority of state energy programs.

## References


[1] E. Miyamoto. Fundamentals of plasma physics and controlled fusion, Transl. From English and. red. V.D. Shafranov, Moscow, 2007, 425 p.

[2] M. Fleischman and S. Pons, *J. Electoral Chem.* **261** *(1989)* 301.

[3] A. Rossi, S. Focardi et al., *J. Nucl. Phys.* 2012. http://www. Journal-of-nuclear-physics.com.

[4] A. Rossi, *J. Nucl. Phys.* 2012. http://www. Journal-of-nuclear-physics.com.

[5] V.D. Kuznetsov et al., *Annales de la Fondation Louis de Broglie* **28**(2) (2008) 173–213.

[6] V.F. Zelensky, 2016-1, Kharkov, KIPT, 2016, Preprint.

[7] G.S. Collins, J.S. Walker and J.W. Norbury, *J. Fusion Energy* **9** (4) (1990).

[8] Vit.M. Bystritsky, V.M. Bystritsky, S.A. Tchaikovsky et al., *Nucl. Phy.* **64** (5) (2001) 920–925.

[9] A. Takahashi, *Proc. J. Condensed Matter Nucl. Sci., The Int. Conf. Cold Fusion-13*, Sochi, Russia, 2007.

[10] A.G. Parkhomov, *J. Emerging Sci. Res.* **11** *(4)* (2016) 58–62.

[11] V.F. Zelensky, Kharkov, 2012, Preprint.

[12] V.F. Zelensky, Problems of atomic science and technology, Ser. Nuclear physics investigations, Kharkov, Apr. 2013.

[13] V.F. Zelensky, V.O. Gamov, A.L. Ulybkin and V.D. Virich, *The Int. Conf. on Cold Fusion-20,* Sendai. Japan, 2016, submitted.

[14] G. Levi Et al., Report:http://amsacta.unibo.it/4084/1/LuganoReportSubmit.pdf. Oct. 6, 2014.

[15] V.F. Zelensky, Problems of atomic science and technologies, Ser. *Phys. Radiation Effect and Radiation Materials Sci.* **2** (56) (1991) 34–45.

[16] B.F. Bush and J.J. Lagowski, *J. Electroanal. Chem.* **304** (1991) 271–278.

[17] E.I. Church and J. Wenser, *Phys. Rev.* **104** (1956) 1382–1386.

[18] M.A. Listengarten, The modern methods of nuclear spectroscopy, M: *Science* (1986) 142–197.

[19] A. Takahashi, *Siena Workshop on Anomalies in Metal-D/H Systems*, Siena, Italy, 2005.

[20] A.S. Davydov, *Physics-Uspekhi (Adv.Physical Sci.)* **34** (1989) 1295.

[21] A.B. Karabut, *The Int. Conf. on Cold Fusion-10,* Cambridge, 2003.

[22] A.G. Lipson, A.S. Roussetski, A.B. Karabut and G.H. Miley, *Proc. J. Condensed Matter Nucl. Sci., The Int Conf. on Cold Fusion-10, Cambridge*, MA, 2003.

[23] M. Gryzinski, *JNR,* 1967, Vol. XVIII, Report No. 810.

[24] M.J. Gryzinski, *Phys. Lett.* **76**A (1980) 28.

[25] M.J. Gryzinski, *Phys. Lett.* **123**A (1987) 170.

[26] M. Gryzinsky, *Nature* **338** (1989) 7121.

[27] M. Gryzinski, *AJP Conf. Proc.* **228** (1990) 717.
[28] A.O. Barut, *Int. I. Hydrogen Energy.* **15** (1990) 907.
[29] J.-P. Vigier, *The Int. Conf. on Cold Fusion*-3, 1992, p. 325.
[30] V.A. Chechin, V.A. Tsarev, M. Rabinowitz and Y.E. Kim, *Int. I. Theoret. Phys.* 3**3** (1994) 617–669, arxiv.org, nucl-th/0303057, 2003.
[31] R.A. Rice, Y.E. Kim, M. Rabinowitz and A.L. Zubarev, *The Int. Conf. on Cold Fusion-4, Theory and Special Topics Papers*, TR-104188-V4, Lahaina, Maui, Hawaii: Electric Power Research Institute, 1994, pp. 41–47.
[32] E. Storms, *The Science of Low Energy Nuclear Reaction, World Scientific, Singapore,* 2007.
[33] S. Szpak and J. Dea, *J. Condensed Matter Nucl. Sci.* **9** (2012) 21–29.
[34] V.V. Pokropivny, Reports of the Ukrainian Academy of Sciences, Vol. 4, 1993, p. 86.
[35] Ya.B. Zeldovich and I.D. Novikov, *Relativistic Astrophysics M.: Science*, 1967, p. 656.
[36] A. Roussetski, *Review and Perspectives*, 2004: http://lenr-canr.org/acrobat/Roussetskicrtrackdet.pdf.
[37] A. Takahashi, *Proc. The Int. Conf. on Cold Fusion-15*, Roma, 2009.
[38] A. Takahashi et al., *Int. J. App. Electromagn. Matter.* **3** (1992) 221.
[39] T. Claytor et al., *Italian Phys. Soc. Conf. Proc.* **33** (1991) 395.
[40] V.A. Romadanov, *Proc. Tenth Int. Conf. on Cold Fusion*, 2003, pp. 325–352.
[41] V.F. Zelensky et al., Problems of atomic science and technologies, *Ser. Nuclear Physics Investigations* **3**(85) (2013) 119–128.
[42] V.F. Zelensky et al., KIPT 1989-61, Kharkov, 1989, 25 p., Preprint.
[43] V.F. Zelensky, V.F. Rybalko, G.D. Tolstolutskaya, A.N. Morozov et al., *Proc. sixth Russian Conf. on Cold Nuclear Transmutation of Chemical Elements (RCCFT-6),* Dagomys, Sochi. 1998.
[44] P.A. Mosier-Boss, S. Szpak, F.E. Gordon and L.P.G. Forsley, *Euro. Phys. J. Appl. Phys.* **46** (2009) 30901.
[45] P.A. Mosier-Boss et al., *Naturwissenschaften* **96** (2009) 135.
[46] P.A. Mosier-Boss, *Euro. Phys. J. Appl. Phys.* **51** (2010) 20901.
[47] J.F. Ziegler and J.P. Biersack, *The Stopping and Range of Ions in Solids,* Pergamon, New York, 1985.
[48] A. Roussetski et al., *Eighth Inter Conf. on Cold Fusion*, Lerici, Italy, 2000.
[49] K.N. Mukhin, *The Experimental Nuclear Physics. An Elementary Particle Physics*, Vol. I, M: Atomizdat, 1974.
[50] K.N. Mukhin, *The Experimental Nuclear Physics. An Elementary Particle Physics*, Vol. II, M: Atomizdat, 1974.
[51] S. Focarfi and A. Rossi, Report, March 22, 2010. http://www.lenr-canr.org/acrobat/FocardiSanewenergy.pdf.

J. Condensed Matter Nucl. Sci. 24 (2017) 168–178

Research Article

# Experimental Device of Cold HD-Fusion Energy Development and Testing (Verification Experiment)

V.F. Zelensky, V.O. Gamov*, A.L. Ulybkin and V.D. Virich

*National Science Center Kharkov Institute of Physics and Technology, Kharkov, Ukraine*

**Abstract**

The work is devoted to experimental check of the intensive cold fusion model (the *chemo*nuclear fusion scenario) developed. For experiments, the experimental device of cold fusion energy in HD–nickel system has been made. To create an "active pseudo-composite" layer (offered in V.F. Zelensky, The Int. Conf. on Cold Fusion-20. Japan, 2016) on the electrode surface, an operating procedure has been developed for processing nickel electrodes of the energy generator. All the phenomena expected in the *chemo*nuclear fusion scenario were observed in the "verification experiment" (except $\gamma$-radiation caused by the bineutron–nickel nucleus interaction). The conclusion is drawn that the *chemo*nuclear fusion scenario successfully explains the process of intensive cold fusion in three fusion systems (2D–, HD– and $({}^{7}\mathrm{Li}{-}{}^{1}\mathrm{H})_{\mathrm{gas}}$-transitive metal), and thus, can form the basis for the development of cold fusion reactors.



*Keywords:* Anomalous $\gamma$-conversion, *Chemo*nuclear fusion, Cold fusion

## 1. Introduction

The work in the new line of scientific research, now known as "cold fusion", was started at the Kharkov Institute of Physics and Technology (KIPT, Ukraine) in 1989, following the publication of information on the phenomenon [1]. In 2012, we published the paper [2], where the two-stage, cluster model of cold fusion was offered. The statements put forth in [2] about participation of virtual photons of anomalous internal $\gamma$-conversion in the fusion process, have provided explanation of cold fusion at a qualitative level. However, the proposed model needed further development.

The point was that the basic version of the scenario on the decisive role of conversion electrons in realization of intensive cold fusion called for experimental verification. That was caused by the fact that this version had been based on the works, the authors of which interpreted their results without considering the participation of the anomalous $\gamma$-conversion phenomenon in the cold fusion process

*E-mail: vgamov@kipt.kharkov.ua.

Considering great importance for understanding the cold fusion phenomenon and its practical use, further studies on both the $\gamma-$radiation yield during cold fusion, and the radioactivity of materials used for the reaction were also needed, because the experimental results on the subject were contradictory in character.

The answers to the questions could be obtained only in the direct experiment, and that has been the task of the present work. The task was performed in two stages. The first stage included creation of the pilot source of cold HD-fusion energy. The second stage was devoted to conducting the "verification experiment". For four years that have passed since publication of our paper [2], the fusion model has received further development in our work [3] and also, the experimental evidence in work [4] and in our present work. Here we report the main results of model verification in the "verification experiment".

## 2. Creation of an Intensive HD-fusion Pilot Energy Source

As a method of cold fusion initiation the electric discharge in gas was used (material of electrodes is nickel) [5]. On the bases of thermo-physical stand of institute [6] gas-discharge installation–KD installation has been created and mounted.

The KD installation is a stainless-steel high-vacuum chamber, equipped with various devices and accessories, which are necessary for cold fusion initiation and conduction of experiments. Installation includes a means for maintaining of high vacuum ($p \leq 10^{-7}$ torr). For storage and metered flow of $D_2$ and $H_2$ chemical pumps based on Zr–Al–Fe-adsorbent are used (they are also necessary for $^4$He and $^3$He clearing).

The water-cooling of the chamber serves as a flowing calorimeter. The $\gamma$-spectrometer Inspector-2000 (Canberra Industries) with the HPGe-detector is used. Outside of the chamber dosimeters for neutron (ALNOR) and total (DTU) radiation level registration are located.

For analysis of chemical/isotopic compositions of gases, the installation is equipped with the MX7304A mass spectrometer, while the gas-discharge ampoule serves for the analysis of optical spectra on the SL40-2 spectrometer.

$D_2$ and $H_2$ are the hydrogen isotopes. Therefore, there have been reasons to expect that the method elaborated in deuterium experiments for initiating intense cold fusion (it consisted in creation of a transition metal-based "pseudo-composite" layer on the electrode surface [5,7]) will be efficient to solve the problem of intensive cold HD-fusion initiation in the $D_2$+ $H_2$ mixture. In our case, the "pseudo-composite" is the nickel-based matrix containing a lot of working gas-loaded cracks, pores and other voids.

The major link of the technological process of obtaining the surface "pseudo-composite" layer under discussion consists in operation of nickel electrodes in the mode of variable-polarity gas discharge.

At operation in the cathode mode, owing to ion implantation, the electrode surface layer is saturated with a mixture of gases to a depth of about 100 Å. Simultaneously with it, the surface layer is exposed to a flow of conversion electrons. Under these conditions, the deeper cathode layers get saturated with the gas due to radiation-stimulated diffusion. With the polarity change, the working gas-filled layer is exposed to both the electrons from the anode discharge current and the flow of conversion electrons. The irradiation by electrons brings about the increase in the quasimolecular clusters concentration, growth of cold fusion intensity and increase of the matrix imperfection [5]. As a result, after repeated changes of polarity, the layer containing a great many cracks, ruptures and other voids filled with the working gas is formed on the electrode surfaces. As it follows from Refs. [5,7], under the action of the electron flow (in this case it is the anode current and the flow of conversion electrons), this sort of "pseudo-composite" becomes the optimum matrix for intensive cold fusion initiation.

In the "verification experiment", the above-described technology of the "pseudo-composite" layer formation on the nickel electrode surfaces and the choice of the appropriate mode of the electric discharge in gas have allowed us to achieve a steady intensive HD-fusion at the KD installation, in the mix of $D_2$ (85%) + $H_2$ (15%). On the basis of the KD installation, the pilot *chemo*nuclear HD-fusion energy source has been developed and tested. The experimentation

**Table 1.** Part of the data from June 6, 2015, experiment.

| Time | $\Delta T$ (°C) | $I$ (A) | $U$ (V) | Water flow rate (100 ml /s) | $P_{\mathrm{elect}}$(W) | $P_{\mathrm{heat}}$ (W) | Efficiency | Efficiency with the calibration account |
|---|---|---|---|---|---|---|---|---|
| 9:36 | 1 | 0.083 | 614 | 79 | 50.96 | 5.29 | 0.10 | 0.11 |
| 10:20 | 6.9 | 0.083 | 589 | 79 | 48.89 | 36.51 | 0.75 | 0.80 |
| 11:24 | 7.6 | 0.109 | 475 | 79 | 51.78 | 40.21 | 0.78 | 0.84 |
| 11:40 | 9.3 | 0.1075 | 466.7 | 79 | 50.17 | 49.21 | 0.98 | 1.05 |
| 12:20 | 11 | 0.1095 | 414.2 | 80 | 45.35 | 57.48 | 1.27 | 1.36 |
| 14:19 | 10.5 | 0.1136 | 281.9 | 80 | 32.02 | 54.86 | 1.71 | 1.84 |
| 14:33 | 10.4 | 0.1147 | 274 | 80 | 31.43 | 54.34 | 1.73 | 1.86 |
| 14:46 | 10.8 | 0.1149 | 288.4 | 80 | 33.14 | 56.43 | 1.70 | 1.83 |
| 16:02 | 14.9 | 0.1414 | 324.5 | 80 | 45.88 | 77.85 | 1.70 | 1.82 |
| 16:16 | 14.4 | 0.143 | 311.9 | 80 | 44.60 | 75.24 | 1.69 | 1.81 |
| 16:31 | 14.2 | 0.145 | 298.7 | 80 | 43.31 | 74.20 | 1.71 | 1.84 |

with this energy source has confirmed the validity of the *chemo*nuclear 2D-fusion scenario for the HD-nickel system, too.

## 3. "Verification Experiment" Results

### 3.1. Measurements of excessive heat generation

In experiment with working mix of $D_2$ (85%) + $H_2$ (15%) the following results are received: the maximum efficiency of a system – 1.8–1.9; excess power – at level of 30–35 W. Table 1 shows a part of the data results, obtained in the June 6, 2015, experiment. Similar results have been obtained in June 3, 2015 and June 7, 2015, experiments.

### 3.2. Changes of electrical parameters of the discharge

In the transition to the mode of intensive cold fusion, as the excess heat yield increased, the resistance of the discharge gap changed from 8–7.5 down to ~2 kΩ. At that, the discharge current at the given electric power of ~50 W increased

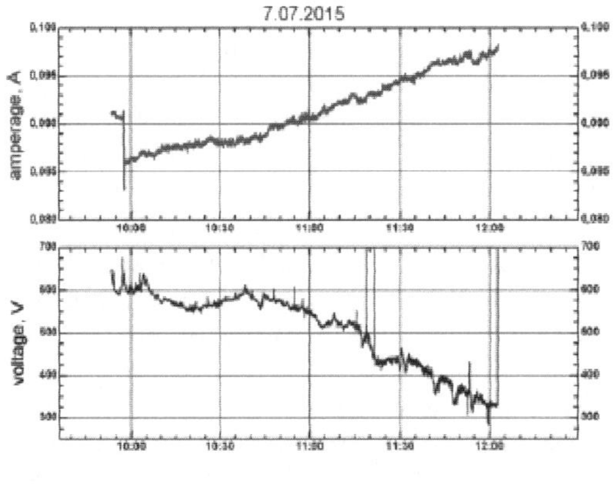


**Figure 1.** Electric characteristics of 7-07-2015 experiment.

**Table 2.** Some products of a bineutron absorption by a nickel isotopes reaction.

| | |
|---|---|
| 1 | $^{59}$Co, stable |
| 2 | $^{61}$Co, $\beta^- T_{1/2} = 1.65$ h, $\gamma = 0.067$ (0.89), $\beta = 1.22$ MeV |
| 3 | $^{62}$Co, $\beta\ ^- T_{1/2} = 1.5$ min |
| 4 | $^{63}$Ni, $\beta^- T_{1/2} = 100$ years, $\gamma = 0$ |

from 83 up to 143 mA, and the discharge gap voltage decreased to the values, at which the discharge could not be initially ignited. The changes in the electrical parameters of the experiment are shown in Fig. 1. Similar results have been obtained in June 3, 2015 and June 6, 2015, experiments.

### 3.3. Absence of induced $\gamma$-radiation

Some supposed products of bineutron absorption by nickel isotopes reaction are presented in Table 2. Neither in the process of experiment nor at the subsequent "flashing" ("flashing", the decay of $\gamma$-activity induced in experiment) is not found a $\gamma$-radiation differs from background level. In Fig. 2 the site of a spectrum in the field of a 0.067 MeV line for a spectrums registered during the discharge, after discharge termination and for a background spectrum ($a$, $b$, $c$ accordingly) is shown. Appreciable differences between spectra are not revealed.

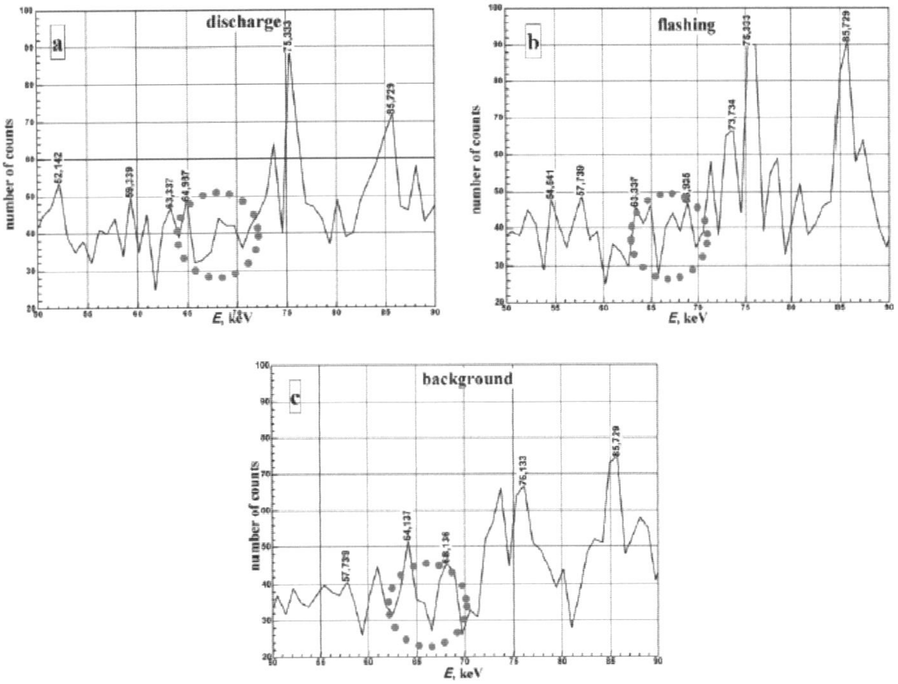


**Figure 2.** $\gamma$-spectral regions in the neighborhood of the 67 keV line for various registration conditions: (a) during the operating time of the discharge (6.07.15); (b) after discharge termination; (c) background.

During carrying out of experiment the sum of $\gamma$-spectrometer count-downs in a range of (50–2000) keV exceeded background activity on $\sim$0.8% that does not fall outside the limits admissible changes of a background level. The activity of radon peaks during the experiment did not exceed the background level. The line of $^{214}$Bi (609.3 keV) decay was not detected. At first sight, it contradicts to our earlier work [8], where $^{214}$Bi activation was reliably registered under similar conditions. In actual fact, we may not compare the results obtained in the present work and in [8], because the activity measurements in [8] were carried out with a low-background stationary $\gamma$-spectrometer at an order-higher measuring sensitivity and at a lower background level (by a factor of $\sim$4) [5].

The signal from ALNOR- (registration of neutron radiation) and DTU- (registration of the total level of radiation) dosimeters corresponds to the natural background level.

### 3.4. $^3$He generation

The analysis of gaseous mixture was carried out in two stages. The first was the analysis of a "zero" spectrum (the analyzer spectrum). The second stage included the analysis of the working gas after its processing by the chemical pump (the chemical pump absorbs reactive gas components such as hydrogen/deuterium/HD, water vapor and carbon oxides; neutral gases are not absorbed, as they do not react with the metal absorber). The gas analysis was carried out with gradual increase of ionization voltage. This approach along with the application of the chemical pump allows distinguishing HD and $^3$He components in the analysis. A similar technique was used in work [9].

In the course of the experiment the isotopic composition of the working gas has changed drastically (Table 3). In the experiment with gas composition of 85% $D_2$ and 15% $H_2$ the ratio of the peak height of mass-3 to the peak height of mass-4 ranged up to 17. In the experiment with low hydrogen content ($\sim$1%) this ratio was equal to 10. This gives grounds to conclude that $^3$He is generated in the experiment. The content of $D_2$ and $H_2$ molecules in the analyzed gas was minimum due to their absorption by the chemical pump adsorbent, and could not affect considerably the $^3$He/$^4$He ratio.

Figures 3 and 4 show the mass-spectrum peaks and their intensities in the range $m/z = 1$ to 4 versus the ionization voltage: in the analyzer chamber (Fig. 3); for the spent gas mixture after processing in the chemical pump (Fig. 4). It can be seen that the curves for $m/z = 3$ in Figs. 3 and 4 have various behavior: in Fig. 3 the curve "intensity" remains practically unchanged with increase in $U_{\text{ionization}}$, whereas in Fig. 4 the "intensity" increases by approximately an order of magnitude. The behavior and the "intensity" level of the curves for $m/z = 4$ are very close (cf., Figs. 3 and 4).

The obtained results allow us to conclude that in our "verification experiment", it is $^3$He that is generated, while the $^4$He concentration remains invariable.

### 3.5. Modifications of electrodes surface

During the operation of electrodes in a mode of intensive cold fusion, the occurrence of specific superficial structures has been revealed. Figure 5 shows two pictures, whose images may be interpreted as the opened blisters. It should

**Table 3.** Results of the gas analysis.

| Dates of experiments | Pressure during the analysis (Pa) | Analyzed gas | Parity of peak intensities of masses 3 and 4 – $I_3/I_4$ |
|---|---|---|---|
| 19-03-2015 | $2.5–4.7\times10^{-4}$ | Vacuum | 2.7–5.5 |
| 09-04-2015 | $7.9–9.6\times10^{-5}$ | $D_2/H_2$ (99%/1%) + chem. pump | 8–9.5 |
| 30-06-2015 | $0.5–1\times10^{-4}$ | Vacuum | 0.7–3.5 |
| 07-07-2015 | $0.5\times10^{-4}$ | $D_2/H_2$ (85%/15%) + chem. pump | 6.4–17 |

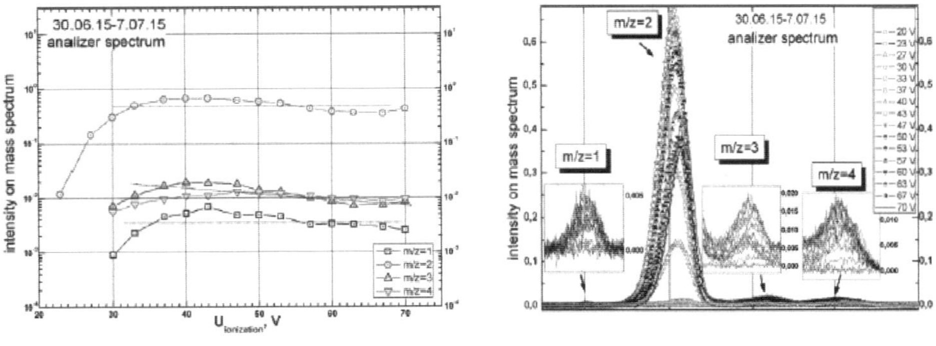


**Figure 3.** Peaks intensity on a mass spectrum in a range $m/z = 1$ to 4 depending on voltage of ionizations (20–70 V) in the gas analyzer chamber (the data from experiments made during 30-06-2015 to 7-07-2015).

be borne in mind that as these images are obtained in the secondary electrons (SEI), the contrast is largely determined by the composition and not of the sample topography. Studies held in the topology of the surface of the backscattered electron detection mode. However, the small size objects (less than 1 $\mu$m) does not allow to apply this method due to its low spatial resolution. Inside the blisters probably disclosed internal relief is observed rather than a smooth surface characteristic of the gas-filled voids, after processing the image.

Based on the foregoing, it is doubtful interpretation of these images as images of gas-filled blisters.

Formation on the surface of shown above structures – structures with an internal relief, well agrees with the mechanism of intense cold fusion considered here (see description responsible for the intense cold fusion matrix, viz "the pseudocomposite", [3]). Studying these structures will provide additional information about the nature of the *chemo*nuclear fusion. Investigations are continuing.

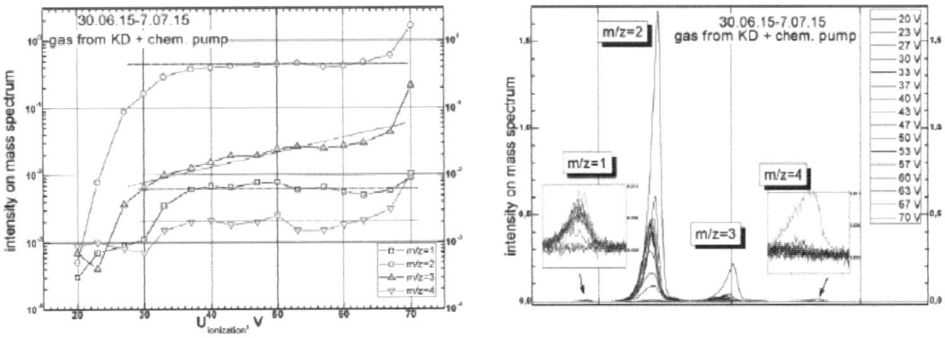


**Figure 4.** Peak intensity of the mass spectrum in the range $m/z = 1$ to 4 versus ionization voltage (20–70 V) for the exit gas from the KD setup after processing in the chemical pump (the data from experiments made during 30-06-2015 to 7-07-2015).

**Table 4.** Nickel isotopic structure in nature, in "ashes:" of HTE-Cat reactor [6] and in present work.

| No. | Isotope | Natural composition (%) | Isotope composition in "ashes" ( % [6]) | Composition on a cathode surface (the present work) (%) | Proton energy in reaction $^{x}Ni + ^{2}n \to$ $\to ^{x+1}Co + p\ (E))$ (MeV) | Columbs barrier $E_c$ (MeV) | $(E_p - E_c)$ (MeV) |
|---|---|---|---|---|---|---|---|
| 1 | $^{58}Ni$ | 67.76 | 0.8 | 66.52 | 10.7 | $\sim 7 \pm 1$ | +3.7 |
| 2 | $^{60}Ni$ | 26.16 | 0.5 | 26.72 | 10.24 | $\sim 7 \pm 1$ | +3.2 |
| 3 | $^{61}Ni$ | 1.25 | 0 | 1.24 | 6.06 | $\sim 7 \pm 1$ | -1 |
| 4 | $^{62}Ni$ | 3.66 | 98.7 | 4.42 | 3.9 | $\sim 7 \pm 1$ | -3.1 |
| 5 | $^{64}Ni$ | 1.16 | 0 | 1.1 | – | – | – |

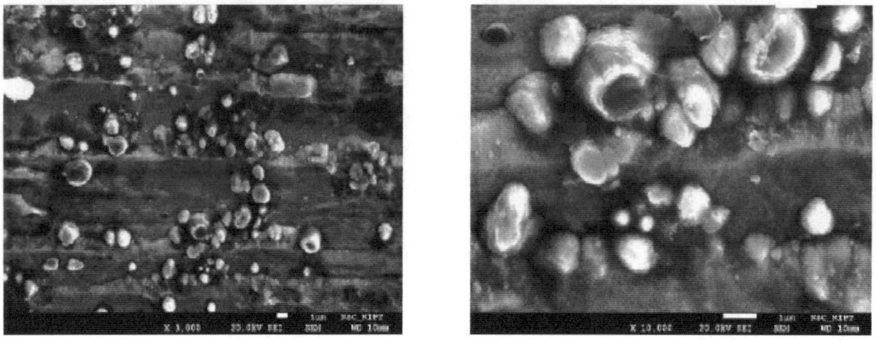

**Figure 5.** Microphotos of a surface of electrodes.

## 3.6. Changes of nickel isotopic composition

The research results for the Ni isotopic composition on the electrode surfaces are presented in Fig. 6 and Table 4. The 20% increase in the $^{62}Ni$ isotope concentration has been found.

In Fig. 7 and Table 4 the data of fuel analysis taken from Ref. [4] is shown. From Fig. 7 and Table 4 it can be

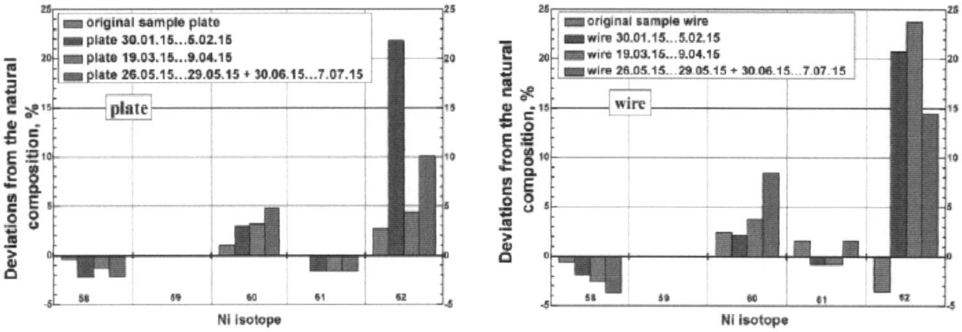


**Figure 6.** Ni isotopic composition deviations in initial samples and on a surface of electrodes (wire is a central electrode, plate is a lateral electrode) relative to a natural composition.

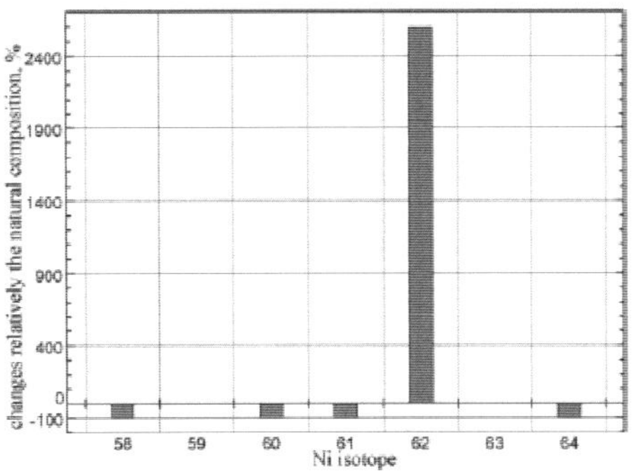


**Figure 7.** Deviations of nickel isotopic composition in "ash" from HTE-Cat reactor [7] relative to a natural composition.

seen that the isotopic composition of nickel observed in [4] has significant changes. On completion of the experiment, the HTE-Cat "ash" contained almost 99% of $^{62}$Ni isotope (cf. its initial content of 3.66%). As to our experiment, the changes in the nickel isotopic composition are not so striking. However, this does not cast any doubt on the reliability of data, since the excess of the $^{62}$Ni isotope content (near 20%) goes far beyond the limits of instrument errors. Such different results are attributed to differences in the operating modes, viz., a considerably shorter duration of our experiment and a lower intensity of process in this case).

## 4. The Discussions of Results

According to the *chemo*nuclear 2D-fusion scenario, one should expect that in the "verification experiment" with the use of the mixture $D_2$ (85%)+ $H_2$ (15%), it is the HD-reaction that will be responsible for the occurrence of intense cold fusion [2,3]. The fusion process in this case should be accompanied by the following nuclear processes.

(a) Considerably strengthened by medium "*chemo*factors" an intercluster (proceeding inside "Gryzinski quasi-molecules" ($D^+$+$e^-$+$H^+$)) HD-reaction:

$$D^+ + e^- + H^+ \rightarrow {}^3_2\mathrm{He} + e^-_{\mathrm{conv}}(\Sigma E = 5.5\ \mathrm{MeV}). \quad (1)$$

(b) Reactions of weak interaction, first of all, generating of a bineutron $^2$n reaction:

$$\mathrm{D} + e^-_{\mathrm{conv}} \rightarrow {}^2\mathrm{n} + \nu_e \quad (2)$$

drastically strengthened in the conditions of intensive cold fusion.

(c) Interaction of bineutron with nickel and working gas nuclei reactions [5].

The major path for a "verification experiment" is the absorption of a bineutron by nickel nuclei reaction:

$${}^x_{28}\mathrm{Ni} + {}^2\mathrm{n} \rightarrow {}^{x+2}_{28}\mathrm{Ni}^* \rightarrow {}^{x+1}_{27}\mathrm{Co} + \mathrm{p}(E). \quad (3)$$

(1) Secondary reactions.

Interactions reactions fast protons from (3) reaction with nickel nuclei (4) and other element of active zone of reactor nuclei.

$${}^x_{28}\mathrm{Ni} + {}^1\mathrm{p} \rightarrow {}^{x+1}_{29}\mathrm{Cu}. \quad (4)$$

Generating copper nuclei with the exception of stable $^{63}$Cu and $^{65}$Cu isotopes decays by one of channels:

$$^{x+1}_{29}\mathrm{Cu} \rightarrow {}^{x+1}_{28}\mathrm{Ni} + \mathrm{e}^{+} + \nu_{\mathrm{e}} \quad \text{(positron decay)}, \tag{5}$$

$$^{x+1}_{29}\mathrm{Cu} + \mathrm{e}^{-} \rightarrow {}^{x+1}_{28}\mathrm{Ni} + \nu_{\mathrm{e}} \quad \text{(electron capture)}. \tag{6}$$

"Products" of nuclear reactions under (a) in a "verification experiment" will be: excess heat without high energy $\gamma$-radiations emission; conversion electrons emission; generating of $^3$He atoms; on a surface of electrodes specific formations will appear (swellings, blisters, craters) etc.

The reactions under (b), (c) should be accompanied by "products" of a bineutron interaction with a matrix and working gas: fast p-, n-nucleons and their clusters d, T, $^4$He, $^3$He yield; generation of anomalous isotopic composition impurities; $\gamma$-radiation of radioactive impurity nuclides and others.

Registered in the conditions of a "verification experiment" under (a), (c) should be: anomalous changes of nickel isotopic composition ($^{62}$Ni concentration growth against burning out of other isotopes); $\gamma$-radiation 0.067 MeV line, $T_{1/2} = 1.65$ h (Table 2).

The intensity of secondary reactions (d) significantly less than intensity of reaction from (a)–(c) and therefore secondary reaction products in "verification experiment", likely, will not register.

One of the main results of the present work has been the creation of the experimental device of intensive cold HD-fusion energy, i.e., the source of thermal energy, which is not accompanied by emission of high-energy radiation. The generator creation is the practical realization of *chemo*nuclear fusion scenario statements, and thus, the very fact of generator creation is the evidence of the scenario validity.

As indicated in [5,7], the main secret of intensive cold fusion lies in the fact that the transition state formed in the process of fusion in the system of co-operating particles, – the "Gryzinski quasimolecule", provides a timely discharge of the fusioned nucleus due to participation of virtual photons and electrons of anomalous $\gamma$-conversion. The last removes the restrictions caused by the problems related to $\gamma$-quantum radiation. In that event, each "Gryzinski quasimolecule" system has its own elemental composition. The fusion process in this case proceeds without rupture of nuclear bonds and demands overcoming the low Coulomb barrier inherent in light nuclei. This accounts for high intensity of cold fusion in all three systems of fusion.

All expected results listed above in a "verification experiment" exception of $\gamma$-activity were observed sufficiently well.

As expected, the experiment has proved the generation of $^3$He atoms (see Table 3). The ratio of the mass-3 peak height to the mass-4 peak height in the experiment with a low hydrogen content ($D_2$ $\sim$99%; $H_2$ $\sim$1%) was approximately equal to 10, while in the experiment with the $D_2$+$H_2$ (85%+15%) mixture this ratio was as high as 17, whereas in nature the ratio is negligibly small. The $^3$He atom generation in the "verification experiment" testifies that the fusion reactions proceed under the given conditions by the intercluster mechanism unrelated to the rupture of the nucleus. In this case, it is the "Gryzinski quasimolecule" ($D^+$+$e^-$ +$H^+$) that serves as an initial ion cluster, and the intracluster reaction (1) is responsible for the HD-fusion process.

The "verification experiment" has also provided evidences in favor of the version that the conversion electrons take part in the cold fusion process. Radical changes in the gas discharge parameters, as the system turns to the mode of intense heat generation (see Fig. 1), testify that the excess heat in the "verification experiment" is taken off by the conversion electrons. The radiation outside the discharge chamber, observed in early experiments by Karabut [10], can be explained in the *chemo*nuclear fusion scenario as the HD-fusion reaction yield of conversion electrons of energy up to 5.5 MeV (see [2]).

The nickel isotopic composition change on the electrode surfaces in the "verification experiment" is one of the most compelling evidences of the validity of the *chemo*nuclear fusion scenario. According to this scenario, the experimentally observed cardinal changes in the Ni isotopic composition are the result of bineutron absorption by the Ni nucleus. This absorption reaction cannot be accompanied by the proton yield from $^{62}$Ni nucleus (Coulomb barrier forbiddenness, see Table 4), and consequently, the $^{62}$Ni isotope accumulation takes place. This was found out in experiment [4] and in our present work.

The "verification experiment" has registered the occurrence of specific formations (swellings, blisters) on the electrode surfaces (see Fig. 5), and this is also in good agreement with the intensive cold fusion mechanism considered in [5,7].

$\gamma$-Radioactivity is the only phenomenon which was expected in the scenario (see Table 2), but was not observed in the "verification experiment". As it follows from Table 2, in accordance with the nickel isotopic composition change, the following reaction should take place in the "verification experiment":

$$^{60}_{28}\mathrm{Ni} + {}^{2}\mathrm{n} \rightarrow {}^{62}_{28}\mathrm{Ni}^* \rightarrow {}^{61}_{27}\mathrm{Co}(\beta;\ T_{1/2} = 1.65\ \mathrm{h};\ E_\gamma = 0.067\ \mathrm{MeV}) + \mathrm{p} \rightarrow {}^{61}_{27}\mathrm{Ni}_{\mathrm{stable}} + \mathrm{p}. \quad (7)$$

When creating the chamber, efforts were made to receive the maximum "signal" from the reaction. The americium-source calibration of the chamber has shown that the 0.067 MeV line intensity in experiment will be suppressed by no more than a factor of 5. In view of this, and also considering the rates of changes in nickel isotopes concentration observed in work [4] and here, one could expect that during the "verification experiment" the 0.067 MeV line would be registered reliably. However, neither in the course of the experiment, nor at subsequent "flashing" any $\gamma$-radiation other than the background radiation was observed. The radioactivity was not found either in the work by the panel of experts in Lugano [4].

At the same time, there is a large number of works, where during cold fusion experimentation the generation of $\beta$-active products was observed and the $\gamma$-radiation was registered [11]. As an example, we will consider our work performed in 2001 [12]. It was indicated there that in the process of pulsed electrolysis of LiO–$D_2O$ solution in the cell having the palladium cathode and the tungsten anode, the generation of $\beta$-active products takes place. After the discharge termination, the electrolysis cell activity in the 250–350 keV energy range was 30% higher than the $\gamma$-background in the spectrometer chamber. In the range between 200 and 2000 keV, the spectral analysis has revealed a number of peaks, the height of which exceeded the background by more than $3\sigma$. The chain of peaks revealed was as follows: 247 keV ($\sim$4.5$\sigma$); 243 keV ($\sim$4$\sigma$); 353 keV (<3$\sigma$); 161 keV (<3$\sigma$). The half-life of the strongest line of the chain ($E = 243$ keV) was determined to be 5 $\pm$ 0.2) days, this being in good agreement with $T_{1/2} = 5.1$ days for the $^{183}$Ta isotope. The above-given $\gamma$-lines peaks were close, in both the energy and the relative intensity, to the lines of the $^{183}$Ta isotope spectrum. This gave us grounds then to conclude [12] that in the conditions of the experiment in the electrolytic cell, along with other radioactive products not identified by us, the generation of $^{183}$Ta isotope occurred.

Out of other similar works, we may mention a detailed research by Savvatimova et al. [8], who investigated the $\gamma-$activity arising during the process of gas discharge experiment in the $D_2$–W and $D_2$–Ta systems. It has been demonstrated by experiments that during the discharge an intensive $\gamma$-radiation arose. The radiation with the same set of lines continued for some time after discharge termination.

Comparison of the radiation spectrum and the target-surface isotopic composition changes during the experiment has provided a reliable identification of the composition of nuclides responsible for the chemical composition change of the electrode surfaces and the $\gamma$-radiation character. As in our experiment, the main result of experiments by Savvatimova et al. [13] is the generation of nuclides lighter than the nuclides of matrices (W and Ta). As an explanation, the authors of [13] have proposed that the obtained results should be attributed to the reactions of W and Ta with deuterium, that being very improbable in those experimental conditions.

At the same time, the results of the both works [12,13] can be adequately explained in terms of the *chemo*nuclear fusion scenario as a result of bineutron absorption by tungsten nuclei:

$$^{182}_{74}\mathrm{W} + {}^{2}\mathrm{n} \rightarrow {}^{184}_{74}\mathrm{W}^{*} \rightarrow {}^{183}_{73}\mathrm{Ta}_{\beta^-} + \mathrm{p}. \quad (8)$$

The seemingly contradictory results of cold fusion experiments may be explained in the *chemo*nuclear fusion scenario reasoning from the following.

The bineutron absorption in both groups of experiments causes the re-build of the nucleus and emission of nucleons or their clusters (see [2,3]). The specific nuclear excitation energy (the ratio of excitation energy to the number of nucleons in the nucleus) at bineutron absorption by the nickel nucleus essentially exceeds the specific excitation energy for tungsten, and the intensity of fusion process in the experiments with Ni is essentially higher than that in experiments with W. On this basis, the absence of $\gamma$-radiation in the Ni experiments can be caused by a deeper reorganization of the nucleus in this case. As it was noted in [5,7], in the matrix microvolume, where the fusion event takes place, the intensity of other phenomena may be drastically changed, too. These questions call for further investigations.

In summary, we note that the present results and the results obtained by the group of independent experts in Lugano (Italy) for the HTE-Cat reactor operation [4] suggest the conclusion that the *chemo*nuclear fusion scenario successfully explains the process of intensive cold fusion in all three (2D–, HD– and $(^{7}\mathrm{Li}\text{-}^{1}\mathrm{H})_{\mathrm{gas}}$–transitional metal) fusion systems, and thus, can serve as a basis for cold fusion reactor development.

## References

[1] V.F. Zelensky, Preprint, Kharkov, 2012.
[2] V.F. Zelensky, Problems of atomic science and technology Ser. "Nuclear physics investigations", Kharkov, Apr. 2013, http://vant.kipt.kharkov.ua/ARTICLE/VANT_2013_3/article_2013_3_76.pdf.
[3] V.F. Zelensky, Preprint KIPT. 2016-1, Kharkov.
[4] V.F. Zelensky, Submitted to *The International Conference on Cold Fusion-20*, Japan, 2016.
[5] V.O. Gamov, A.L. Ulybkin, V.V. Bryk, V.D. Virich, V.P. Ryzhov, V.V. Skripkin and E.V. Glavatskaya, Problems of atomic science and technology Ser., *Nuclear physics investigations*, Kharkov, Apr. 2013. (http://vant.kipt.kharkov.ua/ARTICLE/VANT_2013_3/article_2013_3_119.pdf).
[6] G. Levi, E. Foschi, Bo Hoisteed, R. Pettersson, L. Tegner and H. Essen, Report: http://amsacta.unibo.it/4084/1/LuganoReportSubmit.pdf. Oct. 6, 2014.
[7] V.F. Zelensky, R.F. Polyashenko and A.I. Velikov, Problems of atomic science and technology Ser., Nucl. *Phys. Investigations* **79**(2) (2001) 70–75. http://vant.kipt.kharkov.ua/ARTICLE/VANT_2001_2/article_2001_2_70.pdf.
[8] I. Savvatimova, G. Savvatimov, A. Kornilova, *Proc. The Int. Conf. on Cold Fusion-13*, Dagomys, City of Sochi, 2007.
[9] V.F. Zelensky, V.O. Gamov, V.P. Golchenko, S.G. Boev and V.P. Ryzhov, Problems of atomic science and technology Ser., *Nucl. Phys. Investigations* **95** (1) (2010) 161–164, http://vant.kipt.kharkov.ua/ARTICLE/VANT_2010_1/article_2010_1_161.pdf.
[10] Y. Arata and Zh-Y. Chang, *Cangt. Proc. Japan Acad.* Ser. B **73** (1997).
[11] A.B. Karabut, *10*th *Int. Conf. Cold. Fusion,* Cambridge, 2003.
[12] E. Storms, *The Science of Low Energy Nuclear Reaction,* World Scientific, Singapore, 2007.

J. Condensed Matter Nucl. Sci. 24 (2017) 179–190

Research Article

# Anomalous Excess Heat Generated by the Interaction between Nano-structured Pd/Ni Surface and $D_2$ Gas

Takehiko Itoh*,†, Yasuhiro Iwamura and Jirohta Kasagi

*Research Center for Electron Photon Science, Tohoku University, Sendai, Japan*

Hiroki Shishido

*Department of Quantum Science and Energy Engineering, Graduate School of Engineering, Tohoku University, Sendai, Japan*

**Abstract**

A new experimental setup based on Mizuno's work was introduced in our lab in order to investigate the anomalous heat generation phenomena. Following Mizuno's procedures, we fabricated nano-structured material composed of Pd and Ni by glow discharge on a heater located in the center of a vacuum chamber. The nano-structured Pd/Ni was prepared with $D_2$ before use. Then, we applied electrical power to the heater that was covered with the nano-structured Pd/Ni while evacuating the chamber, and we observed the heater temperature behavior. Next, we introduced $D_2$ gas at about 250 Pa to the chamber while maintaining the heater input. As a result, in three experiments, we observed that heater temperature increases compared to the reference experiment (with no nano-structured Pd/Ni). In particular, in experiments with 7 W input, we observed a123°C heater temperature increase compared to the reference experiment. It can be said that we replicated Mizuno's experiment successfully. Since the heater was covered with nano-structured material, there was a concern that a change in emissivity affected the heater temperature measurement. Numerical calculation was conducted to evaluate the effect of the change of the surface emissivity. It was concluded that even if the emissivity drop dramatically from 0.7 to 0.3 due to the coating of the nano-structured Pd/Ni, the temperature rise would be only 70°C at the most. The postulated emissivity change cannot explain the observed temperature increase of 123°C. These experimental and numerical results suggest that anomalous excess heat was generated by the interaction between nano-structured Pd/Ni surface and $D_2$ gas.



*Keywords:* Anomalous heat, $D_2$ gas absorption, Finite element method, Nano-structured material, Replication

## 1. Introduction

A new division devoted to Condensed Matter Nuclear Reaction (CMNR) was established at the Research Center for Electron Photon Science of Tohoku University in April 2015.

*Corresponding author. E-mail: itoh@lns.tohoku.ac.jp .

†Also at: Clean Planet Inc., Tokyo, Japan.

An experimental apparatus introduced in our lab was based on the experimental device and methods developed by Mizuno [1,2] for the study of anomalous heat generation. Mizuno, Yoshino and their colleagues have been developing excess heat generation methods and devices for a few years. According to their report, Ni and Pd are alternately discharged in a chamber to fabricate nano-structured Pd/Ni. After the baking process, $D_2$ or $H_2$ gas was introduced into the chamber. This process causes D (H) to absorb in the nano-structured Pd/Ni. After that, they observed anomalous heat generation. In one example, they reported as follows. When $D_2$ gas about 100–300 Pa was introduced at 200°C, they observed about 78 W excess power, which was almost as much as the input power. We have begun doing experiments using a similar device and procedures.

## 2. Experiment

### 2.1. Experimental apparatus

We made a device based on information from Mizuno. The chamber and components are almost identical to Mizuno's experimental apparatus. Figure 1 shows a schematic and photograph of our experimental apparatus. The vacuum chamber is made of stainless steel and its volume is 7.5 l. This chamber is evacuated with an oil-free dry scroll pump, ultimately to a vacuum of 1 Pa or less.

Pd electrodes and a ceramic heater are installed inside the chamber. Ni mesh is set around the heater and electrodes. Figure 2 shows the structure of the heater and electrodes. A ceramic heater is installed in the center of the chamber. This ceramic heater is made of alumina of size 25 mm × 25 mm, and has a built-in thermocouple inside. A 0.3 mm diameter Pd wire is wound around this heater so that high voltage can be applied to the wire. A Pd rod ($\varphi$3 mm $L$ = 200 mm) is placed so as to face this heater, and this rod also makes it possible to apply high voltage. A Ni mesh ($\varphi$0.15 mm, 50 mesh) is placed surrounding the heater and Pd rod, and this is grounded to the chamber. As a result, high voltage can be applied between the Pd wire and the Ni mesh, or between the Pd rod and the Ni mesh. The

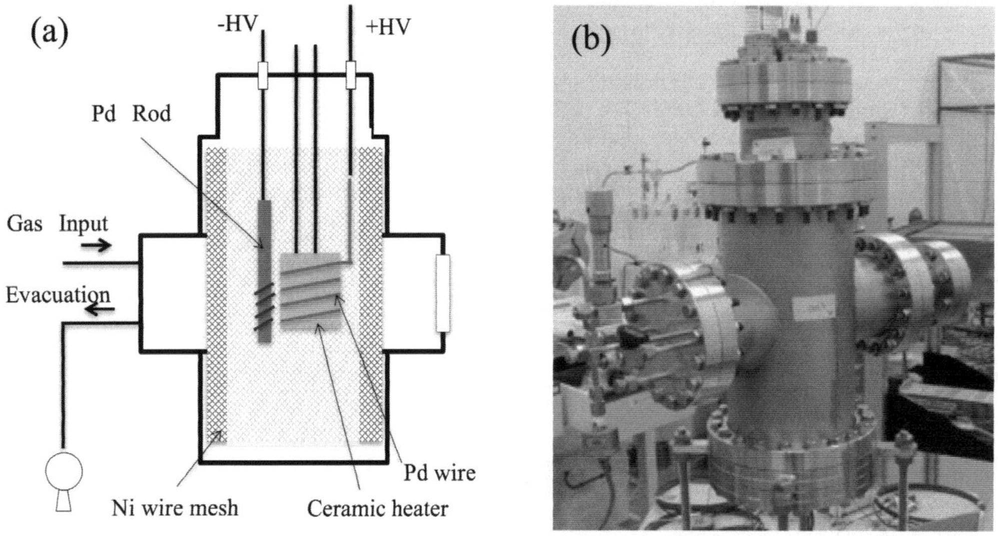


**Figure 1.** Experimental apparatus: (a) schematics of apparatus and (b) photograph of apparatus.

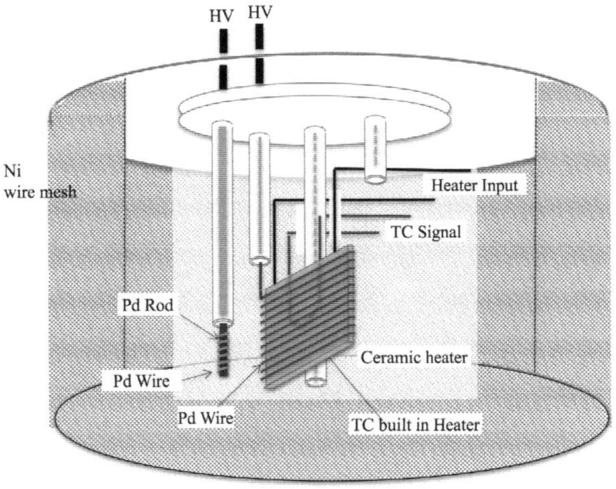


**Figure 2.** Schematic of components around the heater and electrodes.

chamber is covered with Al foil for thermal insulation.

### 2.2. Experimental procedure

Our experimental procedure followed the methods reported by Mizuno. The following three steps are taken.

(1) Reference run without activated nano-structured Pd/Ni material.

(2) Activation treatment: formation and activation of nano-structured Pd/Ni material.

- (a) Sputtering by glow discharge,
- (b) baking,
- (c) $D_2$ absorption.

(3) Foreground run with the activated nano-structured Pd–Ni material.

Determining whether anomalous excess heat was generated by comparing the reference run to the foreground run. Details of each process from (1) to (3) will be explained.

In the reference run, two kinds of measurements were conducted. The first was to measure a temperature of the heater by inputting power to the heater under vacuum condition of 1 Pa or less. The second was to introduce deuterium gas and measure the heater temperature while maintaining pressure of 200–300 Pa.

In activation treatment, the following three processes were carried out. First, the nano-structure was fabricated on the Pd/Ni surface in the chamber, using glow discharge. We performed two patterns of Ar glow discharge (Table 1). We introduced Ar at 10–100 Pa into the chamber. Next, we applied negative high voltage (–1 kV) to the Pd rod, so Pd is sputtered by this discharge. Figure 3 (a) is a photograph of Pd sputtering. We continued this discharge about

**Table 1.** Voltage of electrode for discharge.

| | Pd wire winding heater | Pd rod | Ni mesh |
|---|---|---|---|
| #1 | Float | –HV (1 kv) | Grand |
| #2 | +HV (500–700 V) | Float | Grand |

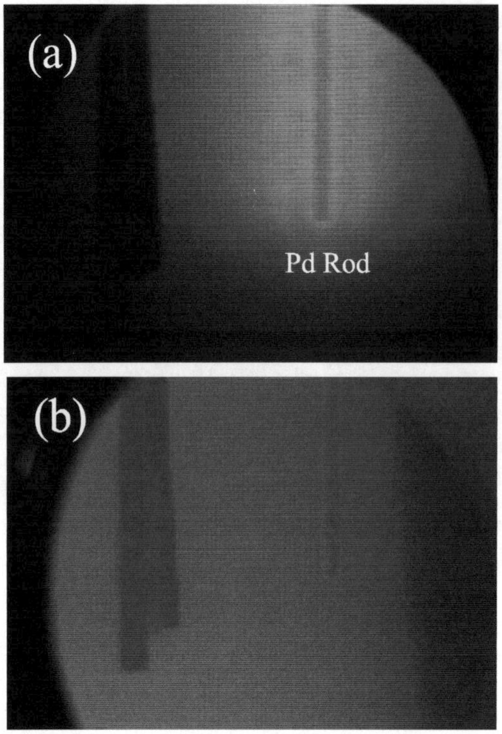


**Figure 3.** (a) Photograph of #1 discharge and (b) #2 discharge.

10,000 s. Next, positive high voltage (+600 to +800 V) was applied to the Pd wire. In this case, Ar ions attack the Ni mesh, so Ni is sputtered (Fig. 3 (b)). This discharge continued about 10,000 s. Table 1 shows the electrode voltage of each discharge. We repeated #1 and #2 discharge alternately three times, so Ni and Pd mixed to form nano-structured Pd/Ni.

Figure 4 (a) and (b) shows the photographs of the electrode before and after experiments. You can see there is a silver colored film on and around a heater after experiments. Figures 4(c) and (d) are the SEM images of the surface of the Pd wire. From the images of (c) and (d), it can be seen that a film covering surface of the wire, i.e., a film formed by sputtering by discharge, is composed of fine particles of several hundred nano-meters in size. According to the Energy dispersive X-ray spectrometry (EDS) analysis, most of the fine particles were Pd and a few contained Ni. We also analyzed other areas and observed similar features. As described above, with this treatment it is possible to fabricate a structure composed of Pd–Ni nano-particles on the heater and its peripheral parts. This structure is similar to the one Mizuno described.

As the second activation treatment, baking was carried out at 100–200°C for about one day to remove water molecules and hydrocarbons from the inner surface of the chamber.

The final activation process is deuterium absorption in the nano-structured Pd/Ni. After baking, $D_2$ gas was introduced into the chamber, which was sealed off at a pressure of 170 Pa, and this state was maintained for more than 12 h.

**Figure 4.** Photograph of electrode and SEM image. (a) Before activation process, (b) after foreground experiment, (c) SEM image of wire and (d) zoom image of (c).

After this process, a foreground run was carried out with the same procedure as the reference run. We will discuss the result in Section 3.

## 3. Results

Figure 5 shows the results of the foreground run of Experiment 1. In this experiment, we applied constant power of 40 W to the heater and maintained this condition. Figure 5 (a) is a reference run (before the activation process). In the reference run, under vacuum conditions (<2 Pa), the heater temperature is 714°C. When $D_2$ gas was introduced at 260 Pa after 1 h, the heater temperature dropped to 587°C from $D_2$ gas cooling. The gas pressure changes in Fig. 5(a) were caused by gas cooling of the heater temperature. As a result, it was found that in the range of 10–300 Pa, there is an effect of several degrees. Figure 5(b) is the foreground run. The chamber was sealed with $D_2$ gas at 170 Pa until the start of the experiment, vacuum evacuation was performed, and power of 40 W was applied to the heater. The temperature of the heater rose to 764°C, which was higher than the temperature of the reference run by 50°C. After 1.7 h, when $D_2$ gas was introduced at up to 260 Pa, the temperature of the heater was 611°C, which was 24°C higher than in the reference experiments. This result suggests that anomalous heat generation is occurring.

We performed Experiment 2 under the same conditions as Experiment 1. Figure 6 shows the result of Experiment 2. The chamber was filled with deuterium and left for 12 h. After that, we applied 40 W of power to the heater while simultaneously evacuating the chamber. The temperature of the heater rose to 760°C, which was 46°C higher than the reference run. This temperature rise was about the same as Experiment 1. After 2 h, when $D_2$ gas was introduced at up to 270 Pa, the temperature of the heater rose to 613°C, which was 26°C higher than in the reference experiments.

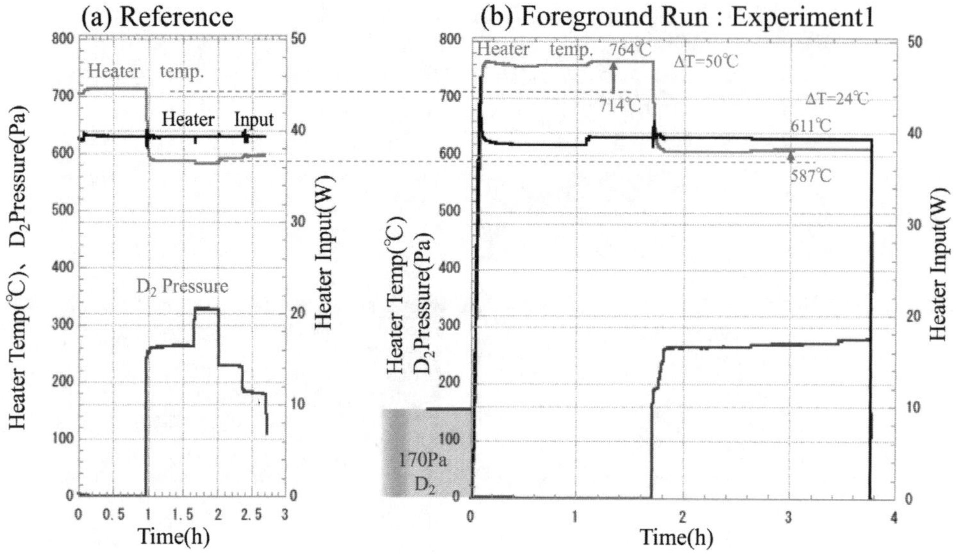


**Figure 5.** Heater temperature, heater input, and $D_2$ pressure of Experiment 1: (a) reference and (b) foreground run. Red line heater temperature, black line heater input and blue line $D_2$ pressure.

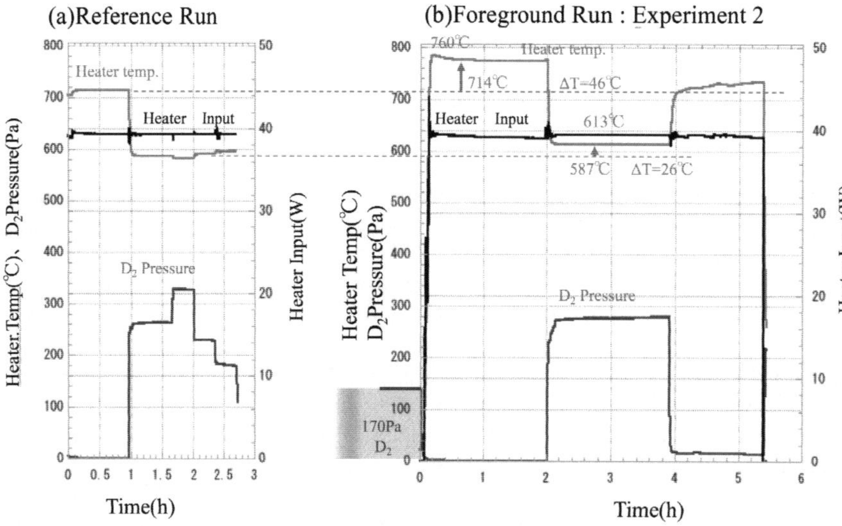


**Figure 6.** Heater temperature, heater input, and $D_2$ pressure of Experiment 2: (a) reference run and (b) foreground run. Red line heater temperature, black line heater input and blue line $D_2$ pressure.

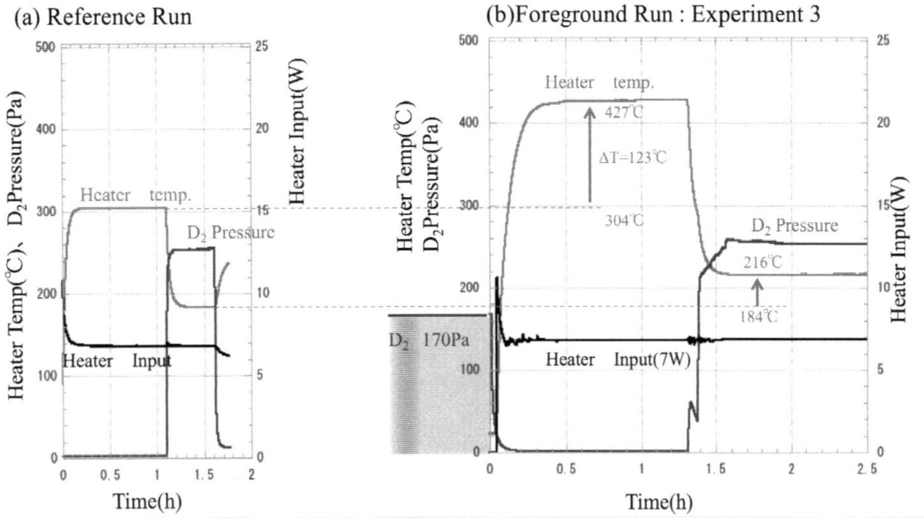


**Figure 7.** Heater temperature, heater input, and $D_2$ pressure of Experiment 3: (a) reference and (b) foreground run. Red line heater temperature, black line heater input and blue line $D_2$ pressure.

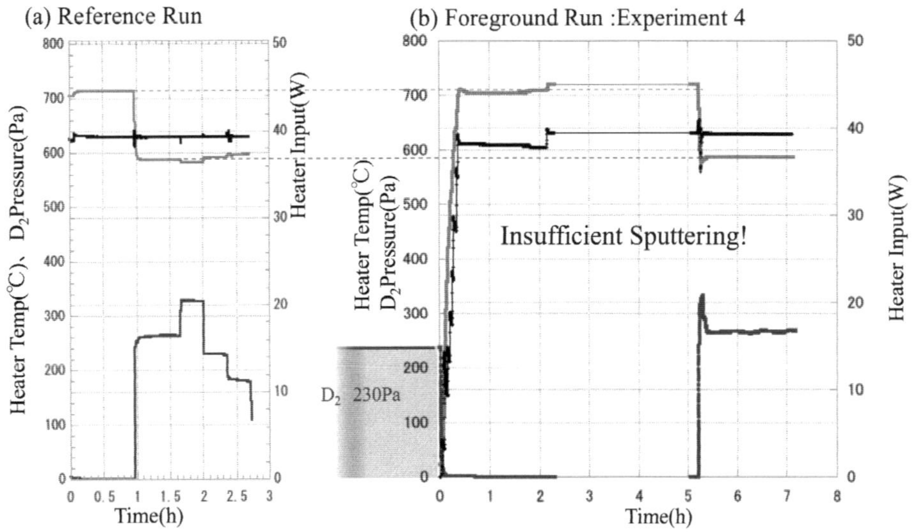


**Figure 8.** Heater temperature, heater input, and $D_2$ pressure of Experiment 4: (a) reference and (b) foreground run. Red line heater temperature, black line heater input and blue line $D_2$ pressure.

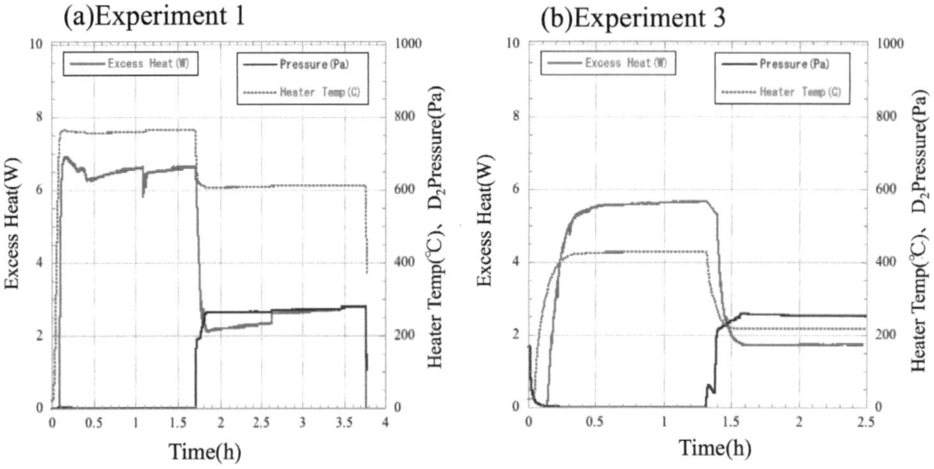


**Figure 9.** Thermal outputs of Experiments 1 and 3. (a) Experiment 1 (heater input 40 W) and (b) Experiment 3 (heater input 7 W). Red solid line Excess heat, red dash line heater temperature and blue line $D_2$ pressure.

In this way, the same extent of temperature rise was observed in the two experiments under the same conditions, so the results were reproduced.

Next, we performed an experiment with input power changed from 40 to 7 W. The results are shown in Fig. 7. In the reference run, under the vacuum condition, the heater temperature reached 304°C, and under the condition with deuterium at 260 Pa, it reached 184°C. On the other hand, in the foreground rum, the temperature in vacuum reached 427°C, which was higher than that in the reference run by 123°C. In addition, when deuterium was introduced, the heater temperature became 216°C, which was 32°C higher than in the reference run.

The next experiment failed to produce excess heat, as is shown in Fig. 8. In this experiment, in the activation process, discharge voltage was insufficient, so sufficient sputtering was not performed. In fact, after the experiment, we examined the heater and the electrode but we found the nano-structured Pd/Ni had not formed. As a result, in the reference run and Foreground run, the temperature of the heater hardly changed and no anomalous heating phenomenon was observed. This suggests that nano-structured Pd/Ni is necessary for abnormal heat generation.

The thermal output of the foreground experiment was evaluated based on the data from the control experiments.

Figure 9(a) is an evaluation of excess heat in Experiment 1. In this experiment, we input 40 W to the heater. In the first half 1.7 h, under vacuum conditions, excess heat is seen at around 7 W. In the second half, 260 Pa of deuterium was introduced, and the excess heat was 2.5 W. Figure 9(b) shows the evaluation of excess heat in Experiment 2, where input power is 7 W. In this case, under vacuum conditions, the excess heat is 5.5 W and the ratio of excess heat to input reaches 80%. When deuterium gas is introduced, the excess heat is 1.6 W.

## 4. Discussion

In this discussion, we will show that the amount of energy in our experiments is much larger than a normal chemical reaction, such as the combustion heat of hydrogen absorbed in nano-structured Pd/Ni.

We assume that the source of excess heat is absorbed deuterium in nano-structured Pd/Ni formed on the heater surface. The EDX analysis shows that the nano-structured film is composed of 99% Pd and 1% Ni, which is formed with a thickness of 10 $\mu$m on the surface of the heater. The amounts of Pd in nano-structured film is approximately 1.414$\times$ $10^{-3}$ mol.

$D_2$ gas absorbed in this material was approximately 1.65 $\times$ $10^{-4}$ mol, which was estimated from pressure decrease of $D_2$ gas which was introduced into the chamber and sealed off before the experiment. This pressure decrease was about 50 Pa in each experiment. Although there are Pd rods and Pd wires in the chamber, we assume that the loading ratio of D/Pd is about 0.23, so only ~1.65 $\times$ $10^{-4}$ mol was absorbed.

If this absorbed deuterium reacts with oxygen for some reason, the total energy of combustion would be 47 J. On the other hand, in the first half of the case of Experiment 1, excess heat of 6 W lasted 1.8 h. This amount of energy is 38,880 J. In the case of Experiment 3, the total energy of excess heat is 19,800 J. These values are much larger than the heat of combustion of deuterium and cannot be explained as a chemical reaction.

Next, we will show that this cannot be apparent excess heat from a radiation change caused by the coating of the Pd / Ni nanostructure film.

In our experiments, since the heater is covered with nano-structured Pd/Ni, there is a concern that the emissivity may vary and affect the heater temperature measurement. We conducted experiments at two temperature ranges of 700°C (Experiments 1 and 2) and 300°C (Experiment 3). Comparing the two experiments, the temperature rise in the low temperature experiment (300°C) was higher than high temperature experiments (700°C). With the effects of radiation, it can be expected that heater temperature rise will increase in the high temperature region, but the experimental results are different. Consequently, we conclude that the observed temperature rise is not related to the effect of radiation.

A numerical calculation was conducted to evaluate the effect of the change of surface emissivity. We performed numerical calculation using the commercial finite element method software COMSOL Multiphysics 5.2 Heat transfer module [3].

The governing equation of heat transfer is described as follows.

$$\rho C_{\mathrm{p}} (\vec{\mu} \cdot \nabla) T + \nabla \cdot q = Q, \quad (1)$$

$$\vec{q} = -k \nabla T. \quad (2)$$

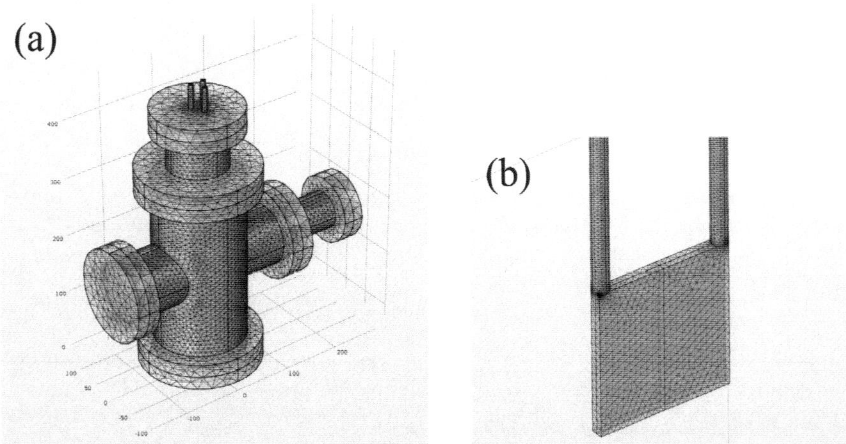


**Figure 10.** Chart of numerical analysis of (a) chamber and (b) heater.

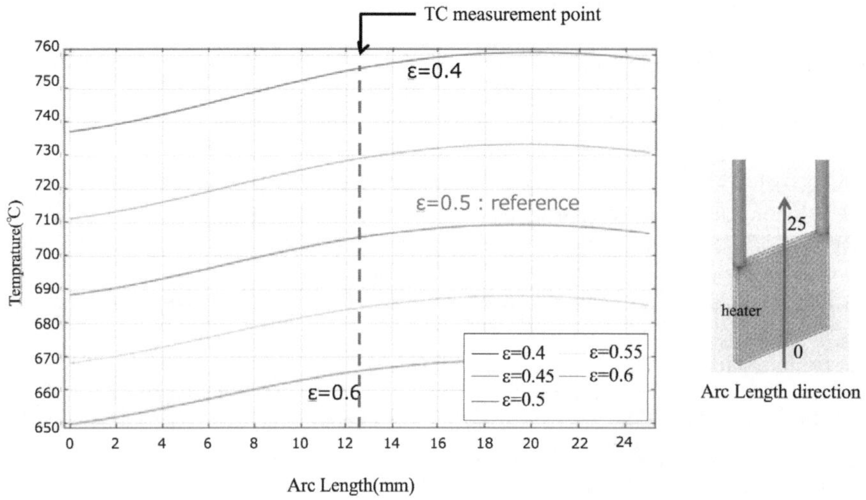


**Figure 11.** Calculation result of temperature distribution of heater. Case of Experiments 1 and 2. (Heater input is 40 W and heater temperature is 700°C.)

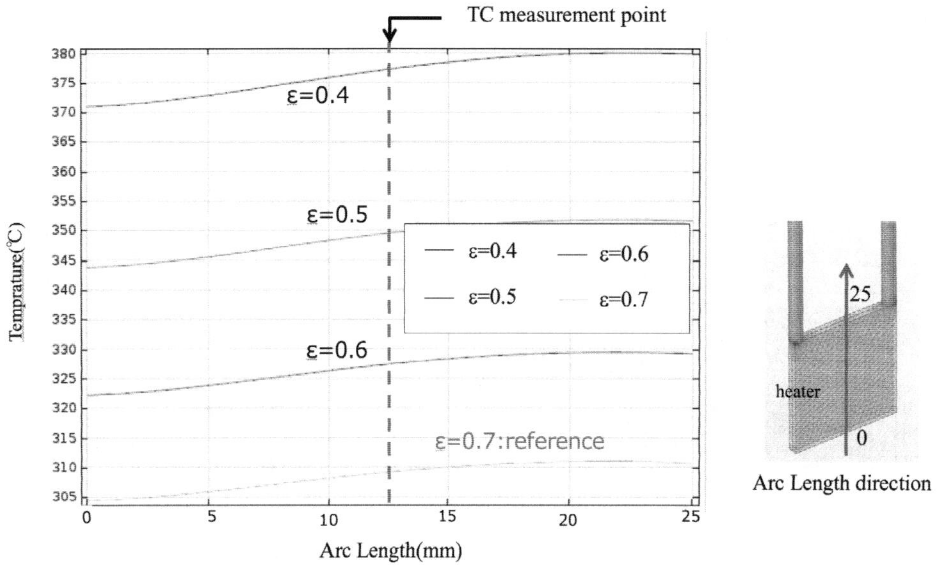


**Figure 12.** Calculation result of temperature distribution of heater, in the case of Experiment 3. (Heater input is 7 W and heater temperature is about 300°C.).

where
$Q$ heat density of heat source
$q$ heat flux
$C_p$ heat capacity of heat medium
$\rho$ density of heat medium
$\vec{u}$ fluid flow vector

As a first step, we simulated the experiments under vacuum conditions (the first half of experiment). In this case, since the fluid flow velocity vector is 0, the term of thermal convection in the first term of Eq. (1) can be neglected, and Eq. (1) includes only the second terms of heat conduction and radiation. The following boundary conditions are applied: adiabatic condition on the outside of the chamber, thermal radiation on the heater surface and inner surface of the chamber written as Eq. (3). A chamber, a heater, and internal members, as shown in Fig. 10 were divided into finite elements of 207,479 free tetrahedrons. We conducted numerical calculations using these procedures.

$$-\vec{n} \cdot \vec{q} = \varepsilon\{G - e_b(T)\} \quad (3)$$

where
$\varepsilon$ total emissivity,
$G$ irradiation,
$e_b(T)$ total emissive power of black body,
$\vec{n}$ normal vector on the boundary.

In the case of Experiments 1 and 2 with 40 W, in the reference run the heater temperature is 714°C (see Figs. 5(a) and 6(a)). On the other hand, since the catalog value is 613°C, the total emissivity $\varepsilon$ is assumed to be 0.5. Based on the above calculation, heat density was $3.0 \times 10^7$ W/m$^3$ by calculation only with the ceramic heater (Fig. 10(b)). Next, a heater was placed in the chamber, and calculations were performed taking into account the effects of heat transfer and irradiation (Fig. 10(a)). As a result, the heater temperature was 706°C, which is in good agreement with the experimental result. In this state, calculations were performed when the heater surface was covered with nano-structured Pd/Ni and the emissivity changed. The results are shown in Fig. 11. According to this, if the emissivity decreases by 0.1 due to the covering of the nano-structured Pd/Ni, the heater temperature is expected to rise by 50°C. For this reason, it is difficult to completely eliminate the temperature rise due to the change in the emissivity, given the temperature rise observed in Experiments 1 and 2.

Next, we consider the case of 7 W input (Experiment 3). In this case, from the heater temperature at 7 W input, the total emissivity was 0.7 and heat density was $5.5 \times 10^6$W/m$^3$, as calculated by the heater alone. Based on this result, calculations were carried out when it was installed in the chamber. The heater temperature was 309°C, which is also nearly the same as the reference run (Fig. 7(a)). As with the case of 40 W, the calculated result of the temperature distribution when emissivity is changed is shown in Fig. 12. According to Fig. 12, even when the emissivity drops dramatically from 0.7 to 0.3 due to the coating of the nano-structured Pd/Ni, the temperature rise is only 70°C at the most. On the other hand, in the experimental results (Fig.7(b)), a temperature rise of 123°C has been observed, which suggests anomalous heat generation.

As described above, at least in the experiment at the 300°C region, we showed anomalous heat generation. At the present time, there are many data analyzing the effects of temperature, pressure conditions and material properties (structure and composition) of this phenomenon.

Based on the above discussion, excess heat observed in our experiments conducted by Mizuno's method cannot be explained by a normal chemical process, which suggests it is a condensed matter nuclear reaction. Since radiation measurement and element analysis were not carried out at the present time, but only thermal measurements, further experiments and analysis will now be performed, detailed data will be acquired and we hope the phenomena will be clarified.

## 5. Conclusion

We tried to reproduce the heat generation experiment reported by Mizuno. We followed the same procedures using same experimental apparatus. As a result, the phenomenon reported by Mizuno could be reproduced.

As a result of evaluating the total amount of excess heat observed in our experiments, we conclude that the total amount of energy was much larger than that of plausible chemical reaction. Since a heater was covered with nano-structured material, there was a concern that change in emissivity affected the heater temperature measurement. Based on numerical calculations, we concluded that the observed temperature increase rise cannot be explained even if we postulate an unrealistic emissivity drop. These experimental and numerical results suggest that anomalous excess heat was generated by the interaction between nano-structured Pd/Ni surface and $D_2$ gas.

## Acknowledgements

The authors would like to thank Dr. Tadahiko Mizuno for the information on his device and procedures, and Prof. Hidetoshi Hashizume and Prof. Noritaka Yusa for their cooperation in the numerical analysis. The Department of Instrumental Analysis Technical Division School of Engineering, Tohoku University, supported our SEM analysis. We also thank Mr. Hideki Yoshino and Mr. Masanao Hattori for their support.

## References

[1] T. Mizuno, Reactant, heating device, and heating method, Patent Application, WO2015/008859 A2.

[2] H. Yoshino, E. Igari and T. Mizuno, Presentation at 2014 CF/LANR Colloquium at MIT, March.21–23, 2014, Massachusetts Institute of Technology, Cambridge, MA, USA.

[3] COMSOL Multiphysics Heat Transfer Module User's Guide, Version 5.2.

J. Condensed Matter Nucl. Sci. 24 (2017) 191–201

Research Article

# Replication Experiments at Tohoku University on Anomalous Heat Generation Using Nickel-based Binary Nanocomposites and Hydrogen Isotope Gas

Y. Iwamura*, T. Itoh† and J. Kasagi

*Research Center for Electron Photon Science, Tohoku University, Sendai, Japan*

A. Kitamura‡, A. Takahashi and K. Takahashi

*Technova Inc., Tokyo, Japan*

**Abstract**

We built a new experimental system to replicate the anomalous heat generation experiments based on the papers by A. Kitamura and A. Takahashi. The system was developed in order to measure precisely heat generation using a flow-calorimetry method with liquid hydrocarbon coolant that enables us to measure at temperatures higher than 373 K. The Ni-based binary nano-composite samples were prepared by the melt spinning method. A fabricated material at Kobe University were separated into two samples. One sample was loaded and tested at Tohoku University and the other at Kobe University, in order to compare the experimental results at the two different places. Two experiments were performed up to now. One was the PNZ4s ($Pd_{0\ 044}Ni_{0\ 31}Zr_{0\ 65}$) with $D_2$ gas experiment and the other is the CNZ5s ($Cu_{0\ 044}Ni_{0\ 31}Zr_{0\ 65}$) with $H_2$ gas experiment. For the PNZ4s with $D_2$ gas experiment, excess heat up to 10 W was observed. The amount of excess energy reached 2.5 MJ and it corresponded to 14.9 eV per absorbed D. CNZ5s ($Cu_{0\ 044}Ni_{0\ 31}Zr_{0\ 65}$) with $H_2$ gas experiment also showed anomalous excess heat ranging from 2 to 5 W. Coincident increase events of the pressure of reaction chamber and gas temperature, which suggested high temperature gas generation in the reactor chamber, was observed many times. The amount of excess energy amounted to 1.9 MJ and the generated energy per hydrogen atom was estimated as 67.8 eV/H. For the both samples subjected to the same fabrication process, results of the present work qualitatively agreed with those of the similar experiment performed at Kobe University. And these observations are supposed to be very difficult to explain by known chemical processes only.


*Keywords:* Anomalous heat, Deuterium gas, Gas loading, Hydrogen gas, Ni-based binary nano-composites, Replication

*Corresponding author. E-mail: iwamura@lns.tohoku.ac.jp.
†Also at: CLEAN PLANET Inc., Tokyo, Japan.
‡Also at: Graduate School of Maritime Sciences, Kobe University, Kobe, Japan.

© 2017 ISCMNS. All rights reserved. ISSN 2227-3123

## 1. Introduction

The Research Center for Electron Photon Science of Tohoku University and CLEAN PLANET Inc. agreed to the establishment of a collaborative research division – Condensed Matter Nuclear Reaction Division, in 2015 [6]. Research on anomalous heat generation started at the Condensed Matter Nuclear Reaction Division (CMNRD) although Iwamura and Itoh have been engaged in the study of transmutation reactions observed in nano-structured Pd/CaO multilayer thin film induced by $D_2$ gas permeation [7–15] for many years at Mitsubishi Heavy Industry and then CMNRD. Replication efforts have been made in two types of experiments as a first step at CMNRD in Tohoku University is the work using nickel-based binary nano-composites with $D_2/H_2$ gas by Kitamura and Takahashi [1–5], and the other is the experiment using nano Pd/Ni fabricated by glow discharge with $D_2$ gas by Mizuno [16,17].

In this paper, replication experimental results on the Kitamura and Takahashi works are described, which were achieved by close cooperation with Technova Inc. and Kobe University. A new experimental system using a flow-calorimetry method was built at Tohoku University, and the experimental results were compared with the results obtained at Kobe University. Similar results were obtained for the samples subjected to the same fabrication process, so it can be said that the replication efforts were successful.

## 2. Experiment

### 2.1. Experimental apparatus

A schematic of our experimental apparatus is shown in Fig. 1. It is basically similar to the apparatus described in Ref. [1], with some improvements. The main features of this system are as follows.

(1) Oil mass flow-calorimetry at high temperature.
(2) Many temperature sensors.
(3) Resistant to outside-temperature fluctuation.

The reaction chamber (RC) that contains the nickel-based binary nanocomposites and hydrogen isotope gas is located in the center of Fig. 1. Heat generation from the RC is estimated by the mass-flow calorimetry, which is widely used in this field, and is considered an accurate measurement method. Our method is standard except that we used oil coolant instead of water coolant. The coolant is input to the RC from the bottom and heated by the RC. Heat generation from the RC is calculated by the temperature difference between the coolant inlet and the outlet. The inlet and outlet oil temperatures are measured by three thermocouples each. The system is intended to accurately measure heat generation at high temperatures above 373 K. A liquid hydrocarbon coolant enables us to use the flow-calorimetry method at temperatures higher than 373 K. The coolant oil is an aromatic hydrocarbon (Barreltherm-400 (BT400); Matsumura Oil Co. Ltd.), which can work up to 600 K. The coolant is driven by a digital liquid tubing pump with a flow rate from 14.0 to 14.8 $cm^3$/min.

$H_2$ or $D_2$ gas is fed from a reservoir through a needle valve precision flow regulator to the nickel-based binary nanocomposite sample in the RC. Pressures in the RC and the reservoir are continuously monitored. Temperature distribution in the RC is measured by four Resistance Temperature Detectors (RTDs) and temperatures along the oil coolant pipe and the stainless-steel gas introduction pipe are monitored by thermocouples. Many temperature measurement points enable us to judge whether observed excess anomalous heat is real or not. The outer chamber is kept at constant temperature with cooling water from the thermobath $23\pm0.1^\circ$C. The entire apparatus is placed in a thermostatic chamber controlled at $23\pm0.1^\circ$C to avoid the influence of outside temperature fluctuations.

A 1 kW sheath heater (#1) is spirally wound on the outer surface of the RC and a 200 W cartridge heater (#2) is located at the central axis of the RC to heat up the sample in the RC. The power to the heater is fed from a finely

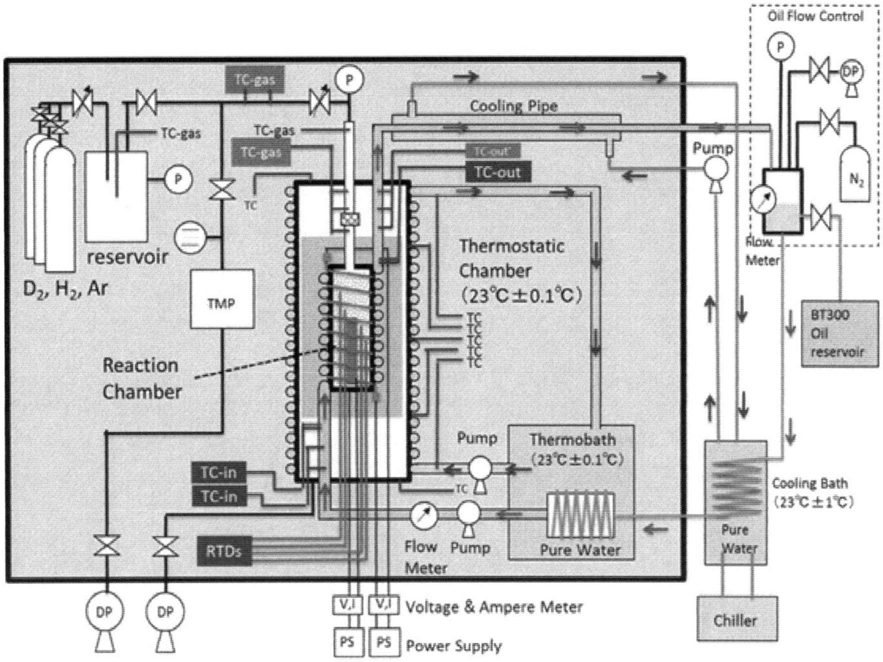


**Figure 1.** Experimental set-up.

regulated DC power supply. The input electric power for every heater is continuously monitored by two independent voltage and ampere meters to avoid mistakes on the input power estimation.

The coolant oil goes to the oil reservoir located outside of the thermostatic chamber and is cooled down to the ambient temperature by a cooling bath. It goes back to the tubing pump through a water bath kept at a temperature of 23±0.1°C.

## 2.2. Sample preparation

The procedure for sample preparation is the same as described in [1,2]. At first, an amorphous mixture of metal elements (Pd, Ni, Zr and Cu) were prepared by the melt spinning method at a Sendai company. By annealing the mixture in air at a temperature of 723 K for 60 h at Kobe University, preferential oxidation of Zr to $ZrO_2$ was expected with a consequent formation of binary-nanoparticles of Pd/Ni or Cu/Ni embedded in it. The sample compositions for the present paper are shown in Table 1. Atomic ratios of Pd:Ni or Cu:Ni were chosen to be 1:7 based on the Technova–Kobe work performed up until now [1–5].

The materials fabricated at Kobe University were separated into two samples. One sample was loaded and tested at Tohoku University in Sendai City and the other at Kobe University in Kobe City, in order to compare the observed experimental results at the two places. The samples CNZ5s and PNZ4s have the same physical and chemical properties as CNZ5 and PNZ4, respectively.

**Table 1.** Sample composition.

| Sample name | Tested at | Weight (g) | Molar fraction (%) | | | | |
|---|---|---|---|---|---|---|---|
| | | | Cu | Pd | Ni | Zr | O |
| PNZ4s | Tohoku University | 109.4 | – | 3.6 | 25.2 | 53.4 | 17.8 |
| PNZ4 | Kobe University | 109.4 | | | | | |
| CNZ5s | Tohoku University | 130.0 | 1.7 | – | 11.6 | 24.5 | 62.3 |
| CNZ5 | Kobe University | 130.4 | | | | | |

### 2.3. Experimental procedure

Calibration or blank runs to measure the heat recovery rate of the experimental apparatus were performed using zirconia beads in the RC before and/or after foreground runs. The diameter of the beads was 1 mm and weight was 1300 g for both experiments.

At the beginning of a foreground run, PNZ4s or CNZ5s sample with the zirconia beads was put into the RC. The RC was evacuated by a turbo molecular pump and then heated up to 200–300°C to remove $H_2O$ or other impurity gas. After the vacuum baking, the RC is cooled down to room temperature. About 1.0 MPa $D_2$ gas for PNZ4s and $H_2$ gas for CNZ5s were stored in the reservoir chamber in advance, then $D_2$/$H_2$ gas was introduced into the RC by opening the needle valve as shown in Fig. 1.

In the case of the PNZ4s sample, $D_2$ gas absorption and heat generation was observed at room temperature due to the presence of Pd. After observing that, we applied electric power to the heaters located at the inside and outer-surface of the RC to increase sample temperature. Data from temperatures, pressures, voltages, currents and a flow rates was logged during the experiments. Based on the data, we estimated the H/D absorption rate and excess heat generation from the samples.

## 3. Results and Discussion

Two experiments were performed up to now. One was the PNZ4s ($Pd_{0\ 044}Ni_{0\ 31}Zr_{0\ 65}$) with $D_2$ gas experiment and the other was the CNZ5s ($Cu_{0\ 044}Ni_{0\ 31}Zr_{0\ 65}$) with $H_2$ gas experiment.

### 3.1. Heat analysis and error estimation

The heat analysis of this system is based on the equation:

$$\eta Q = F_R \rho(T_{ave}) C(T_{ave})(T_{out} - T_{in}), \tag{1}$$

where $\eta$ rate, $Q$ is the heat release rate, $F_R$ is oil flow rate, $T_{ave}$ is the oil density as a function of temperature, $C(T_{ave})$ is heat capacity, $T_{out}$ and $T_{in}$ are the outlet and inlet temperatures of the coolant oil, respectively. Physical data of $\rho(T)$ and $C(T)$ of the coolant oil are already known. As the temperature dependence of $\rho(T)$ and $C(T)$ is linear, we can postulate that $T_{ave}$ is equal to $(T_{out} + T_{in})/2$. $Q$ is expressed as

$$Q = W_1 + W_2 + H_{EX}, \tag{2}$$

where $W_1$, $W_2$ and $H_{EX}$ are the input power of heater 1, the input power of heater 2 and the excess heat power from the RC.

Based on these equations $\eta$ is determined as a function of $(W_1 + W_2)$ by a blank run because $Q$, $F_R$, $\rho(T_{ave})$, $C(T_{ave})$ and $(T_{out} - T_{in})$ are obtained by experimental data. $H_{EX}$ is calculated by a foreground run using the determined $\eta$.

We simplify Eqs. (1) and (2) to estimate experimental error.

$$H_{\mathrm{EX}} = \frac{F_{\mathrm{R}}\rho C}{\eta}\Delta T - W,$$

$$\Delta T = T_{\mathrm{out}} - T_{\mathrm{in}}, \qquad W = W_1 + W_2.$$

Considering that experimental variables are $F_{\mathrm{R}}$, $\Delta T$ the error range of the calculated excess heat is the sum of the fluctuations of the oil flow rate, temperature difference and input electrical power.

$$\delta(H_{\mathrm{EX}}) \approx |\delta(F_{\mathrm{R}})| \frac{\rho C\,\Delta T}{\eta} + |\delta(\Delta T)| \frac{F_{\mathrm{R}}\rho C}{\eta} + |\delta(W)|. \tag{3}$$

Actual experimental data shows that largest contribution to the error of $H_{\mathrm{EX}}$ is of the $F_{\mathrm{R}}$ term and $W$ is the most stable. Considering that experimental variables are $F_{\mathrm{R}}$, $\Delta T$ and $W$, we can assume that error range of the calculated excess heat is the sum of fluctuations of oil flow rate, temperature difference and input electrical. In the case of 80 W input, for example, fluctuations are as follows.

$$|\delta(F_{\mathrm{R}})| = 0.012(\text{ml/min}), \quad |\delta(\Delta T)| = 0.261(\mathrm{K}), \quad |\delta(W)| = 0.031(\mathrm{W}), \quad \delta(H_{\mathrm{EX}}) = 0.260\,(\mathrm{W}).$$

About $\pm 0.26$ W for 80 W input could be the error range for excess heat estimation in our experimental set-up.

### 3.2. PNZ4s

The first experiment was done using the PNZ4s sample at Tohoku University after a blank run. Figure 2 shows the temperature change at four points in the RC during absorption of D into PNZ4s at room temperature. Rapid temperature rises from room temperature could be seen depending on the positions of RTDs when $D_2$ gas was introduced into the RC. RTD1, RTD2, RTD3 and RTD4 are located successively from the bottom as shown in Fig. 2. It could be supposed that PNZ4s sample local density was highest around the RTD2 and RTD3, considering that the weight ratio of PNZ3s to zirconia beads is less than 0.1 and sample powder would be localized somewhere. This might be the reason why peak temperatures are larger at RTD2 and RTD3. The reason for the very fast temperature rise at RTD2 is unknown.

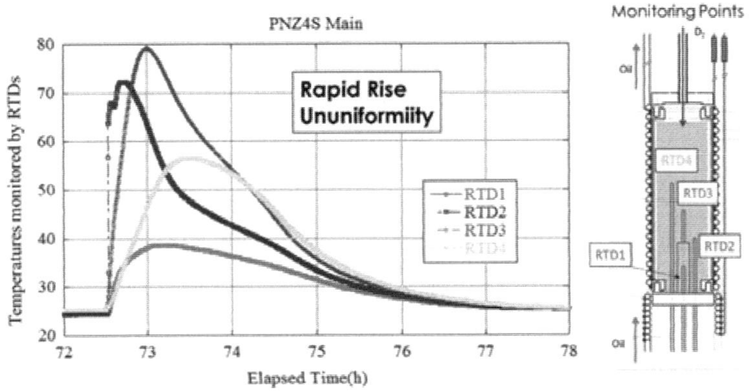


**Figure 2.** Temperature change at four points of the RC during absorption of D into PNZ4s at room temperature.

Figure 3 shows excess heat generation from PNZ4s with $D_2$ gas. The peak power induced by $D_2$ gas absorption at room temperature amounted to 10 W but it did not last for a long time. On the contrary, excess heat power was not so high at elevated temperatures but it continued more than 100 h. The peak ratios of excess heat to input power were about 4% and 5% for 80 and 134 W input, respectively. Integrated excess energies were 0.65 MJ for 80 W input and 1.73 MJ for 134 W input. Therefore, the amount of excess energy released from the RC was about 2.4 MJ.

Based on the pressure measurement at the RC and the reservoir for $D_2$ gas, the total amount of deuterium absorbed into PNZ4s was 1.73 mol. Although it is very difficult to postulate that every absorbed D atoms produced additional energy after the saturation of D-loading, average released energy at elevated temperatures per initial total D-loading is calculated as 14.9 eV/D, which is the lowest estimation of specific energy by one deuterium reaction that might happen. As the long-lasting excess heat was observed after the D-loading was saturated, the real portion of deuterons contributing to the heat generation is regarded to be very small (several orders of magnitude [1–5]) and consequently estimated specific energy might be very large). It is therefore very hard to explain that the generated excess heat was caused by some chemical reactions, as energy generation from known chemical reactions cannot exceed 10 eV/D

### 3.3. CNZ5s

A comparison of temperatures detected by RTDs and E1 between a blank run and an experiment with CNZ5s is shown in Fig. 4 when 134 W of heater power was input to the RC. In this experiment, no input was applied to W2 during the entire course of the test, because the W2 heater was not working due to disconnection-troubles. The thermocouple E1 is located on the top of the RC. Temperature increases of RTD3, RTD4 and E1 from the blank experiment data were observed for the CNZ5s experiment, although there was no difference between RTD1 and RTD2 temperatures compared to those of blank runs. It seems that excess energy is generated at temperatures higher than 220°C based on these observed experimental results.

Pressure of the RC (Pr) and a gas temperature (E2) at the near point of the top of RC are plotted in Fig. 5. Sudden fluctuations of Pr and E2 could be seen during the CNS5s experiment, although no fluctuation could be seen during the blank run and the baking (time-zone 0–150 h) as shown in Fig. 5.

Coincident increase events of Pr and E2 at random timing were observed as shown in Fig. 6. The pressure increase was about 0.02 or 0.03 MPa and the temperature increase ranged from 1°C. These were significant values and cannot

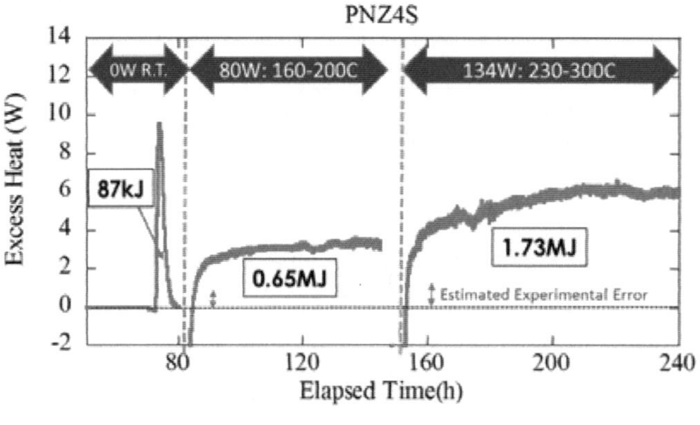


**Figure 3.** Excess heat generation for PNZ4s.

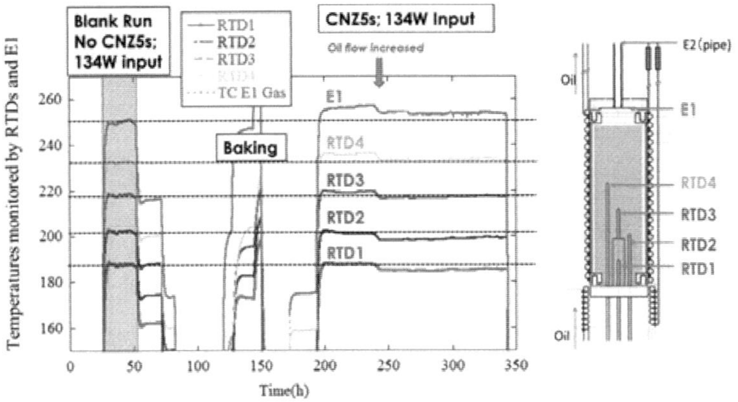


**Figure 4.** Comparison of temperatures detected by RTDs and E1.

be explained as instrument noise, as the other measurement data were stable and only Pr and E2 changes occurred, both at the same time.

A further close-up look at Pr and E2 confirms that Pr and E2 increased simultaneously, as shown in Fig. 7. The time interval for the plotted points for Pr and E2 was 10 s. Considering that the excess energy seemed to be generated in the upper region of the RC, as demonstrated in Fig. 4, these simultaneous increases of Pr and E2 might be due to the generation of high temperature gas. If we assume that high temperature gas was intermittently produced by certain condensed matter nuclear reactions in the upper region of the RC, this assumption will fit the experimental observations.

Excess heat for CNZ5s based on Eqs. (1) and (2) is illustrated in Fig. 8. About 4–5 W and 2–4 W were released during 80 W and 134 W heater power input, respectively. These excess powers are larger than the estimated errors based on Eq. (3). Sudden increase of excess power around 240 h seemed to be induced by the increase of the oil flow rate, but we have no accurate explanation for it.

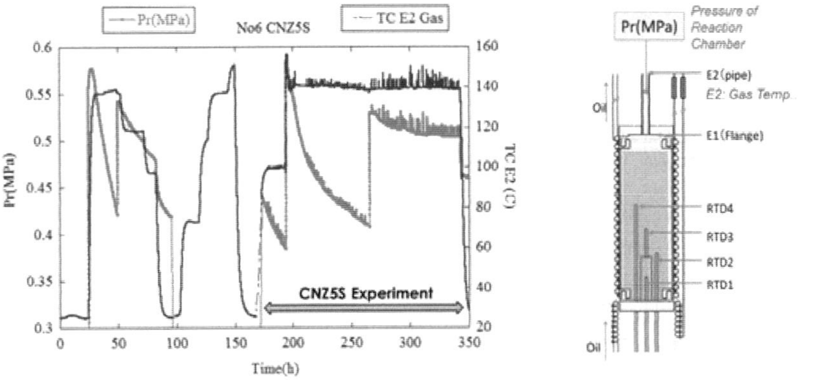


**Figure 5.** Fluctuations of pressure of the RC (Pr.) and a E2 Gas Temp.(E2) during a CNZ5s experiment.

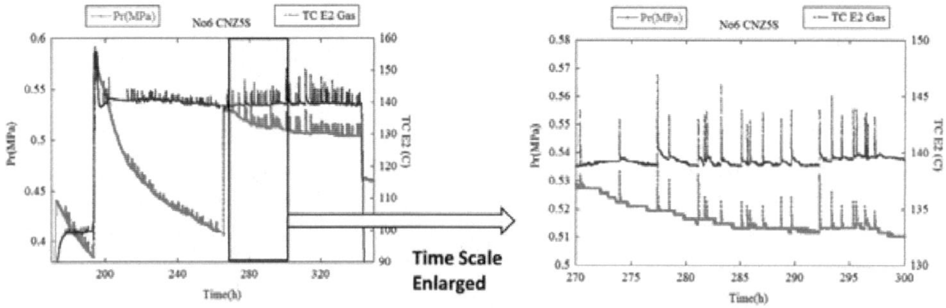


**Figure 6.** Coincident increase of Pr and E2.

All the integrated excess power comes to 1.86 MJ according to Fig. 8. Also estimated H atoms absorbed into CNZ5s during the experiment is 0.29 mol based on the pressure measurement at the RC and the reservoir for $H_2$ gas. Excess energy production is calculated as 67.8 eV/H, although it is very difficult to assume that every absorbed H atoms produced any energy same as D atoms. Also, it is very difficult to explain how the generated excess heat could have been caused by chemical reactions, as 67.8 eV/H energy generation is far larger than usual chemically released energy. This anomalous excess energy generation, in addition to the observed coincident pressure and temperature rises of the RC, suggests that some condensed matter nuclear reactions occurred in the RC.

### 3.4. Comparison between experimental results at Tohoku university and those at Kobe university

Table 2 summarizes experimental results obtained at Tohoku University and those at Kobe University. As previously described, the samples CNZ5s and PNZ4s have the same physical and chemical properties as CNZ5 and PNZ4, re-

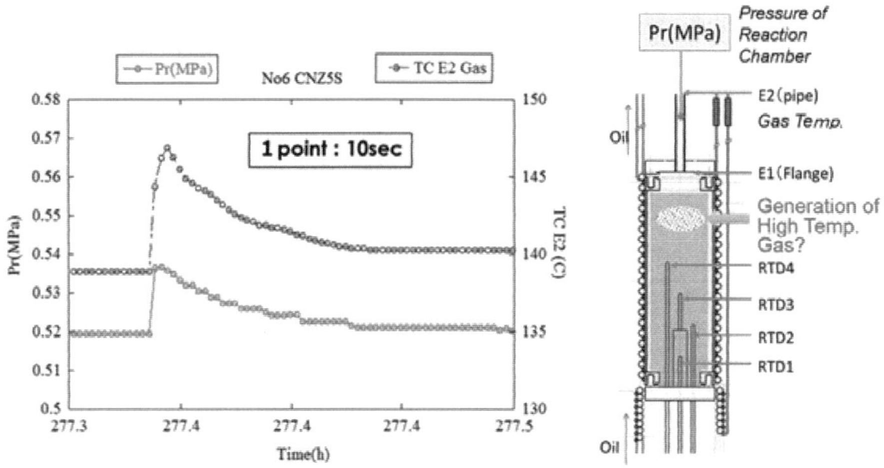


**Figure 7.** Simultaneous increase of Pr and E2 and the hypothesis for it.

**Table 2.** Comparison between experimental results at Tohoku university and those at Kobe university.

| Sample | Composition | Tested at | Released energy at room temp. (eV/D) | Released excess power at elevated temp. (W) | RC temp. (°C) |
|---|---|---|---|---|---|
| PNZ4s | $Pd_{0.044}Ni_{0.31}Zr_{0.65}$ | Tohoku University | 0.57 | 2–6 | 160–300 |
| PNZ4 | | Kobe University | 0.56 | Not estimated due to malfunction | |
| CNZ5s | $Cu_{0.044}Ni_{0.31}Zr_{0.65}$ | Tohoku University | | 2–5 | 150–250 |
| CNZ5 | | Kobe University | | 2–8 | 200–350 |

spectively.

The energy released at room temperature from PNZ4s and PNZ4 were 0.57 and 0.56 eV/D, respectively. These values are very similar. If we look at the released excess power from CNZ5s and CNZ5, they are close to each other. These results show that we successfully replicated experimental results on excess heat at least qualitatively, although there is still not enough quantitative reproducibility.

## 4. Conclusion

A new experimental system to replicate the anomalous heat generation using nickel-based binary nanocomposites and $D_2/H_2$ gas was introduced at Tohoku University. Two experiments were performed up to now. One was the experiment using PNZ4s ($Pd_{0.044}Ni_{0.31}Zr_{0.65}$) with $D_2$ gas, and the other was the experiment using CNZ5s ($Cu_{0.044}Ni_{0.31}Zr_{0.65}$) with $H_2$ gas. Anomalous excess heat generation were observed in both cases, and the experimental results of the present work qualitatively agreed with those of the similar experiment performed at Kobe University. In the case of CNZ5s experiment, events of coincident burst-like increases in the pressure of the reaction chamber and the gas temperature were observed many times, which suggested high temperature gas generation in the reaction chamber. These experimental observations are very difficult to explain by chemical reactions only.

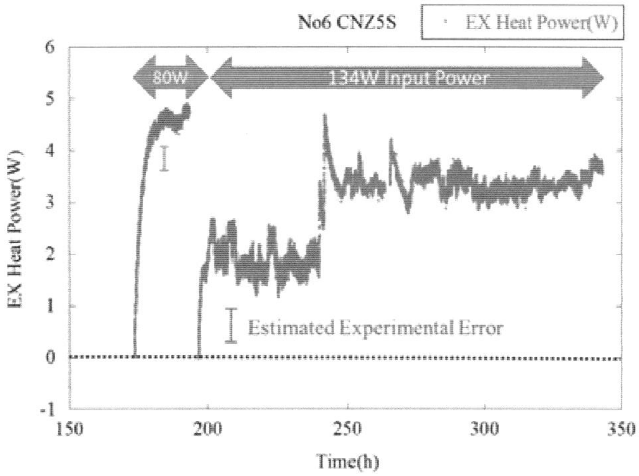


**Figure 8.** Excess heat generation for CNZ5s.

## Acknowledgements

The authors would like to thank Prof. Y. Furuyama (Kobe University), Prof. M. Kishida (Kyushu University), Mr. M. Nakamura, Mr. M. Uchimura and Mr. H. Takahashi (Nissan Motor Co.), Ms. R. Seto and Ms. Y. Matsuda (Technova Inc.), Dr. T. Hioki and Prof. T. Motohiro (Nagoya University) for their supports and valuable discussions. We also acknowledge Mr. Y. Shibasaki (Tohoku University) for his support for establishing experimental system. The authors would like to express appreciation to Technova Inc. for their cooperation in building our experimental apparatus, and we also deeply thank CLEAN PLANET Inc. for its extensive support.

## References

[1] A. Kitamura, A. Takahashi, R. Seto, Y. Fujita, A. Taniike and Y. Furuyama, Brief summary of latest experimental results with a mass-flow calorimetry system for anomalous heat effect of nano-composite metals under D(H)-gas charging , *Current Sci.* **108** (4) (2015) 589–593.

[2] A. Kitamura, A. Takahashi, K. Takahashi, R. Seto 1, Y. Matsuda, Y. Iwamura, T. Itoh, J. Kasagi, M. Nakamura, M. Uchimura, H. Takahashi, T. Hioki T. Motohiro Y. Furuyama and M. Kishida, Collaborative examination on anomalous heat effect using nickel-based binary nanocomposites supported by zirconia, *Proc. ICCF20*, to be published.

[3] A. Takahashi, A. Kitamura, K. Takahashi, R. Seto, T. Yokose, A. Taniike and Y. Furuyama, Anomalous heat effects by interaction of nano-metals and D(H)-gas, *Proc. ICCF20*, to be published.

[4] A. Kitamura, E.F. Marano, A. Takahashi, R. Seto, T. Yokose, A. Taniike and Y. Furuyama, Heat evolution from zirconia-supported Ni-based nano-composite samples under exposure to hydrogen isotope gas, *Proc. JCF16* (2016) 1–16.

[5] A. Kitamura, A. Takahashi, R. Seto, Y. Fujita, A. Taniike and Y. Furuyama, Effect of minority atoms of binary Ni-based nano-composites on anomalous heat evolution under hydrogen absorption, *J. Condensed Matter Nucl. Sci.* **19** (2016) 1–10.

[6] Y. Iwamura, J. Kasagi, H. Kikunaga, H. Yoshino, T. Itoh, M. Hattori and T. Mizuno, The launch of a new plan on condensed matter nuclear science at Tohoku university, *J. Condensed Matter Nucl. Sci.* **19** (2016) 119–126.

[7] Y. Iwamura, T. Itoh and S. Tsuruga, Transmutation reactions induced by deuterium permeation through nano-structured Pd multilayer thin film, *Current Sci.* **108** ( 4) (2015) 628–632.

[8] Y. Iwamura, T. Itoh and S. Tsuruga, Increase of reaction products in deuterium permeation-induced transmutation , *J. Condensed Matter Nucl. Sci.* **13** (2014) 242–252.

[9] Y. Iwamura, T. Itoh, N. Yamazaki, H. Yonemura, K. Fukutani and D. Sekiba, Recent advances in deuterium permeation transmutation experiments, *J. Condensed Matter Nucl. Sci.* **10** (2013) 63–71.

[10] Y. Iwamura, T. Itoh, N. Yamazaki, J. Kasagi, Y. Terada, T. Ishikawa, D. Sekiba, H. Yonemura and K. Fukutani, Observation of low energy nuclear transmutation reactions induced by deuterium permeation through multilayer Pd and CaO thin film, *J. Condensed Matter Nucl. Sci.* **4** (2011) 132–144.

[11] Y. Iwamura, T. Itoh, M. Sakano, S. Kuribayashi, Y. Terada and T. Ishikawa, Observation of surface distribution of products by X-ray fluorescence spectrometry during $D_2$ gas permeation through Pd complexes, *Condensed Matter Nuclear Science*, A. Takahashi, K. Ota and Y. Iwamura (Eds.), World Scientific, Singapore, pp.178–187, 2006.

[12] Y. Iwamura, T. Itoh, M. Sakano, S. Sakai and S. Kuribayashi, Low energy nuclear transmutation in condensed matter induced by $D_2$ gas permeation through Pd complexes: correlation between deuterium flux and nuclear products, *Condensed Matter Nuclear Science*, P.L. Hagelstein and S. Chubb (Eds.), World Scientific, Singapore, pp. 435–446, 2006.

[13] Y. Iwamura, T. Itoh, M. Sakano, S. Kuribayashi, Y. Terada, T. Ishikawa and J. Kasagi, Observation of nuclear transmutation reactions induced by $D_2$ gas permeation through Pd complexes, *Condensed Matter Nuclear Science,* World Scientific, Singapore, pp.339–350, 2004.

[14] Y. Iwamura, M. Sakano and T. Itoh, Elemental analysis of Pd complexes: effects of $D_2$ gas permeation, *Japanese J. Appl. Phys.* **41** (2002) 4642–4650.

[15] Y. Iwamura, T. Itoh, N. Gotoh and I. Toyoda, Detection of anomalous elements, X-ray and excess heat in a $D_2$–Pd system and its interpretation by the electron-induced nuclear reaction model, *Fusion Technol.* **33** (1998) 476–49.

[16] H. Yoshino, E. Igari and T. Mizuno, Presentation at 2014 CF/LANR Colloquium at MIT, March.21–23, 2014, Massachusetts Institute of Technology, Cambridge, MA, USA.
[17] T. Mizuno and H. Yoshino, Confirmation of excess heat generation during metal–hydrogen reaction, *Proc. JCF16* (2016) 17–28.

J. Condensed Matter Nucl. Sci. 24 (2017) 202–213

Research Article

# Collaborative Examination on Anomalous Heat Effect Using Nickel-based Binary Nanocomposites Supported by Zirconia

Akira Kitamura*,†, Akito Takahashi, Koh Takahashi, Reiko Seto and Yuki Matsuda

*Technova Inc., 100-0011 Tokyo, Japan*

Yasuhiro Iwamura, Takehiko Itoh and Jirohta Kasagi

*Research Center for Electron Photon Science, Tohoku University, 982-0826 Sendai, Japan*

Masanori Nakamura, Masanobu Uchimura and Hidekazu Takahashi

*Research Division, Nissan Motor Co. Ltd., 237-8523 Kanagawa, Japan*

Tatsumi Hioki and Tomoyoshi Motohiro

*Green Mobility Research Institute, Institutes of Innovation for Future Society, Nagoya University, 464-8603 Nagoya, Japan*

Yuichi Furuyama

*Graduate School of Maritime Sciences, Kobe University, 658-0022 Kobe, Japan*

Masahiro Kishida

*Graduate School of Engineering, Kyushu University, 819-0395 Kyushu, Japan*

**Abstract**

Hydrogen isotope absorption by nickel-based binary nanocomposite samples has been examined in collaborative work in the new NEDO MHE project. The samples tested so far include Pd0.044Ni0.31Zr0.65 ("PNZ3" and re-calcined "PNZ3r") and Cu0.044Ni0.31Zr0.65 ("CNZ5"). Material characterization by XRD and STEM/EDS has revealed the (continued in next page)


*Keywords:* Anomalous excess heat, Binary Ni-based nano-particles, Elevated temperature, Large hydrogen absorption, Melt-spinning samples, $ZrO_2$ supporter

*E-mail: kitamuraakira3@gmail.com.
†Also at: Graduate School of Maritime Sciences, Kobe University, 658-0022 Kobe, Japan.

© 2017 ISCMNS. All rights reserved. ISSN 2227-3123

(continued from p. 202) existence of crystalline phases of $NiZr_2$, $ZrO_2$, etc., and nano-meter size structures of binary Pd/Ni or Cu/Ni composites in $ZrO_2$ supporter. All samples at elevated temperatures (200–300°C) showed anomalous heat evolution with excess power of 5–10 W for periods of several days, which corresponded to excess energy of 5 keV/atom-D(H) or 0.5 GJ/mol-D(H). This anomalous heat generation phenomenon could be the basis for practical, carbon-free energy devices that do not produce hard radiations.

## 1. Introduction

There has been increasing interest in experiments with hydrogen gas charged nickel-based nano-composite samples for excess energy generation, owing to the higher availability of nickel compared to palladium. A Ni–Cu–Mn alloy thin wire, for example, has been examined extensively by Celani et al. [1]. In addition, a number of entrepreneurs are publicizing their own "products" of nano-fabricated samples on web sites with undisclosed details, and therefore with little scientific corroboration (e.g., [2,3]). Among them, replication experiments of the Rossi-type reactors have been performed by several researchers [4–7], which seemingly show considerable reproducibility of the Rossi method. However, little is known about the accuracy of the calorimetry or the mechanism of the claimed anomalously large energy production.

As reviewed in Ref. [8], the 8-year-long (2008–2015) series of studies on anomalous heat effects by interaction of metal nano-particles and D(H)-gas under the collaboration of Technova Inc. and Kobe University has become the basis of a national research effort sponsored by NEDO, which is hoped to lead to a new $CO_2$ free energy source. The new NEDO project on New Metal–Hydrogen Energy was started in October 2015 with a collaboration of six Japanese institutions, one of which the individual author of the present paper belongs to. The first result by the project is reported in this paper.

In the present work, hydrogen isotope absorption by nickel-based nano-composite samples has been examined in a collaboration using the experimental apparatus installed at Kobe University [8,9,11,13] in order to share scientific understanding of the anomalous heat effects both at room temperature (RT) and elevated temperatures. The system [9] has a reaction chamber with the capacity to hold a sample of 500 $cm^3$, and a flow-calorimetry system capable of working at elevated temperatures up to 300°C with use of a liquid hydrocarbon coolant. The samples tested so far include $Pd_{0.044}Ni_{0.31}Zr_{0.65}$ ("PNZ3" and re-calcined "PNZ3r") and $Cu_{0.044}Ni_{0.31}Zr_{0.65}$ ("CNZ5"), whose D(H)-absorption/heat-generation characteristics are discussed in the present paper.

## 2. Characteristics of Samples

The samples are an amorphous mixture of the metal elements prepared by the melt spinning method, and calcined in air at a temperature of 450°C for 69–100 h, during which preferential oxidation of Zr to $ZrO_2$ is expected with a consequent formation of binary-nano-particles of Pd/Ni or Cu/Ni embedded in it. The sample composition and calcination conditions are summarized in Table 1. The sample designated PNZ3r was subjected to the hydrogen isotope absorption runs at Kobe University, and subsequently re-calcined in air at 450°C for 200 h. The oxygen content was evaluated from the weight difference before and after the calcination/re-calcination. The samples CNZ5s and PNZ4s are from the same lot as CNZ5 and PNZ4, respectively. They were subjected to absorption runs at Tohoku University, whose results are presented in another paper by Iwamura et al. [10]. Atomic ratios of Pd/Ni or Cu/Ni were chosen to be ca. 1/7 based on the experience of the Technova–Kobe contribution [8,9,11,13] to the present work.

ICP-AES and XRD analyses were done at Nissan Motor Co. Ltd., Kyushu University, and Nagoya University independently. Many interesting features including crystalline phases of $NiZr_2$, $ZrO_2$, etc. have been revealed, which will be published independently elsewhere. STEM/EDS analyses done at Kobe University showed features of nano-structure of the samples. Two examples are shown in Fig. 1, showing a tendency of agglomeration of Pd and Ni

**Table 1.** Sample composition and sintering conditions.

| Sample name | Tested at | Weight (g) | Molar fraction (%) | | | | | Duration of oxidation at 450°C (h) |
|---|---|---|---|---|---|---|---|---|
| | | | Cu | Pd | Ni | Zr | O | |
| PNZ3 | Kobe | 95.5 | – | 3.5 | 24.5 | 52.0 | 20.0 | 100 |
| PNZ3r | Kobe | 113.2 | – | 1.7 | 11.6 | 24.5 | 62.3 | 200 |
| CNZ5 | Kobe | 130.4 | 1.7 | – | 11.6 | 24.5 | 62.3 | 60 |
| CNZ5s | Sendai | 130.0 | | | | | | |
| PNZ4 | Kobe | 109.4 | – | 3.6 | 25.2 | 53.4 | 17.8 | 60 |
| PNZ4s | Sendai | 109.4 | | | | | | |

particles together during $D_2$ absorption, runs followed by re-calcination and repeated absorption runs. Some features deduced from the pictures are summarized in Table 2.

## 3. Experimental Procedure

A schematic of the absorption-calorimetry system $C_1$ is shown in Fig. 2. Refer to [9] for detailed description of the system. Calibration of the flow calorimetry with a flow rate of 10 ml/min was done using 1-mm-diam. zirconia balls.

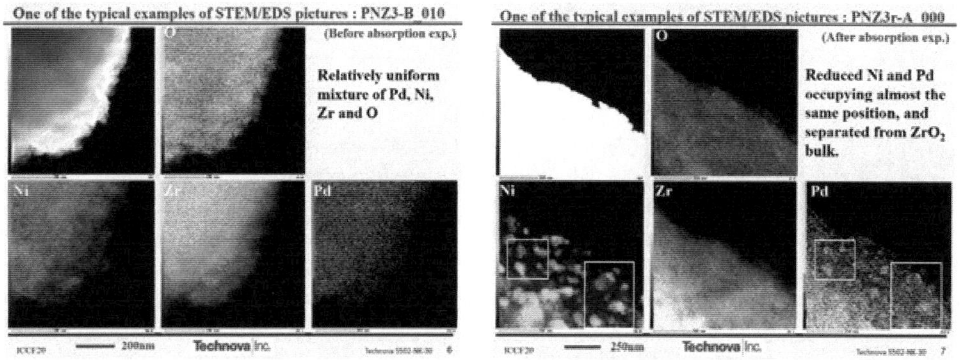


(a) (b)

**Figure 1.** Two examples of STEM/EDS.

**Table 2.** Some features perceived in STEM/EDS pictures.

| | | |
|---|---|---|
| PNZ3 | Amorphous ribbon calcined at 450°C for 100 h | •Mostly, relatively uniform mixture of Pd, Ni, Zr and O, but partly nonuniform distribution of Ni and Pd is recognized<br>•Most Pd and Ni atoms occupy the same position<br>•After absorption runs, $NiZr_2$ decreased, and $ZrO_2$ increased, and some reduced Ni atoms form a block |
| PNZ3r | Used PNZ3 re-calcined at 450°C for 200 h | •The assumed majority is $ZrO_2$+ NiO + PdO<br>•O atoms have increased<br>•NiO and PdO appear to be separated from $ZrO_2$<br>•Nonuniform distribution of Ni and Pd atoms developed further<br>•After absorption runs, NiO and PdO appear to be reduced |

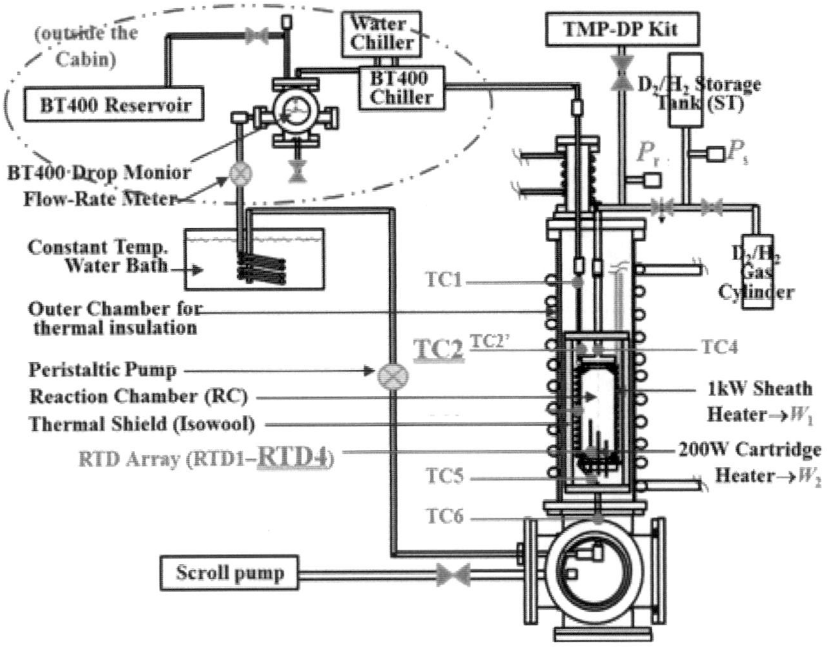


**Figure 2.** Experimental apparatus: Hydrogen isotope absorption-calorimetry system.

The heat conversion coefficient, $dT_{C2}/dW = 2.7°C/W$ or $1.44°C/W$, was obtained at RT or in the temperature range from 200 to 300°C, respectively. The heat recovery rate was calculated as

$$R_h = F\rho C(T_{C2} - T_{C6})/(W_1 + W_2), \tag{1}$$

where $F$, $\rho$ and $C$ are the flow rate, the mass density and the specific heat capacity, respectively, of the coolant BarrelTherm-400 (BT400), Matsumura Oil Co. Ltd., and $W_1$ and $W_2$ the outer sheath (#1) and the inner cartridge (#2) heater power, respectively. The mean value of $R_h = 0.8$ is a little lower than in the case of the $C_1$ system before the addition of the #2 heater due to increased loss of heat through the lead wire. The calibration run also serves as a control run giving reference temperatures, the flow rate of BT400 and heater power for foreground runs using the Ni-based samples. Comparing the temperatures in the foreground and background runs, the excess power is calculated using $dT_{C2}/dW = 2.7°C/W$ or $1.44°C/W$ in RT runs or elevated temperature runs, respectively.

As an example of foreground runs, the temperature and D(H)-loading history in the deuterium-filled PNZ3 runs, D-PNZ3#1 through D-PNZ3#3, and the protium-filled run, H-PNZ3#4, are shown in Fig. 3. Every run has several phase-runs with different combinations of the nominal input power ($W_1$, $W_2$) in watts, as indicated on the extreme right. The run #$n$ is separated by a vacuum baking phase at the end of the preceding run #($n$-1). The hydrogen loading ratio is also plotted in the figure. That is, $L_M \equiv$ D(H)/M, defined as the number of D (or H) atom lost from the gas phase inside the reaction chamber (RC) divided by the number of metal atoms M, where M stands for both for Pd and Ni atoms. That is to say, M denotes (7Ni + Pd)/8 for PNZ3 and PNZ3r, Ni for CNZ5, respectively. The term $L_M$ is calculated from the pressure of RC and of the storage tank (ST), $P_r$ and $P_s$, respectively, and their volumes with a correction for the temperature change based on the Boyle–Charles' law.

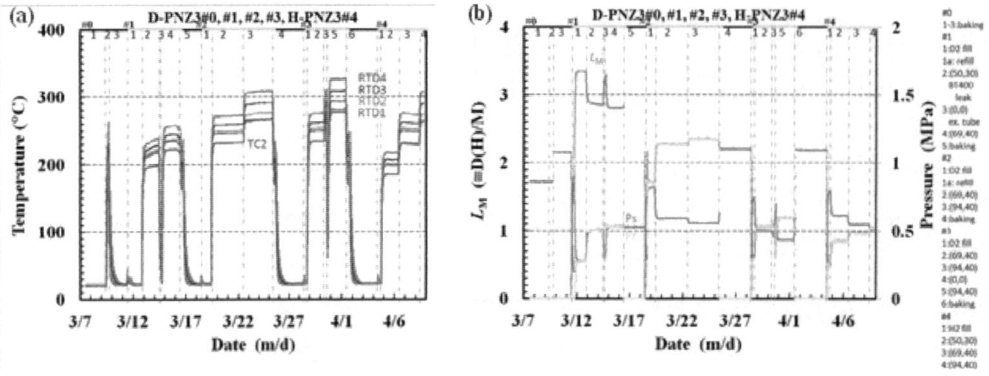


(a) (b)

**Figure 3.** Temperature (a) and D(H)-loading and (b) history of PNZ3 runs.

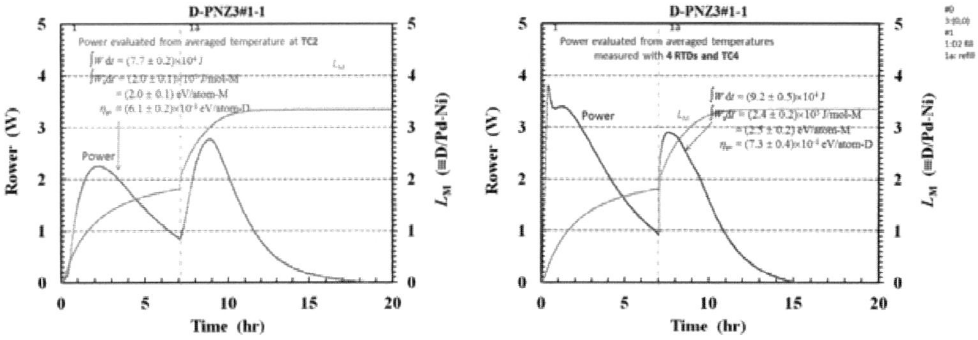


(a) (b)

**Figure 4.** Initial burst of heat recorded on (a) TC2 and (b) RTDav in the PNZ3#1-1 phase at RT.

## 4. Results and Discussion

### 4.1. Heat evolution in RT phases

#### 4.1.1. *PNZ3*

Let us first discuss the PNZ3 sample. In the PNZ3#1-1 phase, when the annealed virgin PNZ3 sample is exposed to hydrogen isotope gas, large initial bursts of heat are observed on the TC and RTD traces. The power recorded on the TC2 and RTDs are shown in Fig. 4. The latter is calculated by using the average temperature of those on RTD1 through RTD4 and TC4, which we call RTDav collectively hereafter. The reason why they have two peaks is simply because the ST was replenished to the initial high pressure at about 7 h after initiation of the run, which increased the heat evolution rate as a result of the increased flow rate of $D_2$ due to the increased pressure in the ST. temperature change based on the Boyle–Charles' law.

The humps are time-integrated in the #1-1 phase to calculate the emerging energy per absorbent atom,

$$E_a = \int_0^{T_1} W_a dt, \tag{2}$$

where $W_a$ is the power per adsorbent atom M, and $T_1$ is the duration of the #1-1 phase. The energy $E_a$ is divided by the saturation value of $L_M$ to give the specific sorption energy averaged in the #1-1 phase,

$$\eta_{av} = \frac{E_a}{L_M}. \tag{3}$$

These are calculated to be $E_a = (2.0 \pm 0.1)$ eV/atom-M and $\eta_{av} = 0.61 \pm 0.02$ eV/atom-D for TC2, or $E_a = (2.5 \pm 0.2)$ eV/atom-M and $\eta_{av} = 0.73 \pm 0.04$ eV/atom-D for RTDav.

It should be stressed that not only Pd but also Ni atoms are contributing to absorption of hydrogen isotopes at RT in the present binary nano-composite system supported by zirconia; this is regarded as a catalytic effect of the minor constituent Pd. The values of $E_a = 2.0$ to 2.6 eV/atom-M and $\eta_{av} = 0.6$ to 0.7 eV/atom-D are rather large in view of the hydrogen absorption energy of about 0.2 eV/atom-Pd and 0.22 eV/atom-D for bulk crystalline Pd. However, this is quite reasonable when we take into account that the XRD analysis of the virgin PNZ3 sample showed the existence of $NiZr_2$ and NiO [11]. Possible reactions to be taken into account other than hydrogen absorption/adsorption by Pd/Ni particles could be oxygen pickup reactions,

$$PdO + D_2(H_2) \rightarrow Pd + D_2O(H_2O) + 1.77(1.63)\,\text{eV/atom-Pd}, \tag{4}$$

$$NiO + D_2(H_2) \rightarrow Ni + D_2O(H_2O) + 0.176(0.033)\,\text{eV/atom-Ni} \tag{5}$$

and the hydrogenation reaction [12] with H replaced by D,

$$NiZr_2 + (4.5/2)H_2 \rightarrow ZrH_2 + ZrNiH_{2.5} + 2.85\,\text{eV/atom-Ni}\,(0.63\,\text{eV/atom-H}). \tag{6}$$

The contribution of these reactions (4)–(6) can account for the large values of $E_a$ and $\eta_{av}$, and also of $L_M > 3$ in the #1-1 phase. A quantitatively exact comparison of the observed $E_a$ and $\eta_{av}$ with the above reaction energies, or a discussion of the existence of anomalous reaction such as nuclear ones, will be possible after the fractional amounts of these phases, $NiZr_2$, NiO and PdO, are known exactly.

As shown in Fig. 3, $L_M$ is reduced to below 1.6 in the #2-1 and later phases after evacuation and baking in the #1-5 phase. The variations of $L_M$, $E_a$ and $\eta_{av}$ in the #$n$-1 RT phases ($n = 1$ to 5) evaluated at TC2 are shown in Fig. 5, and compared with those evaluated from the averaged values of the temperatures recorded on RTDav. Only the #1-1 phase has the exceptionally large values of $L_M$, $E_a$ and $\eta_{av}$. This appears to show that the formation of very stable $ZrH_2$ and possibly oxygen pickup reactions (4) and (5) only occur in the #1-1 phase, and are irreversible, or the $ZrH_2$ does not lose H during the evacuation and baking at temperatures around 300°C. (Hereafter, H stands for both protium and deuterium, unless explicitly declared.)

As can be seen in Fig. 3, $L_M$ depends on temperature rather strongly in the #2-1 and later phases. However, almost the same value of $L_M$ is recovered in the #$n$-1 phases ($n > 2$). The absorption/desorption is reversible and repeatable. However, $L_M$ cannot be decreased below about 1.0 under elevated temperatures up to 330°C. In these phases, formation of hydrides $PdH_x$/$NiH_x$ and $ZrNiH_{2.5}$ could be responsible for the hydrogen absorption. The facts mentioned above imply that hydrogen retention of either $PdH_x$/$NiH_x$ or $ZrNiH_{2.5}$ is dependent strongly on temperature in the range below 330°C, while the other ($ZrNiH_{2.5}$ or $PdH_x$/$NiH_x$) retains H in this temperature range with $L_M \sim 1$.

The specific absorption energies $E_a$ and $\eta_{av}$ in the phases #$n$-1 ($n > 1$) are also large compared with those for the bulk crystalline Pd mentioned earlier. The reason for this could be an anomalous effect including a nuclear one

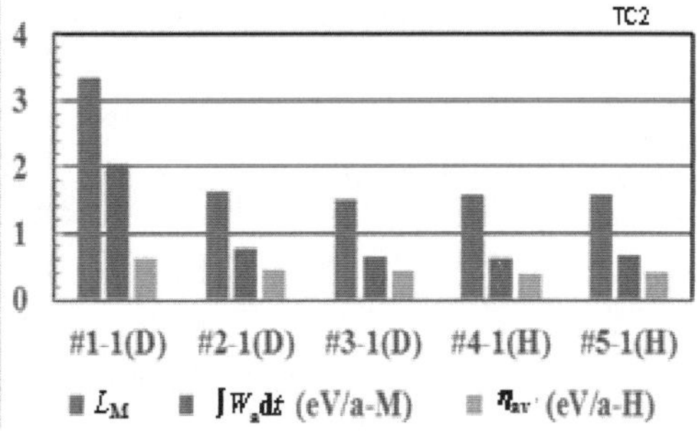


(a)

| | | #1-1(D) | #2-1(D) | #3-1(D) | #4-1(H) | #5-1(H) |
|---|---|---|---|---|---|---|
| $L_M$ | | 3.35E+00 | 1.64E+00 | 1.50E+00 | 1.59E+00 | 1.57E+00 |
| TC2 | | | | | | |
| $\int W_a dt$ | (eV/a-M) | 2.04E+00 | 7.68E-01 | 6.48E-01 | 6.17E-01 | 6.63E-01 |
| $\eta_{av}$ | (eV/a-H) | 6.09E-01 | 4.67E-01 | 4.32E-01 | 3.89E-01 | 4.22E-01 |
| RTDav | | | | | | |
| $\int W_a dt$ | (eV/a-M) | 2.45E+00 | 7.74E-01 | 7.22E-01 | 6.91E-01 | 6.74E-01 |
| $\eta_{av}$ | (eV/a-H) | 7.31E-01 | 4.70E-01 | 4.81E-01 | 4.35E-01 | 4.29E-01 |

(b)

**Figure 5.** Loading ratio and output energy in the first phases at RT (PNZ3#$n$-1) evaluated at TC2 (*left*), which are compared in the right table with those at RTDav.

under hydrogen absorption of Pd/Ni nano-particles. To reach this conclusion, however, the contribution of $ZrNiH_{2.5}$ formation to the absorption energies has to be evaluated quantitatively.

It should also be emphasized here that the values of absorption energies evaluated from TC2 and from RTDav agree within an error of about 20%. The heat evolution, or the sample distribution, is rather uniform in the present system. This justifies the use of RTDav in the calculation of excess power in the elevated temperatures discussed in Section 4.2.

### 4.1.2. *PNZ3r and CNZ5*

Both the PNZ3r and CNZ5 samples showed very little absorption in the RT phases: $L_M \sim 0.1$ with $E_a \sim 0.06$ eV/atom-M and $\eta_{av} \sim 0.6$ eV/atom-D in PNZ3r#1-1, and $L_M \sim 0.2$ with $E_a \sim 0.1$ eV/atom-M and $\eta_{av} \sim 0.5$ eV/atom-H in CNZ5#1-1. It is easy to understand the small values in the sample CNZ5#1-1, since it is established that both Ni and Cu particles do not absorb much hydrogen at RT, in contrast to Pd. The non-zero values of these variables could be related to this sample's ability to absorb substantial amounts of hydrogen at elevated temperatures, as will be discussed in Section 4.2. In other words, the very small values of $L_M$ at RT is simply a result of the Arrhenius law for absorption with an activation energy of around 0.04 eV as inferred by the temperature dependence of the absorption by this sample.

However, the small values in the PNZ3r#1-1 are difficult to understand, since the sample contains Pd. This sample was the PNZ3 re-calcined after finishing the PNZ3#4 run. The re-calcination was done in air at a temperature of 450°C for 200 h, during which hydrogenated Pd/Ni particles should have been reduced and oxidized. The sample appears to have suffered from agglomeration and/or segregation of Pd/Ni particles as seen in Fig. 1. It seems that the change of properties has led to the loss of hydrogen absorption activity of Pd/Ni. The non-zero values here again could be related to the hydrogen absorption property at elevated temperatures, discussed in Section 4.2.

## 4.2. Excess power evolution in the elevated temperature phases

In the elevated temperature phases, where $L_M$ takes the saturated values depending on the temperature, the temperatures $T_{C2}$ at TC2 and $T_{RTDav}$ are compared with the reference ones in the control run. The difference, if any, is divided by the conversion rate, $dT/dW$, and the number of metal atoms M to give the specific excess power, $W_{ex}$ (W/atom-M). This is integrated to give the specific excess energy, $E_{ex}$ (eV/atom-M), which is divided by the increment of $\Delta L_M (T_j)$ during each phase $j$ with duration of $T_j$ to give the specific sorption energy, $\eta_{av,j}$ (eV/atom-H), where the absolute value $|\Delta L_{\rm M}(T_j)|$ is adopted in the denominator. This means that both absorption and desorption are considered to be effective ways to generate anomalous excess power.

$$E_{\rm ex} = \int_0^t W_{\rm ex} {\rm d}t, \tag{7}$$

$$\eta_{{\rm av},i} = \frac{\int_{t-T_j}^{t} W_{\rm ex} {\rm d}t}{|\Delta L_{\rm M}(T_j)|}. \tag{8}$$

Another evaluation of the specific sorption energy, $\eta_{\rm t}$, (eV/atom-H), might be possible by dividing $E_{\rm ex}$ in each run #$n$ by the total amount of hydrogen isotopes having participated in the absorption $L_{{\rm M},n}$ in each run #$n$, and by summing up over the runs,

$$\eta_t = \sum_n \frac{E_{\rm ex} {\rm d}t}{L_{M,n}}. \tag{9}$$

Below in the present paper, $T_{\rm RTDav}$ is used to calculate excess power. This is because TC2 had a problem caused by contact failure at elevated temperatures. The use of $T_{\rm RTDav}$ is justified by the approximate equality of this and $T_{\rm C2}$ shown in Fig. 5, as mentioned earlier.

### 4.2.1. *PNZ3*

Variation of the temperature at RTDav (red) and after correction for the fluctuation of the flow rate of BT400 (green) are shown in Fig. 6(a) in comparison with the reference temperature at RTDav in $ZrO_2$ calibration run (black). The loading ratio $L_{\rm M}$ (orange) is also shown in the figure. As discussed in Section 4.1.1, the loading ratio $L_{\rm M}$ is rather high, possibly due to the formation of $ZrH_2$ (6) and oxygen pickup reactions (4) and (5), and is reduced to 1.6 in the #2-1 and later phases, when thc formation of hydridcs $PdH_x$/$NiH_x$ and $ZrNiH_{2\ 5}$ could be responsible for the absorption.

Figure 6(b) shows the specific excess power $W_{\rm ex}$ after additional correction for input heater power (brown), the integrated excess energy $E_{\rm ex}$ (black square) and the specific sorption energy $\eta_{{\rm av},j}$ (blue circle). In almost all phases the corrected temperature is larger than the reference one, and therefore the excess power is positive. The maximum excess power is about 10 W in the PNZ3#3-3 and #3-5 phases with the input power of $(W_1, W_2) = $ (94 W, 40 W), while the measurement error evaluated in various runs using blank samples is $\pm 1.5\%$ of the input power, or $\pm 2$ W in the present case. The PNZ3#3-4 phase with (0 W, 0 W) was successfully run to confirm the reproducibility of $L_{\rm M} = 1.6$ at RT.

The integrated specific excess energy $E_{\rm ex}$ is several tens of eV/atom-M in each phase, and amounts to $E_{\rm ex} \sim 1.8 \times 10^2$ eV/atom-M ($1.8 \times 10^1$ MJ/mol-M) at the end of the #4 run. The specific sorption energy defined by Eq. (8), $\eta_{{\rm av},j}$, is in the range from $3.2 \times 10^2$ eV/atom-H to $1.6 \times 10^1$ keV/atom-H. These are very large values that cannot be explained by any chemical processes caused by outer shell electrons having binding energies of the order of 10 eV. If we evaluate the specific sorption energy as defined by Eq. (9), we obtain $\eta_{\rm t} \sim 9.8 \times 10^1$ eV/atom-H. This is again one order of magnitude greater than the chemical one. We have to consider the possibility of nuclear origin for the energy production in the present system.

### 4.2.2. *CNZ5*

Similarly, the same variables are plotted in Fig. 7(a) and (b) for the CNZ5 sample. In this sample, only runs with protium gas were tested. This sample also generates excess power of up to 8 W, with $\eta_{{\rm av},j}$ ranging from $2.4 \times 10^{-2}$

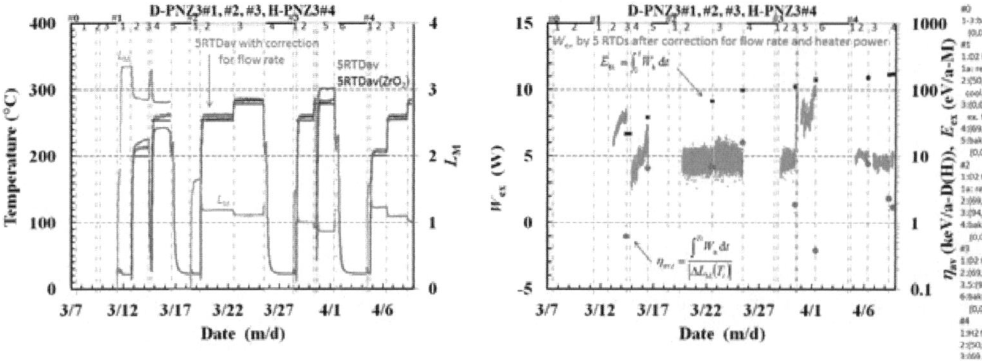


(a) (b)

**Figure 6.** (a) Temperature at RTDav (*red*), the temperature corrected for flow rate fluctuation (*green*), the reference temperature at RTDav (*black*), and the loading ratio $L_M$ (*orange*) in PNZ3#1 through PNZ3#4 runs. (b) The specific excess power $W_{ex}$ (*purple*), the integrated excess energy $E_{ex}$ (*black square*) and the specific sorption energy $\eta_{ex,j}$ (*blue circle*).

to 2.7 keV/atom-H and $E_{ex}\sim5.1\times 10^1$ eV/atom-M (5.1 MJ/mol-M) at the end of the #1 run. The values are rather modest, though not explainable by any known chemical reaction energy.

Very little hydrogen is absorbed in the #1-1 phase, until it starts to absorb a substantial amount of hydrogen at elevated temperatures of up to 200°C. The absorption characteristics of this sample resemble those of the CNS2 sample [13]. The CNS2 is a $CuNi_7$ nano-composite sample supported by mesoporus-$SiO_2$. The sample showed no absorption at RT, and absorbed protium up to $L_M\sim0.9$ in the elevated temperature phases up to $T\sim200$°C. It also produced excess power of $W_{ex}\sim10$ W, $E_{ex} = 3.8\times 10^2$ eV/atom-M and $\eta_{av}\sim1.5\times 10^4$ eV/atom-H in the phases where $L_M$ is saturated. It is inferred therefore that the absorption in CNZ5#1-2 phase and the excess power production is caused by Ni nano-particles with the catalyzing effect of Cu. A small difference between CNS2 and CNZ5 in the absorption at RT, together with the difference of factor 2 in $L_M$, might be a result of the difference in the supporting material. In the present CNZ5 sample, the hydrogenation reaction (6) might be contributing. However, only a small

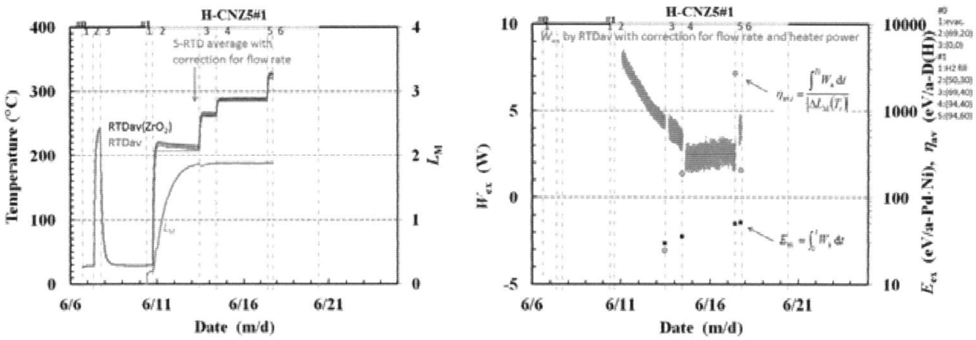


(a) (b)

**Figure 7.** Figures similar to Fig. 6 for CNZ5#1 run.

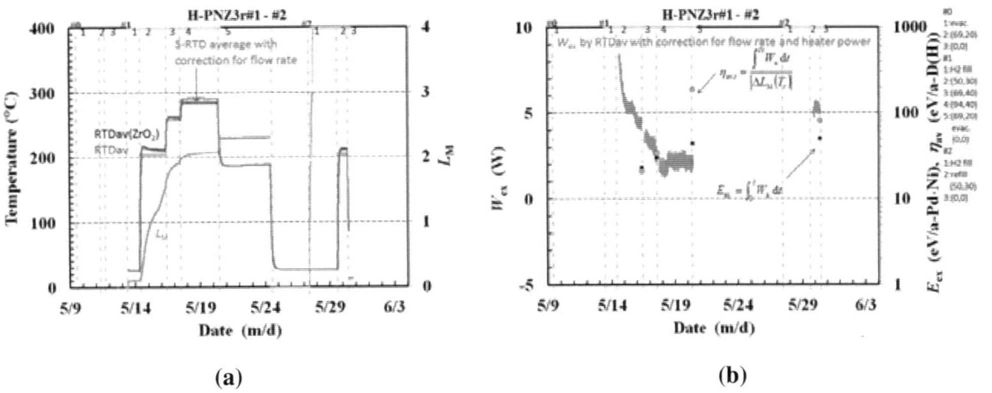

**Figure 8.** Figures similar to Fig. 6 for PNZ3r#1 - #2 runs.

amount of material in the $NiZr_2$ phase was observed in the STEM/EDS pictures. The reason has not been determined yet.

4.2.3. *PNZ3r*

The same variables for PNZ3 and CNZ5 are plotted in Fig. 8(a) and (b) for the PNZ3r sample. This sample was re-calcined after finishing the PNZ3#4 run in air at a temperature of 450°C for 200 h, which caused a drastic change in the absorption property as discussed in Section 4.1.2. The absorption properties both at RT and in the elevated temperature phases are very similar to those in the CNZ5#1 run shown in Fig. 7. However, the evolution of $L_M$ in the PNZ3r#1 run is shifted to the high temperature side by about 50°C compared with that in the CNZ5#1. The non-zero value of $L_M$ in the PNZ3r#1-1 is decreased, and the increase of $L_M$ still continues in the PNZ3r#1-4 phase. This shift might be suggesting the mechanism of hydrogen absorption by these materials at the elevated temperatures; absorption by Ni with catalysis by Pd or Cu, or contribution of the hydrogenation reaction (6). It is not clear at the moment.

This sample also generates excess power of up to 8 W, with $\eta_{av,j}$ ranging from $2.1\times 10^1$ to $1.9\times 10^2$ eV/atom-H and $E_{ex}\sim 5.0\times 10^1$ eV/atom-M (5.0 MJ/mol-M) at the end of the #2 run. The values are very modest, though not explainable by any known chemical reaction energy [8].

At the end, it is noted that the abrupt increase in $L_M$ at PNZ3r#2-1 phase is due to sealing failure at the chamber flange.

## 5. Summary

Hydrogen isotope absorption by nickel-based binary nano-composite samples has been examined in a collaborative work of the NEDO new MHE (metal hydrogen energy) project. The samples tested so far include $Pd_{0.044}Ni_{0.31}Zr_{0.65}$ ("PNZ3" and re-calcined "PNZ3r") and $Cu_{0.044}Ni_{0.31}Zr_{0.65}$ ("CNZ5"). Material characterization by XRD and STEM/EDS has revealed the existence of crystalline phases of $NiZr_2$, $ZrO_2$, etc., and binary-nano-particle structure of Pd/Ni and Cu/Ni in $ZrO_2$ supporter. The results obtained in the absorption runs are summarized in Table 3.

The virgin PNZ3 sample showed very strong absorption with rather large heat output in the #1-1 phase at RT, $L_M\sim 3.3$ and $\eta_{av}\sim 0.6$ eV/D, while in the #$n$-1 ($n>1$) phases after degassing process following elevated temperature runs, $L_M\sim 1.7$ and $\eta_{av}\sim 0.4$ eV/D. The difference could be due to the contribution of $ZrH_2$ formation in the former phase. On the other hand, the CNZ5 sample and the PNZ3r sample showed very little absorption but with comparable

**Table 3.** Summary of the results in comparison with those published earlier.

| Sample | Room temp. | | | | Elevated temp. | | Ref. |
|---|---|---|---|---|---|---|---|
| | #1 | | #2,#3,... | | $L_M$ at 200°C/hump on RTD | Excess power | |
| | $L_M$ | $\eta$(eV/a-D(H)) | $L_M$ | $\eta$(eV/a-D(H)) | | | |
| CNS2 | 0 | – | ~0 | – | 0.9/hump | 10 W (TC2) | [12] |
| PS3 | 2 | 0.7 | 0.7 | 0.4 | – | ~0 | |
| PNZt | (1) | 2 | 2 | 0.4 | – | 6 W /9 W (TC2/RTD) | [10] |
| CNZt | 0.2 | 0.4 | 0.15 | 0.2 | 1.6/hump | 5 W(TC2) | [10] |
| CNZtr | 0.15 | ~0 | 0.15 | ~0 | 2.0/hump | ~0 | |
| PNZ3 | 3.4 | 0.6 | 1.6 | 0.4 | – | 10 W (RTD) | This paper |
| PNZ3r | 0.1 | 0.6 | – | – | 2.0/hump | 8 W (RTD) | This paper |
| CNZ5 | 0.2 | 0.5 | – | – | 1.9/hump | 8 W (RTD) | This paper |
| PNZ4 | 3.5 | 0.6 | 1.7 | 0.4 | – | (Malfunctioning) | |

$\eta_{av}$, $L_M$~0.2 and $\eta_{av}$~0.5 – 0.6 eV/H at RT. Both the CNZ5 and PNZ3r samples at temperatures around 200°C showed strong absorption with $L_M$~2. To determine whether a nuclear process is involved or not, the amount of $NiZr_2$ in the virgin PNZ3 and the virgin CNZ5 samples, and the exact amount of $NiZrH_{2.5}$ in the samples in CNZ5#2-1 phase and PNZ3r#2-1 phases have to be established.

All samples at elevated temperatures (200–300°C) showed anomalous heat evolution: excess power $W_{ex}$~5–10 W for several days, corresponding to excess energy $E_{ex}$~5 keV/atom-D(H) = 0.5 GJ/mol-D(H). This could be the basis for carbon-free energy devices without hard radiation. However, since in practical applications the excess power must continue for an extended time (months or years), further effort to find better materials and better operating conditions is indispensable.

## Acknowledgements

Some of the authors, A. K., A. T., and Y. F. are grateful for Mr. T. Yokose for his assistance in performing the STEM/EDS analysis.

## References

[1] F. Celani, E.F. Marano, B. Ortenzi, S. Pella, S. Bartalucci, F. Micciulla, S. Bellucci, A. Spallone, A. Nuvoli, E. Purchi, M. Nakamura, E. Righi, G. Trenta, G.L. Zangari and A. Ovidi, Cu–Ni–Mn alloy wires, with improved sub-micrometric surfaces, used as LENR device by new transparent, dissipation-type, calorimeter, *J. Condensed Matter Nucl. Sci.* **13** (2014) 56–67.

[2] F. Piantelli, Nichenergy, http://e-catsite.com/2012/06/15/piantelli-moves-closer-to-commercialization/ .

[3] A. Rossi, Leonardo Corporation, http://ecat.com/ .

[4] G. Levi, E. Foschi, B. Höistad, R. Pettersson, L. Tegner and H. Essen, http://www.sifferkoll.se/sifferkoll/wp-content/uploads/2014/10/LuganoReportSubmit.pdf .

[5] A.G. Parkhomov, *Int. J. Unconventional Sci.* **6**(2) (2014) 57–61, ibid. **7**(3) (2015) 68–72, ibid. **8**(3) (2015) 34–39.

[6] S. Jiang, http://ja.scribd.com/doc/267085905/New-Result-on-Anomalous-Heat-Production-in-Hydrogen-loaded-Metals-at-High-Temperature (2015).

[7] J. Cole, http://www.lenr-coldfusion.com/2015/04/16/experiment-generates-apparent-excess-heat/ (2015).

[8] A. Takahashi, A. Kitamura, K. Takahashi, R. Seto, T. Yokose, A. Taniike and Y. Furuyama, Anomalous heat effects by interaction of nano-metals and D(H)-gas, *Proc. ICCF20, J. Condensed Matter Nucl. Sci.* **24** (2017).

[9] For example, A. Kitamura, A. Takahashi, R. Seto, Y. Fujita, A. Taniike and Y. Furuyama, Brief summary of latest experimental results with a mass-flow calorimetry system for anomalous heat effect of nano-composite metals under D(H)-gas charging, *Current Sci.* **108**(4) (2015) 589–593.

[10] Y. Iwamura, T. Itoh, J. Kasagi, A. Kitamura, A. Takahashi and K. Takahashi, Replication experiments at Tohoku university on anomalous heat generation using nickel-based binary nanocomposites and hydrogen isotope gas, *Proc. ICCF20, J. Condensed Matter Nucl. Sci.* **24** (2017).

[11] A. Kitamura, E.F. Marano, A. Takahashi1, R. Seto, T. Yokose, A. Taniike and Y. Furuyama, Heat evolution from zirconia-supported Ni-based nano-composite samples under exposure to hydrogen isotope gas, *Proc. JCF* **16** (2016) 1–16.

[12] P. Dantzer, W. Luo, T.B. Flanagan and J.d. Clewley, Calorimetrically measured enthalpies for the reaction of $H_2$ (g) with Zr and Zr alloys, *Metallurgical Trans.* A **24A** (1993) 1471–1479.

[13] A. Kitamura, A. Takahashi, R. Seto, Y. Fujita, A. Taniike and Y. Furuyama, Effect of minority atoms of binary ni-based nano-composites on anomalous heat evolution under hydrogen absorption, *J. Condensed Matter Nucl. Sci.* **19** (2016) 1–10.

J. Condensed Matter Nucl. Sci. 24 (2017) 214–229

Research Article

# Implications of the Electron Deep Orbits for Cold Fusion and Physics – Deep-orbit-electron Models in LENR: Present and Future

Andrew Meulenberg*

*Science for Humanity Trust Inc., USA*

Jean-Luc Paillet

*Aix-Marseille University, France*

**Abstract**

Relativistic deep-orbit electrons (D-O-Es) have previously been identified as the probable mechanism in the development of low-energy nuclear reaction (LENR) theories based on experimental Cold Fusion (CF) results. This present paper highlights how relativity and the near-field interactions (of deep-orbit electrons with both the nucleus and the lattice) predict accepted and many reported, but not-yet-accepted, CF observations. Included in the former category are: all of the new fragmentation ratios for the D+D ⇒ $^4$He CF nuclear reaction; a high-probability p–e–p ⇒ d reaction; a 'fast' decay process for transitions from excited to ground nuclear states; and a means of transferring excess nuclear energy to the lattice. Included in the latter category are: energy transfer from s-orbit atomic electrons to low-lying nuclear states; the formation of femto-atoms and femto-molecules – a basis for transmutations without the known 'hard' radiation (particulate or photonic) characteristic of neutron activation processes; selective attraction of femto-atoms/molecules to radio-nuclides (nuclear remediation); and the 'preferred' transmutation pathways in CF. Other effects, based on the published deep-orbit models, are predicted. Because of the successes of these models in explaining so much of CF, their mathematical basis is presently being explored beyond previous work(s). The physical bases for, and the consequences of, the mathematical predictions are proposed and described here. These include: special relativity and binding energy; the deBroglie term and spin-axis precessions; the deep-orbit quantum number, $k$; and deep-level splitting from spin–orbit, spin–spin, momentum and magnetic interactions. Theoretical concepts such as: symmetry breaking, 'sequestration', and elementary-particle mass changes to below their rest mass are also addressed. The new results and their interpretation, while incomplete, provide both satisfaction (in resolving prior issues) and surprises (in the magnitude and variety of near-nuclear effects).
.



*Keywords:* Femto-atoms, 500 keV binding energies, Near-field interactions, Nuclear remediation, Relativistic bound electrons, Sub-restmass nucleons, Transmutation

*E-mail: mules333@gmail.com.

© 2017 ISCMNS. All rights reserved. ISSN 2227-3123

## 1. Introduction

For over two decades, relativistic deep-orbit electrons (D-O-Es) have been identified as the probable mechanism for the development of low-energy nuclear reaction (LENR) theories based on experimental Cold Fusion (CF) results. The initial work of Maly and Va'vra [1] in applying this decades-old concept to cold fusion was 'debunked' by Rice, Kim, and Rabinowitz in an exchange of letters to the editor [2] and of papers [3] that performed a critical analysis of all of the CF models of the day. Their criticism was not based on the decades-old mathematical arguments against deep orbits themselves, but on physical principles and concepts: "In this letter, we show that these solutions are unphysical, and thus, these deeply bound energy levels cannot exist." The RKR criticism did include some math differences with the M&V work; however, these were on a different level than earlier criticism of these orbits denying the validity of the anomalous (deep-orbit) solutions of the Klein-Gordon and Dirac equations. Our interpretation of the RKR derivation for the anomalous solution does not deny the existence of the deep orbits (even though RKR claim that it does), but those authors discount their importance. Nevertheless, in their dismissal, they still provide some invaluable information: the contribution from the 'deep-orbit' portion is only two parts per billion. This interpretation is quite different from the 'blind' adherence to the 'pure' mathematics for a $1/r$ Coulomb potential that denies the acceptability of the anomalous solution. *It allows for a cold fusion process to raise that contribution to a 'useful' level.* However, it is doubtful that RKR would share this interpretation.

The RKR letter concludes with two physical bases for rejection: "If such orbits existed, upon collision, they would produce fusion at a much higher rate than muon-catalyzed fusion. So, their (M&V) proposed solution does not get around the nuclear ash problem." The first statement is absolutely correct. It accounts for the non-observability of a deep-orbit electron population. The second statement is false. It is precisely because the deep-orbit electron solves the nuclear ash problem that this concept is so important to CF and why this present paper addresses this and other implications of the deep orbits to CF.

This present paper highlights how relativity and the near-field interactions (of deep-orbit electrons with the nucleus and the lattice) predict accepted and many reported, but not-yet-accepted, CF observations. Included in the former category are: all of the new fragmentation ratios for the D+D ⇒ $^4$He CF nuclear reaction, a high-probability p–e–p ⇒ d reaction, a 'fast' decay process for transitions from excited to ground nuclear states, and a means of transferring excess nuclear energy to the lattice. Included in the latter category are: energy transfer from s-orbit atomic electrons to low-lying nuclear states, the formation of femto-atoms and femto-molecules – a basis for transmutations without the known 'hard' radiation (particulate or photonic) characteristic of neutron activation processes –, selective attraction of femto-atoms/molecules to radio-nuclides (nuclear remediation), and the 'preferred' transmutation pathways in CF.

## 2. Theoretical Concepts

Several theoretical concepts contributing to deep orbit models are described below:

(1) symmetry breaking (phase change),
(2) proximity effects,
(3) 'sequestration' (isolation of effects, [4]),
(4) elementary-particle mass changes to below its rest mass (energy conservation),
(5) Relativistic enhancement of the Coulomb potential (dynamic vs. static).

None of these concepts are esoteric, but neither are they commonly used.

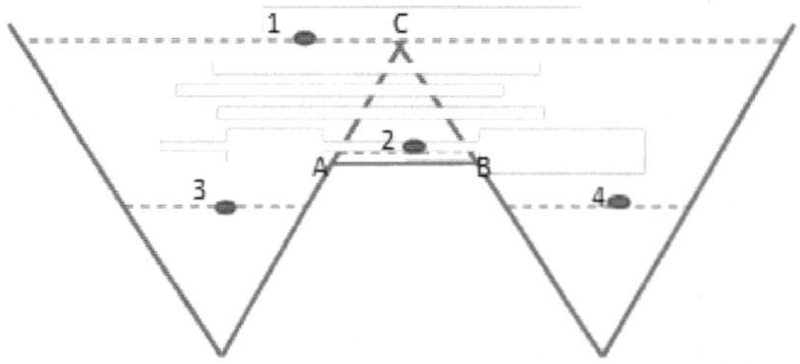


**Figure 1.** Schematic of a potential with energy-level-dependent symmetry.

### 2.1. Symmetry breaking (phase change)

Symmetry breaking is a phenomenon in which small (perhaps infinitesimal) variations of a system at a critical point may decide the system's fate, by determining which branch of a bifurcation is taken. Change in energy level may be such a variation. Figure 1 gives an example of symmetry breaking that applies to the cold fusion phenomenon. Objects at levels 1 and 2 have the same symmetry about the central axis. Moving, from 1 or 2 to 3 or 4, obviously changes symmetry. Nevertheless, moving from region 1 to region 2, while not a change in axial symmetry, still requires a change in energy (slight) and in dimensionality (large) at the critical point (C). On the other hand, if the dotted lines represent discrete states, a change from 1 to 2 requires a not-so-slight change in energy and a much smaller change in dimensionality.

Classically the change from state 1 to state 2 is a continuous process. This gradual change is seen as a phase change. However, change can be gradual or abrupt; both are the same, it is only a matter of degree. Quantum mechanically, the transition from state 1 to state 2 is a quantum jump. However, QM allows a 'mixing' of states 1 and 2 that still represents a phase change through the region between.

Changing Fig.1 to represent a cylindrical potential well in 3-dimensions introduces additional concepts of the critical points. In such a 3-D picture, regions 3 and 4 are joined; but, depending on the nature of the barrier, an electron in the joined region will not generally cross region 2. Electrons orbiting in this joined region will bounce back and forth between the potential walls as they move around the annular region. There are now two additional critical points that have no particular spatial location. At one of them, in angular momentum space, the electron could bounce back and forth between the potential walls, but not move around the annular region. The other, at a specific radius, will have the electron in a circular orbit that never touches the walls.

The 3-D version of Fig. 1 schematically represents the atom; however, the inner repulsive core is now from centrifugal force leading to a pseudo potential with different properties. The state 1 represents the atomic orbitals with the nuclear region as a central point. The state 3+4 represents the deep orbitals where an electron 'sees' the nuclear region as being the whole horizon. The relativistic Coulomb potential (mentioned below) has a repulsive core. In the state 1, the angular momentum would have to be almost zero ($l = 0$) for the electron to pass near to the nucleus. As an example, the s-orbital electrons are the only ones that strongly interact with the nucleus in internal conversion where nuclear energy is transferred directly to an atomic-orbital electron.

In the state 3+4, the deep-orbit electrons cannot have stable orbitals with zero angular momentum, $l = 0$. Such a condition is a critical point and any perturbation would give it rotation and $|l| > 0$. Can it have zero angular momentum within the Heisenberg uncertainty relation (HUR is $\Delta x \cdot \Delta p \geq \hbar$) where $\Delta x$ is within the multi-fermi range? Again, the answer is no. The energy required to obtain the needed momentum would not allow the electron to be bound. (This is an argument often used against the concept of deep-orbit electrons. It is addressed further below. However, the relativistic enhancement of these electrons versus the repulsive core is still to be fully explored.)

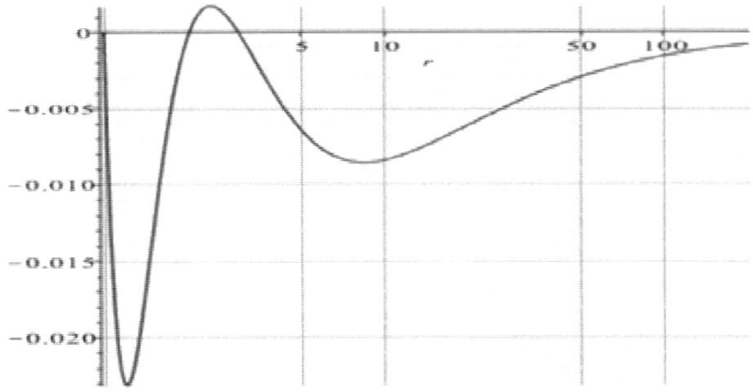


**Figure 2.** Plot of Eq. (1) with values of constants to represent standard and relativistic (proximity) effects.

Change in effective mass is generally considered to be a change in symmetry (e.g., pair production from a photon is 'two fermions from a boson'). Mass is associated with charge (a static electromagnetic, EM, energy). In pair annihilation, both mass and charge change phase to become a photon, an oscillating EM field. Thus, mass change is really a phase change from mass energy to EM energy. We believe that an understanding of this mass change is critical to the acceptance of DD $\to^4$He cold fusion.

## 2.2. Proximity effects

Symmetry change is associated with proximity to critical points. Various interactions have different radial dependencies. The full expression for potential of an electron about a nucleus is of a type:

$$V(r) = C_1/r + C_2/r^2 + C_3/r^3 + C_3/r^4, \tag{1}$$

$$V(r) = C_b\text{.pot.} + \text{centrifugal} + \text{spin–orbit} + \text{from } A^2 \ (A : \text{vector potential}).$$

The first two terms on the right-hand side define the atomic orbitals where the angular momentum $L$ in the centrifugal term, $L^2/2mr^2$, balances the attractive Coulomb potential to produce the minimum in the right-hand side of Fig. 2. As $r$ gets small, proximity effects dominate and the third and fourth terms compete to create a minimum in the left side of the figure, in which the semi-log plot expands the low-$r$ region.

If scaled to the atom, the two minima shown, one provided by the first two terms of Eq. (1) and one by the relativistic effects of the last two terms, are a factor of ~20 apart in the figure. Actual relativistic effects provide the second minimum to be ~4 orders of magnitude below the radius of the first (~5 fm vs. ~50 pm) and more than four orders of magnitude deeper (~500 keV vs. ~10 eV). Thus, the deep-orbit minimum is much more isolated than shown in the figure.

## 2.3. 'Sequestration' (isolation)

The opposite of proximity effects is sequestration. The strong relativistic effects of the deep-orbit interactions, because of sequestration, have negligible effect on low-$Z$ atomic systems. However, they could dominate in the near-nuclear regime. Sequestration does not have to be from spatial isolation. Some sequestration is perceptual or geometric. To an

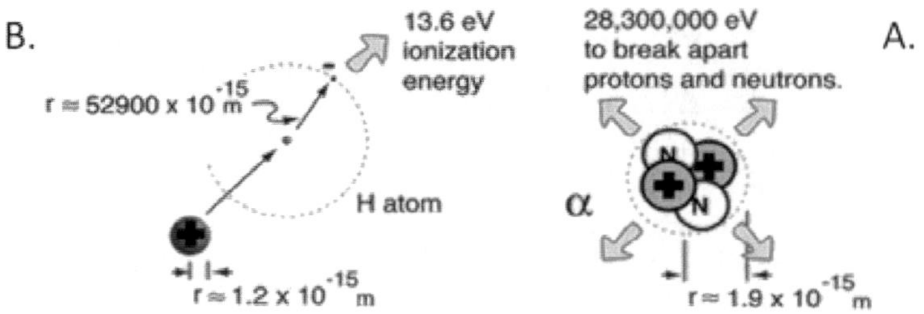


**Figure 3.** Comparison of atomic and nuclear scales and binding energies.

atomic electron, the nucleus is a 'speck' in the distance. To a deep-orbit electron, the nucleus is the constant horizon. The former is sequestered from the nuclear forces.

Relativity might be considered to have a major impact at both ends of the velocity scale. An electron in motion has a magnetic field **B**; a stationary one does not. Any velocity will create both the **B**-field and kinetic energy; however, the energy of the **B**-field is also an incremental increase in electron mass. Velocity removes the sequestration and the input energy goes into both mass (EM) and kinetic energy. Nevertheless, the mass increase is not observable until velocity is near '$c$' because the electron has a starting mass that masks the effect at low velocities.

Some sequestration is not from the spatial isolation (e.g., a $1/f(r)$ dependence); it is isolation from the consequences of proximity. An electron in an s-orbit is more massive when closer to a nucleus (where it is moving faster) because of relativity and the increased bound electro-magnetic fields. An important question for CF is "where does this mass energy come from?"

### 2.4. Mass change to below rest mass (energy conservation)

It is obvious that, when a stationary charge is moved, a portion of the energy creating the motion goes into kinetic energy (KE), as easily measured velocity, and a portion of the energy goes into distortion of the static electric field, **E**. This distortion is the easily measured magnetic field (**B**). Inertia is the resistance to change in velocity, **v**, and acceleration, **a**, is a measure of that change ($\mathbf{a} = \mathrm{d}\mathbf{v}/\mathrm{d}t$). Since $\mathbf{a} = \mathbf{F}/m$, one can see that mass and inertia are closely related. We could extend the concept of mass to that of 'effective mass' to make the relationship complete. In doing so, and in examining the consequences, we are able to get to a better understanding of mass. Some illustrative examples are helpful in this pursuit.

The mass of a nucleus is always less than the sum of the individual masses of constituent protons and neutrons. The difference is a measure of the nuclear binding energy. Nuclear binding energy, $\mathrm{BE_N} = \Delta mc^2$ is generally many MeV. Figure 3 indicates the energy required to ionize a hydrogen atom and to separate all of the nucleons of an alpha particle. The energy difference ($\Delta E = \Delta Mc^2$) in the process of forming a nucleus is often given off as gamma rays ($\gamma$); so we see mass being directly converted into EM radiation. Proton and neutron masses in a nucleus are less than their respective rest masses. The $\Delta M$ is from symmetry breaking and loss of sequestration as two nucleons or nuclei come together and, in a proximity effect, their potentials (nuclear and EM) interact.

Similarly, in the hydrogen atom, H, photo decay of an electron from an excited state results in a loss of mass and this is equivalent to the atomic-electron binding energy, $\mathrm{BE_A}$, or its ionization energy (Fig. 3B). Since the electron moves faster in the lower state, it gains mass in getting there. The emitted photon energy must come from the proton's mass (see Appendix). A phase change begins in the proton and electron as (spatial) sequestration is lost. We have seen that distortion of the static electric fields (of a stationary charge) stores energy expressed in the form of magnetic fields. The interaction of the electron and proton fields is the source of an increase in electron kinetic and mass energy.

The increase in the combined field strength and energy between the electron and proton also is stored energy. Since the electron is orbiting about the proton, the phase change includes not just a distortion of the fields (of stationary to linear velocity), it includes the change from an electro-static to an oscillating-dipole field (of linear to bound motion). In the case of the electron, which is considered to be 100% EM energy, it is expected that the change in fields is both external and internal. The proton, nearly 100% EM energy [4], is also expected to have changes in both external and internal fields. The fields of the relativistic charged quarks inside the proton must respond to the proximate electron's field just as the electron responds to their combined charge field. The result of polarization of the quark charges (another proximity effect) is stored energy. However, the net effect of the presence of the electron field is to lower the bond energy between the quarks (and perhaps between their constituent parts). Thus, the energy (mass) of the proton(s) is reduced enough to provide for the strengthened fields between the proton and electron and the emitted-photon energy.

There is no question of the loss in mass of nucleons as a result of neutron activation. There is probably no reason to ask if some of the nucleons change more than others. There is no question of the loss in mass of atoms resulting from photo-decay. The percent loss in mass of the nucleons is too small to be detected, so there is probably no reason to question if some of the nucleons change more than others. If the electron mass were reduced by the binding energy, spectrographic analysis is sensitive enough to detect a 10 eV change in a 511 keV mass electron. Has this ever been sought?

The decay process to a deep-electron orbit is energetically between that of atomic-electron photo-decay and that of nuclear fusion. However, no experiments have yet shown incontrovertible evidence of either the high-energy ($\sim$1/2 MeV) photo emission from electron decay to the deep orbits or from electron capture preceding an atomic electron fusing with a proton to form a neutron. Nevertheless, if the relativistic effects of a proximity interaction of the Coulomb potential in the nuclear region, or of some other process to absorb the excess potential energy resulting from the change to deep-electron energy levels, is present, then the predicted BE of $\sim$500 keV, plus $\sim$1 MeV (from kinetic energy) per electron would be subtracted from the nuclear mass.

### 2.5. Relativistic effects at small $r$

When an electron gets close to a nucleus and relativistic effects become important, a number of things are different. The virial theorem, which provides the relationship between kinetic and potential energies for stable orbits in a central potential, gives different values for relativistic particles. For a $1/r$ (Coulomb) potential, the non-relativistic relation is KE $=$ −PE/2. This non-relativistic relation is the reason that atomic electrons decaying to lower levels require photon emission. Energy (e.g., photonic, $E_{\mathrm{ph}}$) needs to be released so that the change in potential energy of the two orbits is matched by the change in kinetic and photonic energies, $\Delta$|PE|$=$ $\Delta$KE $+\Delta$E$_{\mathrm{ph}}$. As the velocity approaches the speed of light, KE approaches |PE|. Thus, decay within relativistic orbits requires a much smaller percentage of the transition energy to be dissipated by photon or some other mechanism.

For non-relativistic stable orbits in a $1/r^2$ potential (e.g., in Eq. (1)), the non-relativistic virial theorem states that $\Delta$PE $= -\Delta$KE. As the velocity approaches the speed of light, KE approaches 2|PE|. This means that, for relativistic orbits in a $1/r^2$ potential, it takes extra energy to 'decay' to a lower orbit. This effect continues for higher order potentials. Remembering Eq. (1), we see central potential components up to $1/r^4$. These effects are seen in Fig. 1. One of the consequences is that the requirement for removal of energy from a system (e.g., by photo emission) to allow decay to lower orbits may no longer be valid in the near-field regime. The potential barrier between the two 'wells' may be only a few fermi thick. This means that tunneling between the two would be nearly 100%. However, there remain other barriers.

The first barrier is the angular momentum problem. To even encounter a deep well, an electron must have near-zero angular momentum. This is not the $l = 0$ of the atomic orbitals which allows $l \leq \pm\hbar/2$ for that state. If the outer radius of the deep orbit is $>10^4$ smaller than that of the atomic orbital, then the cross section (just for encounter) is

$>10^8$ smaller. This is close to the RKR-predicted population of the deep orbitals of 2 ppB of the atomic ground state.

The second barrier is that of energy levels within the deep well. The Heisenberg uncertainty relation says that levels can exist in the nuclear region only for electrons with kinetic energies $\sim$100 MeV. Low-energy electrons trapped in atomic orbits will only see the deep levels as a small perturbation in the potential well of their shallow levels. An understanding of what determines an energy level is required to address this issue. This is an issue for another paper.

A third barrier, related to the first two, is one of probability and mechanisms. Assuming that a stable energy level exists within the deep well, what is its nature in terms of energy and angular momentum? These properties must be matched in some way to those of the electrons that could possibly populate them. Not only must the s-orbital atomic electrons that occasionally hit the small near-nucleus-sized target, they must have the correct 'approach' or the encounter would only be that of scattering from a resonant state.

Do we correctly interpret the relativistic quantum mechanics to predict the existence (and nature) of stable orbits in the deep levels? Do the Klein–Gordon and Dirac equations (properly) incorporate *all* of the relativistic effects? Presently, based on the calculated total energy, TE, for an electron of a few keV, both models predict deep-orbits with binding energies, BE, at the 500 keV level. For relativistic electrons, where the values of KE approach that of |PE|, both may be much larger than the static Coulomb potential would predict [5]. Even if both KE and |PE| are in the 100 MeV range, the binding energy could still be as predicted at $\sim$500 keV. This uncertainty in energies is the greatest unknown at the present time.

A few causes of these uncertainties that are presently being considered include:

- Spin–orbit (SO), spin–spin (SS), and magnetic (**B**) effects could greatly increase both the electron kinetic and potential energies, while the binding energy, BE, can remain the same.
- Relativistic and tight-orbit effects still need to be determined for SO, SS, and magnetic-field functions. Anisotropic enhanced forces ($\gamma$ vs. $\gamma^3$, for acceleration perpendicular vs. parallel to velocity vector) result in circularization of deep-electron orbits
- Effects of velocity dependence of mass. Bound (non-emitted) EM field increases electron mass, which increases centrifugal force

## 3. Cold Fusion Observations Explained

Perhaps as a most fortuitous circumstance, the origin of cold fusion was based on the PdD system. It presented a seemingly insurmountable intellectual barrier in that the deuterium–deuterium fusion process was so well known – and known to differ greatly from the proposed CF results. The expected fragmentation products (p, n, t, and $^3$He) of the known D–D reaction were observed at levels orders of magnitude below those commensurate with the excess heat measured. Only later, when the presence of excess $^4$He in CF was incontestable and highly correlated with excess heat [6], could the challenge that "the low-temperature D–D fusion reaction was not a source of excess heat" be met with certainty. The problem of overcoming this well-known reaction with a new physics process required a method of surmounting the Coulomb barrier between two deuterons as well as a model for preventing the known fragmentation of the fusion product [7]. Had CF originated in a hydrogen system (H+H $\Rightarrow$ D + neutrino), the Coulomb barrier would have been the only perceived challenge and many of the other observations of CF and its true nature could not have been modeled.

### 3.1. D+D fragmentation ratios

It has long been 'known', within the CF community, that one (or more) electrons are required to spend a greater than 'normal' time between the nuclei to overcome their Coulomb barrier. Almost no models have been proposed to address the fragmentation issue. To date, only two models are able to explain both the Coulomb barrier penetration and the avoidance of fragmentation in the D–D fusion process [8,9]. Both models utilize lattice confinement to increase the depth of the atomic/molecular potential wells and electron levels. The extended lochon model [10] has electron pairs temporarily bound to one of a pair of deuterons by the lattice **E**-fields oscillating in response to phonon resonance. The linear-H molecule model [11] and Storms' crevice model [12] propose linear and planar-to-linear confinement respectively to increase the electron density between nuclei in a linear, many-atom, H-molecule that is only loosely bound to the confining lattice. The longitudinal-optical phonon modes of the linear molecule (a sub-lattice) have much greater oscillation amplitudes because of their loose binding to the lattice and the enhanced screening from their more-centralized electrons. This, in turn, deepens the potential well about and between the nuclei to further confine the electrons and initiate the fusion process. The two models differ beyond this point as to the details of the fusion process.

The extended-lochon and linear-H (or -D) models propose that the transition of one electron from atomic to deep orbits lowers the proton masses sufficiently to eliminate the neutron+$^3$He fragmentation mode [10], A bound pair of deep-orbit electrons will lower the proton masses sufficiently to also eliminate the proton+tritium fragmentation mode [10]. The proximity of the negative electron(s) to the nucleus also reduces repulsion between the proton pair and this, when added to the reduced-mass effect, raises the proton + $^3$H fragmentation level above the input-energy level. Thus the CF D–D fusion reaction takes place beneath both fragmentation levels and perhaps even beneath the nuclear levels that would permit photo-decay. Even if fusion takes place above these levels, near-field (proximity) electro-magnetic coupling of the relativistic D-O-Es, tightly-bound to both the nucleus and the lattice [13], may still provide the fastest decay mode for removing the excess nuclear energy.

The crevice model proposes (in my interpretation, based on the linear-H model) that the oscillating hydrogen or deuterium atoms in a sub-lattice periodically come close enough together for the excess energy of the fusing nuclei to express itself, but not close enough to initiate fusion. The model's periodic, pulsed, attraction of the adjacent nuclear potential wells and the nuclear Coulomb fields for the bound electrons adds to the atoms' oscillations and this added energy is emitted as photons to be absorbed by the lattice. The nuclear energy of fusion is thus transferred to the many-atom H-molecule before fusion actually takes place.

Aspects of the crevice model have not been accepted because they appear to badly violate accepted physics. The possibility of gradually transferring nuclear energy to the lattice is one apparent violation and the variable lattice spacing of a collapsing multi-H molecule is another. Nevertheless, in 1991 Nobel Laureate Julian Schwinger provided just such a model for nuclear energy transfer [14] (however, he used phonons rather than photons as the transfer medium) and, in 1999, K.P. Sinha provided the quantum mechanical basis for fusion, if the lattice spacing can be varied [15]. Unfortunately, neither of these supporting concepts has been well-received either.

While the presence of screening electrons in some form is generally accepted within the CMNS community as a means of overcoming the Coulomb barrier between nuclei, it is not universally accepted as the only means of doing so. A deep-orbit electron about a proton forms a neutral femto-hydrogen atom H# with a diameter that is not much greater than that of a neutron. The deep orbit automatically brings an electron into the full screening regime so that fusion of a femto-H with another nucleus is quick and nearly inevitable. For this reason, it can be considered as a 'fat' neutron and is a strong transmutant. The p–e–p reaction is known in the solar nucleosynthesis story. Being a 3-body interaction, it is even slower than that of the p–p reaction. However, a femto-H interacting with a proton is essentially a 2-body interaction that gives a very-high-probability p–e–p ⇒ H# + p ⇒ d reaction to explain the light-hydrogen cold-fusion reactions.

### 3.2. A 'faster' nuclear decay process

A decay process for transitions from excited-nuclear to ground states, which is faster than normal gamma-emission processes, would explain many observed CF results. If it were faster than fragmentation processes, it would even explain some of the more unusual CF results. Are such processes possible and, if so, are they available? A very few such process models have been proposed. However, since this is not a review paper and must be limited in length, we will only mention (and reference) how the deep-orbit electrons perform this task.

Gamma decay is a photo-emission process. As such, it is normally associated with a doubly resonant energy and momentum transfer [16] between fixed nuclear-energy levels, with specific angular momentum difference ($\Delta l = 1$), and a transfer medium, the bound electromagnetic (EM) field. At some point, this field energy is also resonant with, and becomes, a free photon, which subsequently may be absorbed in scattering or in another resonant transfer with atomic electrons. The nuclear-energy transfer to a D-O-E is non-resonant [17] because the electron can absorb EM energy from the charged nucleon(s) and transfer that energy to the lattice without ever forming a stable new orbit or a photon. Since an excited nucleon can lose energy to an atomic electron transiting the nuclear region without changing its angular momentum (as in the proximity coupling of internal conversion), it can more readily lose it to a D-O-E, which always is close to the nucleus. Furthermore, since the nucleon and deep-orbit electron natural frequencies are orders of magnitude closer together than are the nucleon to atomic-electron frequencies, it is also possible for the nuclear energy loss not to be quantized. This is a direct EM transfer without a photon's requirement for change in angular momentum [18]. In the case of D+D $\Rightarrow$ $^4$He, both the excited nucleus and the D-O-E(s) have $l = 0$, so $\Delta l = 0$. Thus, no photon is involved. The known reaction of this form is that of internal conversion (IC), which is the preferred decay mode between states of equal angular momenta.

Photo-emission, when not forbidden, is faster than the internal conversion of atomic electrons. However, because of the constant proximity of the D-O-E to the nucleus and their comparable frequencies, the IC rate of D-O-Es may be more than twenty orders-of-magnitude faster than that of atomic electrons. Another critical factor for the high decay rate of an excited nucleus via IC, which has not been previously mentioned, is the high binding energy, BE, of a D-O-E. The BE of s-orbital atomic electrons is 10 eV $<$BE $<$ multi-keV, whereas that of a D-O-E is $>$500 keV. This means that the atomic-electron ejection of normal IC becomes only an excitation (perhaps a large one) of a D-O-E. Thus, the excited deep-orbit electron will remain bound, deliver its excitation energy to the lattice (by various means [13]) and be able to repeatedly absorb more energy from the nucleus. This extension of a well-known concept (IC) to a different orbit explains how the D-O-Es may provide a nuclear-decay path that could be fast enough to reduce, or even eliminate, fragmentation as well as photo-decay in $^4$He or other transmutations involving D-O-Es, even if the nucleon masses were not reduced by their presence. Actual values of relative decay rates still need to be determined.

## 4. Not-yet Fully Accepted Cold-Fusion Observations Explained

After the ICCF-14 conference in 2008, thinking of how DD fusion could lead to $^4$He without fragmentation led to thinking of classical deep transits of perturbed s-orbital electrons [19,20]. Finding that relativistic quantum equations predicted actual deep orbits was a revelation that opened the door to a new theoretical model of cold fusion with far-reaching consequences [10]. Some of these consequences, such as transmutation without any characteristic radiative decay, had been and were being announced in the experimental-CF literature at that time. However, since there was no theoretical basis for such results, which differed so greatly from known neutron-activation data, many of those who had accepted the original CF premise did not credit the new information.

One of the present authors (AM), who was working on theoretical means of overcoming the Coulomb barrier at the time of a workshop on transmutation [21] at ICCF-14, considered the transmutation data to be interesting (if true) but did not feel it would be useful in determining the 'real' processes of CF. However, as the deep-orbit electron model was being developed [10] and matured [17] and transmutation data became more convincing [22], it became obvious that the two models were inextricably linked. The first part of the puzzle was how to determine the possibility and probability of getting deep-orbit electrons. The classical papers on the subject indicated *transient* deep-orbit excursions and thus the possibility, but not necessarily sufficient probability, to solve the CF problem. Furthermore, the transient deep-orbit electrons did nothing to address the transmutation problem.

The relativistic Klein–Gordon and Dirac equations' prediction of deep-electron orbits did provide the means for CF results (including transmutation). Nevertheless, this aspect of the fundamental equations in quantum mechanics has been controversial for over 50 years. As mentioned above, in Section 2.5, there are reasons that the deep-electron levels are suspect. One non-mathematical aspect of the problem is population of the deep levels. If the deep levels *can* be populated, why do not we see evidence of the deep-orbit electrons and their transitions into these levels? Also, why do not all of the atomic electrons fall to these levels and the universe collapse? The answers to these questions are somewhat subtle, but never seem to appear in the literature.

The first reason for the low decay probability to the deep orbits comes most readily from quantum mechanics. The non-zero angular momentum atomic-electron orbitals do not overlap the nucleus (or the deep-electron orbits). Therefore, the transition probability between these states is negligible. The zero angular momentum atomic-electron orbital *does* overlap the nucleus and the zero angular momentum deep-electron orbits. The simple two-level QM calculation of relative populations is of the same order as the transition probability of the highly forbidden zero-zero transition. What is not considered is that the system in a solid-state environment is not two-level. While the thermal transition probability out of the deep level in a star gives a femto-atom lifetime exceeding a century, the lifetime of a femto-hydrogen atom (an atom with a deep-orbit electron) might be less than a pico-second before it fuses with a nucleus in the high-density environment. Thus, outside of the cosmological implications, many predictions using the two-level model are meaningless in the real world. It is sufficient to note that the negligible thermal-excitation rate from the deep orbit, along its calculated ratio relative to the decay rate from the atomic levels, eliminates the possibility of finding electrons in a deep level orbit if there is any other loss mechanism.

Even if cold fusion processes were to increase the decay probability to the deep orbits by a great many orders of magnitude and detection capabilities were available to sensitively measure electrons in these orbits inside a reactor, it is unlikely that the deep-level population could be detected. An isotopic measure of the accumulated transmutation products would be the most sensitive measure of this deep-orbit effect. Detection of radioactivity from transmutation would be the next most sensitive indication of cold-fusion and perhaps of deep-electron-orbit involvement. However, such characteristic radiation has not been observed, for reasons described below. The measure of the thermal output from a cold-fusion reaction is proof of the cold-fusion reaction, but not the process responsible.

A major stumbling block to the general acceptance of CF is the lack of a model for this non-standard process. A similar problem exists for models of populating the deep-orbits. In Section 4.3, two processes are mentioned that have been proposed to produce deep-orbit electrons. Just as the relativistic quantum mechanics equations provided a basis for the deep-electron explanation for cold fusion, relativity, as applied to the highly energetic electrons in close proximity to the nuclear Coulomb potential, provides a basis for the transition of atomic electrons to the deep-orbits.

### 4.1. Transmutation

The D-O-E model, created to explain the D+D $\Rightarrow$ $^4$He CF results, was found to fit the H+H $\Rightarrow$ D results as well (via the p–e–p and p–e–e–p fusion processes) [23,24]. The observed heat generation by CF with protons, not just with deuterons, led to refinements of the D-O-E model to include deep-orbit electron formation of femto-atoms and femto-molecules [25]. These new neutral structures became a basis for transmutation without the known 'hard' radiation (particulate or photonic) characteristic of neutron activation processes [26,27]. The neutral structures are strong transmutants. The hydrogen femto-atom H# becomes equivalent to a 'fat' neutron in terms of its ability to penetrate an atom and nucleus. Likewise, the hydrogen femto-molecule H#$_2$ becomes equivalent to a di-neutron. The $^4$He## femto-atom, as it decays to the $^4$He ground state might be hard to distinguish from a femto-molecule D#$_2$. It acts as a neutral alpha particle until it ejects the D-O-E(s) (if it does so) in its final decay process to ground.

### 4.2. 'Preferred' transmutation pathways in CF

Neutron activation is an extremely well-characterized process. It is a low-incident-energy-particle process that has no Coulomb barrier with which to contend. However, it produces unique transmutations giving radio-nuclides with well-known decay processes that are not observed in CF. We have just mentioned femto-atoms and femto-molecules and referred to them as 'fat' neutrons, di-neutrons, and neutral alpha particles that have the same capacity to penetrate nuclei as do neutrons. How would they differ or mimic the neutron interactions? There are three main differences:

They provide a multi-body decay process rather than the two-body process of neutron activation. One or more components of the femto-particle can be fused with the target nucleus with the remaining components able to carry off the excess energies. This allows a preferred path rather than the forced path of neutron activation. The preferred path is almost always toward the most stable daughter product (which result often mimics the natural isotopic abundance of elements). This preferred path is generally away from fission, if that would have been the normal result of neutron insertion.

Both the PdD and NiH systems have been explored to display details and the consequences of this preference [26,27]. They have one or two deep-orbit electrons present in the incident particle and generally in the daughter product. As in the case of the D+D $\Rightarrow$ $^4$He CF decay process [17], these D-O-Es provide a faster path to nuclear ground levels than photo-emission (gamma decay) and perhaps even faster than fragmentation. They also provide the proximate electron(s) necessary to greatly accelerate the weak interaction required to produce a neutron (e.g., via a p–e–p reaction), if that is the preferred path to greatest stability.

They can cause transmutation without adding kinetic energy to the target nucleus. The ability to add a deep-orbit electron to a nucleus, thereby changing a nuclear proton into a neutron in an accelerated weak interaction, and to carry off excess reaction energy with the other femto-atom/molecule components is unique. This mimics the (n, p) reaction and is a means of reducing/eliminating gamma decay from excited daughter products.

### 4.3. Energy transfer between nucleons and electrons decaying to deep-orbit

Based on unconfirmed reports of NiH-based CF (discounted by many) the D-O-E model contributes some ideas to the formation of conditions necessary for CF. One of the problems with the D-O-E model is the mechanism of populating the deep-orbit levels. The photon decay from H atomic-electron ground states to the deep-orbits is a highly forbidden 0–0 transition. A non-photonic energy transfer has been proposed based on doing work, rather than on photo emission. The lochon-model [28], where the lattice and sub-lattice phonon fields interact to form $D^+$ and $D^-$ pairs, allows work to be done as the charged deuterons are drawn together. While this may not work for transitions to non-relativistic deep orbits in a pure $1/r$ Coulomb potential, it would be possible in practice because of relativity and the additional terms in the potential [5]. Similarly, the linear-H molecule model [24] would provide for a photon-free decay to the deep levels. In both cases, the potential energy of the system decreases with the electron gaining kinetic energy and the deuterons losing mass energy. This part is no different than the atomic-electron case, where photons are emitted to satisfy the virial theorem. However, now the required energy loss goes into atomic or ionic motion and relative position rather than into photonic energy.

### 4.4. Energy transfer between nucleons and deep-orbit electrons

Defkalion reported [29] a non-activity of $^{61}$Ni as the only nickel isotope of the five they tested as pure isotopes (58, 60, 61, 62, and 64) that did not show excess heat generation. While trying to figure out a mechanism to explain the reported isotopic anomalies in these NiH CF results and remembering that atomic electrons can transfer energy with nuclear states by both photonic and non-photonic (e.g., internal conversion) means, one of the present authors (AM) looked at the different nuclei involved. Two differences observed (in http://www.nndc.bnl.gov/chart/) were the high total angular momentum ($J_n = 3/2^-$) of the ground state and the low first excitation level (67 keV) of the odd isotope. The high angular momentum is thought to prevent or retard transmutation via a femto-atom to the next stable element or isotope. The difference in binding energy through a weak interaction to convert a proton into a neutron vs. that through formation of a femto-molecule is an order of magnitude. Both processes involve a deep-orbit electron interacting (binding) with a nucleon.

When Rossi's work, based on the Lugano experiment [30], reported a major isotopic shift to $^{62}$Ni after an extended run starting with a natural isotopic abundance, this appeared to contradict the Defkalion results that would predict an accumulation of $^{61}$Ni. However, in the deep-orbit electron model, the H# reaction with $^{61}$Ni could be a 'choke' point of the proposed transmutation chain along the nickel isotopes. If that isotope became a halo nuclide of $^{61}$Ni ($^{61}$Ni plus the femto-H, H#, forming a femto=molecule), then $^{61}$Ni+H# would, under isotopic-mass analysis, look like $^{62}$Ni (within about 10 MeV). The mass difference results from the difference in binding energy of the halo nuclide vs. that of the fused neutron. If this happened and the pseudo $^{62}$Ni was long-lived in that reactive environment, its final slow decay ($^{61}$Ni+H# into $^{62}$Ni) would leave that later isotope dominant in the residue. The proposed transmutation chain, increasing along the nickel isotopes, would leave the nickel lattice intact to continue the reaction until the end of the natural Ni isotope chain. At this point, the preferred transmutation paths of the heavier nickel isotopes (via $^{62}$Ni + H# and $^{64}$Ni + H#) are no longer to nickel, but to copper and then Zinc. Thus, an apparent contradiction between two experimental results could be resolved to confirm the D-O-E model for both results. If either or both of the experimental claims were to be false, it would not disprove the model. It would simply not confirm it. If either of the results were to be confirmed, then there are specific tests that could confirm the model.

### 4.5. Energy transfer between atomic electrons and nucleons

A second major isotopic shift reported from the Lugano experiment was that of lithium. The report of a dramatic shift between the naturally dominant $^7$Li relative to $^6$Li and the final result with $^7$Li being greatly reduced appears anomalous. However, the 1$^{st}$ excited state of $^7$Li (at 478 keV) permits the possible energy transfer from an s-orbit atomic electron to a low-lying nuclear state. If an atomic electron can give up energy (by photonic or non-photonic means to excite the nucleus) then it can decay to a deep-electron orbit of either the $^7$Li or the H. If the electron is now deeply-bound to the proton, then the resulting femto-H, H#, readily fuses with the $^7$Li to become $^8$Be that quickly fissions into two alpha particles. The resultant $^4$He, as a gas, would probably not be measured in the isotopic analysis of the molten ash. If the deep-orbit electron is bound to the $^7$Li, the resulting hybrid[a] ($^7$Li#) would act as a long-lived $^7$He atom. It would take a very precise mass analysis to distinguish it from $^7$Li and one with very high resolution to separate it from the strong $^7$Li peak of the system. However, as a helium equivalent, a gas, it could escape or be thermally separated from the starting solid and measured. Needless to say, it would not appear in the experimental results reported for the solid ash. Thus, we have a third mechanism (along with the two mentioned in section 4.3) for the transition of an atomic electron into a deep-electron orbit.

### 4.6. Selective attraction of femto-atoms/molecules to radio-nuclides (nuclear remediation)

The ability of CF to not only produce heat from a nuclear source with little or no radioactivity but to reduce other radioactivity that might be present has been a theme that seems to be too-good-to-be-true. Therefore, claims to observe this effect have been generally disregarded. Nevertheless, if it is true, can it be explained? The deep-orbit-electron model, because of the relativistic velocities and extreme accelerations of electrons in femtometer orbits, provides for

[a]While the proposed hydrogen or helium atoms with one or two deep-orbit electrons respectively form femto-atoms, we have not decided on how to name or clearly identify atoms with *both* deep-orbit and atomic-orbit electrons.

intense long-range electric fields at nuclear frequencies [18]. The consequence of these fields and their interactions is the ability of the D-O-Es to transfer energy from the nucleus to the lattice and to interact with nuclear components at many lattice spacings from the source. The resonant-energy transfer from excited nuclei to a D-O-E or the lattice will provide a long-range attractive force between a femto-atom and the radio-nuclide. This force will be much stronger and further-reaching than that of the $1/r^6$ dependent induced-dipole-induced-dipole, interaction between a femto-atom and a stable nucleus. Thus, the small and mobile femto-atom experiences a selective attraction to radio-active nuclei and can become important in nuclear remediation. The D-O-E model is able to explain things that no other model can. There is still concern that some of these phenomena are not real. It will take time to sort out and reproduce the data before refinement and testing of the model becomes possible.

## 5. Summary

Known and proposed physical bases for, and the consequences of, the *mathematical predictions* of deep orbits:
Special relativity leads to:

- deep-electron levels with high binding energy ,
- mass changes (real and effective) for the nucleons and deep-orbit electrons,
- enhanced Coulomb and interaction potentials.

Spin-axis precession(s) leads to:

- the deBroglie wavelength and frequency for linear motion,
- a new higher-frequency 'mode' for wave mechanics of tightly bound electrons,
- a deep-orbit quantum number, $k$, for the new mode,
- deep-level splitting from spin–orbit, spin–spin, momentum, and magnetic interactions.

Effects of deep-orbit electrons (D-O-Es):

- D+D fragmentation ratios – D-O-Es reduce the mass and repulsive energy of nuclear protons to allow D+D $\Rightarrow$ $^4$He$^\#$ as preferred reaction,
- a 'fast' nuclear decay process – proximity coupling of D-O-Es with excited nucleons speeds decay rate, but reduces photonic decay (by both perturbation and competition),
- transmutation – near-total 'screening' of H$^\#$ and $^4$He$^\#$ by D-O-Es gives 'fat' neutrons and 'neutral alphas', which can readily penetrate atoms and nuclei,
- femto atoms – a D-O-E will make an atom with atomic number $Z$ act as one with $Z - 1$, etc.

Effects of D-O-E femto-products:

- femto-atoms can combine to form neutral femto-molecules. Together, with their ions, these form the femto-products,
- p–e–p $\Rightarrow$ d: The 3-body p–e–p problem is reduced to 2-body problem, a proton and a neutral femto-hydrogen atom (p + H#). This greatly enhances the probability of such a reaction,
- D-D = d–e–e–d $\Rightarrow$ $^4$He$^\#$: Lochon mechanism (locally bound electron pair) reduces a 4-body to a 2-body interaction (D–D $\Rightarrow$ D# + D# $\Rightarrow$ $^4$He$^\#$). This greatly enhances the probability of such a reaction,
- 'preferred' transmutation pathways in CF: femto-atoms or femto-molecules break up on entering a nucleus; assuming angular momentum conservation, the multiple options allow 'motion' toward the greatest nuclear stability (the lowest energy state),

- transmutations without 'hard' radiation: femto-products have relativistic D-O-Es that out-compete gamma decay for removing any excess energy from a nucleus.

Effects of relativistic D-O-Es:

- nuclear energy transfer to lattice electrons: relativistic bound electrons have extremely high, bound, EM fields. D-O-Es act like photons as intermediary between nucleus and lattice,
- D-O-E energy transfer to and from nearby nuclei: by proximity coupling and strong EM fields with nucleon-comparable frequencies; (like internal conversion and its inverse),
- selective attraction (nuclear remediation): D-O-E lowering of an excited-nucleus energy will create an attractive force ($F = -\mathrm{d}V/\mathrm{d}x$),
- atomic-electron energy transfer to nucleus: Resonant EM coupling will excite low-energy nuclear levels and allow the decay of atomic electron to deep-orbit levels. This would be the reverse of internal conversion where an atomic electron absorbs nuclear energy.

We have shown, mathematically and physically (in the accompanying ICCF-20 presentations [5,31]), that the prediction of deep-orbit electrons cannot be rejected. These levels are based on relativistic effects associated with the near-field interactions of electrons and nuclei.

If the deep levels really exist and can be populated, they provide the basis for understanding cold fusion and its many experimental observations. They also provide physical concepts that are the basis for new fields of study (femto-physics and femto-chemistry).

## Acknowledgement

This work is supported in part by HiPi Consulting, Ashland, VA, USA; by the Science for Humanity Trust, Bangalore, India; and by the Science for Humanity Trust, Inc., Tucker, GA, USA. Andrew Meulenberg would like to thank David French for an email exchange that triggered an increased recognition of the importance of high binding energy of deep-orbit electrons in the internal conversion mode of nuclear de-excitation.

## References

[1] J.A. Maly and J. Va'vra, Electron transitions on deep Dirac levels I, *Fusion Sci. Technol.* **24**(3) (1993) 307–318, http://www.ans.org/pubs/journals/fst/a_30206.

[2] R. Rice, Y. Kim and M. Rabinowitz, Comments on electron transitions on deep Dirac levels I, *Fusion Technol.* **26**(1) (1994) 110. And *Response* by Maly and Va'vra, *Fusion Technol.* **26**(1) (1994) 111.

[3] R. Rice, Y. Kim, Rabinowitz and Zubarev, Comments on exotic chemistry models and deep Dirac states for cold fusion, *Proc. 4th Int. Conf. on Cold Fusion*, Vol. 4, Dec. 6–9, Hawaii, 1993, p. 4.1, *lenr-canr.org/acrobat/RiceRAcommentsona.pdfý*, http://www.epri.com/abstracts/Pages/ProductAbstract.aspx?ProductId=TR-104188-V4.

[4] F. Wilczek, Origins of Mass, arXiv:1206.7114v2, 22 Aug. 2012.

[5] J.L. Paillet and A. Meulenberg, Advance on Electron deep orbits of the hydrogen atom for LENR, *Proc. of ICCF-20, the 20th Int. Conf. on Condensed Matter Nucl. Sci.*, Sendai, Japan, October 2–7, 2016.

[6] Abd ul-Rahman Lomax, Replicable cold fusion experiment: heat/helium ratio Current Science,25 February 2015, 108–104) http://www.currentscience.ac.in/php/feat.php?feature=Special%20Section:%20Low%20Energy%20Nuclear%20Reactions&featid=10094.

[7] A. Meulenberg and K.P. Sinha, New visions of physics through the microscope of cold fusion invited paper, *17th Int. Conf. on Condensed Matter Nucl. Sci.*, Daejeon, Korea, 12–17 August, 2012, *J. Condensed Matter Nucl. Sci.* **13** (2014) 378–390.

[8] A. Meulenberg, Extensions to physics: what cold fusion teaches, Special Section: Low Energy Nuclear Reactions, Section editors: M. Srinivasan and A. Meulenberg, *Current Sci.* **108**(4) (2015) 25,

http://www.currentscience.ac.in/cs/php/feat.php?feature=Special%20Section:%20Low%20Energy%20Nuclear%20Reactions&featid=10094.

[9] E.K. Storms, *The Explanation of Low Energy Nuclear Reaction*, Infinite Energy Press, Concord, NH, 2014, 365 pages (updated e-version available at Amazon.com).

[10] A. Meulenberg and K.P. Sinha, Tunneling Beneath the $^4$He$*$ Fragmentation Energy, presented at *239th ACS National Meeting*, San Francisco, CA, March 2010, *J. Condensed Matter Nucl. Sci.* **4** (2011) 241–255.

[11] A. Meulenberg, Pictorial description for LENR in linear defects of a lattice, *ICCF-18, 18th Int. Conf. on Cond. Matter Nucl. Sci.*, Columbia, Missouri, 25/07/2013, *J. Cond. Matter Nucl. Sci.* **15** (2015) 106–116.

[12] E.K. Storms, A theory of LENR based on crack formation, *Infinite Energy* **19** (2013) 24–27.

[13] A. Meulenberg, Deep-orbit-electron radiation absorption and emission, *ICCF-18, 18th Int. Conf. on Cond. Matter Nucl. Sci.*, Columbia, Missouri, 25/07/2013, *J. Condensed Matter Nucl. Sci.* **15** (2015) 125–136.

[14] J. Schwinger, Nuclear energy in an atomic lattice, in *The First Annual Conf. Cold Fusion*, Salt Lake City, Utah, 1990.

[15] K.P. Sinha, A theoretical model for low-energy nuclear reactions in a solid matrix, *Infinite Energy* **29** (2000) 5, 29, 54.

[16] A. Meulenberg and K.P. Sinha, Deep-electron orbits in Cold Fusion, *17th Int. Conf. Condensed Matter Nucl. Sci.*, Daejeon, South Korea, 12–17 August, 2012, *J. Condensed Matter Nucl. Sci.* **13** (2014) 368–377.

[17] A. Meulenberg, From the naught orbit to He$^4$ ground state, *16th Int. Conf. on Condensed Matter Nucl. Sci.*, Chennai, February 6–11, 2011, *J. Condensed Matter Nucl. Sci.* **10** (2013) 15–29.

[18] A. Meulenberg and K.P. Sinha, Deep-orbit-electron radiation emission in decay from $^4$He*# to $^4$He, *17th Int. Conf. Condensed Matter Nucl. Sci.*, Daejeon, South Korea, 12–17 August, 2012, *J. Condensed Matter Nucl. Sci.* **13** (2014) 357–368.

[19] T.H. Boyer, Unfamiliar trajectories for a relativistic particle in a Kepler or Coulomb potential, *Am. J. Phys.* **72**(8) (2004) 992–997.

[20] J. Chen, Trajectories for a relativistic particle in a Kepler potential, Coulomb potential, or a scalar field, Senior research thesis, (2005), College of William and Mary,
http://www.wm.edu/as/physics/undergrad/major/seniorhonors/previoustheses/2005/index.php (click on Chen).

[21] George H. Miley, Summary of the transmutation workshop held in association with ICCF-14, *Proc. 14th Int. Conf.*, August 2008, Washington, DC, Vol. 1, pp. 212–216, https://www.iscmns.org/iccf14/ProcICCF14b.pdf

[22] M. Srinivasan, G. Miley and E. Storms, *Low-Energy Nuclear Reactions: Transmutations*, Nuclear Energy Encyclopedia. Ch. 43, S.B. Krivit and J.H. Lehr (Eds.), Wiley, New York, 2011, ISBN 978-0-470-89439-2, http://lenr-canr.org/acrobat/Srinivasanlowenergyn.pdf.

[23] A. Meulenberg and K.P. Sinha, Composite model for LENR in linear defects of a lattice, *ICCF-18, 18th Int. Conf. on Cond. Matter Nucl. Sci.*, Columbia, Missouri, 25/07/2013, Proceedings at http://hdl.handle.net/10355/36818.

[24] A. Meulenberg, Pictorial description for LENR in linear defects of a lattice, *ICCF-18, 18th Int. Conf. on Cond. Matter Nucl. Sci.*, Columbia, Missouri, 25/07/2013, *J. Cond. Matter Nucl. Sci.* **15** (2015) 106–116.

[25] A. Meulenberg and J.L. Paillet, Basis for femto-molecules and -ions created from femto-atoms, *ICCF-19, 19th Int. Conf. on Cond. Matter Nucl. Sci.*, Padua, Italy, 15/05/2015, *J. Condensed Matter Nucl. Sci.* **19** (2016) 202–209.

[26] A. Meulenberg, Femto-helium and PdD Transmutation, *ICCF-18, 18th Int. Conf. on Cond. Matter Nucl. Sci.*, Columbia, Missouri, 25/07/2013, *J. Condensed Matter Nucl. Sci.* **15** (2015) 117–124.

[27] A. Meulenberg, Femto-atoms and Transmutation, 17th International Conference on Condensed Matter Nuclear Science, Daejeon, Korea, 12-17 August, 2012, J. Condensed Matter Nucl. Sci. 13 (2014), 346-357

[28] K.P. Sinha and A. Meulenberg, A model for enhanced fusion reaction in a solid matrix of metal deuterides, *Proc. 14th Int. Conf. Condensed Matter Nucl. Sci.* and *the 14th Int. Conf. Cold Fusion (ICCF-14)*, 10–15 August 2008, Washington DC, p. 633,
https://www.google.com/url?q=http:lenr-canr.org/acrobat/NagelDJproceedinga.pdf&sa=U&ei=YBB4U7_dCpXg8AXlsYGQAg&ved=0CAsQFjAE&client=internal-uds-cse&usg=AFQjCNF7Qd6sCPj15UEA8vj6ZaM7pPKzSw

[29] J. Hadjichristos, M. Koulouris and A. Chatzichristos, Technical characteristics and performance of the Defkalion's hyperion pre-industrial product, *17th Int. Conf. Condensed Matter Nucl. Sci.*, Daejeon, South Korea, 12–17 August, 2012,*J. Condensed Matter Nucl. Sci.* **13** (2014).

[30] G. Levi et al., Observation of abundant heat production from a reactor device and of isotopic changes in the fuel,

*www.sifferkoll.se/sifferkoll/wp-content/uploads/2014/10/LuganoReportSubmit.pdf.*

[31] A. Meulenberg and J.L. Paillet, Physical reasons for accepting the Deep-Dirac Levels, *Proc. of ICCF-20, the 20th Int. Conf. on Condensed Matter Nucl. Sci.*, Sendai, Japan, October 2–7, 2016.

## Appendix: Mass Changes for Atomic Nuclei

Atomic mass change, as a result of photo-emission, is well accepted. Mass change of the proton, but not of the electron, is not a 'given' and is never discussed. However, it can be accommodated in the concept of energy conservation and "which body does work." Consider a hydrogen atom:

- $W = \mathbf{F} \cdot \mathbf{d}$ = force times distance.
- The electrostatic force **F** is the same on both bodies.
  - Both electron and proton contribute to the electric field energy and thus the force. Equally?
  - Force and field energy both increase as the proton and electron move together.
- The proton moves the electron, but the electron has little kinematic effect on the proton.
- The distances moved indicate that more work is done by **E**-field on electron than on proton. However,
- Field energy and electron energy and mass all are known to increase as work is done on electron to bring it closer to proton. What contributes most (or all) of this energy?
- No separate measure of proton mass change has been made. (Is this true?)
- Nevertheless, the proton is the only 'uncommitted' ultimate source of energy for the emitted photon and for the increase in field and electron-mass energy.

QED, the proton mass must decrease as the electron photo-decays toward the ground state.

This concept is critical to the model of "fusion beneath the fragmentation level" [10]. Motion as a whole will increase the electron mass (by relativistic distortion of the Coulomb field and the resultant magnetic-field energy). How does the proton change its rest mass? A strong external field (from the deep-orbit electron) will alter the internal quark relationships of a nucleon (e.g., by polarization and alteration of their interaction distances). While in theory the effect could be both an increase and a decrease in energy levels (e.g., the Stark effect in atomic electrons), in this situation, the lower-energy levels are predominantly occupied to lower the total nucleon energy and mass.

J. Condensed Matter Nucl. Sci. 24 (2017) 230–235

Research Article

# Physical Reasons for Accepting the Deep-Dirac Levels– Physical Reality vs Mathematical Models in LENR

Andrew Meulenberg*
*Science for Humanity Trust Inc., USA*

Jean-Luc Paillet
*Aix-Marseille University, France*

**Abstract**

Limitations to contemporary models of Coulomb and nuclear interactions have previously been identified in the development of low-energy nuclear reaction (LENR) theories based on experimental Cold Fusion (CF) results. However, relativistic quantum mechanics has already provided a means to overcome these limitations. Nevertheless, this 'anomalous' solution has been repeatedly rejected, based on the mathematical formalism of an assumed singular potential, simply because it has not been necessary heretofore in the description of available physical data. The physical realities of a non-singular potential and its interaction with a relativistic electron are not new physics and provide a solid theoretical basis for CF (as indicated in another paper in this conference, and the references therein) and for new fields in femto-physics and nuclear chemistry. This present paper is essentially a reasoned complaint against those critics who proclaim the limitations of the anomalous solution, yet ignore the inapplicability of their chosen potential to the real world where they claim the solution(s) fail. It also castigates those physicists who accept the mathematical limitations as applicable to the real world even when presented with the evidence of assumptions made where they are no longer valid. The paper picks a few of these alleged limitations and shows how relativity and near-field interactions alter predictions based on the mathematical models and approximations used to characterize the conventionally accepted observations. It identifies where the simple models fail, or violate physical reality, and points to the implications of extending the models to better fit the real world and to properly understand the physical mechanism(s) involved.


*Keywords:* Deep-electron orbits, Hydrogen atom, Relativistic equations, Singularities

## 1. Introduction

Few physicists today would question the inadequacy of 19th century physics to correctly calculate many effects when bodies approach the speed of light. And, when a bound s-orbit electron approaches its nucleus, the region of its highest

*E-mail: mules333@gmail.com.

© 2017 ISCMNS. All rights reserved. ISSN 2227-3123

probability density, most will legitimately ignore the effects of its obviously relativistic velocity in that region because it spends so little time there. Yet, when relativistic quantum mechanics predicts new effects, the basic effects are all accepted (even if perhaps misinterpreted) and some of the relativistic effects are ignored. The 'selective acceptance' of a theoretical model's predictions is generally based on confirmation by experiment. However, the prediction of deep electron orbitals (with binding energies into the 0.5 MeV range) by the 'anomalous' solution of the relativistic Klein–Gordon (K–G) and Dirac equations [12] has been rejected in the past on theoretical and purely mathematical grounds. Thus, even if experimental evidence were now to be discovered for these orbits, it would be ignored, rejected, discounted, or attributed to other effects.

Rejection of these deep orbits, based on mathematical considerations, and recent arguments against such rejection are well documented in the present authors' papers on the subject [3–5]. Even when not rejected by mathematical arguments, these deep electron orbits have been previously discounted by the predicted low-population density of such levels [6]. The authors of this early series of discounting papers accepted the anomalous solution, but rejected it as a source of energy for the observed cold fusion results. The rejection by physical consideration of the low population of such levels is revealing in that it shows that some physicists, beyond recent proponents of the deep levels [17] are at least willing to accept their existence and examine possible implications. This examination took place within the framework of the quantum mechanics that predict these levels. The results, while negative for cold fusion, are positive for the deep electron orbit solution. It will be argued below that the conclusions for their rejection of applicability to cold fusion, like that of the mathematical rejections, are based on incorrect assumptions and inadequate physical models.

The present paper seeks to completely eliminate the mathematical arguments against the deep levels by application of physical 'reality checks' that are ignored in the mathematical discussions. It will show that the arguments are based on mathematical models and approximations that do not reflect physical reality and therefore should not be applied to the real world. In particular, two commonly accepted models must be properly extended.

The 'mathematically pure' $1/r$ Coulomb potential gives a non-physical singularity when extended to $r = 0$. The failure of this potential, as it is expressed near to (or within) a distributed charge, is obvious and is described in most elementary college-physics courses. The possible correctness of the singular potential close to a point source and the resultant wave equations is vitiated by the required infinite energy and charge densities of such a source.

The mathematically pure classical and quantum mechanics must be extended into the realm of relativity [4] and to near-field interactions [8]. The mathematics of these extensions can also be pure. Is one model better than the other(s), which should therefore be considered invalid? The relativistic extension of the static Coulomb potential to a dynamic one introduces new ideas into the field of CF.

## 2. A Non-physical Singularity

When Coulomb proposed the $1/r$ dependence of the electrostatic potential, it was based on macroscopic measurements. At the time, there was no concept of extending it to $r = 0$. The 'requirement' that Coulomb's 'law' fulfill all of the mathematical niceties leads to physical absurdities and has obscured important details of the real potential between two charges. One physical absurdity is the resulting singularity of an infinite mass and charge density for every charged particle in the universe. This does not seem to bother the 'opponents' of the anomalous solution of both the relativistic Klein–Gordon and Dirac equations They accept the singular potential and then reject the 'anomalous' solution, dependent on this potential, which they assume has a similar singularity, even for the defined states. For over 50 years the mathematical arguments against the anomalous solution have suppressed interest in its possible physical reality. Until recently, arguments against use of the singular potential appear to be ignored.

Strangely, the common practices in physics for handling singularities and other infinities, e.g. in [9] are also ignored in this case. Use of a non-singular nuclear Coulomb potential, by physical or mathematical modifications, is a trivial change that has been published – and rejected or ignored by the critics. The standard practice of 'cut-offs', a mathematical artifice, to avoid these problems is, for some reason, 'not allowed' in the arguments against the possibility of the anomalous solution being valid. Of course, if incontrovertible proof of the predicted deep-electron orbits were to become available, these same critics would declare themselves experts in the field and using these very techniques would 'prove' the validity of the anomalous solutions (with 'their' modifications).

The problem with the mathematical rejection of the anomalous solution is not with the mathematics itself. It is with the approximations used, and the non-physical assumptions, interpretations and their implications for the solution. The $1/r$ Coulomb potential is so imbedded within many physicists' minds that deviation is unthinkable. We have been taught that it is valid - at least to the nucleus, e.g., [10]. However, the original 'law' must be corrected for source configurations and relativity in the nuclear range. When mathematical rigor, without physical consideration, is applied to the $1/r$ problem at $r = 0$, these details are ignored. This allows a non-physical singularity and its consequences to be introduced into the mathematics. Such singularities are encountered routinely in physical models and 'handled' in various ways. Such treatments for the anomalous solution have been routinely 'rejected' over the decades and, without experimental observation of the deep-orbits predicted by the relativistic quantum mechanics, it is difficult to rationalize such measures. It is precisely under these conditions that near-field and far-field interactions become mixed ('sequestration' fails [11]), symmetry is broken, and 'matter' is created and destroyed. This is the basis for one model of cold fusion that uses the reduction of nuclear mass resulting from energy transfer to the deep-orbit electron to explain D–D fusion to $^4$He rather than to the normal 'hot' fusion paths of fragmentation into protons, neutrons, tritium, and $^3$He [12].

Modern quantum mechanics treats the ambiguities during transitions through these interaction regions by use of the Fock and Stueckelberg model and the uncertainty relation [13]. Energy, mass, and other constants need no longer be conserved during such transitions as long as they are conserved before and after the resultant interactions.

## 3. Solutions Avoiding the Singularity

Khelashvili and Nadareishvili [14] proved that the anomalous solution of the Klein–Gordon equation for the hydrogen atom with a singular $1/r$ Coulomb potential is valid. However, in the same paper (and subsequently affirmed [15 16]), they cannot do the same for the anomalous solutions of the Dirac equations. In addition to the unaddressed higher order terms in $V$ [17], what may be missing in the decades-old argument is that the actual anomalous solutions for the Dirac deep orbits have no electron probability distribution at the singularity (assuming that one physically exists). According to [1], there are no 1s-orbitals in the anomalous solution of the Dirac equations, furthermore, the deep-orbit ground state (2s in their nomenclature) has angular momentum as do all the deep-orbit s-orbital solutions. How can this happen and what does it mean?

In [1], Maly and Va'vra produced a detailed description and tables of the deep orbits, for both the relativistic K–G and the Dirac equations. In [2], the authors included figures of the electron densities for several representative orbits. In this later paper, they were estimating the size of the deep-orbit atoms and were not concerned about arguments against the existence of the deep orbits. However, the information provided in the two papers can be used to indicate that no anomalous solutions of the Dirac equations (for the hydrogen atom with a finite-sized nucleus, thus a non-singular Coulomb-type potential) have any value at $r = 0$ (see next paragraph) If there is no electron density at $r = 0$, there is no singularity in the deep-orbit solution and then all arguments against the deep orbits (if based on a singularity) must be invalid. This result is a step beyond simply using a non-singular potential. As suggested by calculations, but not yet proven, it may be possible (even if unnecessary) to provide this same result with a singular potential.

What is the basis of the non-population at the singular point? The deep level identified as 2s is perhaps better labeled as a 2p' level (with angular momentum of $\hbar$?). The reason for the new notation is related to the new quantum number, $k$, introduced in the Dirac equation solutions. The integer value of $k$, which corresponds to an angular momentum resonance may be positive or negative. However, the deep-orbits all have positive integers, $k$ (corresponding to a positive energy), and the deep-orbit ground state has a value of $k = 1$, which should correspond to a p-, not to an s-, orbital. With no $k = 0$, there is no solution with zero angular momentum. If there is any angular momentum at all, then the electron orbits cannot go to $r = 0$ because the centrifugal barrier is too high. Use of the term "s-orbital," which implies a maximum electron density at $r = 0$ for the $1/r$ potential, and to so identify the deep-orbit ground state is then misleading and could lead to one thinking of the solution as singular.

## 4. Physical Size and Mass of the Electron and Proton

The unmentioned assumptions of the mathematical solution of the relativistic quantum equations leading to the deep orbits include the infinite mass of the point source of a Coulomb potential. We have already addressed the physical absurdity of the singular potential associated with a point source. Here we address the standard correction for the infinite-mass assumption. In elementary quantum mechanics courses, a correction for finite mass nucleus is made to the solution of the Schrodinger equation. This is done after the simplified model of infinite mass source is solved. This correction is based on recognition of the 2-body problem, as existing in the center-of-mass system, being reduced to a single-body problem by introduction of a 'reduced' mass.

Introduction of a finite nuclear mass immediately forces changes to many of the arguments against the deep orbit solutions. First off, the assumed singularity of a point source for the theoretical $1/r$ Coulomb potential does not exist at $r = 0$, the center of mass. The proton is shifted away from $r = 0$ and from the electron. In fact, the singular condition can only exist if the angular momentum of the system is exactly zero and the electron (also assumed to be a point source in the same approximation) is simultaneously at $r = 0$ along with the point source Despite the fact that the wave functions of both nucleus and electron may overlap at $r - 0$, the probability of both of these particles being together at the same time is zero. Thus, the potential is proportional to $1/(r + \varepsilon)$ with $\varepsilon > 0$.

## 5. Physical Approximations Used to Treat Infinities

Many models in physics lead to infinities in one limit or another. To get around this problem, it is common practice to introduce a 'cut-off' to prevent values from going to infinity. Such treatment must somehow be unacceptable to the mathematical physicists who proclaim the invalidity of the anomalous solutions. Perhaps, to be acceptable, such 'fixes' must only be validated by the proposed models better fitting the experimental data when the fixes are included. If so, cold fusion might provide those data. Unfortunately, such data are not yet universally accepted and therefore "any change is unnecessary."

The key to the deep electron levels is relativity [17]. This is also a limit to some extent. Non-relativistic physics is an approximation that is valid under most conditions. In the nuclear range, electron velocities require inclusion of this effect. So, instead of imposing a cutoff such as that imposed by classical quantum mechanics ("no electron orbits can exist below the atomic levels"), it is possible to seek and find the physical limit imposed by, and the consequences of, relativity. Similarly, quantum mechanics has allowed the laws of physics to be 'relaxed' during certain interactions [13]. This limit is as crude as an arbitrary cutoff (and as uninformative). Relativity can provide a much more accurate and informative (physically reasonable) limitation to the classical interaction dynamics. Like quantum mechanics, it also provides a challenge to our perceptions of reality. Cold fusion provides a transition region between the classical atomic and nuclear regimes. It allows physically accessible information to be analyzed and compared with models.

## 6. Relativistic Coulomb Potential and Heisenberg Uncertainty Relation (HUR)

For some people, the fatal flaw in the concept of deep orbits is their failure to obey the HUR. At the predicted KE <1 MeV level of kinetic energies of an electron bound in the low femtometer radius orbits of the near-nuclear environment, the angular momentum is $\sim \hbar/100$ Thus, the HUR is violated by two orders of magnitude, unless the kinetic electron energy is somehow raised to the ~100 MeV level [17] An electron with this energy cannot be bound in the *static* Coulomb potential of a single proton. This argument is hard to contest without going against what has become a fundamental tenet of quantum mechanics One of the present authors (JLP) has found a possible answer in effects such as the near-field magnetic interactions between the nucleus and the electron, that raise the kinetic energy of a tight-bound electron into the 100 MeV range [17,18]

It is known that the deep orbits predicted by the Klein–Gordon and Dirac equations are a result of relativity [5]. We have been seeking the physical mechanism that satisfies both the singularity and the HUR limitations of the anomalous solutions. Early-on in the study, it was recognized that there was a correction needed to the virial theorem as it is applied to deep orbits [19]. Orbital requirements in a central potential are defined by the virial theorem. For non-relativistic orbits in a $1/r$ Coulomb potential, the virial theorem requires that the kinetic energy of the bound particle be one-half of the magnitude of its potential energy. The relativistic correction to the virial theorem predicts that as the particle velocity approaches that of light, this ratio of the energies approaches unity (i.e., KE $\Rightarrow$ |PE|). While this correction has major implications for the creation of a deep-orbit-electron population, it resolves neither the singularity nor the HUR problem of the solution.

The most recent contributions [5] to resolving the HUR problem are also presented at this conference [17]. Relativity not only introduces mass energy into the Hamiltonian, thus providing the deep orbits, it also affects (greatly increasing) the attractive forces between the proton and electron as the average electron velocity approaches a significant fraction of c. In the low-femtometer range of orbits, the Coulomb potential is no longer a $1/r$ potential and, with magnetic interactions, bound-electron kinetic energies can exceed 100 MeV. This unexpected turn of events resolves the HUR issue very nicely. However, it introduces some other questions, such as "where do the >100 MeV energies come from?" and "how do these relativistic electromagnetic energies compare with the nuclear energies (fractional quark size) that are so much greater than the static Coulomb energies normally considered?"

## 7. Conclusion

The use of mathematics in physics is essential. However, it is a tool. Any dogmatic adherence to its 'rules' and predictions, which may be based on improper or inadequate formulation and approximations should be avoided. The mathematics needs to be guided by the physics, before it can properly guide physics into (or out of) new regions of interest. Neither physics nor mathematics should be considered sacrosanct. This paper has addressed a few of these issues associated with the deep electron orbits predicted by relativistic quantum mechanics.

Physics has said that the Coulomb potential varies as $1/r$. Unguided mathematics then states, categorically, that any solution of an equation using this potential must be rejected unless it is zero at $r = 0$ because it is otherwise singular at that point. Both are correct in their frame of reference; but, they are both wrong when applied to the special case near $r = 0$. We have shown that the singularity in the $1/r$ potential is non-physical and therefore mathematical solutions that include the singularity must be 'adjusted' so that they are no longer singular and rejected. The means of such adjustment are well known in physics and some have even been applied to the case of interest here.

We believe that the dilemma of cold fusion theory is resolved in the acceptance of the deep-orbit electrons. Because of the uncertainties in the influence of some of the physical interactions (e.g., spin–spin) at these femtometer dimensions, actual values for the deep-orbit radii are still under investigation. Nevertheless, within the wide range of approximations and assumptions made in the calculations, most variations are not great and the implications for CF are unchanged.

In the process of exploring (with an open mind) the mathematics and physics of this near-field region, important things beyond CF are being discovered. The most recent of these is the effect of relativity on the Coulomb potential energy for deep-orbit electrons. With energies greatly in excess of the known nuclear binding energies and with the greatly enhanced interaction between the bound electron and the charged-quarks of a nucleon, the theoretical relationship between the strong-nuclear force, the electromagnetic force, and the weak interaction may be altered. It would appear that cold fusion is operating in the transition region between the atomic and the nuclear forces and is providing experimental data that will illuminate many aspects of physics today.

## Acknowledgement

This work is supported in part by HiPi Consulting, Ashland, VA, USA; by the Science for Humanity Trust, Bangalore, India; and by the Science for Humanity Trust, Inc, Tucker, GA, USA.

## References

[1] J.A. Maly and J. Va'vra, Electron transitions on deep Dirac levels I, *Fusion Sci. Technol.* **24**(3) (1993) 307–318, http://www.ans.org/pubs/journals/fst/a_30206.

[2] J. Maly and J. Va'vra, Electron transitions on deep Dirac levels II, *Fusion Technol.* **27** (1995) 59–70.

[3] J.-L. Paillet and A. Meulenberg, Arguments for the anomalous solutions of the Dirac equations, *J. Condensed Matter Nucl. Sci.* **18** (2016) 50–75, http://viXra.org/abs/1506.0177.

[4] J.-L. Paillet and A. Meulenberg, Relativity and electron deep orbits of the hydrogen atom, http://viXra.org/abs/1606.0259, RNBE-2016, 19–20 March 2016, Avignon (France), *J. Condensed Matter Nucl. Sci.* **21** (2016) 40–58.

[5] J.-L. Paillet and A. Meulenberg, Special relativity: the source of electron deep orbits, *Found. Phys.* **47**(2) (2017) 256–264.

[6] R. Rice, Y. Kim, Rabinowitz and Zubarev, Comments on exotic chemistry models and deep dirac states for cold fusion, *Proc. 4th Int. Conf. on Cold Fusion*, Hawaii, Vol. 4, p. 4.1, Dec. 6–9, 1993, *lenr-canr.org/acrobat/RiceRAcommentsona.pdf.*

[7] J. Naudts, On the hydrino state of the relativistic hydrogen atom, arXiv:physics/0507193.

[8] A. Meulenberg, Deep-orbit-electron radiation absorption and emission *ICCF-18, 18th Int. Conf. on Cond. Matter Nuclear Science*, Columbia, Missouri, 25/07/2013, *J. Cond. Matter Nucl. Sci.* **15** (2015) 125–136.

[9] C. Cohen-Tannoudji, B. Diu and F. Laloë, *Quantum Mechanics*, Volume 2, XIIC, Wiley, New York, 1977.

[10] R.P. Feynman, R.B. Leighton and M. Sands, *The Feynman Lectures on Physics*,2nd edition, Vol. 3, Addison–Wesley, USA, 1965.

[11] F. Wilczek, *Origins of Mass*, arXiv:1206.7114v2 22, Aug. 2012.

[12] A. Meulenberg and K.P. Sinha, Tunneling Beneath the $^4$He$*$ Fragmentation Energy, *J. Condensed Matter Nucl. Sci.* **4** (2011) 241–255.

[13] M. Davidson. Theories of variable mass particles and low energy nuclear phenomena, *Found. Phys.* **44**(2) (2014) 144–174.

[14] T Nadareishvili and A. Khelashvili, Some problems of self-adjoint extension in the Schrodinger equation, arXiv:0903.0234.

[15] A. Khelashvili and T. Nadareishvili, On the existence of additional (hydrino) states in the Dirac equation, arXiv:1601.01957.

[16] A. Khelashvili and T. Nadareishvili, Dirac reduced radial equations and the problem of additional solutions, arXiv:1610.07936.

[17] J.-L. Paillet and A. Meulenberg, Advance on electron deep orbits of the hydrogen atom for LENR, *Proc. of ICCF-20, the 20th Int. Conf. on Condensed Matter Nucl. Sci.*, Sendai, Japan, October 2–7, 2016.

[18] A.O. Barut and J. Kraus, Resonances in e+–e – system due to anomalous magnetic moment interactions, *Phys. Lett.* B **59**(2) (1975) 27.

[19] A. Meulenberg, J.-.L. Paillet, Implications of the electron deep orbits for cold fusion and physics, *Proc. of ICCF-20, the 20*th *Int. Conf. Condensed Matter Nucl. Sci.,* Sendai, Japan, October 2–7, 2016.

J. Condensed Matter Nucl. Sci. 24 (2017) 236–243

Research Article

# Fundamental Experimental Tests toward Future Cold Fusion Engine Based on Point-compression due to Supermulti-jets Colliding with Pulse (Fusine)

Ken Naitoh,* Jumpei Tuschiya, Ken Ayukawa, Susumu Oyanagi, Takuto Kanase, Kohta Tsuru and Remi Konagaya

*Faculty of Science and Technology, Waseda University, Tokyo, Japan*

**Abstract**

Our previous reports based on theoretical considerations and supercomputer simulation showed the possibility that super multi-air jets of gases such as air or deuterium colliding with pulse (K. Naitoh, patent: 2012-519298 (2010)) lead to self-compression over 60 MPa and 2000 K at single point around the reacted center, at maximum. This may bring about a more stable occurrence of cold fusion. This approach due to supermulti-jets will also cause an insulation effect because of encasing, which will result in less heat loss from the reactor walls. Based on this, we developed three types of prototype engine reactors using the supermulti-jets colliding with pulse. In the present report, we show some fundamental experimental data for one of the three prototype engine reactors, derived now, before we plan to begin testing for cold fusion.


*Keywords:* Experiment, Pulse, Reactor, Simulation, Supermulti-jets colliding, Theory

## 1. Introduction

A very impressive and mysterious phenomenon called cold fusion has been reported over 20 years [1–4]. Among many studies, it was found that the use of nano-particles is an effective means of stabilizing the reaction [5].

Further studies were conducted, which also indicated stability of the reaction [6–8]. However, the level of production energy is still small.

Our thermo-fluid dynamic theory [9,10] clarified the inevitability of sizes of various particles generated in phenomena from subatomic to macroscopic levels, which include atoms produced by cold fusion and liquid fossil fuel droplets atomized in combustion engines. The theory indicates the possibility that strong impact due to instantaneous pulse of pressure and temperature on particles including nano-particles for cold fusion may produce stable excess heat release in far greater amounts than combustion can produce.

*E-mail: k-naito@waseda.jp

© 2017 ISCMNS. All rights reserved. ISSN 2227-3123

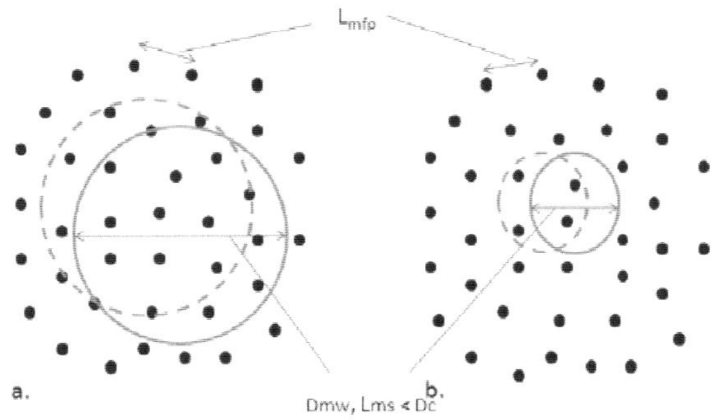


**Figure 1.** Stochasticity and indeterminacy generated by a mesoscopic window size smaller than that of continuum mechanics. a: weak stochasticity in density averaged in a larger window; b: stronger stochasticity in density averaged in a smaller window (Mean free path: Lmfp; Mesoscopic window size for averaging (MW): Dmw; Minimum scale dominating the phenomenon: Lms; Continuum assumption window: Dc).

## 2. Weakly Stochastic Field Theory based on Fluid Dynamics

While statistical mechanics usually uses a probability density function defined and averaged in the space of molecular speeds, the Boltzmann equation is also based on averaging a large volume in Euclid space because of the continuum assumption, which leads to the form of a deterministic partial differential equation for a stochastic field [11].

Here, the basic idea is that physical quantities such as velocity and density are intentionally averaged in a spatial window on a scale smaller than that for continuum mechanics (mesoscopic window: MW), in order to resolve the minimum scale (Lms) dominating the phenomenon between molecules and the continuum, which is also mesoscopic window size for averaging (Dmw). When the smaller spatial window for averaging, including a small number of molecules, moves right and left, the number of molecules inside the window varies and also the physical quantities are a little vague or indeterminant (Fig. 1).

As a result, a mesoscopic equation in the form of a stochastic partial differential equation for the field is obtained with stochastic fluctuations, in which stochasticity comes from the discontinuity of molecules in space.

The traditional macroscopic governing equation in continuum mechanics for the physical quantity $f(t, x_i)$ such as density, fluid velocities, and temperature defined in four-dimensional space of time $t$ and space $x_i$ is generally written in the form of

$$L\, f(t, x_i) = Q(t, x_i) \tag{1}$$

with a partial difference operator $L$ on time and Euclid space and a source term $Q$.

The mesoscopic equation averaged in a smaller averaging window (mesoscopic window size: Dmw) can be written as

$$L\, \bar{f}(t, x_i) = \bar{Q}(t, x_i) + \varepsilon(t, x_i), \tag{2}$$

where $\bar{f}$, $\bar{Q}$, and $\varepsilon$ denote the physical quantities averaged on the minimum scale dominating the phenomenon for $f$ and $Q$ and a stochastic fluctuation, respectively.

Weak vagueness of $\bar{f}$ also has the advantage that a smaller averaging window resolves the smaller scale of physical fluctuations and shows the possible ranges of the natural phenomena. Vagueness brings a solution, albeit a little vague because of $\bar{f}$ that includes indeterminacy. This vague solution is welcome, because the deterministic macroscopic governing equation averaged on the traditional scale for the continuum, which is too large to resolve the minimum scale dominating the phenomenon, cannot show any solutions. The important point is that the indeterminacy of $\bar{f}$ reveals another advantage in the form of the level of instability in the phenomenon and the range of the solution, which is important for scientific and engineering applications. It is stressed that complete random motion differs from that of the vague solution with weak indeterminacy [16,17].

The quantum mechanics of the Schrodinger and Klein–Gordon equations is based on an indeterminacy principle [12]. The presence of electrons is given in a certain area, not at a deterministic point. This vague viewpoint with indeterminacy shows an outline of a possible solution. We can obtain vague solutions in exchange for abandoning determinant ones.

Thus, this new approach to solving several problems with weakly stochastic partial differential equations can be said to be at the triple point of the indeterminacy principle in the Schrodinger equation, the stochasticity principle in the Langevin equation [11], and deterministic field theory in the Boltzmann equation [11]. In short, it represents the fusion of indeterminacy, determinism, and stochasticity.

In nature, discrete molecules in space are essentially discontinuous, which leads to stochasticity, while phenomena are relatively continuous in time. Thus, averaging for time can be regarded as being consistent with the observation window for spatial averaging.

## 3. Nano-particles Explained by the Weakly Stochastic Theory

An important momentum equation (Eq. (3) ) [9,10,15]

$$\frac{\mathrm{d}^2}{\mathrm{d}t^2}X(t) = -A(t)\left[\frac{\mathrm{d}}{\mathrm{d}t}X(t)\right]^2 - B(t)\,X(t) + \varepsilon(t) + \xi(t) + Q_2(t) \tag{3}$$

with initial deformation speed of $\mathrm{d}X(t)/\mathrm{d}t_{t=0} = Uo$

describes the deformation rate $X(t)$ of a particle (a nano-particle or a cluster) consisting of atoms like palladium for time $t$,while $A(t), B(t), \varepsilon(t), \xi(t),$ and $Q_2(t)$ denote nonlinear function of $X(t)$ for convection inside the particle, nonlinear function of $X(t)$ for contact surface force, stochastic term related to the small number of sub-particles

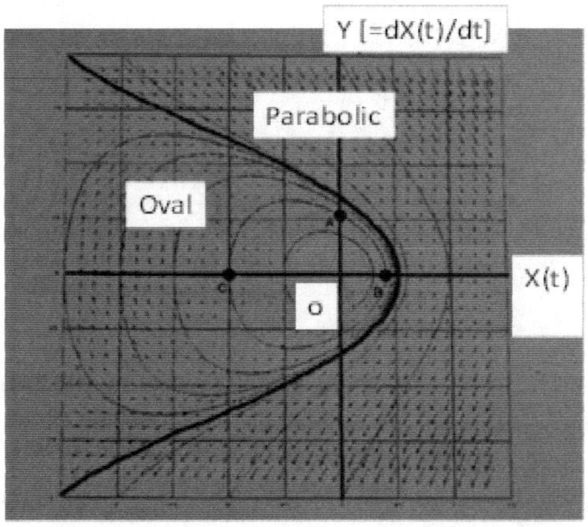


**Figure 2.** Trajectories of the solution for Eq. (3) ($Y = \mathrm{d}X/\mathrm{d}t$ plotted against $X$).

(atoms) explained Section 2, random force due to collisions of the other type of particles such as hydrogen (H) or deuterium (D), and another particle (another nano-particle or a cluster) consisting of atoms like palladium connecting just before breakup or just after collision, respectively.

Next, $A(t)$ and $B(t)$ in Eq. (3) can be set as fixed constants, because cold fusion will occur in a very short time. Then, $\varepsilon(t), \xi(t),$ and $Q_2(t)$are eliminated. Equation (3) is still nonlinear because of the first term on the left-hand side. The analytical solution of Eq. (3), which we fortunately found, shows that increasing input energy causes a drastic transition from oval-type of oscillation implying small deformation of $X(t)$ to parabolic divergence implying breakup of particle due to infinitely large deformation (Fig. 2).

Equation (4) then shows the critical condition of initial deformation speed $Uo$ (minimum input energy at the initial stage), after which the breakup of the particle occurs.

$$Uo^2 > \frac{B}{2A^2}. \tag{4}$$

The theory based on Eqs. (3) and (4) brings two important sources of knowledge.

First, the present theory can clarify the reason why decreasing sizes of particles such as palladium will result in a little more stable occurrence of cold fusion, as proposed by Prof. Yoshiaki Arata. When the input energy is smaller

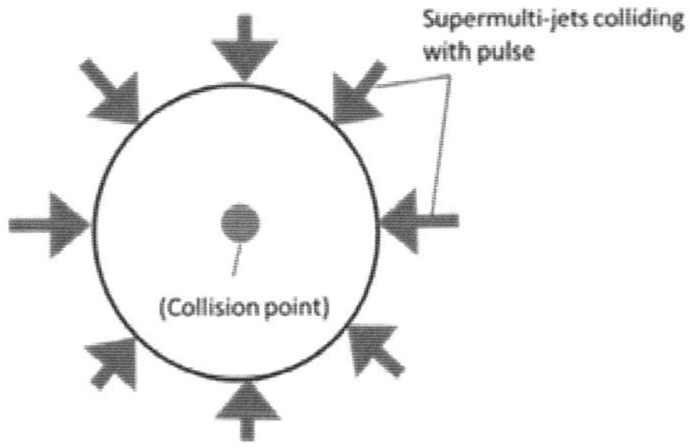


(a)

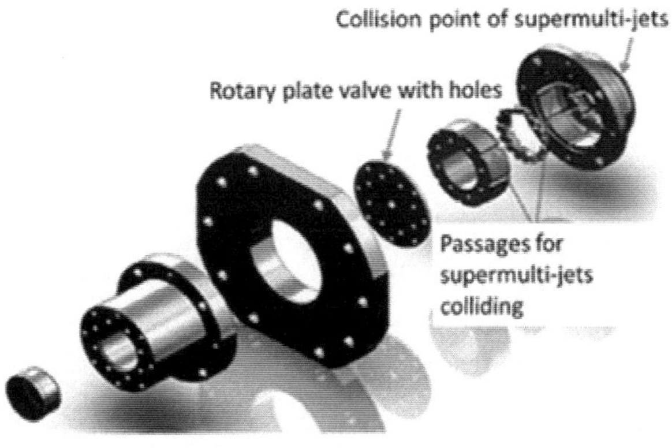


(b)

**Figure 3.** Point compression due to supermulti-jets colliding with pulse. (a) Principle of the point compression due to supermulti-jets colliding with pulse. (b) Actual system of supermulti-jets colliding with rotary valve for generating pulse.

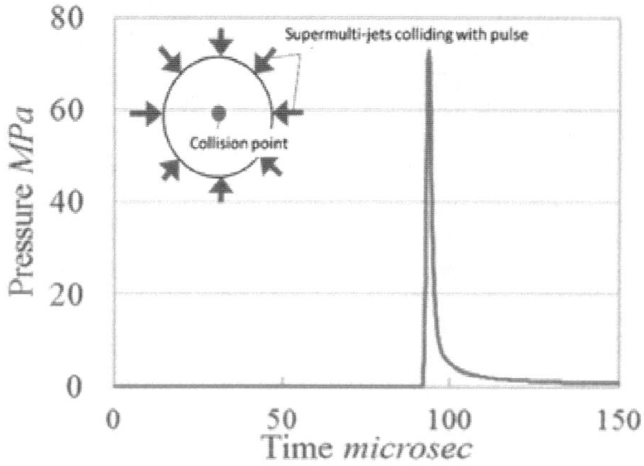


**Figure 4.** Pressure history computed for supermulti-jets colliding with pulse.

than that for breakup, the stochastic term $\varepsilon(t)$ due to small number of sub-particles, which appears for nano-particles, often gives an effect surpassing the critical condition of breakup.

## 4. An Approach toward Higher Energy Generation Based on the Supermulti-jets Colliding with Pulsation

Second, the above theoretical consideration and numerical simulation on a supercomputer shows a concrete methodology for making the cold fusion phenomenon more stable. An important point of the methodology is supermti-jets of gases such as deuterium with pulse, which collides at the center of reactor (Fig. 3). This is because point-compression due to this supermulti-jets colliding with pulse gives particles larger deformations in the form of an increase of the initial force $U$o shown in Eqs. (3) and (4) or $\xi(t)$, which will lead to a stable cold fusion reaction.

**(a)** **(b)**

**Figure 5.** Prototype engine reactor with 14 nozzles for supermulti-jets (*Left*: reactor. *Right:* Chemical reaction occurrence in the prototype engine reactor).

The super multi-air jets of gases such as air or deuterium colliding with pulse, i.e., collision of jets of transonic- or supersonic- speeds, lead to self-compression over 60 MPa and 2000 K at a single point around reacted center, at maximum (Fig. 4). This compression level will be larger than that in traditional cold fusion systems, whereas it wil be much less than for high temperature fusion systems. This may bring about more stable and frequent occurrences of cold fusion. This approach due to supermulti-jets will also cause an insulation effect because of encasing, which will result in less heat loss from the reactor walls.

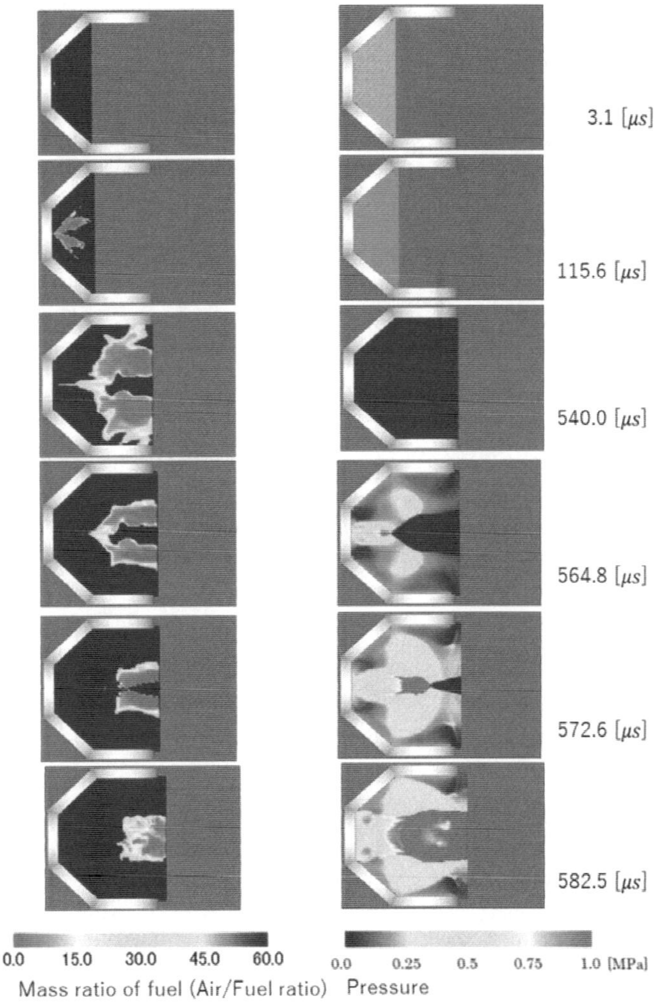


**Figure 6.** Unsteady three-dimensional computation of the supermulti-jets and fuel supplied into the Fusine. Injection cone angle is 70°. Injection speed of fuel is 135 m/s. Piston diameter is 39 mm. Piston stroke is 40.5 mm, Piston speed is 12,500 rpm. Intake pressure due to turbo-/super-charged system is 3.0 bar. Computational grid: 150× 120× 120 points.

We developed three types of prototype engine reactors using the supermulti-jets colliding with pulses. One of them is shown in Fig. 5. In the present report, we show some fundamental experimental data collected before testing cold fusion with one of the three prototype engine reactors. First, we measured the compression level due to supermulti-jets colliding in the case of atmospheric air. From the data, we got evidence that self-compression is possible. Second, we added a hydrocarbon fuel around the reactor center, i.e., around the compression point, which resulted in a chemical reaction with large increases of pressure and temperature (see the right-hand side of the figure in Fig. 5 as an example) [13,14]. High thermal efficiencies and nearly complete air insulation effect were obtained from the combustion experiments [13,14,19,20].

## 5. Concrete Design of the Fusine System

The engine reactor system must be extended to trigger cold fusion in a future cold fusion engine (a "Fusine"), because solid particles such as palladium are necessary. Figure 6 shows unsteady three-dimensional computational results for the supermulti-jets colliding having 17 nozzles with pulsation around the reactor center and also fuel injected from the left position of the reactor [18]. Injection of fuel starts at microsecond zero, while the supermulti-jets of the other vapor reactant enter into the reactor about 550 $\mu$s later. A high pressure region over 1 MPa can be observed around the reactor center after 564.8 $\mu$s, while fuel is also compressed into the reactor center due to the supermulti-jets after 564.8 $\mu$s. Thus, by using the present numerical model [18], we can find optimum conditions including injection timings of $H_2/D_2$ gases and nanoparticles such as palladium, injection pressure, and injection cone angle (width of fuel injected). A special injection system of solid particles designed by using the numerical model [18] can be developed with durability.

## 6. Conclusion

Most of the studies on cold fusion have been done with closed reactors until now. Thus, the open reactor systems shown in Figs. 3 and 5 will be important for the actual implementation of cold fusion phenomenon. We have started to make a concrete reactor system including one with nanoparticles, and we will also perform fundamental experiments. In the near future, we will report the results.

## Acknowledgements

This paper is part of the outcome of research performed under the JSPS grant for research projects (25630072). Sincere thanks are also due Mr. Kan Yamagishi worked in the Naitoh laboratory in Waseda University until March 2016 and also the members of the Naitoh Laboratory for their help.

## References

[1] M. Fleischmann, S. Pons and M. Hawkins, *J. Electroanalytical Chem.* **261** (2) Part 1 (1989) 301–308.
[2] M. Fleischmann, S. Pons, M.W. Anderson, L.J. Li and M.Hawkins, *J. Electroanalytical Chem.* **287** (2) (1990) 293–348.
[3] G.H. Miley and J.A. Patterson, *J. New. Energy* **1–3** (1996) 5–30.
[4] http://newenergytimes.com/v2/conferences/LENRConferenceProceedings.pdf.
[5] Y. Arata and Y. Zhang, *J. High Temp. Soc.* **34-2** (2008) 85–93.
[6] A. Kitamura, T. Nohmi, Y. Sasaki, A. Taniike, A. Takahashi, R. Seto and Y. Fujita, *Phys. Lett.* **A373** (2009) 3109–3112.
[7] T. Hiroki, N. Sugimoto, T. Nishi, A. Itoh and T. Motohiro, *J. Condensed Matter Nucl. Sci.* **13** (2014) 223–233.
[8] A. Kitamura, A. Takahashi, R. Seto, Y. Fujita, A. Taniike and Y. Furuyama, *Current Sci.* **108** (4) (2015) 589–593.
[9] K. Naitoh, Quasi-stability theory: revealing various atomic breakups and cold fusion, *Proc. of ICCF17*, South Korea, 2012.

[10] K. Naitoh, A spatiotemporal structure: common to subatomic systems, biological processes, and economic cycles, *J. Phys. CS* **344** (2012) 1–18.
[11] J.O. Hirschfelder, C.F. Curtiss and R.B. Bird, *Molecular Theory of Gases and Liquids*, Willey, New York, 1964.
[12] L.D. Landau and E.M. Lifshitz, *Quantum Mechanics*, 3rd Edition, Butterworth-Heinemann Elsevier, Oxford, 1981.
[13] K. Naitoh, J. Tsuchiya, D. Ikom, T. Nakai, S. Oyanagi, T. Kanase, T. Okamoto, Y. Tanaka, K. Ayukawa and R. Konagaya, SAE paper 2016-01-2331 (2016).
[14] K. Naitoh, K. Ayukawa, D. Ikoma, T. Nakai, S. Oyanagi, T. Kanase and J. Tsuchiya, SAEpaper 2016-01-2337 (2016), doi:10.4271/2016-01-2337.
[15] K. Naitoh, Cyto-fluid dynamic theory, *Jpn. J. Indust. Appl. Math.* **18-1** (2001) 75–105.
[16] K. Naitoh and H. Shimiya, Stochastic determinism capturing the transition point from laminar flow to turbulence, *Jpn. J. Indust. Appl. Math* **28** (1) (2011) 3–14.
[17] K. Naitoh, *Gene Engine and Machine Engine*, Springer, Japan, Tokyo, 2006, 1–261.
[18] K. Yamagishi, S. Onuma, S. Ohara, K. Hasegawa, K. Kojima, T. Shirai, T. Kihara, K. Tsuru and K. Naitoh, SAE Technical Paper 2016-01-2334, 2016, doi:10.4271/2016-01-2334.
[19] K. Naitoh S. Ohara, S. Onuma, K. Kojima, K. Hasegawa and T. Shirai, SAE Technical Paper 2016-01-2336, 2016, doi:10.4271/2016-01-2336.
[20] R. Konagaya, S. Oyanagi, T. Kanase, J. Tsuchiya, K. Ayukawa, K. Kinoshita, J. Mikoda, H. Fujita and K. Naitoh, SAE Techncal paper, 2017.

J. Condensed Matter Nucl. Sci. 24 (2017) 244–251

Research Article

# Observation of Anomalous Production of Si and Fe in an Arc Furnace Driven Ferro Silicon Smelting Plant at levels of Tons per day

C.R. Narayanaswamy*

*The Silcal Metallurgic Ltd., Coimbatore 641004, Tamil Nadu, India*

**Abstract**

In the period 1978–2002, The Silcal Metallurgic Ltd., a Coimbatore (India) based company, was engaged in the production of ferro silicon alloy deploying a 12 MVA "Submerged Carbon Arc" powered smelter. During a 11-week long non-stop round the clock operation of the plant in 1995, daily feed of raw materials was: Quartz (33.4 ton), charcoal (with fixed carbon content of 13.2 ton) and scrap steel (5.1 ton) while the daily output production of Fe–Si alloy (73.5% Si) was 24.75 ton. From the total weights of Si and Fe in the input feed and assuming 100% recovery of the metals, the daily output alloy production could at best have been only 20.5 ton. However to our surprise throughout the 11-week period the total daily Fe–Si alloy (with 73.5% Si) output was consistently 24.75 ton, corresponding to a daily "anomalous" excess metal production of 4.25 ton of Fe–Si alloy. The only source of Si entering the smelter furnace was the quartz raw material and that of Fe was the scrap steel (except for minor additional amounts of Fe originating from the steel casing of the consumable Söderberg carbon electrodes). Very careful vigil of the weights of daily input feed of raw materials and output alloy drained out as also the electrical energy consumption was maintained. It was evident that roughly 20% more metal than could be accounted for from the input feed was being produced and consequently we have been obliged to come to the conclusion that anomalous quantities of Si (2.8 ton/day) and Fe (1.45 ton/day) were being synthesized during the smelting process. Discussions with researchers involved in the Cold Fusion/LENR field have suggested that a likely explanation for the anomalous metal production could be the occurrence of transmutation reactions between nuclei of C and the O stripped from the $SiO_2$ during the chemical reduction process. It is speculated that the intense varying magnetic fields generated by the kilo-amp levels of alternating current (AC) driving the arcing between the three gigantic carbon electrodes and the carbonic hearth of the furnace, in the 2000°C temperature environment, could have somehow catalyzed transmutation reactions to occur, very similar to the transmutations reported in laboratory scale "Carbon-Arc experiments" first revealed by George Oshawa in 1964. But the more puzzling aspect of our observations is that there was no evidence of release of the expected massive amounts of nuclear energy that should have accompanied the postulated transmutation reactions based on the atomic masses of the nuclei involved.


*Keywords:* Carbon Arc, Energy balance, Fe–Si alloy, Transmutation

*Former Managing Director, E-mail: cr_narain@yahoo.co.in.

© 2017 ISCMNS. All rights reserved. ISSN 2227-3123

## 1. Introduction

Silcal Metallurgic Ltd. was incorporated as a private limited company in the southern Indian industrial town of Coimbatore in 1978 and attained commercial production in 1980. It deployed the well-known ferro silicon smelting technology involving high current "Submerged Electric Arc furnaces". This electro-thermic manufacturing process for ferro-silicon alloy is known to be highly power intensive since the temperature in the reaction zone has to be maintained at around 2000°C. References [1–3] give comprehensive overviews of this technology. The Silcal furnaces had ratings of 5 and 12 MVA and were operated on a round-clock basis. They deployed the traditional Söderberg self-baking carbon electrodes in steel casing, which have been successfully used for over a century [4]. The 5 MVA furnace was used for the production of low carbon silico manganese while the 12 MVA furnace was dedicated to the production of Fe–Si alloy of 70–75% Si grade.

## 2. Brief Remarks on the Plant and Operation

Raw materials used for the production of Fe–Si alloy are low alumina content quartz ($SiO_2$) of 98–98.8% purity, steel scrap and wood charcoal with low ash content which served as the reducing agent. Quartz was sourced directly from selected mines in the state of Tamil Nadu, which are known to have low alumina content. On receipt of the consignment at the plant site, dust and fines were screened out and the stock stored outdoors. Wood charcoal on arrival was tested for moisture and fixed carbon and screened to separate fine dust and placed in indoor/outdoor storage. Steel scrap was stored in an outdoor yard. All received raw materials were analyzed for purity at the in-house testing lab and the data carefully archived.

The screened raw materials were taken using a conveyer system to the third floor of the furnace and stored in separate over head bunkers. Each of the three raw materials were weighed according to a computerized batching system and transferred into charging buckets running on monorails in the second floor. Charging buckets then discharged the premixed raw materials into the furnace every 10–15 min through chutes. Shift-wise consumption of all raw materials was totalized to obtain daily (24 h) consumption data.

The molten alloy product was drained through one of the three tap holes at the bottom of the furnace every 2–2.5 h into tiltable "teeming ladles" mounted on rail tracks. The teeming ladles were then emptied into large stationary heat resistant cast iron trays to a thickness of approximately 50 mm. Next day, during the day shift the solidified Fe–Si slabs were manually broken into small pieces, weighed and packed into 40 kg bags for domestic consumers or in 1 ton jumbo bags for export. Each batch of Fe–Si was individually analyzed adopting standard procedures prevalent in this industry.

The voltage applied to the three electrodes was 3-phase alternating current, typically in the 100–200 V region, using step down transformers to convert from 110 kV/11 kV/ to furnace voltage from grid supplied power. Arc currents were in the 30–60 kA region. The arc is struck between the 1 m diameter vertically mounted steel encased consumable Söderberg electrodes and the floor of the carbon hearth. Both the carbon of the self-baking electrodes and its steel casing are consumed in the smelting process, the consumption being 50–60 kg/ton of Fe–Si. Details of the procedure adopted to replenish the electrodes online without interrupting furnace operation is discussed in [1–4]. The whole process is slagless and the only product is molten Fe–Si alloy drained from the bottom and carbon monoxide (CO) effluent gas which burns at the top of the furnace, combining with the atmospheric oxygen to become $CO_2$ which is released through a stack after scrubbing as per applicable environment regulations.

The 12 MVA furnace was operated round the clock at variable ratings from 7 to 12 MVA, depending on the availability of power. Various charge mix ratios and operating electrical parameters were experimented with in order to arrive at the optimum conditions required for achieving 73–74% silicon content alloy. Systematic records of the total weight of the raw material feed consumed every day, as also the total weight of the product alloy tapped out daily was maintained. Cumulative daily consumption of electrical power was also recorded. A maximum daily production of 27.5 ton of product alloy was achieved when the furnace was operated under full load conditions. The company was very successful and made good profits, supplying high quality products to both local and export markets.

## 3. Energy Requirement for Chemical Reduction of $SiO_2$

Plant records show that to produce 1 kg of Silicon content in the product alloy, about 10.92 kWh of electrical energy is consumed, on 100% recovery basis. This observation also tallies with the expected energy consumption estimate based on theoretical considerations of the chemistry of the reduction reaction which is endothermic. In our plant the product alloy contained 73–74% silicon content. Taking an average value of 73.5%, 735 kg of Si would be present in each ton of product alloy (balance being iron). Power consumption for producing 1 ton of alloy thus works out to $10.92 \times 735 = 8027$ kWh. However, dissolving iron into molten silicon is exothermic; rough estimate of this energy release for 265 kg of Iron dissolving into 735 kg of silicon is around 75 kWh of heat energy. Thus production of 1 ton of ferro silicon alloy of 73.5% Si content would require a net energy input of $8027–75 = 7952$ kWh. (Ref. [3] also quotes a similar figure.) The relevance of discussing chemical reaction energy consumption considerations will become apparent later in the paper.

## 4. Remarks On Inconsistencies Observed between Weights of Input Feeds and Output Products During a 11-week Run in 1995

During early 1995, the furnace was operated continuously on a round the clock basis at a rating of between 8.5 and 8.75 MVA with a daily power consumption of 1,68,000 kWh/day. The feed mixture composition of the raw materials which was not changed throughout the 11-week period was: (a) Quartz – 33.4 ton; accounting for its purity of 98.7% the actual weight of $SiO_2$ in the feed stock works out to 33 ton. The weight component of Si in this therefore works out to 15.4 ton. (b) Daily iron feed was 5.1 ton. Minor additional contributions from the quartz and electrode casing have been included in this. (c) Fixed carbon input from charcoal was set as 13.2 ton based on chemistry considerations for reduction of 33 ton of $SiO_2$. The actual weight of charcoal fed on any given day was however higher depending on the moisture content of the particular feedstock of charcoal. Thus the total weight of the main raw materials, namely $SiO_2$, Fe and C consumed daily was 51.3 ton.

Based on the above raw materials consumption, the maximum possible daily production of Fe–Si alloy at 100% recovery of Si for an alloy of 73.5% silicon should have been not more than 20.5 ton. However, in comparison with the actual daily production of Fe–Si alloy was consistently 24.75 ton with a 73.5% content silicon. This implies a total "anomalous" excess metal (silicon plus iron) production of 4.25 ton/day. Net silicon content in product alloy works out to $24.75 \times 0.735$ ton $= 18.2$ ton compared to 15.4 ton in the input feed. Excess silicon was thus 2.8 ton corresponding to18% increase. Likewise net weight of Fe in product alloy works out to $24.75 \times 0.265 = 6.55$ ton compared to 5.1 ton that was input, implying excess iron of 1.45 ton or 28.4% weight gain.

## 5. Discussion and Remarks

Although we had been observing anomalous excess production of Si and Fe ranging from 200 to 400 kg/day right from 1985 onwards, we were not sure whether these were due to errors in weighing or could be attributed to anomalous generation of Si and Fe. However, following the consistent and repeated observation of about 4.25 ton of daily excess metal production over the 11-week round the clock run in 1995, we were convinced beyond doubt that anomalous transmutation processes are indeed occurring, pointing to the existence of new science. It was only after this that we went public and released our findings in a press briefing in 1999.

We are fully aware that our claims of ton level transmutations will be met with intense skepticism. The first most obvious possible source of doubt leading to incorrect conclusion is errors in weighing of input feedstock and output alloy produced. All raw materials on arrival are weighed in an electronic weigh bridge and stored indoor/outdoor after screening out the fines and dust. In each shift, quartz, charcoal and steel scrap are shifted from ground level bunkers, to storage bins in the top of the furnace floor through conveyors, after a final fines screening. The discharges from the bins are weighed in a computerized batching and weighing system and recorded before feeding into the furnace through chutes. The finished product in the packing floor are weighed in precision weighing balances after size reduction and loaded in bags for shipments to customers. These weighing machines are made by reputed manufacturers such as Avery, whose technicians visit the factory periodically to calibrate and check their accuracy as per industrial regulations. It should be remembered that as operators of a commercial production plant we are most conscious of maintaining proper records of material balance from an accountancy point of view. Payments are involved for raw materials received and products dispatched to customers. Shift wise consumption of raw materials, power used and total daily production and also logs of operational plant parameters were monitored and recorded accurately. There were thus no chances for any errors to occur in the measurement of quantities of input raw materials or output finished product.

The other doubt often voiced is that somehow additional amounts of Si and Fe could have entered the furnace without the knowledge of the plant management. One critic for example suggested that may be the quartz used was not 100% $SiO_2$ but may have been partly in the form of SiO in which case the weight fraction of silicon in the quartz would be more than 46.7% (28/60) as assumed by us. Critics argued that this could explain the anomalous appearance of "additional" Si. The main argument against this criticism is that SiO is actually a gas and there is no question of its being a contamination in the quartz raw material. References [1–4] clearly point out that quartz as mined is primarily in the form of $SiO_2$

During the 11-week continuous nonstop run of the furnace, when excess production of 4.25 ton/day was attained, the daily power consumption and load was maintained at a constant level of 1,68,000 kWh/day. In the presently adopted technology it is impossible to produce 1 ton of 73.5% grade Fe–Si alloy at less than 7952 kWh power consumption, as discussed earlier. But in our case in the presence of transmutation also the total power used remained the same in spite of the weight of product alloy having increased, resulting in the "specific power consumption" dropping to (1,68,000/24.75) = 6788 kWh /ton of product alloy, during the entire period of 11 weeks. This observation can be taken to be indicative of the fact that the additional Si production did not come through chemical reduction processes, but must have arisen as a result of some other cause (see Appendix A).

Some amount of SiO gas does get formed during the smelting process if operating conditions are not optimized and this escapes along with CO. One can clearly observe the presence of SiO in the effluent gas stream from the blue tinge in the flame caused by CO burning to $CO_2$. In our material balance considerations we have not accounted for any loss of Si through the effluent gas and have assumed 100% recovery of Si. If SiO escape is taken into account, the quantum of transmuted Si would actually work out to be even more.

A significant additional qualitative observation we made was that whenever nuclear transmutation processes appeared to be taking place, the heat radiated in the furnace top floor where CO burns to form $CO_2$, was noticeably less implying that the quantum of CO reaching the top of the furnace was lesser.

## 6. Possible Relevance of the Carbon Arc Experiment

In quest of an explanation of these anomalous observations, the author was advised (in 1999) to meet the then Director of the Indira Gandhi Centre for Atomic Research (IGCAR) at Kalpakkam, Tamil Nadu, who himself was a distinguished Metallurgist. This Director, however, cautioned us that mainstream Science has no ready explanation for these results and suggested that I meet one Dr. Mahadeva Srinivasan, formerly of the Bhabha Atomic Research Centre (BARC) Mumbai, India, who had been closely following the progress of a new field of research called Cold Fusion/LENR. It is only after discussions with Dr. Srinivasan in the year 2000 that we became aware of the existence of LENR. It was Dr. Srinivasan who first introduced us to the so-called Carbon Arc experiments [5–8] in which anomalous generation of Si and Fe had earlier been reported by many researchers. Indeed, the carbon arc experiment itself was pioneered by a Japanese researcher by the name of George Oshawa in 1964 and the iron reportedly generated in such experiments has come to be known as George Oshawa steel in [9].

At ICCF 20 conference where our paper was presented, coincidentally there was also a paper presented by George Egely [10,11] of Hungary which described a variation of the Carbon Arc experiment in which also anomalous quantities of iron appears to have been generated. Instead of arcing between carbon electrodes Egely has however used carbon powder carried in a silica crucible and subjected it to microwave heating. Presence of iron could be observed readily by using a magnet placed above the post reaction powder mixture.

Our attention was also recently drawn to a patent titled "Silicon Extraction Method" that has been awarded in Ukraine in January 2007 to Bolotov [12] for what appears to be a somewhat analogous method of producing Si from the fusion of the nuclei of the neighboring elements of copper and phosphorous. The method involves application of electric currents with density more than $10^{11}$ A/m$^2$ under temperatures below 1050$^\circ$ C. However, there does not appear to be any arcing involved in this process but rather high pressure is employed. Details are available in the patent disclosure [12].

## 7. Puzzle of the Missing Nuclear Energy

A worth noting feature of the Silcal observations was that there was no dramatic change whatsoever in the energy dissipation. Using the estimated energy release values of 17.13 MeV/atom of Si or 49.58 MeV per atom of Fe given in Appendix B, for the postulated nuclear transmutation reactions, it can be shown that corresponding to 4.25 ton of metal transmutation, the power generated should have been the equivalent of the total thermal power generated by hundreds of 1000 MWe nuclear power stations. However, in our plant there was no evidence of such massive amounts of nuclear energy being released throughout the 11-week period, giving a handle to the skeptics to question our claims of ton level elemental transmutations. In this context it is worth noting that nobody in published LENR literature (to the best of our knowledge) has established a clear correlation between the quantum of transmutation products generated in carbon arc and the expected nuclear heat release based on atomic mass considerations. On the other hand neither has any publication claimed that the Carbon Arc experiment violates Einstein's $E = mc^2$ dictum. Thus if indeed the Silcal transmutation claims are confirmed it would clearly point to the operation of new Science wherein transmutation could be occurring without the accompaniment of the expected nuclear energy release.

In the context of these remarks the arguments of Daniel Szumski elaborated in his "Least Action Nuclear Process" (LANP) Theory appear relevant. We learnt about Szumski's work through his paper presented at ICCF 20 conference [13]. Szumski who has taken great pains to analyze in detail the transmutation observations of George Miley (see www.LeastActionNuclearProcess.com) argues that both endothermic and exothermic nuclear reactions can and do occur concurrently in LENR experiments, partly or wholly cancelling out net energy release. In fact he has referred to some experimental observations of Mizuno wherein transmutations have reportedly been observed by him not accompanied by energy release. Szumski is thus not at all surprised by our observation of "energy neutral" transmutation reactions.

## 8. Remarks on Possible Future Studies

Unfortunately Silcal operations were severely crippled by power cuts ranging from 30 to 100% and frequent power interruptions from 1996 onwards till 1999. Following a dispute regarding violation of Power Tariff agreement with the state government and withdrawal of exemptions from power cut for our industry, the Silcal plant had to be shut down in 2002 and the company wound up in 2010. This was an inevitable consequence of the fact that cost of electrical power consumed forms a substantial component of the net cost of production of Fe–Si alloy by the smelting process.

What has been has been observed by us in our plant may be considered to be only partial transmutation of the C and O present in the reaction zone into Si and Fe. As such we were making attempts to attain 100% transmutation of all the carbon and oxygen nuclei present in the input feed, from 1995 to 2002. For the last 15 years, we have been contemplating various ways of improving the technology, hoping to move towards achieving 100% transmutations. We do believe that we now have the design and operational parameters for such an improved plant design. If we had carried out isotopic analysis of the produced Fe–Si alloy it would have given us very valuable clues. We deeply regret not having thought of doing this those days. We are meanwhile continuing efforts to try and set up a new plant where these transmutation results could be replicated once again.

The objective of this paper is to share our findings with the LENR community, with the hope that operators of similar plants elsewhere in the world, especially in Norway, could be encouraged to look for the occurrence of anomalous production of Si and Fe in similar submerged arc furnace smelting plants. On our part we shall be most happy to share our results and experience in a true scientific spirit with anyone interested.

## Acknowledgements

The author is most grateful to Dr. Mahadeva Srinivasan for sharing his knowledge of the LENR field, especially of anomalous elemental transmutations and his guidance in moving forward in quest of understanding these puzzling observations. I am indebted to him for his help in writing this paper.

## References

[1] A.G.E. Ribiette, *Electric Smelting Processes*, Griffin, London, 1973.
[2] A. Riss and Y. Khodorovsky, *Production of Ferro Alloys*, Translated from Russian by I.V. Savin, Foreign Languages Publishing House, Moscow.
[3] M. Tangstad, *Ferrosilicon and Silicon Technology in Handbook of Ferro Alloys: Theory and Technology*, First Edition, 2013, M.M. Gasik (Ed.), Butterworth–Heinemann, Elsevier, UK.
[4] J.W. Richards, *The Soderberg Self-Baking Continuous Electrode,* 1920. Early publications of the Lehigh Faculty. Paper 277. http://preserve.lehigh.edu/ .
[5] R.A. Monti, Low energy nuclear reactions: the revival of alchemy, *Proc. Int. Conf. Space and Time*, St. Petersburg, Russia, 2001, p. 178.
[6] M. Singh, M.D. Saxena, V.S. Dixit and V.B. Kartha, Verification of the George Oshawa experiment for anomalous production of iron from carbon arc in water. FusionTechnol., 1994, **26**, 266–270.
[7] R. Sundaresan and J.O.M. Bockris, Anomalous reactions during arcing between carbon rods in water, *Fusion Technol* . **26** (1994) 261–265.
[8] X.L. Jiang and L.J. Han, Anomalous element production induced by carbon arcing under water, *Proc. 7th Int. Conf. on Cold Fusion*, F. Jaeger (Ed.), Eneco Inc., Salt Lake City, UT, 1998, pp. 172–179.
[9] http://www.nuenergy.org/transmutation-of-carbon/ and http://www.levity.com/alchemy/nelson2_3.html.
[10] G. Egely, Transmutation by dust fusion, *Infinite Energy*, Issue 130, Nov./Dec. 2016.
[11] G. Egely, Influencing radioactivity via transmutation, *Proc. of ICCF 20 conf.*, under publication in *JCMNS*(2017).
[12] B.V. Bolotov, Ukraine Patent No 19900, Silicon Extraction Method, Issued 15 January 2007 byMinistry of Education and Science Ukraine, State Department of Intellectual property.
[13] D.S. Szumski, Laws of nature are precise and reproducible, *JCMNS, Proc. of ICCF 20 Conference* (2017).

**Appendix A: Ferro–silicon 73.5% (Balance Iron) Production Parameters: Recovery Percentages and Electrical Power Consumption Per Ton of Products**

Electrical power consumption per kg of silicon produced is 10.92 kWh at 100% recovery for 735 kg of silicon in 1 ton of alloy produced, power consumption is 735 × 10.92 is 8027 kWh 8027 – 75 = 7952 kWh in regular operation. If the recovery of Si is lesser than 100% the corresponding power consumption values will be higher as given below:

| | | | | | |
|---|---|---|---|---|---|
| 100% | 99% | 98% | 97% | 96% | 95% |
| 7952 | 8032 | 8114 | 8198 | 8283 | 8371 |

Typical energy demand in kWh and recovery percentages of various grades of Fe–Si per ton of alloy in European plants [3]

| | Fe–Si 45 | Fe–Si 65 | Fe–Si 75 |
|---|---|---|---|
| Electrical energy | 4800 | 7400 | 8800 |
| Recovery % | 98–99 | 92–94 | 91–93 |

**Appendix B: Atomic mass data and computation of expected nuclear energy release in transmutations reactions**

Amu data used for computations:

| | | |
|---|---|---|
| 2 Helium | 3He 3.016029 | 0.000137 |
| | 4He 4.002603 | 99.999863 |
| 6 Carbon | 12C 12.000000 | 98.93 |
| | 13C 13.003355 | 1.0 |
| 8 Oxygen | 16O 15.994915 | 99.757 |
| | 17O 16.999132 | 0.038 |
| | 18O 17.999160 | 0.205 |
| 14 Silicon | 28Si 27.976927 | 92.2297 |
| | 29Si 28.976495 | 4.6832 |
| | 30Si 29.973770 | 3.0872 |
| 26 Iron | 54Fe 53.939615 | 5.845 |
| | 56Fe 55.934942 | 91.754 |
| | 57Fe 56.935399 | 2.119 |
| | 58Fe 57.933280 | 0.282 |

| Suggested nuclear reaction leading to production of Si | | | | |
|---|---|---|---|---|
| $_6C^{12}$ | + | $_8O^{16}$ | => | $_{14}Si^{28}$ |
| 12.000 | | 15.9949 | | 27.9765 (amu) |
| Total amu of LHS = 27.9949 | | | | |
| Total amu of RHS = 27.9765 | | | | |
| Therefore mass lost = 0.0184 amu (means energy released) | | | | |
| Multiply by 931 Mev to convert to Mev units | | | | |
| Energy release is 17.13 MeV per nuclear reaction (exothermic) | | | | |

| Suggested nuclear reaction path for generation of iron ($Fe^{56}$) | | | | |
|---|---|---|---|---|
| 2 $_6C^{12}$ | + | 2 $_8O^{18}$ | ⇒ $_{26}Fe^{56}$ | + $_2He^4$ |
| 2 × 12.00 | | 2 × 17.999160 | 55.9349 | 4.002603 (amu) |
| Total amu of LHS = 59.99076 | | | | |
| Mass lost is 0.05326 amu | | | | |
| Total of RHS = 59.9375 amu | | | | |
| In MeV units = 49.58 MeV (exothermic) | | | | |
| 2 Helium | | 3He 3.016029 | | 0.000137 |
| | | 4He 4.002603 | | 99.999863 |
| 6 Carbon | | 12C 12.000000 | | 98.93 |
| | | 13C 13.003355 | | 1.07 |
| 8 Oxygen | | 16O 15.994915 | | 99.757 |
| | | 17O 16.999132 | | 0.038 |
| | | 18O 17.999160 | | 0.205 |
| 14 Silicon | | 28Si 27.976927 | | 92.2297 |
| | | 29Si 28.976495 | | 4.6832 |
| | | 30Si 29.973770 | | 3.0872 |
| 26 Iron | | 54Fe 53.939615 | | 5.845 |
| | | 56Fe 55.934942 | | 91.754 |
| | | 57Fe 56.935399 | | 2.119 |
| | | 58Fe 57.933280 | | 0.282 |

J. Condensed Matter Nucl. Sci. 24 (2017) 252–257

Research Article

# Physical Model of Energy Fluctuation Divergence

K. Okubo* and K. Umeno†

*Graduate School of Informatics, Kyoto University, Kyoto, Japan*

**Abstract**

In this paper, we propose a new classical model in which energy fluctuation diverges. In detail, for certain parameter ranges, kinetic energy diverges since the momentum obeys the Cauchy distribution. This phenomenon will be applied to the cold fusion since jumping over the potential wall is essential to cold fusion.


*Keywords:* Cauchy distribution, Chaos, Energy fluctuation divergence

## 1. Introduction

It is well known that the second moment of a probability variable $x$ diverges such that

$$\langle x^2 \rangle = \int x^2 p(x)\, \mathrm{d}x = \infty,$$

when $x$ obeys a power-law density function defined as

$$p(x) = O\left(\frac{1}{x^{1+\alpha}}\right), \qquad x \ll 1, \quad 0 < \alpha < 2.$$

Then, the fluctuation of the probability variable $\left\langle (x - \langle x \rangle)^2 \right\rangle$ also diverges. When momentum obeys a power-law density function, the fluctuation of kinetic energy diverges. Then energy fluctuation divergence will play an important role in jumping over a potential wall. One can read research about the divergence of fluctuation or large fluctuation of energy [1,2]. In this paper, we propose a classical chaotic model in which power-law density appears for certain parameter ranges and energy fluctuation can be observed.

*E-mail: okubo.kenichi.65z@st.kyoto-u.ac.jp.
†E-mail: umeno.ken.8z@kyoto-u.ac.jp.

## 2. Classical Chaotic Model

We consider a classical dynamical system in which a power-law distribution can be observed owing to the chaotic structure. In general, chaos is characterized by initial condition sensitivity. That is, the distance between two orbits grows exponentially although the initial points of two orbits are very close. To realize a system in which power-law densities appear, we apply the fact that when a probability variable $\alpha$ obeys the uniform distribution on $\left(-\frac{1}{2}, \frac{1}{2}\right)$, another probability variable $\beta$ defined as $\beta = \gamma \tan(\pi\alpha)$, $\gamma > 0$ obeys the Cauchy distribution $f_\gamma(\beta)$ defined as

$$f_\gamma(\beta) = \frac{1}{\pi}\frac{\gamma}{\beta^2 + \gamma^2}.$$

Then, from a simple calculation, the variance of $\beta$, $\langle\beta^2\rangle$ diverges. Then if momentum $p$ obeys the Cauchy distribution, then the expectation value of kinetic energy $\left\langle\frac{p^2}{2}\right\rangle$ and its fluctuation

$$\left\langle\left(\frac{p^2}{2} - \left\langle\frac{p^2}{2}\right\rangle\right)^2\right\rangle$$

diverge. From this, we propose a four dimensional classical Hamiltonian in order to realize the power-law distribution defined by

$$\begin{aligned} H(p_1, q_1, p_2, q_2) &= \tfrac{1}{2}\left(p_1^2 + p_2^2\right) - \varepsilon \log\left|\cos\left\{\pi(q_1 - q_2)\right\}\right|, \\ V(q_1, q_2) &= -\varepsilon \log\left|\cos\left\{\pi(q_1 - q_2)\right\}\right|, \end{aligned} \tag{1}$$

where $\varepsilon$ is a perturbation parameter and

$$p_1, p_2 \in R, \quad q_1, q_2 \in I_{\delta_N}, \quad I_{\delta_N} = \left(-\frac{1}{2} + \frac{\delta_N}{\pi}, \frac{1}{2} - \frac{\delta_N}{\pi}\right).$$

$\delta_N$ satisfies the condition such that

$$\frac{2\left(1 - \frac{2\delta_N}{\pi}\right) - 2\varepsilon(\Delta\tau)\left[\tan\left(\frac{\pi}{2} - \delta_N\right) - \tan\left(-\frac{\pi}{2} + \delta_N\right)\right]}{1 - \frac{2\delta_N}{\pi}} = N,$$

where $N$ is a natural number. Figures 1 and 2 show the shape of potential $V(q_1, q_2)$ for $\varepsilon = 2$ and $\varepsilon = -2$, respectively. Although, the potential $V(q_1, q_2)$ is artificially constructed in order to generate the power-law distribution, this potential is periodic and in condensed matter physics periodic potentials are often treated [3,4].

From this Hamiltonian, one obtains the canonical equation such that

$$\begin{aligned} \dot{p}_1 &= -\frac{\partial H}{\partial q_1} = \pi\varepsilon \tan\left[\pi\left\{q_1 - q_2\right\}\right], \quad \dot{q}_1 = \frac{\partial H}{\partial p_1} = p_1, \\ \dot{p}_2 &= -\frac{\partial H}{\partial q_2} = -\pi\varepsilon \tan\left[\pi\left\{q_1 - q_2\right\}\right], \quad \dot{q}_2 = \frac{\partial H}{\partial p_2} = p_2. \end{aligned} \tag{2}$$

By using first order symplectic integrator [5] defined as

$$\begin{aligned} p_i(n+1) &= p_i(n) - \frac{\partial H}{\partial q_i}\left(p_1(n), q_1(n+1), p_2(n), q_2(n+1)\right), \quad i = 1, 2, \\ q_i(n+1) &= q_i(n) + \frac{\partial H}{\partial p_i}\left(p_1(n), q_1(n), p_2(n), q_2(n)\right), \quad i = 1, 2, \end{aligned} \tag{3}$$

one obtains four dimensional time discrete model $T_\varepsilon$ as follows.

$$\begin{pmatrix} p_1(n+1) \\ q_1(n+1) \\ p_2(n+1) \\ q_2(n+1) \end{pmatrix} = T_\varepsilon \begin{pmatrix} p_1(n) \\ q_1(n) \\ p_2(n) \\ q_2(n) \end{pmatrix} = \begin{pmatrix} p_1(n) - \varepsilon \tan\left[\pi\left\{q_1(n+1) - q_2(n+1)\right\}\right] \\ p_1(n) + q_1(n) \mod I_{\delta_N} \\ p_2(n) + \varepsilon \tan\left[\pi\left\{q_1(n+1) - q_2(n+1)\right\}\right] \\ p_2(n) + q_2(n) \mod I_{\delta_N} \end{pmatrix}, \tag{4}$$

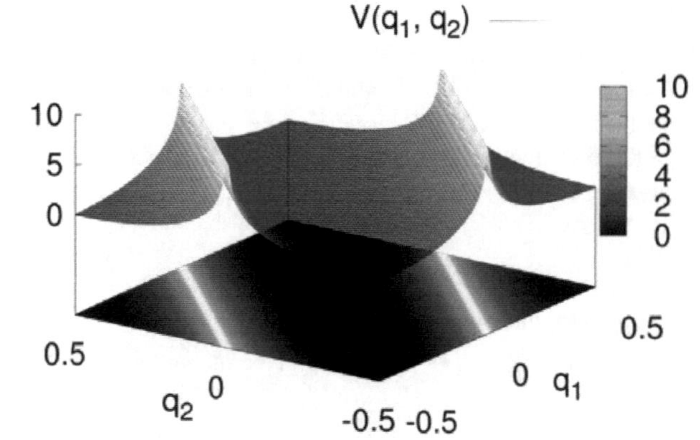


**Figure 1.** The shape of potential $V(q_1, q_2)$ for $\varepsilon = 2$. $V(q_1, q_2)$ diverges where $\cos\{\pi(q_1 - q_2)\} = 0$.

where the operation $\bmod I_{\delta_N}$ is defined such that

$$\begin{array}{l} x \mod I_{\delta_N} = x - n\left(1 - \frac{2\delta_N}{\pi}\right), \\ -\frac{1}{2} + \frac{\delta_N}{\pi} + n\left(1 - \frac{2\delta_N}{\pi}\right) < x \leq \frac{1}{2} - \frac{\delta_N}{\pi} + n\left(1 - \frac{2\delta_N}{\pi}\right), \quad n \in Z. \end{array} \tag{5}$$

This map preserves the sum of moment such as $p_1(n) + p_2(n) = \cdots = p_1(0) + p_2(0)$. Then, the probabilistic property for $p_2$ is the same as that of $p_1$. One obtains chaotic orbits for almost all initial points for $\varepsilon < 0$, $\frac{2}{\pi} < \varepsilon$.

According to [6], one can prove that $\{ q_1(n) - q_2(n) \}$ obeys the uniform distribution on $I_{\delta_N}$ and have mixing property owing to the chaotic structure when the parameter satisfies the condition $\varepsilon < 0$, $\frac{2}{\pi} < \varepsilon$. When we set $N$ as

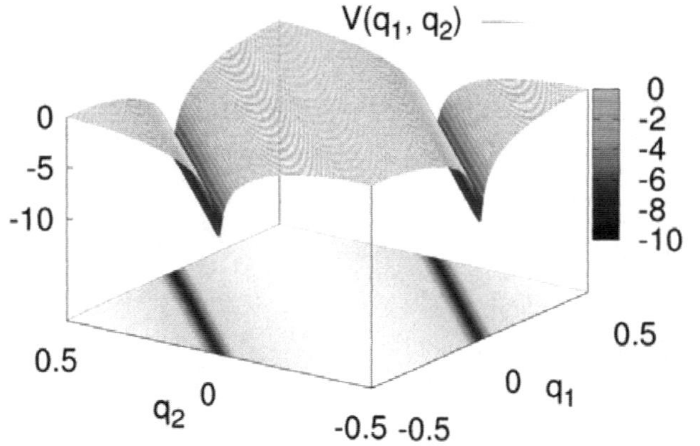


**Figure 2.** The shape of potential $V(q_1, q_2)$ for $\varepsilon = -2$. $V(q_1, q_2)$ diverges where $\cos\{\pi(q_1 - q_2)\} = 0$.

$N \gg 1$, variable $\{ q_1(n) - q_2(n) \}$ are considered to distribute uniformly on $\left(-\frac{1}{2}, \frac{1}{2}\right)$. Then the Cauchy distribution appears. That is, the time series $\{x_n = -\varepsilon \tan[\pi \{q_1(n) - q_2(n)\}]\}$ obey the Cauchy distribution $f(x)$ denoted as

$$f(x) = \frac{1}{\pi} \frac{|\varepsilon|}{x^2 + |\varepsilon|^2}, \tag{6}$$

when the parameter satisfies the condition $\varepsilon < 0$, $\frac{2}{\pi} < \varepsilon$. The momentum $p_1$ and $p_2$ are denoted by sum of variables $\{x_n\}_{n \geq 0}$ such that

$$p_1(n) = p_1(0) + \sum_{i=1}^{n-1} x_i,\ p_2(n) = p_2(0) - \sum_{i=1}^{n-1} x_i. \tag{7}$$

In this system, time series $\{x_n\}_{n \geq 0}$ have mixing property when the parameter satisfies the condition $\varepsilon < 0$, $\frac{2}{\pi} < \varepsilon$. Then, according to [7], momentum $p_1$ and $p_2$ obey the stable distribution in the condition. If each element of $\{x_n\}_{n \geq 0}$ are independent of the other elements, then $p_1$ and $p_2$ obey certain Cauchy distribution $\bar{f}(x)$ and the expectation value of kinetic energy $\langle K_1 \rangle$ and $\langle K_2 \rangle$ defined as

$$\langle K_1 \rangle = \left\langle \frac{p_1^2}{2} \right\rangle = \int_{-\infty}^{\infty} \frac{p_1^2}{2} \bar{f}(x)\, \mathrm{d}x, \qquad \langle K_2 \rangle = \left\langle \frac{p_2^2}{2} \right\rangle = \int_{-\infty}^{\infty} \frac{p_2^2}{2} \bar{f}(x)\, \mathrm{d}x, \tag{8}$$

diverges even though an initial condition of the kinetic energy $K_1(0)$ and $K_2(0)$ are finite.

We investigate whether $\{p_1(n)\}$ obey the Cauchy distribution at a finite time $n$. Figure 3 shows the result of numerical simulation for a distribution with $\{p_1(100)\}$ for $\varepsilon - 1$ where the number of ensemble $M$ is as $M = 10^6$.

When the momentum $\{p_1(100)\}$ obey the Cauchy distribution whose scale parameter is $\gamma$, a new probabilistic variable $\{y\}$ defined as

$$y \equiv \frac{p_1(100) - \mu}{\sqrt{\sigma^2}}$$

obey $g(y)$ such that

$$g(y) = \frac{1}{\pi} \frac{\gamma\sqrt{\sigma^2}}{\left(\sqrt{\sigma^2} y\right)^2 + \gamma^2}, \tag{9}$$

where for finite number of ensemble of $\{p_1(100)\}$, $\mu$ and $\sigma^2$ correspond to their average and variance, respectively. Figure 3 shows the density function with $\{y\}$ and $g(y)$ where $g(y)$ is fitted by least squares method.

Then from the fitted density function $g(y)$, one obtains the estimated scale parameter $\hat{\gamma}$ such as $\hat{\gamma} = 98.1 \approx 1.00 \times 100$. If the time series $\{x_n\}_{n \geq 0}^{99}$ are independent, the estimated scale parameter $\hat{\gamma}$ satisfies the relation as $\hat{\gamma} = 100$. Then the result as $\hat{\gamma} = 98.1 \approx 1.00 \times 100$ means that the times series $\{x_n\}$ are almost independent. Therefore, this result suggests that not only $\{x_n\}$ but also $\{p_{1,2}(n)\}$ obey the Cauchy distribution.

## 3. Time Evolution of Energy

In Section 2, we introduced the property of this chaotic model, in which it was proven that the expectation value of kinetic energy diverges in the conditions described. In this section we show the time evolution of total energy in this system defined as follows.

$$E(t) = \frac{p_1^2(t)}{2} + \frac{p_2^2(t)}{2} - \varepsilon \log |\cos\{\pi (q_1(t) - q_2(t))\}|. \tag{10}$$

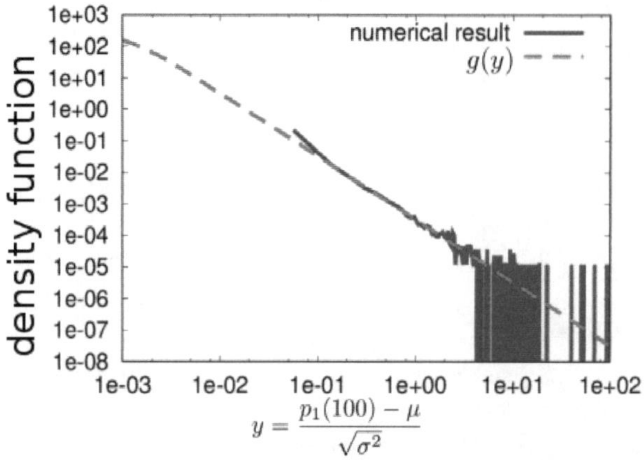


**Figure 3.** The density function of $y$ composed of $p_1$ (100) for $\varepsilon = 1$. A solid line shows the result of the numerical simulation and a broken line shows a density function $g(y)$ fitted by least squares method. The value of average and variance for finite number of data are about $\mu \approx -7.13 \times 10^5$ and $\sqrt{\sigma^2} \approx 7.01 \times 10^5$ points *M* is $M = 10^6$. Estimated parameter value is $\hat{\gamma} = 98.1 \approx 1.00 \times 100$.

According to [8], by discretization, a system does not conserve energy although the Hamiltonian (1) is an autonomous system. According to [9], the energy fluctuates focusing on the initial value. From Fig. 4, the energy is not so large until $n \approx 2 \times 10^2$ and the intermittent burst occurs around $n \approx 3 \times 10^2$. Thereafter the energy begins to fluctuate wildly.

## 4. Conclusion

In this paper, a classical chaotic model has been proposed. It has been proven that this model has mixing property and momentum $\{p_1, p_2\}$ obey the stable distribution for $\varepsilon < 0, \frac{2}{\pi} < \varepsilon$ [1]. By numerical simulation, one can see

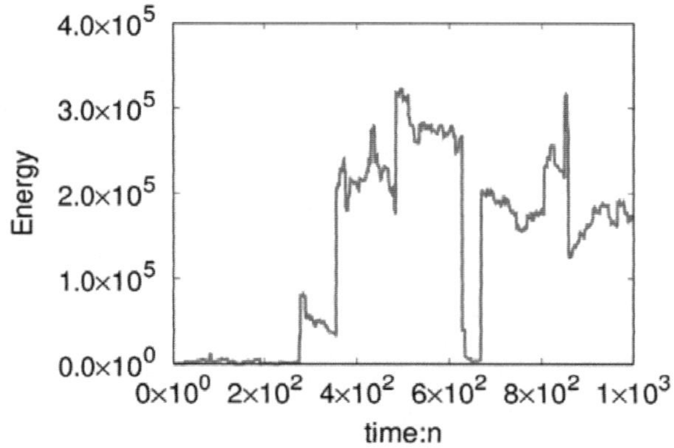


**Figure 4.** The time behavior of the total energy for $\varepsilon = 1$. Although the initial total energy is about 12.1, it fluctuates wildly and one can observe intermittency. Initial condition is $(p_1, q_1, p_2, q_2) = (0.3,\ 0,\ 4.91,\ 0)$.

$\{p_1, p_2\}$ obey the Cauchy distribution. From this result, we have confirmed that the variance of $\{p_1, p_2\}$ diverges and the expectation value of the kinetic energy $\langle K_1 \rangle$ and $\langle K_2 \rangle$ also diverges. Thus, this model can simply explain the phenomena in which the energy fluctuation divergence occurs although the initial energy is finite. This model can simply explain the phenomenon in which jumping over the potential wall occurs and can be a toy model which can explain cold fusion.

## References


[1] V.I. Vysotskii and S.V. Adamenko, *J. Condensed Matter Nucl. Sci.* **8** (2012) 91.

[2] V.I. Vysotskii, M.V. Vysotskyy and S.V. Adamenko, *J. Exp. Theoret. Phys.* **114** (2012) 243.

[3] B. Hu, B. Li, J. Liu and Y. Gu, *Phy. Rev. Lett.* **82** (1999) 4224.

[4] A.M. Lacasta, J.M. Sancho, A.H. Romero, I.M. Sokolov and K. Lindenberg, *Phy. Rev.* E **70** (2004) 051104.

[5] H. Yoshida, *Phys. Lett.* A **150** (1990) 262.

[6] K. Okubo and K. Umeno, arXiv:1610.01254v2 (2016).

[7] I. A. Ibragimov, *Theory Probab. Appl.* **7** (1962) 349382.

[8] G. Zhong and J.E. Marsden, *Phys. Lett.* A **133** (1988) 134.

[9] K. Umeno and M. Suzuki, *Phys. Lett.* A **58** (1993) 2644.

J. Condensed Matter Nucl. Sci. 24 (2017) 258–277

Research Article

# Advance on Electron Deep Orbits of the Hydrogen Atom

Jean-Luc Paillet*

*Aix-Marseille University, France*

Andrew Meulenberg†

*Science for Humanity Trust Inc., USA*

**Abstract**

In the previous works, we discussed arguments for and against the deep orbits, as exemplified in published solutions. So we considered the works of Maly and Va'vra on the topic, the most complete solution available and one showing an infinite family of EDO solutions. In particular, we deeply analyzed their second of these papers, where they consider a finite nucleus and look for solutions with a Coulomb potential modified inside the nucleus. In the present paper, we quickly recall our analysis, verification, and extension of their results. Moreover, we answer to a recent criticism that the EDOs would represent negative energy states and therefore would not qualify as an answer to the questions posed by Cold Fusion results. We can prove, by means of a simple algebraic argument based on the solution process, that, while at the transition region, the energy of the EDOs are positive. Next, we deepen the essential role of Special Relativity as source of the EDOs, which we discussed in previous papers. But the central topic of our present study is an initial analysis of the magnetic interactions near the nucleus, with the aim of solving important physical questions: do the EDOs satisfy the Heisenberg Uncertainty relation (HUR)? Are the orbits stable? So, we examine some works related to the Vigier–Barut Model, with potentials including magnetic coupling. We also carried out approximate computations to evaluate the strength of these interactions and the possibilities of their answering some of our questions. As a first result, we can expect the HUR to be respected by EDOs, due to the high energies of the magnetic interactions near the nucleus. Present computations for stability do not yet give a plain result; we need further studies and tools based on QED to face the complexity of the near-nuclear region. For the creation of EDOs, we outline a possibility based on magnetic coupling.


*Keywords:* Deep electron levels, LENR, Magnetic interactions, Relativistic quantum physics, Singular solutions

## 1. Introduction

The question of the existence of electron deep orbits (EDO) for the hydrogen atom has led to a number of works and debates.

*E-mail: jean-luc.paillet@club-internet.fr
†E-mail: mules333@gmail.com

© 2017 ISCMNS. All rights reserved. ISSN 2227-3123

In previous works [1,2] we answered to main criticism found in the literature, which are of a mathematical nature. Also, in this paper, we only list them with short summaries of our solutions and of the origin of the EDOs as solutions of relativistic equations for atomic H (essentially the relativistic Schrödinger equation – also called the Klein–Gordon equation – and the Dirac equation for a single particle in a potential).

Recently, we faced a new and unexpected argument against the deep orbits, concerning the sign of their energy levels. As a consequence, we present here a straightforward algebraic reasoning, based on known methods to solve the radial Dirac equation for atomic H, which allows us to show that deep orbits have positive energy.

With this perspective, it is interesting to see how Special Relativity plays a central role in the theoretical genesis of the EDOs, when considering only the Coulomb potential. Regarding this subject, we open a little discussion about the singular solutions in a non-relativistic context. One can ask what is precisely the element, in Special Relativity, that is the source of EDOs. For the present, and from a more precise analysis [3], we can think that a quadratic relation involving the energy $E$, eigenvalue of the Hamiltonian, and the potential energy, inserted into the equation, plays a dominant role.

However, as the heart of our present paper, we now analyze the role of various kinds of magnetic interactions inside the atom, for the EDOs. For example: is the Heisenberg uncertainty relation satisfied by electrons so confined near the nucleus? That is effectively a very important question, of physical nature and touching an essential pillar of the Quantum Mechanics. So, we recall most known magnetic interactions in a simple atom H, and examine works explicitly including several magnetic potentials in the Hamiltonian of the atom. Do not forget the Dirac equation includes (in a hidden way) the spin-orbit interaction for electron spin, but no interaction with the nuclear spin; and the Schrödinger equation does not include any spin considerations (Note: While it is generally accepted that the Schrödinger equation does not consider spin, one can find in the literature [4] arguments suggesting that it represents a particle in a spin eigenstate.) Our first study on the magnetic interactions and on the works based on the Vigier-Barut model, lead us to make some correlations with other works and results. After developing this subject, we made several approximate computations, in order to estimate the energies of the interactions, which, as expected, are very high when approaching the nucleus.

Partial results given at the end of this study are hopeful, especially about the HUR.

## 2. Deep Orbits Obtained as Solutions of Relativistic Equations for Atom H

In this section, we quickly recall our analysis, verification and extension of works about EDOs obtained by relativistic quantum equations, and exposed in previous papers [5,6] in a developed way. While looking for the solutions of the radial equations by an analytic method, there is a parameter generally denoted $s$ that occurs during the process. This parameter appears in an exponential term of form $\rho^s$ included in the used ansatz, and $s$ has to satisfy a quadratic relation on the coefficients of series included in the ansatz.

In the case of the relativistic Schrödinger equation, this quadratic relation for $s$ is the following: $s(s+1) + \alpha^2 - l(l+1) = 0$, where $l$ is the usual angular number and $\alpha$ the coupling constant. This equation admits two roots: $s = -1/2 \pm [(l+1/2)^2 - \alpha^2]^{1/2}$. For the Dirac equation, the condition on $s$ is the following: $s^2 - k^2 + \alpha^2 = 0$, where $k$ is the specific Dirac angular number.

This equation also admits two roots: $s = \pm(k^2 - \alpha^2)^{1/2}$. For both cases of relativistic equations, when one takes the "plus" sign for $s$, one obtains the usual "regular" solutions, whereas with the "minus" sign one obtains the so-called "anomalous" or "singular" solutions.

We recall here the expressions of energy levels of these singular solutions. For the Schrödinger, it reads:

$$E_{\mathrm{S}} = mc^2 \left[ 1 + \frac{\alpha^2}{\left( n' + \frac{1}{2} - \left[ \left( l + \frac{1}{2} \right)^2 - \alpha^2 \right]^{\frac{1}{2}} \right)} \right]^{-1/2}, \tag{1}$$

while for the Dirac equation, we have:

$$E_{\mathrm{D}} = mc^2 \left[ 1 + \frac{\alpha^2}{\left( n' - \sqrt{(k^2 - \alpha^2)} \right)^2} \right]^{-1/2}. \tag{2}$$

In previous works, we noted that all energy values given by $E_{\mathrm{S}}$ or $E_{\mathrm{D}}$ do not correspond to deep orbits. In fact, the deep orbits are obtained only when the angular number ($l$ or $|k|$) is equal to the radial number $n'$. Note that for the Dirac equation, one takes the absolute value of $k$, because $k$ can be positive or negative, but only $|k|$ is involved in the computation of the $E_{\mathrm{D}}$.

Moreover, the EDOs solutions have the following properties:

- The total energy $E_{\mathrm{S}}$ (or $E_{\mathrm{D}}$) $\ll mc^2$, which implies the binding energy $|\mathrm{BE}|$ is near the rest mass energy, more precisely, for H, $|\mathrm{BE}| > 509$ keV for the $E_{\mathrm{D}}$ solutions, and $|\mathrm{BE}| > 507$ keV for the $E_{\mathrm{S}}$ solutions. This difference is only due to the fact that the Schrödinger equation admits an additional solution, for $n' = l = 0$, whereas $n' = k = 0$ is not possible for the Dirac equation, since $k$ cannot be equal to 0.
- When $n'$ increases, for $E_{\mathrm{S}}$ or $E_{\mathrm{D}}$, the absolute value $|\mathrm{BE}|$ of the binding energy increases.
- Moreover, the condition $|k| = n'$ (or $l = n'$) separates the positive- and negative-energy solutions. More precisely:
- When $|k|$(or $l$) > $n'$, $E_{\mathrm{S}}$ or $E_{\mathrm{D}}$ correspond to "negative energy" states.
- When $|k|$ (or $l$) < $n'$, the energies are positive and their values are near those of regular solutions for $s > 0$. We called them "pseudo-regular" solutions.

The question of the sign of the energy for the EDO is analyzed in a more developed way in Section 3.

### 2.1. Possible solutions to the arguments against the EDO states

Here, we only summarize our previous studies, about answers to mathematical arguments against the EDOs.

In the literature, there was essentially the following criticism:

(1) The wavefunction has a singular point at the origin
(2) The wavefunction cannot be square integrable
(3) The orthogonality of eigenvectors cannot be satisfied
(4) There is a paradoxical relationship between the coupling constant $\alpha$ and the binding energy.

For the point #4, we concluded the result in question is obtained in the context of an ill-defined system. For the other points, we reported the deep orbit solution Klein–Gordon equation, given in [7] in the case $l = 0$ is square integrable, and according to works [8], it satisfies the orthogonality conditions and boundary conditions. Nevertheless, in the other cases, and in particular the Dirac equation, the difficulties #1 to #3 are related to the singularity of the Coulomb potential.

Then, we analyzed the works of [9], where the authors remove the singularity by taking into account the nucleus is not a point, and look for solutions with a modified potential inside the nucleus. We summarize the computation process used:

- To choose a radius $R_0$, so-called "matching radius", delimiting two spatial domains: an "outside" one, where the potential is expressed by the usual Coulomb potential, and an "inside", where a chosen potential, different of the Coulomb potential has no longer a singular point at $r = 0$. Of course, the outside Coulomb potential and the inside potential have to be connected at the matching radius, while satisfying continuity conditions.
- To solve the system of radial Dirac equations for the "outside potential", i.e. the Coulomb potential, which gives outside solutions consisting of two components.
- To solve the system for the chosen inside potential, which gives the "inside solution" consisting of two components.
- To satisfy continuity conditions while connecting the respective components of inside and outside solutions.

In fact, during our analysis, we had to look for a more complex ansatz [5] for the inside solutions to satisfy more complete continuity conditions at the matching radius. Next we computed the electron probability density for the wavefunctions, and values of the mean radius $\langle r \rangle$ of the deep orbits, for increasing values of $k(= n')$.

### 2.2. Results on EDOs obtained as solutions of Dirac equation with a corrected potential at the origin

Numerical computations were computed by choosing the matching radius $R_0 = 1.2$ F for atomic H. For the inside solutions, the ansatz includes polynomials of degree 6. The outside solutions are those indicated in [9], i.e. they are built with hyper-geometrical series determined by an analytic method as indicated in [10]. We made computations of the mean radius for increasing values of $k$, to show the progression of the $\langle r \rangle$ as a function of $k$. So, for $k$ respectively equal to 1,2,3,10,20, we respectively obtain $\langle r \rangle = 6.62, 1.65, 1.39, 1.22, 1.2$ in F. These results confirm and extend those of [9]. Moreover they show $\langle r \rangle$ decreases when $k$ increases: it is consistent with |BE| increasing when $k(= n')$ increases.

Further computations were carried out, while changing the values of some parameters: e.g. the precision of our ansatz, the form of the nuclear potential, the chosen matching radius $R_0$. Note that the choice of the matching radius has to depend on the charge radius of the nucleus: for a greater Z, one has to take a greater value for $R_0$. With these conditions, we observed the values of $\langle r \rangle$ are nearly independent of the parameters except for the value of the matching radius $R_0$.

Next, we considered two related facts:

- On the one hand, we have some lack of dependence of the solutions on the nuclear electrostatic potential [11].
- On the other hand, while computing in an approximate way the values of BE corresponding to the obtained values of $\langle r \rangle$, we found some discrepancy with the values deduced from the algebraic expression $E_D$. In fact, these values, as in the tables of [12], are obtained with an equation in pure Coulomb potential. To correct the discrepancy, we used a method of iterative computation with convergence, which leads to the following results [5]: for $k = 1, 2, 3, 10$, we respectively obtain BE = - 56 keV, - 275 keV, -301 keV, -320 keV, corresponding respectively to $\langle r \rangle = 12, 1.65, 1.39$, and 1.22 F.

We can observe two things:

- There is an important gap between the result associated with $k = 1$, and those associated with values of $k > 1$.
- The value obtained for $k = 1$ , i.e. BE $= -56$ keV and $\langle r \rangle = 12$ F, seem to be similar to a result obtained by [13] with a very different method, indicated in Section 5.2.1

## 3. What is the Sign of the EDOs Solutions? We Show below it is Positive

As a consequence of our previous publications, it has been stated that EDOs represent in fact negative energy states. We have looked for and present, in this section, a clear solution to this issue.

First, we can note that the final steps to compute the energy levels of the solutions of relativistic equations lead to a relation defining the square of the energy $E$, i.e. $E^2 = \langle\text{expression}\rangle$. In these conditions, one has two alternatives for the expression of $E$, either to take the positive square root of $\langle\text{expression}\rangle$ or to take its negative square root. But most of time, in the textbooks, the positive root is systematically taken without discussion, since one implicitly supposes that the energy of regular energy solutions cannot be negative.

What about "singular solutions" such as EDOs? One knows the solutions of Dirac equation applied to electrons in an atom form two continuums: the electrons (positive energy continuum) and the positrons (negative energy continuum). It is the cause of the well-known Brown–Ravenhall disease [14] (also referred as the "continuum dissolution" problem), which happens while applying the Dirac equation for at least two atomic electrons. On double excitation of a pair of correlated electrons, one electron can end in the negative energy continuum (positrons), while the other lands in the positive energy continuum, the total system energy being retained. As a consequence, the number of such electron-positron states for an initial electron pair is infinite, and any energy level of the bound electron pair system is infinitely degenerate. Of course, as noted by Rusakova [15], this problem appears only in many-body cases. Nevertheless, it is quite legitimate to ask the question of the sign of the EDO energies, and as we indicated above, this sign cannot be deduced from the relation $E^2 = \langle\text{expression}\rangle$. In fact, if we look more attentively at the solution process of Dirac equation, we can observe the energy $E$, as the variable in an eigenvalue problem, is included in some parameters of the radial equations, after putting it in a pure numerical form. Most interesting: $E$ appears in these parameters without square. For example, in [16], the author ends up to the following relation:

$$\frac{E\gamma}{\hbar c} = \alpha(n' + s), \tag{3}$$

where $\gamma$ is the coupling constant (usually noted $\alpha$), while $\alpha$ is here a positive numerical constant, $s$ the usual parameter defined by $s = \pm(k^2 - \gamma^2)^{1/2}$, taken here with the sign "+" (for regular solutions), and $n'$ the radial quantum number. With $s$ always positive, as $\gamma^2 \ll k$, the author says "…$\langle$the above relation$\rangle$ incidentally shows that $E > 0$". But, in fact, one can more generally say: if the sign of $s$ is not fixed, $E$ has the sign of "$n' + s$".

Nevertheless, one can find another method to solve the Dirac equation, where one obtains the same result without needing to look for any auxiliary numerical parameters in the solution process. We refer here to the well-known method used in [10] and based on the transformation of the system of first order differential radial equations into a second order differential Kummer's equation. The solutions are in form of confluent hyper-geometrical functions. The final step to obtain the expression of the energy $E$, is carried out by solving the following algebraic relation:

$$\frac{1}{2}\alpha\left\{\sqrt{\frac{mc^2+E}{mc^2-E}} - \sqrt{\frac{mc^2-E}{mc^2+E}}\right\} = (n' + s)\ , \tag{4}$$

where $n'$ and $s$ have the same meaning as above, but $\alpha$ is the coupling constant. It is easy to see that the left side has the same sign as $E$, and thus $E$ has the sign of $(n' + s)$.

Now, we return to the sign of EDOs energy solutions. The EDOs are obtained when $|k| = n'$ and $s < 0$, i.e. $s = -(k^2 - \alpha^2)^{1/2}$. As $\alpha \ll 1$, one has $n' + s = |k| - (k^2 - \alpha^2)^{1/2}$ and $|k|$ is always greater than $(k^2 - \alpha^2)^{1/2}$, thus $n' + s > 0$, we can affirm *the energy E is positive*.

Furthermore, we have, as already mentioned in our previous papers:

- If $|k| > n'$, the energy $E$ is $< 0$, so it does not concern physical solutions.

- If $|k| < n'$, the energy $E$ is $> 0$. We have named "pseudo-regular" [1] the corresponding solutions, because they have energy levels near the regular atomic solutions. We can conclude that the case $|k| = n'$ separates the positive- and negative-energy solutions and it falls on the positive-energy side.

## 4. Role of Special Relativity for the Existence of EDOs

### 4.1. Special Relativity generates EDOs

In the recent papers [3,6], we reveal and analyze the essential role of Special Relativity for the possibility of EDOs among "singular" (i.e. not regular) solutions of quantum equations.

From the beginning of our study of these singular solutions, including the works of M&V [9,12], we had observed a pure algebraic argument, which allows us to sort the EDOs in the set of singular solutions. As recalled in Section 3, the energy levels $E$ of the singular solutions of Dirac equation, are given by the algebraic expression (Eq. (2)), again written here:

$$E = mc^2 \left[ 1 + \frac{\alpha^2}{\left(n' - \sqrt{(k^2 - \alpha^2)}\right)^2} \right]^{-1/2} . \tag{5}$$

In the case where $|k| = n'$, one can easily show the expression of $E$ is drastically reduced into $E \sim mc^2\alpha/2n'$. This leads to a binding energy BE $= E - mc^2 \sim -mc^2(1 - \alpha/2k)$, whose absolute value is close to the rest mass energy $mc^2$. This is the signature of deep orbits. One can observe a similar result in the case of the relativistic Schrödinger equation: the EDOs are obtained when the quantum angular $l$ and radial $n'$ are equal. But, to find a deeper reason behind this pure algebraic fact, we have compared the two versions, relativistic and non-relativistic of the Schrödinger equation, as extracted from [16]. Doing this, we observed they have the same form, and only differ by the occurrence of a term including the coupling constant $\alpha$. It is sufficient here to give the expression of the relativistic version, under dimensionless form, which reads:

$$\frac{1}{\rho^2}\frac{\mathrm{d}}{\mathrm{d}\rho}\left(\rho^2 \frac{\mathrm{d}R}{\mathrm{d}\rho}\right) + \left[\frac{\lambda}{\rho} - \frac{1}{4} - \frac{l\,(l+1) - \alpha^2}{\rho^2}\right] R = 0. \tag{6}$$

In fact, the non-relativistic version does not include the term $\alpha^2/\rho^2$. Of course, the numerical parameter noted $\lambda$ and occurring in this equation, has not the same expression as the similar one occurring in the non-relativistic version.

So we have "tracked" the term $\alpha^2/\rho^2$ including the very small "salt grain" $\alpha^2 \sim 5.3 \times 10^{-5}$ brought by Relativity into the Schrödinger equation, which makes all the difference. Then we finally found it comes from the following "energy" factor, used while building the relativistic equation:

$$(E - V)^2 - (mc^2)^2 = E^2 - m^2c^4 - 2VE + V^2. \tag{7}$$

The left-hand-side expression is itself an intermediate expression coming from the relativistic expression of the total energy, after replacing $E$ by $(E - V)$ to introduce the action of the "exterior" Coulomb field on the electron, into the equation in covariant form with minimal coupling. More precisely, the term $\alpha^2/\rho^2$ originates from the far-right term above, $V^2$: the square of the Coulomb potential energy. On the other hand, we can observe that $\alpha^2$ also occurs in the algebraic expression of the level energies for singular solutions, both for the (relativistic) Schrödinger equation and for the Dirac equation, and it is the source of the EDOs (by a dramatic simplification, as explained above). Moreover, we note the two following things:

(1) Expression (7) contains in a relativistic way, i.e. in quadratic form, the energy of the electron in a central Coulomb field minus the rest mass energy. So, the insertion of the energy term is under quadratic form $(E - V)^2$, instead of a linear form $(E - V)$ for the non-relativistic case. We can observe a kind of "entanglement" of $E$ and $V$ due to the non-linear form of expression of the relativistic (total) energy.
(2) The term $V^2$ corresponds in fact to a dynamic relativistic correction of the Coulomb potential $V$. This correction (see e.g. [17,18]), which is also provided by the Dirac equation (in a less visible way), leads to an effective "dynamic" electric potential given by the general following formula:

$$V_{\text{eff}} = V(E/mc^2) - V^2/2mc^2. \tag{8}$$

If considering relativistic electrons, this expression can be rewritten under the following form:

$$V_{\text{eff}} = \gamma V + V^2/2mc^2. \tag{9}$$

Here $\gamma$ denotes the well-known relativistic coefficient, i.e. $\gamma = (1 - \nu^2/c^2)^{-1/2}$.

It is not obvious, from both expressions of $V_{\text{eff}}$, to see one has always a strengthening of the normal Coulomb potential, i.e. $|V_{\text{eff}}| > |V|$. Nevertheless we have shown in [3] that it is true in the case of quasi-circular orbits, more precisely:

- $|V_{\text{eff}}| > |V|$ and $V_{\text{eff}}$ is attractive: i.e. $V_{\text{eff}} < 0$.
- $|V_{\text{eff}}|$ quickly increases as a function of $|V|$, with a parabolic behavior in $|V|^2$ when $|V| \to +\infty$.

Moreover, while computing $V_{\text{eff}}$ from results [5] concerning EDOs in atomic H, approximated by considering almost circular orbits for comparison purposes, we can see the following:

- When $k$ increases, and as the mean radius of the orbits decreases, then $\gamma$ increases and $|V_{\text{eff}}|$ increases too.
- The ratio $|V_{\text{eff}} - V|/\text{KE}$ increases with the kinetic energy KE. So, it seems we have a kind of entanglement between mass, kinetic energy, and potential energy, as Special Relativity mixes total energy and potential energy in a non-linear way: it is a similar observation as we noted above, about the effect of the non-linear energy term $(E - V)^2$.

Here we illustrate this result, by giving only two examples about EDOs extracted from [6]. We indicate approximated values of $\langle r \rangle$, $V$, $\gamma$, $V_{\text{eff}}$, the corresponding potential energy shift $|\Delta V| = |V_{\text{eff}} - V|$ and the kinetic energy KE:

- $k = 1 \Rightarrow \langle r \rangle \sim 6.6$ F, $\gamma \sim 1.2$, $V \sim -218$ keV, $V_{\text{eff}} \sim -222$ keV, $|\Delta V| \sim 4$ keV, KE $\sim 120$ keV.
- $k = 10 \Rightarrow \langle r \rangle \sim 1.2$ F, $\gamma \sim 2.7$, $V \sim -1.17$ MeV, $V_{\text{eff}} \sim -1.79$ MeV, $|\Delta V| \sim 620$ keV, KE $\sim 854$ keV.

### 4.2. What about the singular solutions in the non-relativistic case?

Now, after having recalled how one can see how Special Relativity induces the existence of EDOs, we examine the case of the singular solutions in the non-relativistic case, essentially while considering the (non-relativistic) Schrödinger equation. There exist different (and equivalent) forms of this equation. Here, we consider the form used in [16]: the radial equation (16.7) having, as indicated above, the same form as (6), but without the term $\alpha^2/\rho^2$. The parameter noted $\lambda'$, instead of $\lambda$ in (6) to avoid confusion, is defined by the following expression:

$$\lambda' = (e^2/\hbar)(m/2|E|)^{1/2}. \tag{10}$$

The ansatz used to solve the Schrödinger equation involves, as usual, a factor of the form $F(\rho) = \rho^s L(\rho)$, where $L(\rho)$ is a series in powers of $\rho$. It leads to a condition on the parameter s for convergence of the solutions when $r \to +\infty$, which reads:

$$s(s+1) - l(l+1) = 0. \tag{11}$$

This equation in $s$ has two solutions:

- If taking the root $s = l$, one has the regular well-known solutions.
- Whereas, if taking $s = -(l+1)$, one obtains singular solutions.

In the non-relativistic context, the total energy does not include the rest mass energy and is negative; moreover it represents the binding energy too. The values of energy values are expressed by means the following relation:

$$|E| = mc^2\alpha^2/2\lambda'^2. \tag{12}$$

As showed in [6], it is not possible to put $l = n'$ to obtain EDO's, because this would give a null denominator. But this only shows one cannot apply, to the non-relativistic equation, the recipe used to find EDOs with the relativistic one. Here we can indicate, in a simple way, the necessity of the relativity to obtain EDOs. Based on the solution method used in [16] and even if the recurrence relations for the case $s = -(l+1)$ are different from the case $s = l$, it clearly appears that the series $L(\rho)$ must terminate to provide the convergence of the solutions for $\rho \to +\infty$. This implies $\lambda'$ has to be an integer, as in the regular case. Under these conditions, we can see the maximum (absolute) value of the binding energy is obtained for the minimal possible value of the integer $|\lambda'|$ . If this minimal value is 1, we have in fact $|E| = mc^2\alpha^2/2$, where we recognize the expression of the Rydberg energy ~13 eV, i.e. the ground state energy associated with the Bohr radius of 53 pm.

In conclusion, we have the following result: *with the classical Schrödinger equation for atomic H, one has no EDO.*

Putting together the positive result deduced from the relativistic case with the negative result deduced from the non-relativistic case, we can "surmise" the following global result: Special relativity is sufficient to obtain EDOs, and reciprocally, the existence of EDO needs Special relativity, i.e. Special relativity is necessary for EDOs. That can be written in a more symbolic way: Special Relativity ⇔ EDO. Of course, as we suppose the electron is in the Coulomb electric potential of a nucleus, we wrote only "surmise".

### 4.3. Discussion

In the literature, one can find some results about singular solutions with a non-relativistic Schrödinger equation, but it seems they require a Coulomb potential modified by an *external confining action*, as e.g. in [19], to really obtain deep orbits. Indeed, in that article, the author considered a "compressed" atom in a confinement potential described by a finite potential step at some given radius. In a further article [20], the same author obtains a continuum spectrum of singular bound states, by solving the radial Schrödinger equation in Coulomb field in integral form (integral on closed paths in the complex plan), by using the Laplace method. Those singular wave-functions could contribute to screen the nuclear charge in a zone of radius $10^{-4}\times$ Bohr radius, i.e. ~5.3 F. That paper does not give the mean radius of the electron orbits, nor of their binding energy, but yields probabilities of population of the deep states leading to a compact "pico-atom", depending on the total energy $E$. That probability seems to go from $\sim 2\times 10^{-6}$ to $6.6\times 10^{-5}$.

It is very difficult to compare those results with the EDOs obtained by the relativistic equations. Nevertheless, that result can be related to the one of [21] for a computer simulation of hydrogen atom behavior in a flow of free electrons: during this simulation, the formation of "mini-atoms" is observed, with a smallest size of ~10 F. The incoming free electrons have a kinetic energy of ~3.6 eV, which is the average energy of conduction electrons in palladium. If confirmed, this study could help to suggest a way for reaching EDO states, while taking into account strong magnetic interactions (see Section 5) near the nucleus.

## 5. Survey on the Magnetic Interactions

At atomic levels, the magnetic interactions such as spin–orbit interactions or spin-spin interaction, have little effect on the energy levels of bound electrons, leading only to very small energy shifts. Their essential interest is to break the degeneracy of energy levels and split the spectral lines. Nevertheless, at deeper levels, they can become very strong and even be dominant over the Coulomb interaction. Below, without claiming to be exhaustive, we recall some elements about the magnetic interactions; then we analyze several works on the magnetic interactions, the methods to account for their actions at short distances, and the difficulties encountered in these questions.

### 5.1. Elements on magnetic interactions

We recall some simple and necessary elements of knowledge about magnetic interactions in an atom. We only consider the atomic H, or D (when we explicitly indicate it).

#### 5.1.1. *Spin–orbit interaction*

When taking into account the spin of a bound electron, one has the well-known Spin–Orbit (SO) coupling: it has the greatest observable effect, since it is involved in the "fine-structure" effect, a splitting of the spectral lines, and it partially breaks the degeneracy of the atomic energy levels. The Spin-Orbit interaction is the cause of only a part of the energy correction leading to the fine-structure effect, as further corrective terms are involved: a kinetic energy relativistic correction at first-order in $v^2/c^2$ and the Darwin term. In fact, these three corrective terms are automatically included (but hidden) in the Dirac equation. We do not expose specifics showing this result, as it is in most QM textbook and courses. We recall that, in fact, one considers the relative orbital movement of the nucleus in the electron reference frame, which gives a magnetic field "seen" by the electron. So, on account of the magnetic moment associated with the electron spin, this relative nuclear magnetic field causes a spin precession (Larmor), expressed in a quasi-classical way by a potential energy. But one has also to take into account the further Thomas precession, having a pure relativistic nature and acting in the opposite direction of the precession. If the electron is only slightly relativistic, the energy of Thomas precession is half the Larmor precession.

The general quantum expression of the spin-orbit interaction reads:

$$H_{\mathrm{SO}} = \frac{1}{2m^2c^2}\left(\frac{1}{r}\frac{\partial V}{\partial r}L\cdot S\right), \tag{13}$$

where $L$ and $S$ respectively denote the electron angular momentum and spin operators.

As $V$ is the Coulomb potential energy, by replacing $V$ by its expression as a function of $r$, one obtains:

$$H_{\mathrm{SO}} = \frac{\mu_0 e^2}{8\pi m^2}\frac{1}{r^3}L\cdot S = \xi(r)L\cdot S. \tag{14}$$

Moreover, the scalar operator $L\cdot S$ commutes with the total angular moment $J = L+S$ and the potential expressed by (14) commutes with $L^2$, $S^2$, $J^2$ and $J_z$. By using known rules on composition of angular moments and Clebsch–Gordan coefficients, one shows $L\cdot S$ can be expressed by means of the eigenvalues $j, l$, and $s$ (fixed $= 1/2$). If $l \neq 0$, one can write:

$$L\cdot S = \frac{\hbar^2}{2}\left[j\left(j+1\right) - l\left(l+1\right) - s\left(s+1\right)\right] = \frac{\hbar^2}{2}\left[j\left(j+1\right) - l\left(l+1\right) - 3/4\right]. \tag{15}$$

Here $j = l \pm 1/2$ and

- If $l = 0$, there is no SO interaction.
- For any value of $l > 0$, one has two cases:
- $j = l - 1/2 \Rightarrow L \cdot S = -(\hbar^2/2)(l+1)$, that gives an attractive potential, since $\xi(r)$ is $> 0$.
- $j = l + 1/2 \Rightarrow L \cdot S = +(\hbar^2/2)l$, giving a repulsive potential.

To obtain numerical values for the SO potential energy as function of $\langle r \rangle$, one replaces the physical constants occurring in $\xi(r)$ by their values in a chosen unit system, e.g. the SI standard. Then the average energy $\langle E_{\rm SO} \rangle$ for $l = 1$ reads:

$$\langle E_{\rm SO} \rangle \sim (1.71 \times 10^{-53}/\langle r^3 \rangle)\ \mathrm{J} \sim (1.07 \times 10^{-34}/\langle r^3 \rangle)\ \mathrm{eV}.$$

As an example, we give a quick computation: for the 2p orbital (of atomic H), where $\langle r \rangle \sim 10^{-10}$ m, $\langle E_{\rm SO} \rangle \sim 1.06 \times 10^{-4}$ eV. But we can easily see that, for a very small radius, the energy becomes very high. For example, with $\langle r \rangle \sim 10^{-13}$ m $= 100$ F, one has $\langle E_{\rm SO} \rangle \sim 106$ keV. Note this generalization is made, by supposing the rules for angular moments can be extended for any radius. Moreover, when high energies are concerned, we have to take into account relativistic corrections. Finally, we remark the following things:

- The SO coupling is based on the relative motion of the proton with respect to the electron, but one considers the *angular momentum operator L of the electron* to compute it.
- The composition of angular moments, not totally developed here, leads to a computation of the mean value of the energy by using the eigenvalues while avoiding the integration of the operator with a wavefunction.

### 5.1.2. *Magnetic interactions involving the nuclear spin*

In taking into account the nuclear spin, we have further magnetic interactions, which are involved in the so-called "hyperfine structure" of the atomic spectral lines and further breaking of the degeneracy of the energy levels. The analysis of these magnetic interactions is complex, but we only need deal with the energy. Therefore, we can make some simplifications while evaluating orders of magnitude of the coupling energies. For example, there is a spin–orbit interaction $\mathrm{S_pO}$ concerning the nuclear spin and the (direct) orbital motion of the electron represented by $L$. It can be repulsive or attractive, depending on the relative directions of the nuclear spin and of the electron angular momentum. As the magnetic moment of the proton is $\mu_{\rm p} \sim 1.41 \times 10^{-26}$ SI units, whereas for the electron it is $\mu_{\rm e} \sim 9.3 \times 10^{-24}$ SI units, we have a ratio $\mu_{\rm e}/\mu_{\rm p} \sim 670$. Under these conditions, for any combination of type (repulsive/attractive) of interactions SO and $\mathrm{S_p}$ O, the coupling energy of $\mathrm{S_pO}$ can be neglected relative to SO.

For the same reason, another magnetic interaction with the nuclear spin, the spin–spin interaction, which is much weaker than the SO interaction, will be neglected when we take into account SO. Nevertheless, if $L = 0$ one has neither SO nor $\mathrm{S_pO}$ coupling. Thus, one has to take into account the interactions between the nuclear and electron spins. We have the following expression, including the respective magnetic moments $M_{\rm p}$ and $M_{\rm e}$ of the proton and the electron, due to their spins:

$$H_{\rm SS} = -\frac{\mu_0}{4\pi}\left[\frac{1}{r^3}\left(3\left(M_{\rm p}\cdot\hat{r}\right)\left(M_{\rm e}\cdot\hat{r}\right) - M_{\rm p}\cdot M_{\rm e}\right) + \frac{8\pi}{3}M_{\rm p}\cdot M_{\rm e}\delta\left(r\right)\right], \tag{16}$$

where $\hat{r}$ denotes a radial unit vector.

As the spins are involved in the magnetic moments, one has to consider a combination of both electron and proton spin operators $S_{\rm e}$ and $S_{\rm p}$. By introducing the total spin $S = S_{\rm e} + S_{\rm p}$, one can write the following relation:

$$S_{\rm e}\cdot S_{\rm p} = 1/2(S^2 - (S_{\rm e})^2 - (S_{\rm p})^2) = (\hbar^2/2)\left[s(s+1) - 3/2\right], \tag{17}$$

where the only possible values of $s$ are $s = 1$ ("triplet" state) and $s = 0$ ("singlet" state). So, we have two cases:

- $s = 0 \Rightarrow S_e \cdot S_p = -(3/4)\hbar^2$.
- $s = 1 \Rightarrow S_e \cdot S_p = +(1/4)\hbar^2$.

Here, we wrote only a little part of the computation (which is in most QM books/courses) and we note that the case $s = 0$ gives an *attractive* interaction whereas $s = 1$ a *repulsive* one. In fact one can use the following relation for the energy $E_{SS}$ of the spin–spin interaction in the attractive case: $\langle E_{SS}\rangle = -(3/4)A(r)$ where, for $\langle r\rangle = a_0$ (the Bohr radius), $A(a_0)$ is proportional to $1/a_0^3$ and has a value $\sim 5.87 \times 10^{-6}$ eV for the fundamental level $1s$ of the hydrogen atom. In these conditions, to make approximate computations (Section 6), we extrapolate this result to a general expression of the spin-spin interaction energy $E_{SS}$ for any radius $\langle r\rangle$:

$$\langle E_{SS}\rangle \sim (1 \times 10^{-55}/\langle r^3\rangle)\ \mathrm{J} \sim (0.64 \times 10^{-36}/\langle r^3\rangle)\ \mathrm{eV}. \tag{18}$$

As for the SO coupling and when high energies are concerned, we have to take into account relativistic corrections. At high speed considered in the observer reference frame, the direction of the spin appears to realign itself closer to that of the translational velocity vector [22] which is normal to the orbital angular momentum and is time varying, with an average value approaching zero as $v \Rightarrow c$: this should imply a weakening of both the $L \cdot S$ of the SO term and the $S_e \cdot S_p$ and $1/\langle r^3\rangle$ dependence of the energy $E_{SS}$.

### 5.1.3. *Case of a term in 1/$r^4$*

In the next section, we identify a term including the square of the magnetic vector potential, i.e. $\mathbf{A}^2$, generally neglected at atomic levels. It occurs, e.g., while developing the hyperfine structure Hamiltonian or a Pauli equation. In both cases, it comes from a term of the form $(\mathbf{p} \pm q\,\mathbf{A})^2$/m. As it includes the vector potential, it is a term of magnetic interaction: it expresses a diamagnetic energy. We give further the expression used to compute this energy term.

## 5.2. The Vigier–Barut model

In this section, we speak about works referring to the "Vigier–Barut model", but we first specify the origin of this naming.

### 5.2.1. *Works of Barut, with a "more complete" Dirac equation*

The Vigier–Barut model and the works related to this model, derive from works of Barut, e.g. [23,24]. In these articles, the author looks for an analytic solution of the Dirac equation for a charged lepton with anomalous magnetic momentum (AMM) in Coulomb potential. Note that including the AMM into the Dirac equation, which contains already the "normal" magnetic momentum of electron, is in fact a way to take into account a second order QED contribution: more precisely a one-loop correction to the fermion magnetic moment. For doing this, the author uses a Dirac equation, where the vector potential is completed by a term introducing the electron AMM and expressed by means of the EM tensor $F\mu\nu$ and Dirac matrices $\gamma$. As pointed out by Barut, the AMM of the electron is small relative to the normal magnetic moment, but it is of the same order-of-magnitude as the magnetic moment of the proton. After numerous and complex transformations, he obtains a second order differential radial equation. This equation has a form similar to a radial Schrödinger equation, with a specific dynamic potential $V$ expressed as sum of inverse powers of the radius $r$, plus a term $k^2$ depending on $E^2$, where $E$ represents the energy eigenvalue. More precisely, the differential equation reads:

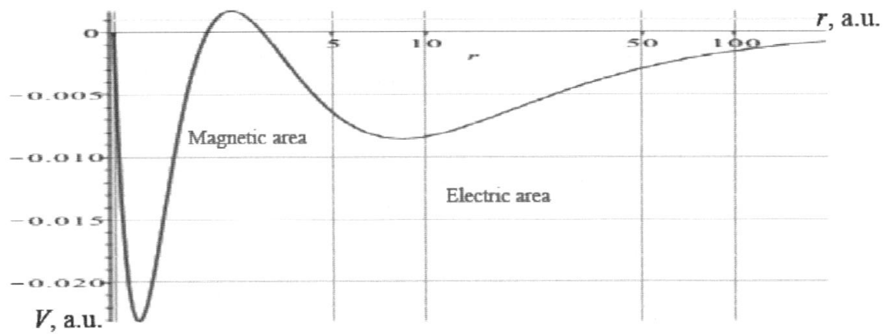


**Figure 1.** Potential with two wells.

$$\left(\frac{\mathrm{d}^2}{\mathrm{d}r^2} - V\left(r\right) + k^2\right)\psi = 0. \tag{19}$$

The potential $V$ is expressed by

$$V(r) - A/r + B/r^2 + C/r^3 + D/r^4. \tag{20}$$

Here *A, B, C* and *D* are built from physical constants, the mass or the reduced mass, $E$, and initial parameters of the equation, written in natural units, i.e. with $\hbar = c = 1$. In the second cited paper, these coefficients are constant for any given eigenvalue $E$. We have to note the occurrence of $E$ in a non-linear form that is the "signature" of Relativity

Note that in the "dynamical" potential $V(r)$ computed by Barut, the expressions of the coefficients $B$, $C$ and $D$ are complex, energy-dependent, and difficult to interpret physically.

But, on the other hand, in the works associated with the so-called Vigier–Barut model:

- The term in $1/r^2$ has clear meaning of the well-known "centrifugal barrier" energy.
- The term in $1/r^3$ represents magnetic interactions, such as those described in Section 5.1:
  - either spin–orbit, available only if the angular quantum $l$ is not null. It is included in the Dirac equation. It can be attractive or repulsive, depending on the total angular number $j$
  - either spin–spin, involving the nuclear spin, which is not included in the Dirac equation but can be involved in the Pauli equation. It can be attractive (case of "singlet" state) or repulsive (case of "triplet" state).
- The term in $1/r^4$ is the least known. It has meaning as a "diamagnetic" term and involves the vector potential **A**. It is always repulsive and can be deduced from the Pauli equation. Some explanations are given in Section 5.1.3 and especially in the body of Section 5.2.2.

In fact, Barut does not apply Eq. (19) to the atomic H, but to a coupled system (e+, e–) called "positronium." Doing this, he finds two potential wells as pictured in Fig. 1, and he looks for resonance states of positive energy. Note $V$ reaches a maximum at the huge energy $V_{\max}$ of order 35 GeV, when the radius becomes very small and orders-of-magnitude smaller than the region corresponding to the regular (atomic) positronium bound state. Then Barut raises the abscissa axis to $V_{\max}$ to treat the decay of the resonances. In conclusion, he summarized the following result: there are new resonant states at distances of order $\alpha r_0$ of very large positive mass–energy of order GeV; here, $r_0$ denotes the classical electron radius ~2.8 F, which gives $\alpha r_0$ ~0.02 F. Moreover, he says the used model could describe the electron–proton system and give (very) deep bound electrons, while noting the proton (finite) size leads to a further effect. However, it is difficult to imagine a deep electron resonance being at a distance 0.02 F from the origin, since it would be inside the proton.

In another paper [25], the author uses a very different method to treat tight molecules and finds an inter-atomic binding energy of ~50 keV. On the other hand, in [26] he extends his model, by taking into account the short-ranged strong magnetic forces, to the high-energy physics, including strong and weak interactions.

Note an article of Behncke [27], where the author analyzes the properties of the Dirac equation for an electron in a spherically symmetric electric potential $V$, under a general linear covariant form including the electron AMM. While proving the Hamiltonian $H$ is essentially self-adjoint, it appears the AMM term induces a repulsive force, regardless of the sign of $V$, which forces the wavefunctions in the domain of $H$, to vanish at the origin. This latter result is indicated in [28], under the form of a further term in $V'$ in the usual radial Dirac operator.

### 5.2.2. *Works associated with the Vigier–Barut model*

Regardless of the relativistic origin of the model, we speak now about some works carried out in a non-relativistic framework: we can cite the research of Vigier [29], Samsonenko et al. [13], Dragic et al. [30], on possible tight orbits under the Bohr level for atomic H (or D), and even for tight molecules. Of course, this list does not exhaust the concerned subject. All these works are attempts to explain LENR results by screening effects in "tight" orbits, in particular the initial experiment of Pons and Fleishmann [31], which was recent at the time. We set aside the work of Ozcelik et al. [32], which does not aim at finding tight orbits, but is referred to by the previous cited works, because it presents an exact solution of the radial Schrödinger equation for inverse-power potentials. But, as its purpose is different and concerns mainly the interatomic interactions, the solution method contains some restrictions on the coefficients of the potentials, which limit the possibilities of application for the intra-atomic magnetic interactions. All these works have in common the use of a potential $V$ defined as a sum of inverse powers, similar to the expression (20), with constant coefficients $A, B, C$ and $D$.

For the first three cited papers, the starting point is a two-body Hamiltonian for the electron and the nucleus, taking into account the magnetic moments of the two particles and including, as usual, the momentum **P**. This Hamiltonian is inserted into a Schrödinger-like equation. So interactions such as spin–spin and spin–orbits can be explicitly expressed. In fact, the obtained equation is an extension of the (non-relativistic) Pauli equation to a two-body system of charged particles. Next, algebraic transformations and developments applied to the equation lead to a new equation including a potential $V$ expressed as in (20). Note that, while defining the reduced mass, the authors of [30] explicitly talk about a proton, which means they only handle the atomic H. In [29], the author explicitly speaks about hydrogen and deuterium, and in [13] the authors speak about particles without specifying them, but as they refer to the work of [29], one can suppose their study can be applied to H or D.

We can see, in the potential $V$, the term in $1/r$ is the Coulomb potential and the term in $1/r^2$ is the classical term in $l(l+1)$ expressing the centrifugal barrier. The term in $1/r^3$, characteristic of the magnetic interactions spin–spin and spin–orbit, are similar in the different works. The term in $1/r^4$ could seem weird, but by coming back to the initial Hamiltonian, we can see it comes from terms equivalent to $(\mathbf{P}_\mathrm{i} - q_\mathrm{i}\mathbf{A}_\mathrm{j})^2/\mathrm{m}_\mathrm{i}$ ; $(\mathbf{P}_\mathrm{i} - q_\mathrm{i}\mathbf{A}_\mathrm{j})^2$ is the square of what Feynman names "dynamical momentum" or "p-momentum" [33], including the coupling of one particle with the EM field of the other particle. The expression of the vector potential produced by a magnetic dipole moment **m** reads:

$$\mathbf{A}\,(r) = \frac{\mu_0}{4\pi}\frac{\mathbf{m}\times\hat{r}}{|\mathbf{r}|^2}. \tag{21}$$

Here $\hat{r}$ is the unit vector in the direction of the radial vector **r** and $\mu_0$ is the vacuum permeability. So, the square of the p-momentum generates a term in $1/r^4$. Note that, to take into account the magnetic interactions in a symmetric way, the Hamiltonian must include the p-moment of each particle (e.g. electron and proton) and the dipole–dipole interaction terms.

One can ask what is the physical meaning of the square $A^2$, included in the term $D/r^4$. But the complete term $q'^2\mathbf{A}^2/m$, having the dimension of an energy, is considered [34] as expressing a diamagnetic energy. But it would be interesting to deepen its physical meaning. Moreover, we can also cite an article [35] about the significance of $\mathbf{A}^2$ in QFT context, but it exceeds the framework of our paper.

Here, it is not superfluous to say some words about the vector potential: It allows us to greatly simplify the writing of relations and equations of Electromagnetism; the electric **E** and magnetic **B** fields can be directly derived from **A**, the Coulomb gauge is simply $\nabla \cdot \mathbf{A} = 0$ and the Maxwell's equation in this gauge is $\partial_\mu \partial^\mu \mathbf{A} = 0$, the Lorenz gauge can be written $\partial_\mu A^\mu(x) = 0$ and so the Maxwell's equation in Lorenz gauge can be written $\partial_a \partial^a A^\mu$=0. On the other hand, the concept of vector potential has a proper scientific interest, whereas in the past, famous physicists supposed it was only an abstract mathematical tool without physical reality. Nevertheless, it became necessary to explain some observations in the EM theory, e.g. the well-known and strange phenomenon called Aharonov-Bohm effect [36], which shows the local fields **E** and **B** do not contain full information on the EM field.

To return to the term $D/r^4$: this term is generally neglected, as having a very small value at the atomic levels. But $D/r^4$ has an important role in the Vigier–Barut model: as $D$ is $> 0$, $D/r^4$ acts as a repulsive core at small distances of the nucleus, while the Coulomb potential $A/r$ is attractive at large distances. $B$ is always $> 0$, since $B/r^2$ represents the centrifugal barrier, and $C/r^3$, including magnetic interactions, can be $> 0$ or $< 0$. With these conditions, if $C < 0$, we have the following succession of potentials, intentionally written in the order of the decreasing powers of $r$:

$$D/r^4 \ \text{[repulsive]}, \quad C/r^3 \ \text{[attractive]}, \quad B/r^2 \ \text{[repulsive]}, \quad A/r \ \text{[attractive]}.$$

For some combinations of values of the coefficients, if $V(r)$ has three zeros, i.e. there are three different values $r_1$, $r_2$ and $r_3$ such that $V(r_1) = V(r_2) = V(r_3) = 0$, *then V(r) includes two wells*:

- The first well, for the small values of the radius $r$, corresponds approximately to a particular zone where the magnetic interactions are dominant: one can call it the "magnetic region" as in [37]
- The second well, occurring when $r$ increases toward the Bohr radius, corresponds to a zone where the electric Coulomb potential is dominant: the "electric region" [37]

In Fig. 1, we represent a pure abstract mathematical exercise to show, in a simple way, that some combinations of coefficients can give rise to two local minima for the plotted curve of a potential like $V$. This curve does not correspond to the computation of an actual physical case. Here, as well as in [29] and in these conditions, the values on the axes have no physical units. Plotting actual values, with even a limited range from 1 eV to $\pm 1$ MeV against 1 nm to 1 F, is a bit awkward. It would have to be done on a modified log-log plot (>7 $\times$ 6 cycles). Therefore, we only illustrate such a situation in Fig. 1 by plotting, in semi-log, a simple simulation of a representative potential $V$ satisfying the above conditions of Eq. (20).

In fact, the methods used in the cited works are slightly different, so some terms can be neglected because of the complexity of the computations and the different approximation methods (e.g. the "hard core" approximation is used in [30]). The analytic methods meet numerous difficulties, which lead to solving the equation numerically. With these circumstances, as written in [13]: "there is a set of contradictory estimates of energy values and quantum-orbit radii based on the starting Hamiltonian." It is not a criticism of these works, because they are important and difficult studies in a field that has been barely explored, if at all: the domain of the magnetic interactions near to the nucleus.

We note Vigier [29] also develops a theoretical study for computation of new tight orbits, based on the causal mathematical formalism of L. deBroglie [38] and Bohm [39]. Unfortunately he does not give numerical results on these tight orbits. He indicates that he would continue this study in a future article, which seems to have never been published. On the other hand, his work [29] includes a long study about a three-body problem, by considering possible new "tight" molecules, i.e. one noted $H_2^+$ and another $D_2^+$, constituted from two nuclei aligned with an electron between them, while the nuclei rotate around the electron. It can be compared to a similar work of Barut [25], where the author also considers linear molecules, but in a triangular structure, where the electron oscillates in the plane perpendicular to the line joining the nuclei. For these tight molecules, Vigier computes *numerical results* for energies of ground states and he obtains ~28.1 keV for $H_2^+$ and ~56.2 keV for $D_2^+$. Barut indicates similar size order of energies, e.g. ~50 keV for $D_2^+$.

To conclude with the numerical results, we can see

- In [13], the authors indicate a tight state of energy ~40 keV. We recall that, as it is a study in non-relativistic context, this value is that of the total electron energy and the binding energy too.
- In [30], the authors give numerous results for the spectra of the hydrogen atom, at different energy levels and for different values of quantum numbers L, S, and J. Nevertheless, the indicated energies are similar to classical atomic values, i.e. a few eV. In fact, they did not find tight orbits with high energies of order keV.

In conclusion, we can note the results concerning tight orbits in these works seem to correspond to a result we found [5] for the "ground" level of EDOs, i.e. for $k = 1$, when applying a special process for coherence between the energy and the value of the mean radius, as indicated in Section 2.2: $|\mathrm{BE}|$ ~56 keV for $\langle r\rangle$ ~12 F.

## 6. Physical Questions About EDOs, and First Attempts of Solution by Computations

Two important questions have not yet been resolved in our previous papers:

(1) Can the EDOs satisfy the HUR, a pillar of QM?
(2) Are the EDOs stable solutions? That is to say, do they have a lifetime sufficient to induce a fusion or transmutation process?

### 6.1. The question of the *HUR*. Relativistic confinement energies

Consider an electron confined near the nucleus, in a region delimited by a mean radius $\langle r\rangle$ of order ~2 F, the size order of EDOs for atomic H. We compute an estimate of the kinetic energy of the electron: from HUR, $\Delta|\mathbf{p}|\Delta r \geq \hbar/2$. It is usual to approximate the uncertainty in $|\mathbf{p}|$ by $\Delta|\mathbf{p}| \sim \hbar/\langle r\rangle$ and to take for $\Delta|\mathbf{p}|$ an estimate of the momentum $|\mathbf{p}|$ of the confined particle. So we let $|\mathbf{p}| \sim \hbar/\langle r\rangle$, which gives $|\mathbf{p}| = 5.27 \times 10^{-20}$ SI units for the given $\langle r\rangle$. On the one hand, one sometimes finds a reasoning based on the classical kinetic energy for a particle in a Coulomb potential $V$, $E - V = \mathbf{p}^2/2m$; then one can deduce the value of $E - V \sim 1.5 \times 10^{-9}$ J $\sim 9.5$ GeV: one obtains an actual "hadronic" energy. On the other hand, as it makes no sense to talk of non-relativistic models in the nuclear region, consider the relativistic expression $E - V = [\mathbf{p}^2c^2 + m^2c^4]^{1/2}$. Then, with the previous value of $|\mathbf{p}|$, one has $|\mathbf{p}|c = 1.58 \times 10^{-11}$ J, while the electron rest mass $mc^2 \sim 8.2 \times 10^{-14}$ J. So, as $m^2c^4 \ll \mathbf{p}^2c^2$, one can write $E - V \sim |\mathbf{p}|c$ and KE $\sim E - V - mc^2 = 1.57 \times 10^{-11}$ J $\sim 98$ Mev, i.e. ~100 times smaller than the value found by using a non-relativistic formula. The mathematical reason of this difference is well-known [40]: from the relativistic expression of $E - V$, we see the kinetic energy increase linearly with $|\mathbf{p}|$ for large momenta, whereas in non-relativistic mode it increases as $|\mathbf{p}|^2$

These are rather coarse computations, but the result is very interesting, because energy of about 100 Mev can be almost realized by the accepted spin–spin interaction alone (Section 5), which is still much weaker than the accepted spin–orbit interaction. Indeed, for $\langle r\rangle$ ~2 F, we have $E_{\mathrm{SS}} \sim 1 \times 10^{-55}/(2 \times 10^{-15})^3 \sim 1.3 \times 10^{-11}$ J $\sim 81$ MeV. With such values, one can think *the HUR is practically satisfied by the electrons on EDOs,* thanks to magnetic coupling. We have yet to make more precise computations, by taking into account more parameters of interactions near the nucleus, as well as precise effects of relativity, which tends to bend the spin into the direction of the motion of the particle.

### 6.2. Question on the stability of the EDOs

An approximate but classical way to verify the stability of a particle state is to show it corresponds to a local minimum of total energy of the particle. For doing this, we start from the relativistic expression of energy $E$, and we replace the momentum $|\mathbf{p}|$ by $\hbar/r$, in order to include the HUR, and then we add a term $V$ to $E$, to represent a potential energy. So, we have the following expression:

$$E_H = \sqrt{\frac{\hbar^2 c^2}{r^2} + m^2 c^4} + V. \tag{22}$$

Next we look for a sum of potential terms as in the Vigier–Barut model, to obtain a local minimum of $E_H$.

- In particular, we include in $V$ a repulsive term in $1/r^4$, which, as indicated in Section 5, is essential to obtain a potential well near the nucleus with a repulsive hardcore. This term is computed by means of the following expression, deduced from the electron vector potential and including usual physical constants in SI units:

$$V_4 = K_4/r^4, \quad \text{where } K_4 = \left(\frac{\mu_0}{4\pi}\right)^2 \frac{\mathbf{e}^4 \hbar^2}{4m_e^2 m_p} \sim 1.3 \times 10^{(-71)} \text{ SI}. \tag{23}$$

There is a symmetric term built with the proton vector potential, but it can be neglected in relation to the former one.

- One considers terms in $1/r^3$, for magnetic interactions, and the classical term in $1/r$ corresponding to the electric Coulomb potential energy $V_1 = -\alpha c\hbar/r$. So, at this stage, we have $V = V_1 + V_3 + V_4$, where $V_3$ is yet to be determined.

Initially, as the expression of $E_H$ already includes a momentum-like term (under the square root), we do not include a centrifugal barrier term in $V$, because it already corresponds to a part of the Laplace operator in spherical coordinates, which introduces the dependence on the angular momentum $l$. Of course, by doing this, one partly eliminates the role of $l$.

On another hand, if we compute the potential energy $V_3 = E_{\text{SO}}$ associated with the spin–orbit interaction in attractive mode, while supposing $l = 1$ and applying the usual quantization rules, we obtain $E_{\text{SO}} \sim -13$ GeV for an electron at a distance $r \sim 2$ F, i.e. in the expected region of the EDOs. Such a huge value seems not physically reasonable. Moreover we obtain the deep local minimum of $E_H$ at $r \sim 0.001$ F, i.e. totally inside the nucleus: it has no physical meaning since the used expressions for computing are no longer correct when reaching within the nuclear surface. Then, we tried to introduce the repulsive centrifugal barrier term in $1/r^2$ into $V$, in thinking it can have a restraining effect on the value of $E_H$ near the nucleus. But we obtain again the deep local minimum at $r \sim 0.001$ F. We can also remark, for each value of $n'$, the energy levels of EDO solutions of Dirac equation differ very little from those of the Schrödinger equation, although the Dirac equation takes the SO interaction into account: so one can think that $E_{\text{SO}}$ does not act for the EDOs. Moreover, the SO coupling requires more hypotheses to be satisfied than the SS coupling: the spins are intrinsic parameters and the rules of spin addition are independent of the context. Thus, we do not take into account $E_{\text{SO}}$, but rather the potential energy $E_{\text{SS}}$ corresponding to the spin–spin interaction in attractive mode. So, with $V_3 = E_{\text{SS}} \sim (1.04 \times 10^{-55}/\langle r^3 \rangle)$ SI, we put $V = V_1 + V_3 + V_4$ in $E_H$.

Here, we give some numerical results obtained with this potential:

- The deep local minimum of $E_H$ is at $r$ ∼0.17 F, i.e. again inside the nucleus, albeit much less deep than with $E_{S0}$
- There is a local maximum of $E_H$ of 42 MeV, at $r$ ∼3 F.
- At $r = 2$ F and assuming the orbit to be quasi-circular, we have $\gamma \sim 150$ and the kinetic energy KE ∼ 76 MeV.

At this time, and after numerous further computation trials, we can observe that the repulsive diamagnetic potential $V_4$ is too weak to "push" the local minimum outside the nucleus. Moreover, if we computed an "optimal" value of $K_4$, to obtain a local minimum for $r \sim$ a few F, it would be of size order $10^{-68}$, i.e. three orders greater than the present value. For example, with $K_4 \sim 3.4 \times 10^{-68}$ the deep local minimum is at $r$ ∼2.6 F. Of course, all these computations are made in a context where several further effects due to relativity and the nearness of the electron to the nucleus are not taken into account. For example:

- By simply considering the finite size of the nucleus and electron, one can think that the magnetic potentials, dominant in this region, overlap and decrease while approaching the nuclear "surface" defined by the charge radius, thus tending toward a finite limit.
- The tightly bound ultra-relativistic electron has its spin axis precessing about its velocity vector, which is normal to its orbital angular-momentum vector, so closely that the L ·S and orbit-averaged $S_n \cdot S_e$ terms go toward zero.
- Moreover, QED effects such as electron self-energy and vacuum polarization are certainly strengthened in the extreme EM field near the nucleus. So one can expect a cloud of virtual lepton pairs (e+, e–) to surround the electron, causing a strong screening effect of the interactions.

With these effects, a potential $V$, highly attractive near the nucleus, could stop increasing (in absolute value) when approaching the nuclear surface and would decrease within: so one would have a potential well leading to a local minimum of the energy. Resonances between these effects would lead to the predicted quantized orbits and deep-energy levels.

Of course, such a qualitative reasoning requires confirmation by rigorous mathematical proof.

## 7. Conclusion, Open Questions and Future Work

Several questions are investigated here, in the continuation of our previous work.

- We quickly recalled our studies on the EDOs of the atomic H as solutions of relativistic quantum equations, in particular our review of the method of Coulomb potential corrected at the origin [9] which, we believe, counters all mathematical arguments against the EDOs and which we have extended with new results about general properties of EDOs.
- Then, from methods used to find deep orbits as singular solutions of relativistic equations, we have discussed a recent and subtle criticism of the sign of the energy levels of these solutions, which is of course a crucial question. So we showed, in a simple way, that the *EDOs have a positive energy E*. More precisely, we can see that the equality condition on angular and radial numbers, $|k| = n'$ (case of the Dirac equation) or $l = n'$ (Schrödinger equation), generates the EDOs as a particular class included in the set of singular solutions. This condition separates the positive energy solutions from the negative energy ones: if $|k| > n', E < 0$, whereas for $|k| \leq n', E > 0$ including the case $|k| = n'$ which characterizes the EDOs.
- Next we gave new elements about the role of the Relativity for the deep orbits: from an analysis of singular solutions of the non-relativistic Schrödinger equation, we surmise Relativity is not only a source of the deep orbits, but is necessary to obtain them. Nevertheless, the question is not completely resolved if one considers atoms "compressed" by external potentials.

- But the heart of our paper is devoted to a study of the magnetic interactions near the nucleus, with the aim to solve two further important questions, of physical nature:
- Can the EDOs satisfy the HUR, which is a pillar of QM?
- Can we guarantee their stability?

As the topic of the magnetic interactions between the electrons and the nucleus has multiple aspects and is the subject of numerous works, we began here with our first study on the most-known expressions of these interactions, e.g. spin–orbit and spin–spin coupling. Our motivation is based on the fact that these interactions, very weak at the atomic levels, become very strong and even dominant near the nucleus. Thus, they have to be taken into account for the EDOs. To this end, we analyzed some works of Barut, carried out in a relativistic context, and some studies related to the well-known Vigier–Barut model (VB) developed in a non-relativistic context. These works aim to find tight-orbit solutions with high binding energies.

- We noted that Barut found resonances for positronium, for a very short inter-particle distance ~0.02 F and huge energy of order GeV. The author expected the extension of his method (based on a Dirac equation "completed" by the AMM of the electron) to the atom H, but we did not find this extension in the literature, and it seems very difficult on account of the finite dimension of the proton, whereas in positronium the particles are considered "point-like". Nevertheless, in another article of Barut, we found a result on tight (H or D) molecules with high binding energy ~50 keV. The non-relativistic works on the VB model use Schrödinger-like or Pauli-like equations with a potential in the form of a sum of inverse powers of the radial distance. There are few numerical results about deep orbits in the literature, but we noted results for a tight state with energy $\sim -40$ keV and for tight molecules with – 28 keV ($H_2$) and – 56 keV ($D_2$). These results have a size near the result (Section 2) we found with a special iterative process [5] for coherence of the energy w.r.t. the radii.
- Next we presented some computations based on magnetic interactions. As result, we can observe the *HUR, expressed relativistically, can certainly be satisfied by EDOs*, thanks to the high energies of these interactions: in particular the spin–spin coupling is sufficient. Our computations are carried out in a very simplified context and so they lead to an approximate conclusion; nevertheless, it is already a hopeful result.
- Computations were also carried out to look for a local minimum near the nucleus, to provide stability for EDOs. Unfortunately, the local minimum tends to take place inside the nucleus, as the repulsive potential in $1/r^4$ is much too weak to overcome, in the EDO region, the strong attractive potentials we computed. So, we have to continue to work on this question, by taking into account further interactions and effects occurring near the nucleus and by using adequate relativistic quantum tools. In particular, in a QED context, the Two Body Dirac Equation of Dirac's constraint dynamic [41], used in several works by Crater, e.g. [42], allows one to express, in a consistent relativistic way, numerous magnetic interactions at short distances, under the form of "quasi-potentials". Moreover it gets around the Currie–Jordan–Sudarshan "non-interaction theorem" [43], about the bad habit of adding simple potentials to obtain a relativistic Hamiltonian dynamics.
- There is also a further open question: how can we obtain deep orbits from atomic H (or D)? Of course, it seems difficult to believe such a transition could happen spontaneously, for free atoms, e.g. in the spatial vacuum. Nevertheless we can think about the following possible process in condensed matter: from a computer simulation [21] cited in Section 4.3., mini-atoms could be generated from H atoms confined in a metallic net and exposed to a beam of slow quasi-free electrons (from the conduction band) set in motion by low-energy electric arcs. In the simulation, these mini-atoms are unstable; but, when the electrons approach the nucleus, they could be "hooked" by strong magnetic interactions, such as spin–spin coupling, and so they could remain bound. Of course, one has to mathematically confirm this qualitative reasoning, perhaps by scattering computations.

## Acknowledgement

This work is supported in part by HiPi Consulting, Ashland, VA, USA; by the Science for Humanity Trust, Bangalore, India; and by the Science for Humanity Trust Inc., Tucker, GA, USA. The authors would like to thank the reviewer for his judicious advices allowing to improve the comprehension of our work.

## References

[1] J.L. Paillet and A. Meulenberg, Arguments for the anomalous solutions of the Dirac equations, *J. Condensed Matter Nucl. Sci.* **18** (2016) 50–75.

[2] J.L. Paillet and A. Meulenberg, Basis for electron deep orbits of the hydrogen atom, *Proc. of ICCF19, 19th Int. Conf. on Cond. Matter Nuclear Science*, Padua, Italy, 13–17 April 2015, *J. Condensed Matter Nucl. Sci.* **19** (2016) 230–243.

[3] J.L. Paillet and A. Meulenberg, Special relativity, the source of the electron deep orbits, *Found. Phys.* **47** (2) (2017) 256–264.

[4] R. Gurtler and D. Hestenes, Consistency in the formulation of the Dirac, Pauli, and Schrödinger theories, *J. Math. Phys.* **16** (1975) 573–784.

[5] J.L. Paillet and A. Meulenberg, Electron deep orbits of the hydrogen atom, *Proc. 11th Int. Workshop on Hydrogen Loaded Metals*, Airbus Toulouse, 15–16 Oct. 2015, *J. Condensed Matter Nucl. Sci.* **23** (2017) 62–84.

[6] J.L. Paillet and A. Meulenberg, Relativity and electron deep orbits of the hydrogen atom, *Proc. of the 1st French Symp. RNBE-2016 on Cond. Matter Nucl. Sc.*, Avignon, 18–20 March, *J. Condensed Matter Nucl. Sci.* **21** (2016) 40–58.

[7] J. Naudts, On the hydrino state of the relativistic hydrogen atom, arXiv:physics/0507193v2.5, 2005.

[8] T. Nadareishvili and A. Khelashvili, Some problems of self-adjoint extension in the Schrödinger equation, http://arxiv.org/pdf/0903.0234.pdf.

[9] J.A. Maly and J. Va'vra, Electron transitions on deep Dirac levels II, *Fusion Sci. Technol.* **27**(1) (1995) 59–70, http://www.ans.org/pubs/journals/fst/a_30350.

[10] S. Fluegge, *Practical Quantum Mechanics*, Vol. 2, Springer, Berlin, 1974.

[11] R.T. Deck, J.G. Amar and G. Fralick, Nuclear size corrections to the energy levels of single-electron and - muon atoms, *J. Phys. B: At. Mol. Opt. Phys.* **38** (2005) 2173–2186 [27].

[12] J.A. Maly and J. Va'vra, Electron transitions on deep Dirac levels I, *Fusion Sci. Technol.* **24** (3) (1993) 307–318, http://www.ans.org/pubs/journals/fst/a_30206.

[13] N.V. Samsonenko, D.V. Tahti and F. Ndahayo, On the Barut–Vigier model of the hydrogen atom physics, *Phys. Lett. A* **220** (1996) 297–301.

[14] G.E. Brown and D.G. Ravenhall, *Proc. Roy. Soc. Lond. A* **208** (1951) 552–559.

[15] L. Irina Rusakova, Yurii Yu. Rusakov and Leonid B. Krivdin, Theoretical grounds of relativistic methods for calculation of spin–spin coupling constants in nuclear magnetic resonance spectra, *Russ. Chem. Rev.* **85**(4) (2016) 356–426.

[16] L.I. Schiff, *Quantum Mechanics*, 3rd Edition, McGraw-Hill, New York, 1968.

[17] S.V. Adamenko and V.I.Vysotskii, Mechanism of synthesis of superheavy nuclei via the process of controlled electron–nuclear collapse, *Found. Phys. Lett.* **17** (3) (2004) 203–233.

[18] S.V.Adamenko and V.I. Vysotskii, Evolution of annular self-controlled electron–nucleus collapse in condensed targets, *Found. Phys. Lett.* **34**(11) (2004) 1801–1831.

[19] V.K. Ignatovitch, A missed Solution for an Atom – a Gate toward cold nuclear Fusion, *Infinite Energy Magazine* **117** (2014) 33–36.

[20] V.K. Ignatovitch, Singular bound states and cold nuclear fusion, *Phys. J.* **2** (1) (2015) 6–11.

[21] A.I. Goncharov and V.A. Kirkinskii, Theoretical modeling of electron flow action on probability of nuclear fusion of deuterons, *Proc. of ICCF12, 12th Int. Conf. on Cond. Matter Nuclear Science*, Yokohama, A. Takahashi, K.-I. Ota and Y. Iwamura (Eds.), World Scientific, Singapore, 2006.

[22] R.D. Klaubert, *Student Friendly Quantum Field Theory*, 2nd Edn., Sandtrove Press, Fairfield, Iowa, 2013.

[23] A.O. Barut and J. Kraus, Resonances in $e+ - e^-$ System due to anomalous magnetic moment interactions, *Phys. Lett.* **59B** (2) (1975) 175–178.

[24] A.O. Barut and J. Kraus, Solutions of the Dirac equation with Coulomb and magnetic moment interactions, *J. Math. Phys.* **17** (4) (1976) 506–508.

[25] A.O. Barut, Prediction of new tightly-bound states of $H_2^+$ ($D_2^+$) and cold fusion experiments, *Int. J. Hydrogen Energy* **15** (12) (1990) 907–909.

[26] A.O. Barut, Stable particles ad building block of matter, Int. Centre for Theoretical Physics Lib., Miramare-Trieste, IAEA, 1979.

[27] H. Behncke, The Dirac equation with an anomalous magnetic moment, *Math. Z* **174** (1980) 213–225.

[28] B. Thaller, *The Dirac Equation*, Springer, Berlin, 1992.

[29] J.P. Vigier, New hydrogen(deuterium) Bohr orbits in quantum chemistry and cold fusion processes, *Proc. ICCF4*, Hawaii, 1994. http://www.epri.com/abstracts/Pages/ProductAbstract.aspx?ProductId=TR-104188-V4. Section 7, pp. 1–25.

[30] A. Dragic, Z. Maric and J.P. Vigier, *Phys. Lett.* **A237** (1998) 349–353.

[31] M. Fleishman, S. Pons and M. Hawkins, *J. Electroanal. Chem.* **263** (1989) 301.

[32] S. Ozcelik and M. Simsek, *Phys. Lett.* **A152** (1991) 145–149.

[33] R.P. Feynman, *The Feynman Lectures on Physics*, Vol III, Chapter 21, Web site, Feynman-Leighton-Sans, California Institute of Physics.

[34] W.H.E. Schwarz, *An Introduction to Relativistic Quantum Chemistry, Relativistic Methods for Chemists*, Chapter 1, Challenges and Advances in Comp. Chememistry and Physics Series, J. Leszczynski (Ed.), Vol. 10, M. Barysz (Ed.), Yasuyuki Ishikawa, Springer.

[35] F.V. Gubarev, L. Stodolsky and V.I. Zakharov, On the significance of the vector potential, *Phys. Rev. Lett.* **86** (11) (2001) 2220–2222.

[36] A.M. Stewart, Role of the non-locality of the vector potential in the Aharonov–Bohm effect, https://arxiv.org/abs/1209.2050.

[37] N.V. Samsonenko, F. Ndayaho and A. Ousmane Manga, Systems of particles and nuclei bounded by magnetic interactions (Barut Model), Presentation at *Proc. JINR*, Dubna, 14 April 2016.

[38] L. de Broglie, *Nonlinear Wave Mechanics*, Elsevier, Amsterdam, 1960.

[39] Bohm, *Phys. Rev.* A suggested Interpretation of the Quantum Theory in Terms of Hidden Variables I, II, **85**(2) (1952) 166–179 and 180–193.

[40] B. Thaller, The Dirac Operator, in Relativistic Electronic Structure Theory: Part 1, Theoretical and Computational Chemistry No. 11, Chapter 2, P. Schwerdtfeger (Ed.), Elsevier, Oxford, pp. (2002)239-105

[41] Two-body Dirac equations, https://en.wikipedia.org/wiki/Two-body_Dirac_equations.

[42] H.W. Crater, R.L. Becker, C.Y. Wong and P. Van Alstine, A detailed study of nonperturbative solutions of two-body Dirac equations, Oak Ridge National Laboratory (ORNL), Central Research Lib., Martin Marietta, D.O.E, Dec. 1992.

[43] D.G. Currie, T.F. Jordan and E.C.G. Sudarshan, Relativistic invariance and Hamiltonian theories of interacting particles, *Rev. Mod. Phys.* **34** (1963) 350–375.

J. Condensed Matter Nucl. Sci. 24 (2017) 278–283

Research Article

# Evidence for Nuclear Transmutations in Ni–H Electrolysis

K.P. Rajeev* and D. Gaur

*Department of Physics, Indian Institute of Technology Kanpur, Kanpur 208016, India*

**Abstract**

We report an easily reproducible Low Energy Nuclear Reactions (LENR) electrolysis experiment on an Ni–H system. An electrolytic cell with a Ni cathode and a Pt anode with an aqueous ($H_2O$ *not* $D_2O$) solution of $K_2CO_3$ as electrolyte was used for the experiment. A dc-power supply capable of supplying up to 5 A current was used to drive the electrolysis with the typical current being around 4 A and the applied voltage about 100 V. After running the electrolysis for about two weeks with a daily run time of 8 h on an average, a part of the nickel electrode, which by this time had become black, was taken for elemental analysis using Energy Dispersive Spectroscopy (EDS). The EDS analysis showed that a whole host of new elements had appeared on the cathode such as: K, Fe, Cu, O, Rh, Zr and Pb. The apparent concentration of the elements varied from just over a percent for K and Pb to about 10% for Rh and 20% for Cu. The sample was then taken to a Time of Flight Secondary Ion Mass Spectrometer (ToF-SIMS) for isotopic analysis. This analysis also showed that a set of new elements had appeared on the nickel wire which were: K, Si, Mg, Mn, Zn, O & Rh. The Ni itself was found to have the following isotopes: $^{58}$Ni, $^{60}$Ni and $^{62}$Ni with isotope concentration ratios differing significantly from the ratio found in naturally occurring Ni. It was clear that in the electrolyzed Ni the concentrations of $^{60}$Ni and $^{62}$Ni had gone down compared to the concentration of $^{58}$Ni. This is a clear indication that nuclear reactions had taken place during electrolysis.
 ISSN 2227-3123

*Keywords:* Electrolysis, LENR, Ni–H systems, Transmutation

## 1. Introduction

Since the announcement of the discovery of low energy nuclear reactions (LENR) by Fleischmann and Pons in 1989, there has been a large number of attempts to replicate their results [1]. There have been many successful replications of excess heat in palladium–heavy water systems (Pd–$D_2O$ or Pd–D) even though the detection of excess heat is not always easy, especially if the effect is small [2]. On the other hand, finding transmutation products provide undeniable evidence of nuclear reactions and there has been large number of instances of this in the Pd–D and other systems [3]. In the light of this it would be useful to find an easily reproducible experiment to convince any interested researcher of the truth of the LENR phenomena. The Ni–$H_2O$ (or Ni–H) system fits the bill, being very easy to set up and run and seems to work the first time unlike the Pd–D system which seems to require considerable coaxing before it will reveal

*E-mail: kpraj@iitk.ac.in.

© 2017 ISCMNS. All rights reserved. ISSN 2227-3123

the LENR phenomenon. There have been many reported studies on the Ni–H system that have reported excess heat as well as transmutation products [3].

## 2. Experiment

### 2.1. Experimental apparatus

A schematic of the electrolytic cell used is shown in Fig. 1. The experiments were done with a Ni (99.99%) cathode and a Pt (99.999%) anode in a 1 l glass beaker with the electrode diameters being 0.75 mm (Ni) and 0.5 mm (Pt). The electrolyte was $K_2CO_3$ (minimum purity 99%; the maximum limits of impurities were: iron: 0.001%, sodium: 0.1%, chloride: 0.03%, nitrate: 0.01% and sulphate: 0.02%, the remaining being moisture) dissolved in about one liter of distilled water with an approximate 1/2 molar concentration to begin with. The molarity of the solution was maintained between 1/2 and 1 by adding distilled water to the beaker as needed. Metallic impurities in $K_2CO_3$ were less than 0.1%. To prevent accidental combustion within the electrolytic cell the evolving hydrogen and oxygen gases were allowed to escape from the cell through different vents.

### 2.2. Experimental procedure

A dc current less than 5 A at voltages up to 100 V was used to drive the electrolysis. Because of the relatively large power dissipated in the electrolyte it would be boiling vigorously and losing water which was replenished by adding distilled water as required.

A single Ni electrode was used for the electrolysis experiments. The tip of the electrode about 3–4 mm in length was cut-off at various stages of the experiment for EDS analysis to monitor the progress of the nuclear reactions. The duration of electrolysis was 100 h. Detailed analysis including both EDS and SIMS were done only on a piece of the Ni electrode subjected to the full 100 h of electrolysis and is reported here. EDS analyses were done on the pieces Ni electrode cut off at earlier stages to check whether LENR process had started and, if so, how it was progressing.

We were unable to carry out a detailed elemental analysis of the electrolyte before and after the experiment for lack of suitable facilities.

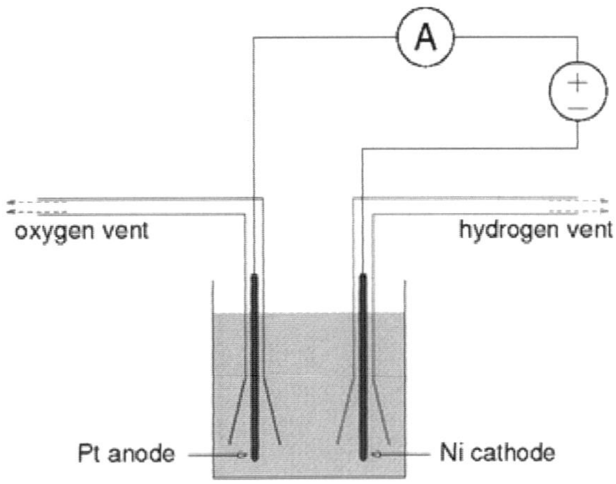


**Figure 1.** Schematic of the electrolytic cell used for the experiment. The electrolyte was $K_2CO_3$ dissolved in distilled water.

## 3. Analysis technique: EDS

Energy Dispersive Spectroscopy (EDS) is a surface spectroscopic technique [4]. It is usually associated with electron microscopy. Some of the electrons used for imaging knock out core electrons from the surface atoms giving rise to characteristic X-ray emissions from the excited atoms. The X-rays are analyzed to determine the elements and their concentration.

### 3.1. Analysis technique: ToF-SIMS

Secondary ion mass spectrometry (SIMS) is a technique used to analyze the composition of solid surfaces and thin films by sputtering the surface of a specimen with a focused primary ion beam and collecting and analyzing the ejected secondary ions [5]. The mass/charge ratios of these secondary ions are measured with a mass spectrometer to determine the elemental, isotopic, or molecular composition of the surface to a depth of, typically, a few nanometers.

For the determination of concentration of isotopes in a sample "multiple collector inductively coupled plasma spectrometry (MC-ICP-MS)" is a far more accurate technique than SIMS [6]. But SIMS has an advantage over MC-ICP-MS in that it determines the isotope ratios at the surface of a solid sample down to a few nanometers while MC-ICP-MS can only determine isotope ratios in a sample which is dissolved in some solvent. But of late spectrometers have started becoming popular which uses laser ablation (LA) to dislodge ions from the surface to be analyzed and thus can analyze isotopic concentration of surfaces very well. For lack of easy access to such an LA-MC-ICP-MS machine we decided to do our analysis on a SIMS machine.

Time of Flight (ToF)-SIMS is a particular type of SIMS where the isotope determinations are done by measuring the transit time of ions which are accelerated according to their charge to mass ratio through a certain region of space [5].

## 4. Results and Discussion

After a run of a total of about 100 h the Ni electrode was analyzed for new elements using EDS. See Fig. 2 for a typical spectrum. Evidence for a whole host of new elements can be seen with substantial amounts of new metals such as copper (20%), rhodium (10%) and zirconium (6%) and iron (2.5%). Interestingly Ni concentration is only 3.1% in this particular spectrum! It is to be pointed out that the spectra obtained from different regions of the sample differ in the observed elements and their concentrations. EDS data show that new elements have appeared on the cathode. To confirm that nuclear transmutations have taken place in the Ni wire we carried out mass spectrometry on the Ni wire using a ToF-SIMS. Even though, inadvertently, the sample was cleaned lightly with an emery paper to expose the shining nickel metal underneath before being analyzed in the ToF-SIMS machine, the results turned out to be very significant.

The SIMS analysis found various elements on the wire surface as expected such as K, Si, Ca, Zn, Mg, Mn, Rh, Zr and Pb. We confined our isotopic analysis to nickel for which we had a standard available as a piece of unused Ni wire. The isotopic concentrations of nickel found in its natural state are: $^{58}$Ni: 68%, $^{60}$Ni: 26%, $^{61}$Ni: 1.1%, $^{62}$Ni: 3.6% and $^{64}$Ni: 0.9% [7]. For the electrolyzed Ni the isotope analysis was done only for $^{58}$Ni, $^{60}$Ni and $^{62}$Ni because we got clear signals from these isotopes only.

The SIMS analysis was done on two regions (each 10 $\mu$m diameter) of the Ni sample. Figure 3 shows the SIMS signals from Ni isotopes from region I of the sample. In Table 1, we have shown the isotope ratios of Ni observed in nature as well as what we observed in our electrolyzed samples. It is clear that the isotopic ratio in the electrolyzed Ni is not what is found in the natural state of Ni. This fact strongly suggests that nuclear transmutation does take place during electrolysis. Isotopic ratio of Ni is modified by the electrolysis process with $^{60}$Ni and $^{62}$Ni being significantly depleted with respect to the more abundant $^{58}$Ni.

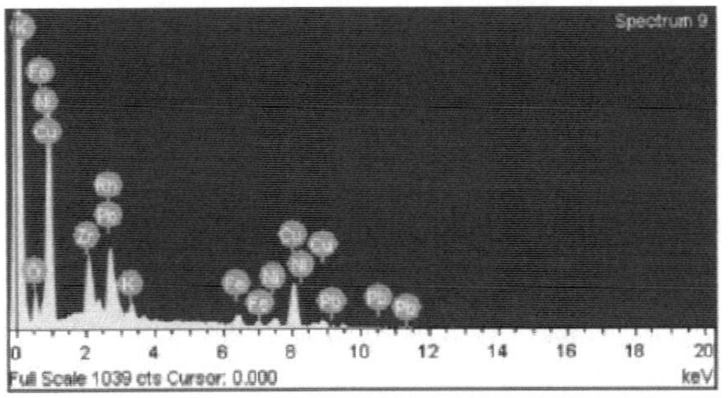


| Element | App Conc. | Intensity Corrn. | Weight% | Weight% Sigma | Atomic% |
|---|---|---|---|---|---|
| O K | 7.27 | 0.8426 | 11.52 | 1.09 | 36.87 |
| K K | 1.47 | 1.1026 | 1.77 | 0.29 | 2.33 |
| Fe K | 2.50 | 1.0083 | 3.31 | 0.59 | 3.03 |
| Ni K | 3.14 | 1.0259 | 4.08 | 0.81 | 3.56 |
| Cu L | 20.29 | 0.5632 | 48.10 | 1.36 | 38.78 |
| Zr L | 6.10 | 0.7398 | 11.00 | 0.79 | 6.18 |
| Rh L | 10.03 | 0.7885 | 16.98 | 1.12 | 8.45 |
| Pb M | 1.79 | 0.7402 | 3.23 | 0.83 | 0.80 |
| Totals | | | 100.00 | | |

**Figure 2.** EDS spectrum of a certain region of a Ni cathode after about 100 h of electrolysis.

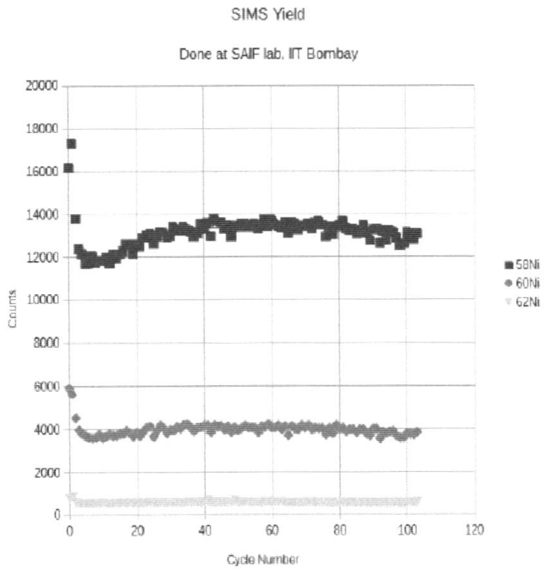


**Figure 3.** SIMS Yield from Region 1. Each cycle lasts 20 s. Total data acquisition time is about 0.5 h.

**Table 1.** Comparison of isotope ratios of natural and electrolyzed nickel.

| Isotope ratios | Natural nickel | Electrolyzed nickel | |
|---|---|---|---|
| | | Region I | Region II |
| $^{58}$Ni/$^{60}$Ni | 2.596(1) | 3.3(1) | 3.5(1) |
| $^{58}$Ni/$^{62}$Ni | 18.73(1) | 24(1) | 27(2) |

More work needs to be done to systematize the findings such as:

(1) What happens if the electrolysis process is carried out at a different voltage? Is there a threshold voltage? Please note that this work was done at around 100 V.
(2) What are the isotope numbers of the other elements found on the Ni cathode? Are they all stable nuclides? Are some of them radioactive?
(3) How do the isotope ratios evolve with the duration of electrolysis?
(4) Does the electrolyte have any effect on the transmutation process? Will the use of a different electrolyte result in different isotope ratios for Ni?
(5) Would an external applied magnetic field have any effect on the outcome?
(6) Are there any hazardous neutron or $\gamma$-ray emissions during the process? No monitoring for such radiation was done during the experiments reported here as it is believed generally that LENR reactions do not produce any hazardous radiations. But still, this is a very important safety consideration and monitoring for such radiations should be done in LENR experiments as a matter of routine.

One of the observations made during the course of this work was that Ni–H electrolysis always seem to produce new elements at the cathode. The transmutation process seems to start right away, without delay, once the electrolysis is started. We could see this from the EDS analysis of a piece of the Ni wire subjected to only a few hours of electrolysis, but not reported here. Thus Ni–H systems seem friendlier than the Pd–D systems in which the LENR processes start with a large time delay, if at all.

Another strange observation is that the isotope $^{62}$Ni which has the highest binding energy per nucleon of all the nuclides got depleted during our experiment! This raises the question of whether we are dealing with an endothermic process here.

## 5. Conclusion

It is easy to reproducibly induce nuclear transmutations in Ni–H systems and a large number of transmutation products are seen on the cathode on elemental analysis. Mass spectrometric analysis shows that isotope ratios of Ni are modified in the Ni cathode which is an undeniable indication that nuclear transmutations can be induced by low energy electrolytic processes.

### Acknowledgements

Many thanks to Mr. Ramashray for technical help during the experiments and to Dr. M. Srinivasan for his untiring encouragement and also for presenting this work at ICCF20.

### References

[1] M. Fleischmann and S. Pons, *J. Electroanal. Chem.* **261**(2A) (1989) 301.

[2] E. Storms, *The Science of Low Energy Nuclear Reaction: A Comprehensive Compilation of Evidence and Explanations About Cold Fusion*, World Scientific, Singapore, 2007.

[3] M. Srinivasan, *Current Sci.* **108** (2015) 624 and the references therein.

[4] J. Goldstein, D.E. Newbury, P. Echlin, D.C. Joy, C.E. Lyman, E. Lifshin, L. Sawyer and J.R. Michael, *Scanning Electron Microscopy and X-ray Microanalysis*, 3rd Edition, Kluwer/Plenum, New York, 2003.

[5] A. M. Spool, *The Practice of TOF-SIMS: Time of Flight Secondary Ion Mass Spectrometry*, Momentum Press, New York, 2016.

[6] [6] F. Albarède, P. Telouk, J. Blichert-Toft, M. Boyet, A. Agranier and B. Nelson, *Geochimica et Cosmochimica Acta* **68** (2004) 2725.

[7] IUPAC, *Pure Appl. Chem.* **70** (1998) 217.

J. Condensed Matter Nucl. Sci. 24 (2017) 284–295

Research Article

# Helium Measurements From Target Foils, LANL and PNNL, 1994

Roger Sherman Stringham*
*First Gate Energies, Kilauea, HI 96754 , USA*

**Abstract**

The alphas formed in Ti and Pd target foils were measured as $^4$He in a mass spectrum, MS, analysis at the Pacific Northwest National Laboratory, PNNL, a lab that specializes in $^3$He and $^4$He measurements. The exposed bcc crystal lattice Ti target foil, TF, measured an average of $39 \times 10^{12} \pm 1.4$ trapped $^4$He atoms. The fcc crystal lattices of Pd, Ag, Ni, and Cu target foils, particularly the two measured fcc Pd TF lattices produced $\sim 0.35 \times 10^{12}$ trapped $^4$He atoms. This helium level was just above its background level, and 1/100 that of the bcc Ti TF. In the fcc palladium TF, SEM crater volume measurements show that most of the alphas are ejected into the circulating $D_2O$, where they were measured in the gas phase. In 1994 the samples were run at LANL and measured at PNNL. The helium was measured by melting small TF pieces cut from the active center zone. Then each piece was placed in a crucible, and melted under vacuum, releasing all gases including background helium. These collected gases from one piece were pretreated; then were measured. The data was converted to total $^4$He atoms trapped in the active zone of the TF lattice. A total of 24 measurements for $^3$He and $^4$He on three TFs found no $^3$He but $^4$He in all 12 measurements. SEM photos of single ejecta sites were combined with calculations of old data that used knowledge gained from 23 years of cavitation experience that enhances the old data.


*Keywords:* Cavitation, Gamma, $^4$He, Lattice, Mass spectrum

## 1. Introduction

EQuest Science, a small research laboratory in California, USA, first visited Los Alamos National Laboratory (LANL) in 1993, with the MI 20 kHz reactor, and then returned the next year for further work with a new reactor, the MII 20 kHz, Fig. 1. The new reactor was the same except with an added calibration resistance heater. For three weeks, from mid-April into May 1994, EQuest [1], ran experiments day and night in a collaborative effort with Dale Tuggle and Tom Claytor and their staff in LANL's Tritium Lab and equipped with mass spectrum, MS. We had the support of LANL's scientific staff, including advice and instrumentation. Also, assistance from Naval Research Lab, NRL from George Chambers, and Dave Nagel. The three weeks of nearly constant experimental runs produced much data. A MS filament failure during the first experiment at the Tritium Lab, induced Tom Passel of Electric Power Research Institute, EPRI, to negotiate a contract with PNNL, for a delayed MS analysis [2,3] of the collected sample gases and

*E-mail: firstgate@earthlink.net.

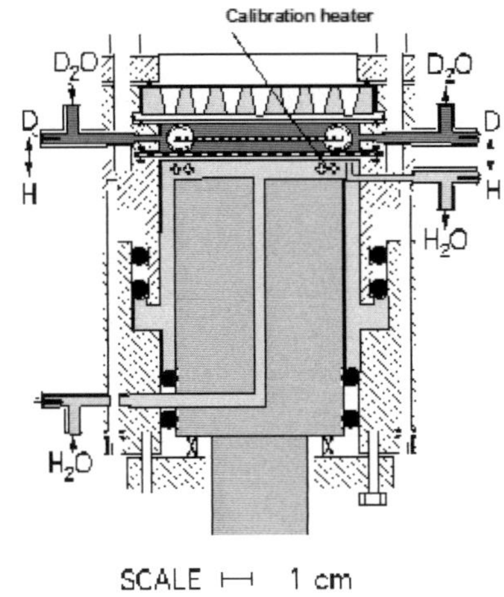


**Figure 1.** The M-II 20 kHz, a dual resonating reactor with two concentric flow systems.

metal TF stored at the EQuest facilities for 4.5 months after their collection. Among these stored gas samples and the corresponding TF were the samples the Ti 3-A, a 195-minute cavitation run (run at LANL 29 and 30 April 1994) from data recently found is highlighted, and was part of the PNNL report [2] along with 2 Pd TF samples. There were signs that the gas sample volumes were compromised, but the 3 TF were not affected by any problems before and after the experiment was shut down.

Power input to the above sonofusion reactor, Fig. 1, is similar to ion beam implantation experiments [4] and studies of the nuclear reactor containment materials damage studies from the 1980s to the present [5,6]. These papers discuss the swelling lattice and voids filled with D+ and/or $^4$He. Analysis of collected samples show nm bubbles in grain boundaries at extreme pressures from irradiation of nuclear reactors containment [7,8]. This is similar to the damage measured by NRLs George Chambers's in his X-ray of diffraction study of an exposed Pd TF in 1994, that showed similar swelling and lattice changes compared to an unexposed edge piece from the same Pd TF [9]. The reactor in Fig.1 shows that it is divided into two co-axial reactors, one on top of the other. The driving horn powers the lower $H_2O$ cavitation system at 20 kHz, and was coupled through the 40-mil stainless steel disk to drive the smaller $D_2O$ TF reactor. This dual reactor set up was necessary because the Ti horn surface was damaged by the cavitating $D_2O$. Dividing into two systems by the separator allowed the horn powered $H_2O$ system and an acoustic coupled $D_2O$ system, to solve the problem. The collapsing bubble in the cavitating $D_2O$ was adjusted to the correct resonance settings. *Argon Pressure, Temperature, and Cavitation power* are the working parameters producing a $z$-pinch jet in a magnetic squeeze by its accelerating sheath electrons [10] that implanted the TF with an accelerating micro ion plasma.

This paper is somewhat disjointed by time and circumstances. In 1993 R. George negotiated a working visit to LANL at Los Alamos, New Mexico. A year later in April, EQuest returned. Unfortunately, the MS in the tritium laboratory burned out a filament before the first sample was run, but we continued producing data at LANL for two more weeks. The samples were stored at the EQuest laboratory for 4.5 months, when Tom Passell made arrangements with PNNL in Richland, Washington to analyze some of our LANL samples. Many of the liquid and gas samples were contaminated during storage in the EQuest lab except for the TFs where $^3$He and $^4$He could be trapped for years. One year ago, data was discovered in a file that was never analyzed. The data was very revealing after so many years and was first presented at the ICCF-20 2016 in Sendai Japan, where Francesco Celani asked the question, "What was the ratio between background and the measured $^4$He"?

To make this cavitation make sense, a model was created many years ago to explain a path to the creation of an alpha in a TF lattice. The model is constantly changing. The model follows: Excess plasma electrons, where an electron–deuteron pinched jet plasma pulse approached the target foil. Where the plasma implanted first the plentiful more mobile electrons, followed by deuterons is in an atomic scale. Temperatures and time frames were drawn together as plasma particles M cluster implanted into the TF lattice. In the sub-picoseconds, the electrons surrounded the less mobile deuterons and squeezed clusters of 2–20 deuterons that were in the M cluster $\sim$ 25 nm below the surface of the TF. M cluster, the high-density cluster, compressed at 20 000 K preventing interaction between D+ and electrons. The interaction produces a D atom destroying the cluster. Before the M cluster's natural destruction, two deuterons will be forced together via the attractive image forces between like charges to produce an alpha with no gamma [11]. Or a BEC oriented boson containing D+ and Cooper pair, CP, where the deuterons occupy same energy level, behaving as a single atom, are squeezed to densities of muon fusion and no gamma. The speculation is that as the distance between deuterons decreases an alpha fusion occurs, and the expected gamma was not emitted [12]. These are two possible paths, one for the bcc Ti lattice and one for fcc lattice. The plasma particles involved, deuterons and electrons, were one fermion in radius, and the short time frame add to their possible existence as paths. Also, SEM photos show ejecta sites of the smallest size, from numerical study that form a minimum base size ejecta. The binding energy forming an alpha, $E = 3.83 \times 10^{-12}$ J, is close to the minimum base size energy that produces the smallest crater that has a diameter and depth of 50 nm, producing an ejecta of similar energy [13]. The M cluster is an entity that is the bridge to the alpha.

## 2. Experimental LANL

At LANL the MII dual reactor set up was constructed with two concentric flow systems. At the top of the dual reactor is a small, with a 36 ml volume, that was the reactor chamber that held the TF. The Ar saturated $D_2O$ flowed through it, shown in Fig. 1. The bottom system, the larger of two systems, produced the cavitation in $H_2O$, with the heat systems 2-inch diameter Ti acoustic horn transducer in the 100 ml of flowing $H_2O$. It was discovered that the Ti horn was damaged by cavitating $D_2O$, and it was remedied by this configuration. Each system had its own circulation controls, pressures, gauges, reservoirs, and heat exchangers. The experimental procedure for experiments with target foils followed this path, with exposure times ranging from 3 to 24 h.

Figure 1 shows the M-II 20 kHz, a dual resonating reactor with two concentric flow systems: $D_2O$ flow in dark gray, $H_2O$ in light gray. The white crosses in the $H_2O$ are the ARI calibration heater. The larger dotted line is the 40-mil stainless steel disk separator between the two systems, and the smaller dotted line is the $5 \times 5 \times 0.01$ cm Ti target foil. The Ti 3-A target foil was centered and sealed with a squeezed Teflon ring. The horn and cavitating $H_2O$ system is coupled to the 40-mil through the ss disk to power to the top $D_2O$ system. The gray area in the center is the 20 kHz commercial transducer from heat systems, a Ti horn, cooled by circulating $H_2O$, light gray. The bolted down top piece consists of a clamping ring that holds a 1/2 inch thick Quartz disk window or the shown 1/2 inch Al disk with vertically conically drilled holes on top of 40-mil FEP transparent seal. This allows for direct observation of flowing $D_2O$, dark gray, in the top experimental reactor. The set-up allowed for photos, UV irradiation, and videos, when using the 1/2 " Quartz window. The cylindrical body consists top and bottom components that are hatched and bolted together, and are sealed with "O" rings, black dots. The support of pumps and flow distribution systems are not shown. The $D_2O$ $\Delta$T and flow measurement give the power out.

The cavitating $D_2O$ exposure process of TF Ti 3-A took place in the top $D_2O$ chamber of the MII reactor, where the $D_2O$ circulation rates and Ar pressures were adjusted in a closed system that passed the circulating $D_2O$ through its cooling coil in and out of the 16-liter heat exchanger. The circulation rate of $D_2O$ was 3.2 ml/s and flowed over the 40-mil stainless-steel separator, where the TF was located (see Fig. 1). On the other side of the separator, where the

$H_2O$ pump picked up the hot $H_2O$ flowing at a rate of 13 ml/s. through the center of the acoustic horn. The hot flowing $H_2O$ was transported through its cooling coil in and out the 16-liter $H_2O$ heat exchanger. The $H_2O$ circulation was pressurized from 70 to 100 Psig with $N_2$ gas, where the acoustic horn driver was suppressing the formation of $H_2O$ cavitation bubbles by the $N_2$ high pressure. The 40-mil thick separator removed the potential damage to the horn in the bottom system of the dual MII reactor that passed a 20 kHz acoustic signal through the 40-mil separator, into the $D_2O$ producing a jet plasma pulse implanting into the TF. This experiment ran for 105 min and the $H_2O$ bubbler broke shutting down the experiment for 30 min for a bubbler replacement. The experiment was restarted and ran for another 90 min and broke again shutting down the experiment, with the $D_2O$ system still intact. The running pressures were for the $D_2O$ system the input pressure of Ar was 45 Psig and for the $H_2O$ system was 90 Psig. This gave a total run time of 195 min. There was no damage to the circulating $D_2O$ system.

The system was shut down and samples of gases, $D_2O$, and target foil, were moved from the experimental system. These samples were stored at the EQuest lab. in Mountain View, CA, waiting for further instructions. It was fortunate when EQuest was informed about funding negotiated by Tom Passell and others, that EQuest had an agreement with Rockwell International, in Canoga Park, CA and PNNL in Richland, WA for mass spectrum analysis of the LANL samples. NPPL took over the task of the analysis of the TFs. The TFs and liquid and gas samples were removed and shipped to the PNNL MS facilities, where the TF helium analyses were done.

## 3. Target Foil Information

The calorimetry data was compromised by the breakdown of the $H_2O$ flow system for the Ti 3-A run. The target foil that was removed from the MII reactor was viable and was stored with the other sample volumes and TF for 4.5 months at the EQuest's laboratory in Mountain View, CA. Then selected sample volumes and TFs were shipped to NPPL for MS measurements. The three target foils that measured Ti 3-A, Pd 9, and P13a from the recovered data are shown in Table 1. The TF flow cartoon diagram, Fig. 3, for Ti 3-A, also applies to the general process of all three TFs.

Three TFs were measured by PNNL MS. 12 measurements were made on pieces of TF were cut from the active zone of each TF, about 1/3 of the TF total area (see Fig. 3). Each of the 12 pieces cut was analyzed separately. A piece was placed into ceramic crucible and the TF piece melted, driving off gases under vacuum conditions for MS measurement the $^3$He an 4 isotopes.

Old and more recent experiments show the way to interesting comparisons with respect to SEM photos of bcc Ti TF [14,15]. Ti tubes of 700 $\mu$m long and 1 $\mu$m diameter that rose above the surface of the TF appear hollow. Most of the SEM electron beam passes through the tube leaving a shadow on the Ti TF surface. The tubes are common and very fragile, and would have a short lifetime in the $D_2O$ cavitation field. So, the tubes' ability to survive would be a few acoustic cycles. Only the newest would survive when the power is turned off. These fragile tubes would add to the debris of the bcc Ti TF.

These target foils produced ejecta sites equivalent to the *nE* energy, where n is the number of events, of single and multiple alpha events in the low frequency range of 20 kHz. In the high frequency of RF the single event craters dominate. These craters are $\sim$ 50 nm in diameter with about the same energy as a single event, alpha, $E$ of $3.83 \times 10^{-12}$ J. These single events are shown in SEM photos [13]. These ejecta sites for fcc TF not only show single events but also measure multi events at 20 kHz that will destroy a Pd TF in ten minutes of cavitation exposure. These have been viewed in videos [16]. The spectrum of crater event diameters cover 0.050–10 $\mu$m diameter craters in the surface events of fcc TFs (see Fig. 2B). The most prevalent event was the 50 nm diameter crater at a frequency of 20 kHz in cavitating $D_2O$ in the MII reactor, Fig. 1. The fcc TFs produce the same products, heat and alphas, but their final paths are different. The bcc Ti TF Ti 3-A retains $^4$He within the lattice, and the measured fcc Pd TFs ejects most of the alphas from its lattice at low and high frequencies, see Table 1. The TF lattice, where the M cluster exists for a sub-picosecond, will either eject the alpha, fcc type TF lattice atoms, Pd, Cu, Ni, and Ag, and heat into the circulating

$D_2O$, or trap the alpha that is measured as $^4$He in the bcc Ti TF lattice. The two crystal structures of bcc Ti and fcc Pd TFs, with different crystal structures show different cavitation results.

## 4. Experimental PNNL

The TFs were delivered to Brian Oliver for PNNL specialized mass spectrum laboratory, MSL, analysis about 4.5 months after shutting down the working visit at LANL. The PNNL MSL criteria was for MS measurements of exposed TF active zones' area. Brian M. Oliver had no knowledge of the TFs origin or the exposure, but as in the past applied MSL methodology of $^3$He and $^4$He measurements. Brian Oliver cut and weighed the four pieces from each TF lattice. The emphasis of this paper is on the three TFs and 24 MS measurements made by Brian, searching for $^3$He and $^4$He. Twelve measurements from the active zone showed no $^3$He in any of the three TFs. Twelve measurements from the active zone all showed $^4$He in the TFs.

PNNL MS measurements were of three TF shown in Table 1. All TFs measurements were in the activity zone showed $^4$He measurements were greater than background levels. The dimensions of all TFs were $50 \times 50 \times 0.01$ cm$^3$. Figure 3 shows the activity zone where each piece was cut and weighed for a typical TF measurement. four pieces each released gases that were collected by Brian Oliver for his MS measurements for helium isotopes. The natural background levels were determined from measurement of the edge pieces in rows 4, 8, and 12 in Table 1. The Ti 3-A TF, where the three center cut foil pieces showed three different determinations for $^4$He in the active zone, see Fig. 3 and Table 1. And the cut edge piece, see Fig. 3, measured for the background $^4$He in the active zone. The data shows the three consistently high results for bcc Ti 3-A TF using the NPPL MS technique for trapped helium in foils. The data collected by this measurement was obtained through a very elaborate protocol of MS standards, blanks, vacuum pumping. and vacuum heating [2]. The PNNL evolved gases were pre-treated so the MS only measured the helium isotopes. After pretreatment, the first gases through the LN cooled column were $^3$He and $^4$He. The helium isotopes were identified by their mass charge difference by the MS. $^3$He and $^4$He are compared to standards. Ti 3-A TF showed high $^4$He concentrations above the background level of the edge pieces. Following the identical process, the two fcc

**Figure 2.** (A) The scale of 2A is shown and the scale of 2B is shown in the diameter of largest of the small condensed Pd spheres $\sim 1$ $\mu$m in diameter. The SEM photo 2A reveals layers of $TiO_2$ spherical nodules, in the range of 50 nm. This leads to very colorful radiant layers of TiOx on the surface of the bcc crystal of a Ti TF exposed to 20 kHz cavitation. SEM photo 2A by J. Wheeler of Evans Surface Sciences. (B) 2B with about 1/1000 resolution as surface with a few large craters and the TF covered with smaller craters in the sub-micron range. Recently it was observed that target foils of different crystal structures respond to $D_2O$ cavitation in unique ways. 2A, Ti forms a heavy oxide and traps the alpha in a bcc structure. 2B, ejects the Alfa and is a fcc structure. SEM photo by John Dash of Portland State University.

Pd TF showed low $^4$He concentration very close to the background levels, and are shown in Tables 1 and 2.

A SEM photo of a typical exposed Ti TF surface is shown in Fig. 2A, where the active zone is covered with Ti$Ox$ in a multi colored surface that was built up in thin layers of $TiO_2$ after exposure at 20 kHz cavitation exposure. The bcc Ti TF shows in the SEM photo packed layers of small nodules, about 50 nm in diameter, of $TiO_2$ that gives its bright colored standing wave pattern. No evidence of ejecta site production, see first two paragraphs in Section 6. There is one exception suggested, bursting nano tubes [14]. The bcc Ti TF path was very different from fcc Pd and Ag, Ni, and Cu fcc metal TFs. The many SEM photos of fcc lattice crystals exhibit a size-spectrum of ejecta site craters shown in the SEM photos of a large ejecta crater, Fig. 2B. SEM photos of other fcc lattice TFs, also have similarities showing small condensed metal spheres less than 1 $\mu$m in diameter located inside their craters [17].

## 5. PNNL $^4$He Data

The purpose of the PNNL measurements was to demonstrate if there was any trapped $^4$He in $D_2O$ cavitation exposed TF lattices from EQuest 1994 LANL projects. Brian Oliver, Mass Spectroscopist, at NPPL in 1994 followed an elaborate protocol [2] for measurement of the helium isotopes in the metal Ti and Pd TF . The samples were delivered to Brian Oliver at PNNL, Richland, WA for MS measurement. The process used by Brian Oliver at PNNL mass spectroscopy facility was to extract and measure gas trapped in metal lattice target foils exposed to cavitating $D_2O$. Three pieces were cut from the center of the TF exposed active zone, and weighed, Fig. 3; Table 1, column 1. The fourth piece was cut from the edge of the TF and was used to determine $^3$He and $^4$He background.

The crucible was heated in vacuum and cooled preparing for the 7.63 mg cut piece. The piece in the crucible was evacuated of all gases, and then heated at known rate (see Fig. 3). The evolved gases from the piece passed into a 0.25 inch packed charcoal column with hydrogen getters. The column was cooled with liquid nitrogen. The remaining eluting gases were held for the proper time. $D_2$ and T were removed as well as condensables, air, $CO_2$, Ar, etc. Then the ionized gases were passed into the MS magnetic field and sorted for mass 3 and 4 ions, and the peak size determined the helium atom peak count. All the pieces were the same except that the TF densities were different. The density for the Ti TF 4506.0 mg/cm$^3$ and Pd TF 12023.0 mg/ cm$^3$. are used in the calculations.

Converting Brian Oliver data, the procedure was to cut a piece from the TF active zone. [2] It weighed 7.63 mg. Next the volume of the active zone 0.0833 cm$^3$ $\times$ density of Ti in mg/ cm$^3$ $=$ 37500 mg. See the above paragraph. The ratio of these mg values is R= 492. See Table 1; row 1, column 3. $R$ is the multiplying factor for converting PNNL $^4$He measured data into Ti TF $^4$He content 42.5 $\times$ $10^{+12}$ atoms, Table 1, column 4. Follow this procedure for each piece and Ti 3-A TF and the 2 Pd TF and Table 1 data is filled, using the proper density.

These procedures and protocols for helium measurements were developed over many years. All $^4$He data from PNNL were in units of $10^{+9}$ atoms, and is shown in Table 1 in column 2. The measured $^4$He for this example piece,

was measured at 86.4 $\times$ $10^9$ atoms and no $^3$He atoms were found as mentioned earlier. From the PNNL measurement the $^4$He atoms (see column 2 of Table 1) must be multiplied by the ratio of masses $R$ in column 3 to convert the atoms of $^4$He measured to the $^4$He atoms in the active zone of the three TFs. $D_2O$ cavitation affects the active zone, pink area Ti TF, Fig. 3.

Figure 3 shows a cartoon diagram shows PNNL MS and results for TF Ti 3-A.

Table 1 shows The result of the PNNL MS measurements showed the presence of $^4$He trapped in the Ti TF lattice [2]. The gases that were trapped in the lattice of melted TF pieces were released. The TF's background $^4$He atoms from the edge piece were subtracted. They were not in the zone of activity of the TF. The cavitation activity was confined to 1/3 the total TF volume. This was a PNNL MS measurement of the $^3$He and $^4$He isotopes only.

(This measurement includes background $^4$He content and the trapped $^4$He in the lattice that will be subtracted to yield the alphas produced.)

The data shows the numbers of $^4$He atoms measured from the Ti lattice trapped alphas during its $D_2O$ cavitation

**Table 1.** MS analyses of target foil Ti and Pd 4506 mg/cm$^3$, D, or Pd at 12 020 mg/ cm$^3$.

| | Foil mass (mg) | Measured $^4$He $10^{+9}$ atoms | Ratio $R$ active mg/piece mg | $R \times$ mg weight $^4$He MS total $\times$ $12^{12}$ atoms |
|---|---|---|---|---|
| *Centre cut sample* Ti 3-A | | | | |
| Center 1 | 7.63 | 86.4± 2 | 492 | 42.5 |
| Center 2 | 4.46 | 47.9± 2 | 842 | 40.3 |
| Center 3 | 5.21 | 51.9± 2 | 720 | 37.4 |
| Edge | 7.18 | 2.6± 1 | 523 | 1.36 |
| *Centre cut sample* Pd 9 | | | | |
| Center 1 | 9.01 | 6.1± 1 | 111 | 0.678 |
| Center 2 | 8.64 | 5.01± 1 | 116 | 0.580 |
| Center 3 | 9.33 | 6.4± 1 | 107 | 0.687 |
| Edge | 5.78 | 2.1± 1 | 173 | 0.364 |
| *Centre cut sample* Pd 13a | | | | |
| Center 1 | 8.25 | 6.2± 1 | 121 | 0.753 |
| Center 2 | 10.92 | 6.8± 1 | 91 | 0.624 |
| Center 3 | 10.48 | 10.4± 1 | 95 | 0.994 |
| Edge | 6.03 | 2.4± 1 | 166 | 0.399 |

*Mass uncertainty is ± 0.02 mg.
The $^3$He was the first out of the column followed by $^4$He. No $^3$He was found.

exposure of 195 min.

Summary of the MS PNNL measurements of TFs cut pieces average trapped $^4$He atoms are listed in Table 2.

It was recognized that Ti TF was different from the other TF in cavitating $D_2O$ as there were no observed ejecta sites and there was a gray powder visible. Also, the other fcc TF with SEM photos showed ejecta sites, and matched crystal lattice group with observed ejecta sites. The several SM photos of bcc Ti show the ejecta tubes [14]. Table 2 shows red for bcc and yellow for fcc. The different crystal types may help to explain other differences noted in the text. Alphas or $^4$He are produced in sub-surface M clusters and remain in the TF lattice of Ti, see Table 2 red. fcc TF lattices in a cavitation system ejected alphas produced in sub-surface M clusters, see Table 2 yellow [13,19].

The fcc Pd TF are showing some small residue of $^4$He in its lattice that just exceed the error measurements that may point to interruptions on an atomic scale in the action of a damaged ejecta mechanism in the M cluster in the fcc lattice.

The bcc Ti TF is showing 39 times background $^4$He. This poses the question, was the Ti TF a static or a dynamic system? The static system would be a constant build-up of $^4$He in the lattice during the 195 min of the experiment. The dynamic system would be an equilibrium of helium produced in the lattice and helium lost from the lattice. If a helium flux exists it was of an unknown flow rate. There is a short treatment of this in Section 6.

The bcc Ti foil was unique showing no crater evidence via SEM photos. However, in SEM photos of fcc metal TF craters ranged from 0.05 to 10 $\mu$m. See Fig 2A and 2B. There are SEM photos that show 1 $\mu$m bcc Ti ejecta tubes on or near the TF surface [14,15]. Were these phenomena related to ejecta sites? SEM photos of fcc TF, Pd, Ag, Ni and

**Table 2.** The calculated MS Measured Lattice Trapped $^4$He Results from LANL and PNNL 1994.

| Target foil sample | $^4$He total $10^{+12}$ atoms | $^4$He back ground $10^{+12}$ atoms | $^4$He - BG $10^{+12}$ atoms |
|---|---|---|---|
| Ti-3A | 40± 2 | 1.4± 1 | 39± 2 |
| Pd 9 | 0.65± 0.15 | 0.36± 0.15 | 0.29± 0.15 |
| Pd13a | 0.79± 0.15 | 0.40± 0.15 | 0.39± 0.15 |

Cu, show craters with diminishing sizes as the input frequency increased [16]. At 20 kHz cavitation inputs produced the largest events, but harmonic modes may be responsible for single events [13]. Also, one must consider the driven oscillations of TF as it became the source for resonating and collapsing bubbles as presented in ICCF 20 at Sendai, Japan, 2016 [19].

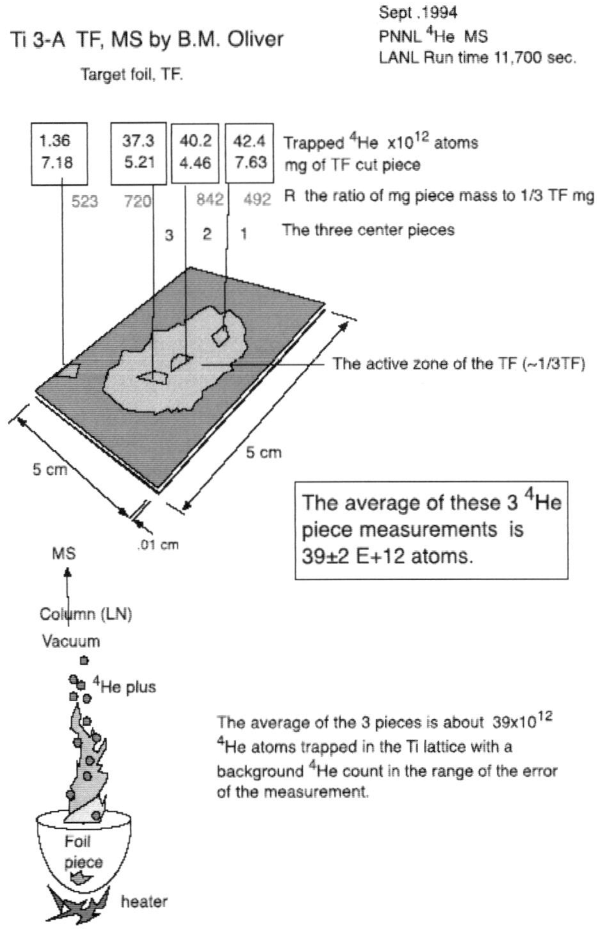


**Figure 3.** A cartoon of the first row of Fig. 1 showing the MS process of Brian Oliver's PNNL methodology and measurement of the helium isotopes with the final average measurement of $39\pm 2 \times 10^{+12}$. A weighed piece Ti TF is heated in a crucible under vacuum, and the evolved gases were passed through a $0.25 \times 12$ inch LN cooled column, that quickly elutes $^3$He and $^4$He into mass spectrum device measuring 4,250 $^4$He column 3 of the table.

## 6. Discussion

In SEM photos, there is a possible correlation between fcc ejecta and bcc micro tubes that were bursts of TF atoms. The very delicate hollow 1 $\mu$m tubes of bcc Ti SEM photos show burst of its lattice atoms ending in TF 1 $\mu$m tubes. These tube events might be compared to the ejecta from the ejecta sites in fcc Pd and other fcc lattices. The SEM photos of bcc Ti TF show Ti tubes that were partially transparent to an SEM electron beam. They produced shadows during the SEM photo process, and indicated the hollow nature of the bcc Ti tubes [14,19]. And their fragility shortened their lifetime to a few acoustic cycles, and led to a high production rate of lattice atomic debris. These 1 $\mu$m tubes, Ti or $TiO_2$, could originate from near surface M cluster events that were produced in a bcc Ti crystal TF that was a fusion alpha burst, ejecting the Ti TF lattice as 1 $\mu$m tube into the cavitating $D_2O$. Can alphas be ejected via these delicate short-lived Ti tube networks, with their debris circulating in the $D_2O$? This $^4$He will find its way to the Ar gas phase above the Ar saturated $D_2O$ for measurement.

For sub-surface M cluster activity, there was a concentration of about one He for each 50 Ti lattice atoms in the Ti TF surface active zone. For the purpose of this of discussion, the active surface TF was one atom deep. When the experiment was turned off, a flux of $^4$He atoms ended, and the atoms were trapped. Either the Ti lattice had trapped $^4$He from a steady additive situation, or from a flux rate of $^4$He in the bcc Ti TF lattice where helium was free to move out of the lattice. The measured 39 $\times$ $10^{+12}$ atoms of $^4$He represented only helium trapped in the lattice at shutdown, and not counted was the $^4$He lost from the Ti lattice during the running time.

The $^4$He results pose a question. Where were the gammas [12] ? An interesting comparison, following the expected path of gamma radiation from the number of $^4$He atoms measured over the Ti 3-A experiment's running time of 11,700 s is 3.4 $\times$ $10^{+9}$ counts/s representing the expected $^4$He flux. This compared to the known flux from 1 g of radium at 3.7 $\times$ $10^{+10}$ counts/s that is 1 Curie by definition, and their energy is 10 MeV for radium gammas. The comparison shows that potential gamma radiation expected from $^4$He is a factor of 10 lower than for the 1gm of radium but double the intensity of the radium and about 0.5 MeV. No gamma radiation count was detected from experiments while we were present working at LANL, and none was reported. This means that there must have been another path for giving up heat formation of an alpha to the lattice. This is just an idea and shows that the expected gamma energy chose another path or mechanism to dissipate the binding energy of the alpha that was faster than the usual gamma. It must be faster and new and excludes the usual path to gamma. Suggested paths are image charge [11] and the BEC [18] in the M cluster. The foil is gone now. All that is left is the NPPL MS data trail that tells part of the story. It was during the sub-picosecond in the Ti-3A M cluster that the M cluster alpha was created [13,14]. The location of $^4$He is based on speculation for the purpose of discussion. So if the radiation from the active zone surface area is 1 atom thick the ratio is about 1–50 Ti atoms.

The PNNL MS measurements of $^4$He were given in values of $10^{+9}$ units. That was the way the PNNL MS instrument was set. Also, it supposes that alphas were a product. Ti 3-A and Pd 9 and Pd 13 TF $^4$He values were worth reporting. They informed us about reproducibility with the three pieces from the center cut from the activity zone of the bcc Ti TF, Table 1. Because of the good reproducibility the magnitude is believable. The confidence levels of the PNNL MS data lived up to its reputation. The data shows that the fcc Pd TF sub-surface M cluster bursts into a single event shown in the TF activity zone. These single events produced crater ejecta velocities can be found from SEM photos of single event ejecta sites $E = 0.5\,mv^2$ [13]. The binding energy for alpha is $\sim 4 \times 10^{-12}$ J and from the fcc Pd ejecta site is 3 $\times$ $10^{+3}$ m/s and mass ejected Pd atoms 1.2 $\times$ $10^{-19}$ kg and energy is 4.4 $\times$ $10^{-12}$ J. This is equivalent to the energy of single alpha producing events of 4 $\times$ $10^{-12}$ J. In the fcc Pd TFs there was a low level of $^4$He atoms trapped in the lattice as low steady state lattice population, and were not ejected, remaining in the lattice at low levels close to the background level (see Table 2). When the experiment was turned off there was perhaps a flux of $^4$He that was left trapped in the lattice matrix of the Pd foils. Or was the $^4$He just a continual build-up of $^4$He atoms in the surface lattice of the Pd TFs? The same question can be applied to the bcc Ti TF. If there was a continuous flux, that means the number of helium atoms produced was larger than in an accumulation method over time, and the system used to calculate the Ti 39 $\times$ $10^{+12}$ atoms was a low result of the total potential atoms produced. In the flux system, a build-up of $^4$He was ejected into the cavitating $D_2O$ and migrated to the gas phase, where helium atoms could be collected and measured. In the Ti 3-A experiment the steady state temperature profile dropped off during the system failure during the 30-minute repair. Then the system recovered to the same steady state temperature. The experiment continued until the final breakdown and stopped any further activity in Ti 3-A TF.

Sample TFs that generated craters did not lose all their $^4$He atoms. Some were trapped in the fcc TF lattices of Pd, Ni, Ag, and Cu. The cavitating $D_2O$ implantation is a destructive process. On top of this cavitation damage was a faster much smaller in size and more powerful alpha event process in the M cluster resulting in the lattice ejecting some target foil. Several videos of this phenomena show the glistening small areas of TF viewed through 0.5-inch-thick Quartz window looking at the $D_2O$ cavitation vaporization of the target foil's continuous erosion [16]. Decoupling from the driving piezo supports the idea that the foil was also a generator of implanting bubble jets in recent RF stimulated cavitation producing measured acoustic beats [19].

$^4$He ejected atoms pass from the fcc TF into the Ar saturated cavitating $D_2O$, and were found and measured in the Ar gas above the circulating $D_2O$ [2]. The bcc Ti target foil is unique, showing an altered mechanism, with no large ejecta site craters as measured in 20 kHz SEM photos of TF, Fig. 2A. Helium residing in the bcc Ti TF lattice as measured in PNNL MS was different as TF M clusters left $^4$He in the TF lattice. The energy of formation of 3.83 $\times$ $10^{-12}$ J per 1 alpha, is speculated to originate from implanted deuterons in an M cluster that often produces a single alpha event at high frequencies at 2 MHz [17,20], but also many single events at low frequency. Multiple events produced in fcc TF increased in the number of alpha energy events in proportion to their crater size. The $^4$He presence and containment were found in the active zone of the TF volume (see Fig. 3). Investigation by SEM Tomography analyses is needed to examine in the interior of the bcc Ti TF to confirm the existence and location of the lattice bubbles or atoms which would settle this speculation, but it has not been done yet on TFs in First Gate Energy's possession. These are the unique characteristics of Ti TF at 20 kHz, showing less surface damage, and a relatively high count of trapped $^4$He in the bcc Ti, TF crystal lattice. For a Ti TF crystal lattice, sh $\rightarrow$bcc, change phase at 1200 K reverts back to the bcc if it was not already there [20]. This contrasts with the fcc crystal lattice TF, particularly the two Pd TFs above that were measured and showed alpha ejection events, single and multi. These fcc Pd TFs showed small amounts of $^4$He collected from experiments measured by PNNL MS. A wide distribution of ejecta from single to large multi events were measured with 20 kHz devices and in the higher frequencies [19]. At 1600 kHz Pd TF, the major ejecta population was the single alpha ejecta and its fcc ejecta debris [13,17].

## 7. Conclusion

It is difficult to produce good data. The expensive analyses required were not done and few opportunities come along to make measurements that stimulate new work. This is one of those good opportunities. In storage in the First Gate's Lab are 60 exposed target foils collected over 27 years that have never been looked at or tested by SEM, MS, or other methods. Over a period of 23 years and many experiments, a new appreciation for the cavitation process was realized as new results and data altered the thinking of old experiments. This is an opportunity to evaluate data using new cavitation knowledge. The stored gas sample cylinders were contaminated, but the TF samples were not, because all activity in the TF lattice active zone were frozen when the power to the experiment was shut down. The contents and lattice changes remained frozen in their TF crystal lattices. Metal TF foils remain unchanged from that day forward, including the two Pd and Ti 3-A target foils that would retain helium isotopes in their crystal lattices until melting to give up their helium data information. Target foil lattices are less vulnerable in the storage than the valved off 50 $cm^3$ cylinders of liquids and gases.

The irradiation of the TF by nanometer plasma jets of deuterons can produce high pressure, 100 GPa, in bubbles in TF lattices [6]. The fcc lattices of TFs Pd, Ni, Cu, and Ag, are more favorable to the production of surface eruptions in the form of craters, in the fcc TF (forming surface cavities, ejecta sites) than in bcc TF lattices. Ti bcc and other bcc Zr crystals favor the production of small bubbles or single atoms trapped in the bcc lattice [21]. Is there a parallel between Ti tubes and Pd ejecta sites as both appear very explosive? There have been no SEM photos of any tubes in the fcc TF lattices.

There have been several experiments involving bcc Ti lattice helium production, no gamma production mentioned, but neutron counts measured heat production equivalent to $10^{15}$ helium atoms produced in 2.7 h [4] in the Russian paper in 2015. This has a connection to the 1960s–1980s. Many accelerator experiments were done with beams of deuterium and helium ion collisions with lattices of various metal targets that found helium present trapped in nm size bubbles at 100 GPa with deuterons and He; not a bad fit with sonofusion. In most cases the helium and or deuterium ions were accelerated into various target foils [4,6,13]. A very powerful petawatt laser device is described as accelerating ions into thin fcc Al TF, where a pre-pulse creates surface ions on an fcc Al TF. The ions are accelerated back into the TF surface in time frames of femtoseconds; also, not a bad fit for the sonofusion model [8]. Using a bcc Ti TF would be very interesting as it would accelerate deuterons into a stable $TiD_2$ TF with the production of a possible alpha event, where the $^4$He would stay, much like sonofusion events in Ti 3-A. This is a proposed path to the helium found in Ti 3-A target foil data. Energy density in smaller bubbles may have higher energy densities. Keep the energy densities as high as possible. All particles are found in the creation of stars, but here only their shadows exist (on paper).

## Acknowledgments

The facilities at Los Alamos National Labs and the people at the Tritium Lab, Tom Claytor and Dale Tuggle and staff. The facilities at PNNL's MS Lab, Brian Oliver. EPRI in Palo Alto, CA, Tom Passell negotiated funding. The support of SRI and its Energy Research Center under the direction Michael McKubry and Fran Tanzella, and many others: Gene Mallove, Jean Paul Biberian, Scott Chubb, etc. connected to the field of CMNS (Condensed Matter Nuclear Research). Over 23 years later, with help from Julie Wallace, Kim Nunlist, Dick Raymond, and Malcolm Fowler.

## References

[1] EQuest Sciences, and later First Gate Energies, a small research company dedicated to cavitation studies and sonofusion lead to measurable fusion events. The company founded by Dick Raymond, Larry Klein, and Roger Stringham, 1989.

[2] B.M. Oliver, 1995 Helium analysis of metal and gas samples, U.S. Department of Energy, Pacific Northwest National Laboratory, Battelle Memorial Institute, Richland, WA, May 1994.

[3] B.M. Oliver, J.G. Bradley and H. Farrar , Helium concentration in the earth's lower atmosphere, *Geochim. Cosmochim. Acta* **48** (1984) 1759.

[4] O.D. Dalkarov, M.A. Negodaev and Ruetskii, Investigation of heat release in the targets during irradiation by ion beams, *Russian Acad. of Sci.* **8** (2015), ISSN 2410-4914.

[5] P.B. Johnson and D.J. Mazy, Helium gas bubble superlattice in copper and nickel, *Nature* **281** (1970) 359–360.

[6] A.N. Zhiganov and A.Y. Kupryazhkin, Grain-boundary diffusion and solubility of helium in submicrocrystalline palladium, *Tech. Phys.* **50** (2005) 1026. doi:10.1134/1.2014533.

[7] G.L. Copeland, R.W. Hobbs, G.L. Hoffman and J.L. Snelgrove, Perform of Loe-Enriched U3Si2-aluminm dispersion fuel elements in the Oak Ridge Research Reactor, NCL Collective store, Argon National Lab, 1987, p. 6.

[8] T.Z. Esirkepov et al., Prepulse and amplified spontaneous emission effects on the interaction of a petawatt class laser with thin solid targets, arXiv:0568v1 (2013).

[9] G. Chambers, NRL X-ray diffraction studies of exposed Pd target foil at LANL, 1994.

[10] M.A. Liberman, J.S. De Groot, A. Toor and R.B. Spielman, *Physics of High-Density Z-Plasmas*, Springer, Berlin, 1999.

[11] N.M. Lawandy, Interaction of charged particles on surfaces, *Appl. Phy. Lett.* **95** (2009) 234101-1-3.

[12] J. Yoon, *The Curious Story of the Muon-Catalyzed Fusion Reaction*, Stanford University, 2016.

[13] R.S. Stringham, Conservation of E and M, single cavitation heat events, *J. Condensed Matter Nucl. Sci.* **15** (2015) 55, www.iscmns.org/CMNS/JCMNS-Vol15.pdf.

[14] R.S. Stringham, Sonofusion produces tritium that deacys to helium three, *Proc. ICCF 15*, V. Violante and F. Sarto (EDs.), , Oct. 5– 9, 2009, p. 57. Some parts need a rewrite.

[15] G. Mogilevsky, Q. Chen, A. Kleinhammes and Y. Wu, The structure of multilayered titania nanotubes based on delaminated anatase, *Chem. Phys. Lett.* **460**(4–6) (2008).517–520. DOI: 10.1016/j.cplett.2008.06.063.

[16] 20 MHz cavitation input to M LL reactor 0.5 inch Quartz window of target foil in circulating DOD. Videos of exposed TF shows the destruction of the FCC lattice of Pd; the first 1993 at EQuest Lab, second at LANL in 1994, and third at Stanford Research International in 1995.

[17] R.S. Stringham, When bubble cavitation becomes sonofusion, *JCMNS* **6** (2012) 1–12.

[18] A. Wenz and M. Kronerett, //www.physi.uni-heidelberg.de/∼ fschney/2008SS-Preseminar/BEC_BCS.pdf ; Physikalisches Institute Heidelberg (2008).

[19] R.S. Stringham, EM resonance of a piezo antenna produces heat, helium, *ICCF20,* Sedai, Japan, Co-charis Yashiro Iwamura and Jirohta Kasagi (2016).

[20] Over a period of 27 years three different input frequencies were used; 20 kHz, 46 kHz, and 1600 kHz and as the frequency increased the average ejecta size dropped essentially to 1 alpha/per event.

[21] A. Arjhangmehr, and S.A.H. Feghhi, Evolution pattern of collision cascades in *bcc* V with different grain boundary structures: an atomic scale study, Published online: 26 Apr 2017.

J. Condensed Matter Nucl. Sci. 24 (2017) 296–300

Research Article

# Plasmonic Concepts for Condensed Matter Nuclear Fusion

Katsuaki Tanabe*

*Department of Chemical Engineering, Kyoto University, Kyoto, Japan*

**Abstract**

We propose and numerically investigate a scheme to provide high-density optical or electromagnetic energy to fuel materials for condensed matter nuclear fusion. Surface plasmons in metal nanoparticulates strongly interact with electromagnetic fields at their resonance, resulting in an intense focusing of the incident energy around the metal surface. It is therefore possible to take the advantage of such high energy concentration, under laser or electric power input, for instance by simply coating the conventional Pd-based fuel materials with noble-metal nanoparticles or nanoshells. This field-enhancement effect can be a powerful method to boost the fusion-triggering electromagnetic or optical energy to significantly improve the reaction rate and experimental reproducibility, and to create low-D-load fusion systems. Importantly, this enhancement scheme is applicable to both gas- and liquid-phase systems, and furthermore, not only for laser/optical power, but also for the conventional electrolysis systems due to the equivalency between light and electromagnetic fields. We also show that the conventional fusion-catalyzing metals of Pd, Ni, and Ti themselves intrinsically exhibit a certain degree of field enhancement for their nanostructured form. Therefore, we point out that this plasmonic effect may have been unknowingly produced particularly in the electrolysis-type fusion experiments reported so far. This field-enhancement phenomenon thus could be a clue to solve the mystery of the energy supplied to overcome the gigantic Coulomb barrier to produce the fusion reaction observed with visible rates, as well as a powerful tool for further technical progress.


*Keywords:* Electromagnetic field enhancement, Laser, Metal nanoparticles, Metal nanoshells, Nanophotonics, Plasmonics, Power/energy density

## 1. Introduction

The intensity and density of the triggering energy supplied to activate the nuclear fusion reaction are key factors to produce a smooth and reproducible initiation of the reaction. The incident electric power onto Pd samples in liquid [1,2] and gaseous [3,4] systems reported in the field of condensed matter nuclear fusion has been generally around 1 W/cm$^2$ or less. Note that we estimated the chemical power, instead of an electric power, from the temperature increment and heat capacity of the sample for Ref. [4]. Semiconductor or solid-state (e.g., YAG) lasers, on the other hand, are able to strike power densities several orders of magnitude greater. This approach aims an initial local ignition of nuclear fusion reaction enabling the generated heat there to trigger subsequent reactions throughout the fuel material.

*E-mail: tanabe@cheme.kyoto-u.ac.jp

© 2017 ISCMNS. All rights reserved. ISSN 2227-3123

Moreover, the plasmonic field enhancement effect [5–7] induced by metal nanoparticles can concentrate the incident optical energy even further, as explained below.

## 2. Calculation Methods

To numerically examine the plasmonic energy focusing effect, we did the following calculations. We calculate the field-enhancement factors, which are the intensity ratios of the fields around the object to those in the absence of the object, or the original incident fields, for spherical metal nanoparticles in $D_2$ and $D_2O$. These calculations are based on the classical electromagnetic field theory in the quasistatic limit [6,7] to numerically show how much energy can be concentrated out of the incident optical or electric power. The electrostatic calculations conducted in this paper are valid for particle sizes in the range of 10–100 nm, for which the phase retardation is negligible throughout the particle, the field enhancement will be largest, and metal nanoparticles and nanoshells therefore become most applicable, as discussed in Section 3. The empirical complex dielectric functions of metals and of the surroundings on frequencies listed in [7,8] are used for the computation in this paper. The dielectric functions of $D_2$ and $D_2O$ were assumed equal to those of the air and $H_2O$, respectively. The calculation of the field-enhancement factors was carried out for spherical nanoparticles of Pd, Ni, and Ti. The method used to calculate the field enhancement factors is described in [7]. In short, the field-enhancement factor is calculated as

$$\eta \equiv \frac{\left|\vec{E}\right|^2}{\left|\vec{E_0}\right|^2} = \left|1 + 2\frac{\varepsilon_1 - \varepsilon_m}{\varepsilon_1 + 2\varepsilon_m}\right|^2, \tag{1}$$

where $\vec{E}$ is the maximum static electric field around the metal nanoparticle, $\vec{E_0}$ is the original uniform electric field in the absence of the nanoparticle, and $\varepsilon_1$ and $\varepsilon_m$ are the frequency-dependent complex permittivities or dielectric functions of the sphere and the surrounding medium, respectively. Note that this field-enhancement factor is defined as the ratio of field *intensities* and not field *magnitudes*.

We also conduct field-enhancement-factor calculations for $Ag/SiO_2$ nanoshells, which are subwavelength-scale heterogeneous concentric spherical particles with $SiO_2$ cores and Ag shells. In these calculations, we vary the diameter ratio of the inner material to the outer material, $f$, to understand the differences with the metal-nanoparticle case as well as the resonant-frequency dependence on $f$. The calculations in this part are also carried out in the quasistatic limit with the scheme described in [7]. The resulted field-enhancement factor is

$$\eta = \left|1 + 2\frac{(\varepsilon_2 - \varepsilon_m)(\varepsilon_1 + 2\varepsilon_2) + f^3(\varepsilon_1 - \varepsilon_2)(\varepsilon_m + 2\varepsilon_2)}{(\varepsilon_2 + 2\varepsilon_m)(\varepsilon_1 + 2\varepsilon_2) + f^3(2\varepsilon_2 - 2\varepsilon_m)(\varepsilon_1 - \varepsilon_2)}\right|^2, \tag{2}$$

where $\varepsilon_2$ is the dielectric function of the outer spherical shell. Ag has the highest electrical conductivity among the whole metal elements and therefore exhibits the highest field enhancement for its nanoshells [7,9]. Hence, we choose Ag as the nanoshell material for the calculations.

## 3. Results and Discussion

Figure 1 shows the calculated spectra of field enhancement factors around nanoparticles of Pd, Ni, and Ti, which have been conventionally used for deuterium-containing fuel materials in the field of condensed matter nuclear fusion, in $D_2$ (Fig. 1(a)) and $D_2O$ (Fig. 1(b)) under the electrostatic approximation. Local energy enhancement around 10 times is seen for a wide range of optical frequencies, through visible to near infrared and beyond. These nanoparticles thus concentrate optical or electromagnetic energy in their vicinity like antennae. Among the whole metal elements,

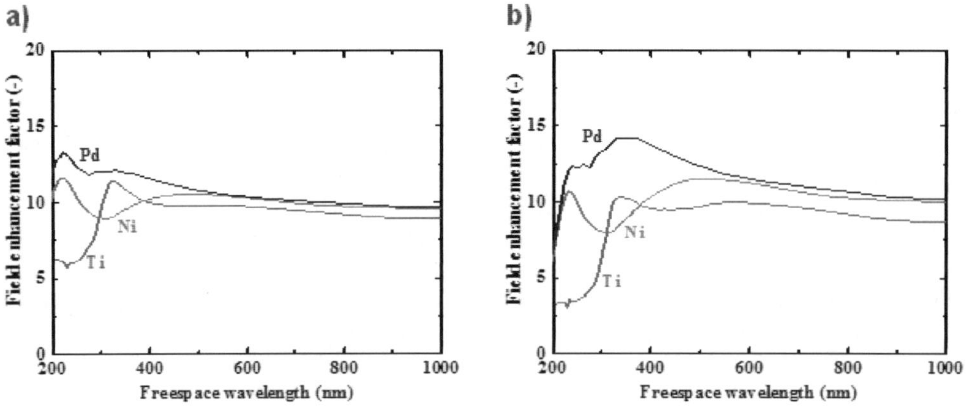


**Figure 1.** Electromagnetic field-enhancement factors around Pd, Ni, and Ti nanoparticles in (a) $D_2$ and (b) $D_2O$.

Al and the noble metals Ag, Au, and Cu are known to exhibit distinctively higher field enhancement factors than other metals due to their high conductivities [7,9]. A certain degree of field enhancement, however, is still attainable even for the common metals used for condensed-matter fusion, as seen in Fig. 1. Importantly, a certain number of the electrolysis-type condensed matter nuclear fusion experiments reported so far may actually have unknowingly benefited from this plasmonic local energy enhancement effect. Note that the field-enhancement-factor spectra (peak positions, intensities) are independent of particle size under the quasistatic limit but are valid for particle diameters of 10–100 nm in this calculation [7]. The peaks seen in these spectra are associated with the resonance or surface mode, characterized by internal electric fields with no radial nodes. The electrostatic calculations shown in this part and the following nanoshell part are valid for particles and shells smaller than the wavelengths at which the phase retardation is negligible throughout the objects. Also, the dielectric functions of materials used for the computation in this paper were empirical values for bulk materials, whose validity is debatable when the material sizes become smaller than 10 nm. The validity becomes problematic due to the electron mean free path limitation or scattering of conduction electrons off particle surfaces [6,10,11]. Therefore, the computation results for optical wavelengths under the quasistatic approximation are valid for metal particles and shells with diameters of 10–100 nm. Metal particles both smaller and larger than these limits exhibit broader plasmon resonances and smaller field enhancements due to surface scattering losses and radiative losses or electrodynamic damping, respectively [10,11]. The choice of particle sizes (10–100 nm) in this paper is therefore most suitable for plasmon-enhanced photonic applications due to the largest field enhancements. This size aspect should therefore be also accounted for the optimizing design of the deuterium-containing metal composite materials for condensed matter fusion.

Figure 2 shows the spectra of field-enhancement factors around Ag nanoshells with $SiO_2$ cores with varied $f$ in $D_2$ and $D_2O$. We can take advantage of such gigantic energy concentration and resonant-frequency tunability with $f$, for instance by simply coating the conventional Pd-based fuel materials with metal nanoshells. Once an initial nuclear fusion reaction occurs in the energetic highly concentrated "hot spot" region around a metallic nanoparticulate, a gigantic amount of heat locally generated by the nuclear reaction induces subsequent reactions around the region by supplying the activation energy, and thus effectively initiates heat-mediated chain reactions to spread throughout the fuel material. The local energy focusing scheme we propose in this paper thus significantly increases the probability of the initial nuclear reaction even if the total power irradiated into the fuel material is the same, and therefore may

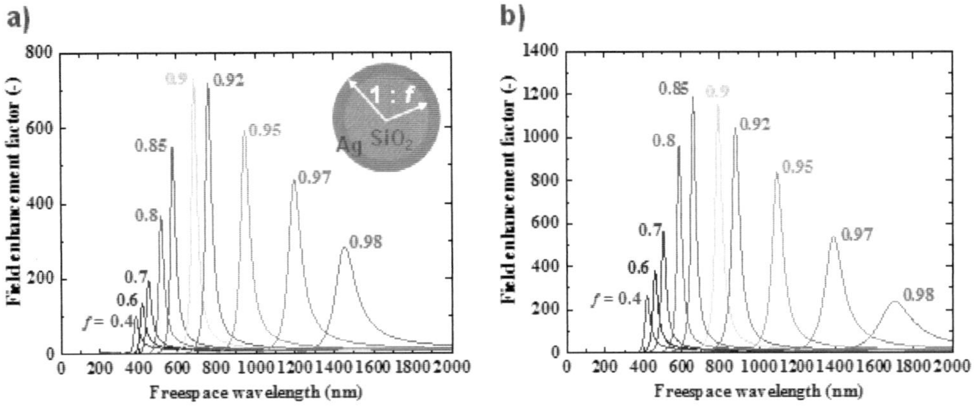


**Figure 2.** Electromagnetic field-enhancement factors around $Ag/SiO_2$ nanoshells in (a) $D_2$ and (b) $D_2O$.

effectively reduce the threshold input power. This enhancement approach may enable fusion with significantly lower deuterium loading than reported so far. Note that this metal-nanoparticle/shell-induced energy focusing scheme is applicable not only with lasers but also to the conventional electrolysis-type condensed-matter fusion, since the electromagnetic field enhancement is equivalent for both systems. Our approach may be also beneficial for laser fusion [9]. This sort of optical or electromagnetic model, incidentally, is simple and flexible to connect to other models, and may construct a comprehensive model by combining with transport and reaction submodels [12], for instance. The wide tunability of the resonant frequency to $f$ seen in Fig. 2 allowing it to be matched with the wavelength of the incident laser or electric field offers a significant advantage in the use of nanoshells over solid nanoparticles. This tunability as well as the large field enhancement can be understood to be a result of the strong interaction between the plasmons for a metal sphere and a metal-dielectric hollow cavity, whose hybridization forms a metal shell [13,14]. Some of the plots in this paper incidentally coincide with those in [7] due to the assumed equivalency in the dielectric functions between $D_2$ and the air, and between $D_2O$ and $H_2O$. The reproduction permission has been granted by the American Chemical Society; copyright 2008. This paper aims to give a conceptual understanding of the nature of the concentration of electromagnetic fields around nanoparticles/nanoshells, so we assumed that the case for a single stand-alone spherical nanoparticle/nanoshell geometry with high symmetry is suitable, due to its relatively straightforward computation and analysis. Therefore, in the present work, we show calculations only for spherical nanoparticles and nanoshells for the sake of simplicity in a preliminary investigation, but ellipsoidal ones would provide even higher field enhancement factors around their tips, because the sharper curvature of the metal/dielectric interfaces allows the electromagnetic field to concentrate further. This is known as the "lightning-rod effect" [6,15,16]. Also, surface plasmons located in between multiple metal nanostructures, or the so-called "gap plasmons," are known to have distinctive characteristics [17,18]. The shape and interparticle/shell effects will be discussed elsewhere.

## 4. Conclusion

In this work, we have proposed and numerically investigated a scheme to provide high-density optical or electromagnetic energy to fuel materials for condensed matter nuclear fusion. Plasmonic metal-nanoparticle field enhancement can be utilized by combining with laser or electric power input, for instance, by simply coating the conventional

Pd-based fuel materials with noble-metal nanoparticles or nanoshells. This field-enhancement effect can be a powerful method to boost the fusion-triggering electromagnetic or optical energy to significantly improve the reaction rate and experimental reproducibility as well as create low-D-load fusion systems. Importantly, this enhancement scheme is applicable both for gas- and liquid-phase systems, and furthermore, not only for laser/optical power, but also for the conventional electrolysis systems due to the equivalency between light and electromagnetic fields. We have also shown that the conventional fusion-catalyzing metals of Pd, Ni, and Ti themselves intrinsically exhibit a certain degree of field enhancement for their nanostructured form. Therefore, we point out that this plasmonic effect may have been unknowingly applied, particularly in the electrolysis-type fusion experiments reported so far. This field-enhancement phenomenon thus could be a clue to solve the mystery of the energy supplied to overcome the gigantic Coulomb barrier to produce the fusion reaction observed with visible rates, as well as a powerful tool for further technical progress.

## Acknowledgement

This work was partially supported by the Thermal and Electric Energy Technology Foundation.

## References

[1] M. Fleischmann, S. Pons and M. Hawkins, *J. Electroanal. Chem.* **261** (1989) 301.

[2] S.E. Jones, E.P. Palmer, J.B. Czirr, D.L. Decker, G.L. Jensen, J.M. Thorne, S.F. Taylor and J. Rafelski, *Nature* **338** (1989) 737.

[3] T. Mizuno, T. Akimoto, K. Azumi, M. Kitaichi and K. Kurokawa, *Fusion Technol.* **29** (1996) 385.

[4] Y. Arata and Y, Zhang, *J. High Temp. Soc. Jpn.* **34** (2008) 85.

[5] R.H. Ritchie, *Phys. Rev.* **106** (1957) 874.

[6] C.F. Bohren and D.R. Huffman, *Absorption and Scattering of Light by Small Particles*, Chapters 3 and 5, Wiley-VCH, Weinheim, 1983.

[7] K. Tanabe, *J. Phys. Chem.* C **112** (2008) 15721.

[8] E.D. Palik (Ed.), *Handbook of Optical Constants of Solids*, Academic Press, Orland, 1985.

[9] K. Tanabe, *Jpn. J. Appl. Phys.* **55** (2016) 08RG01.

[10] M. Moskovits, *Rev. Mod. Phys.* **57** (1985) 783.

[11] E. Hao and G.C. Schatz, *J. Chem. Phys.* **120** (2004) 357.

[12] K. Tanabe, *Heliyon* **2** (2016) e00057.

[13] A.L. Aden and M. Kerker, *J. Appl. Phys.* **22** (1951) 1242.

[14] E. Prodan, C. Radloff, N.J. Halas and P. Nordlander, *Science* **302** (2003) 419.

[15] P.F. Liao and A. Wokaun, *J. Chem. Phys.* **76** (1982) 751.

[16] M.B. Mohamed, V. Volkov, S. Link and M.A. El-Sayed, *Chem. Phys. Lett.* **317** (2000) 517.

[17] K.H. Su, Q.H. Wei, X. Zhang, J.J. Mock, D.R. Smith and S. Schultz, *Nano Lett.* **3** (2003) 1087.

[18] D.F.P. Pile, T. Ogawa, D.K. Gramotnev, Y. Matsuzaki, K.C. Vernon, K. Yamaguchi, T. Okamoto, M. Haraguchi and M. Fukui, *Appl. Phys. Lett.* **87** (2005) 261114.

J. Condensed Matter Nucl. Sci. 24 (2017) 301–311

Research Article

# Controlled Electron Capture: Enhanced Stimulation and Calorimetry Methods

Francis Tanzella*

*SRI International, Menlo Park, CA, USA*

Robert Godes, Rogelio Herrera and Cedric Eveleigh

*Brillouin Energy Corporation, Berkeley, CA, USA*

**Abstract**

The Controlled Electron Capture (CEC) method has been extended to use faster rise and fall time pulses, hydrogen isotope gas based systems at temperatures up to 650°C, and more precise and accurate calorimetry relative to results presented earlier. Our isoperibolic (IPB) cell/calorimeter is operated as an isothermal compensation type calorimeter. Potential sources of error in this system are discussed as well the methods used to minimize them. In power compensation mode the cell is held at a constant temperature using a heater power feedback system and constant power pulses or DC power steps are added to the system, resulting in a reduction of heater power. The relationship between this heater power reduction and DC power passed along the reactor core yields a calibration curve at different temperatures that allows us to evaluate how much output power increased during a given stimulation pulse. The IPB cell/calorimeter was stimulated by commanding different pulse widths at constant amplitude with the pulse power held constant by appropriately varying the pulse repetition rate. At 250–300°C the ratio of output power increase to input pulse power varied from 1.0 to over 2.0 depending on the pulse width at constant input power. That ratio was always 1.0 at all pulse widths attempted at 600°C. These results have been seen tens of times. The amount of excess power was also dependent on the composition of the gas and the metal alloy coatings on the core. The outer layer of the core was always pure Ni. The composition of a multilayer metal–dielectric metal coated core was chosen to allow for reasonable hydrogen solubility and mobility at 300°C. The results of various experiments are discussed Importantly these results presented here ignore the heater power necessary to maintain temperature and the losses in the pulse generator, which can be several times greater than either the stimulation power or power gain.



*Keywords:* Controlled electron capture, Hydrogen, Isoperibolic, Nickel, Power compensation

*E-mail: francis.tanzella@sri.com.

© 2017 ISCMNS. All rights reserved. ISSN 2227-3123

## 1. Introduction

Since August 2012, SRI International (SRI) has been performing tests on two different versions of Brillouin Energy Corp.'s low-energy nuclear reactors (LENR) [1]. We have operated these reactors independently in an attempt to verify results that Brillouin has found with these reactors and others like them. We have also monitored and advised Brillouin on the results from reactors operated by Brillouin in their own laboratory. This report documents the results obtained by studies in SRI's laboratory, as well as verification and validation of results obtained in Brillouin's laboratory over the past nine months. Brillouin has indicated that it has designed the control systems in its reactors to drive the underlying physics of LENR, as described in its Controlled Electron Capture Reaction (CECR) hypothesis [2]. The CECR hypothesis explains how scientists at Brillouin believe their reactors generate controlled LENR reaction heat. Our study did not attempt to prove or disprove Brillouin's CECR hypothesis.

The systems tested and described in this report consist of three parts — cores, reactors, and calorimeters. The cores are the reactive components of the system. The reactors provide the environment and stimulation that causes the cores to produce reaction heat. The calorimeter is used to measure the thermal efficiency and absolute heat produced by the core-reactor system. The calorimeter was designed by both SRI and Brillouin personnel to be perfectly matched to the reactor. The results from two of these reactors are described in this report.

SRI has brought over 75 person-years of calorimeter design, operation, and analysis experience to this process. We have used our expertise in LENR calorimetry to validate the results summarized herein. Brillouin's system design utilizes compensation calorimetry, in which the core and reference temperatures are held constant by varying the input heater power while applying different types of stimulation that also input power to the reactor/calorimeter.

## 2. Experiment

### 2.1. Design

The cores consist of a metal substrate, which in some configurations includes a heater and thermocouple with several spray-coated layers. Generally, these coatings alternate between a hydrogen-absorbing metal and an insulating ceramic. One example is shown in Fig. 1. Other designs may have more or fewer layers. All of the layers are porous, allowing the gas(es) in the reactor chamber access to all coatings. There is a heater and thermocouple in the center of the core. The power to the heater is measured directly from the voltage and current supplied by the direct current (DC) power supply.

A photograph of the reactor/calorimeter system is shown in Fig. 2. The system is contained in an acrylic container filled with argon gas to minimize the probability of a hydrogen–oxygen reaction from any $H_2$ that might leak from the system. A schematic diagram of the reactor/calorimeter system is shown in Fig. 3. In a traditional isoperibolic calorimeter, the reactor temperature is distributed along a massive thermal block (inner block) surrounded completely

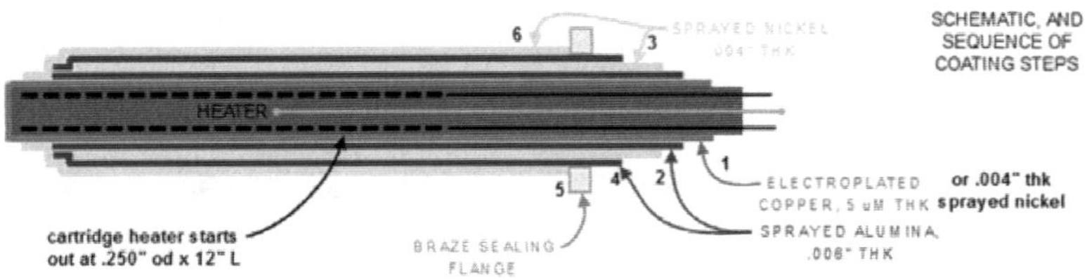


Figure 1. Example of Brillouin's fourth-generation hydrogen hot tube cores.

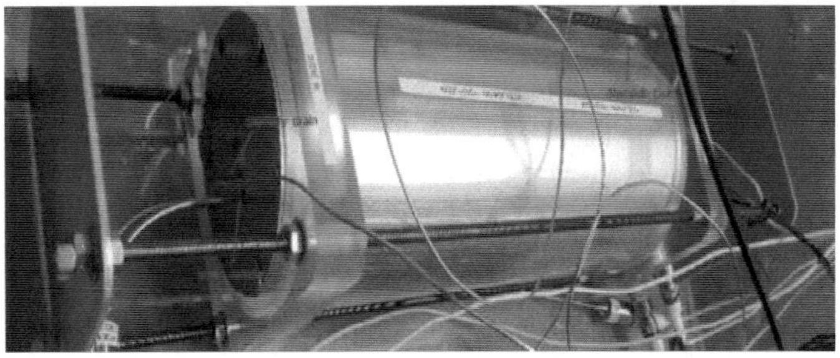

**Figure 2.** Photograph of the reactor/calorimeter system.

by a thick insulating layer, which itself is surrounded by another thermally conductive metal mass (outer block). This latter block is kept at a constant reference temperature.

Referring to the labeled parts of Fig. 3, the core (4) is centered in and insulated from a metal sheath (1). This core/sheath combination, together with the electrical connections (15), comprises the reactor. An annular copper or stainless steel block (3) is in intimate contact with the reactor sheath and contains a thermowell (2) and thermocouples and acts as the inner block. This copper block is surrounded by an annular ceramic insulator (14). Surrounding this insulator is an aluminum shell (5) with a thermowell and thermocouples. This shell, kept at constant temperature by flowing temperature-controlled water between it and the outer acrylic sleeve (12), serves as the outer block. Argon gas is circulated through the chamber outside of the calorimeter.

## 2.2. Measurement

The outer active layer is stimulated by sending pulses through the outer layer or layers and returning electrically through the innermost layer. The nature of the pulses is such that its current travels primarily on the surface of the metal in contact with the ceramic (the "skin effect"). This effect is caused by the very fast rise time of the pulses. An

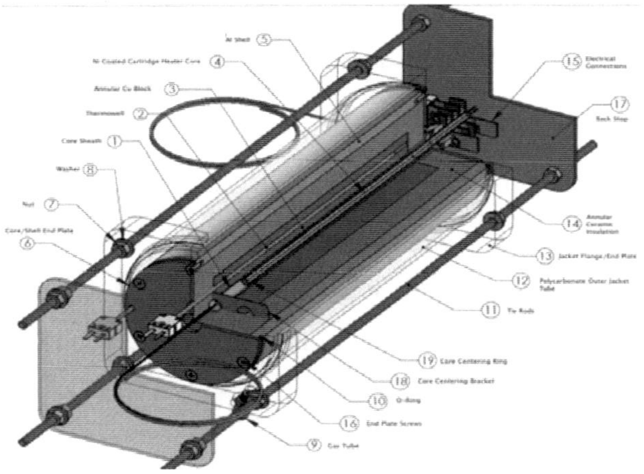


**Figure 3.** Schematic diagram of the isoperibolic hydrogen hot tube reactor/calorimeter.

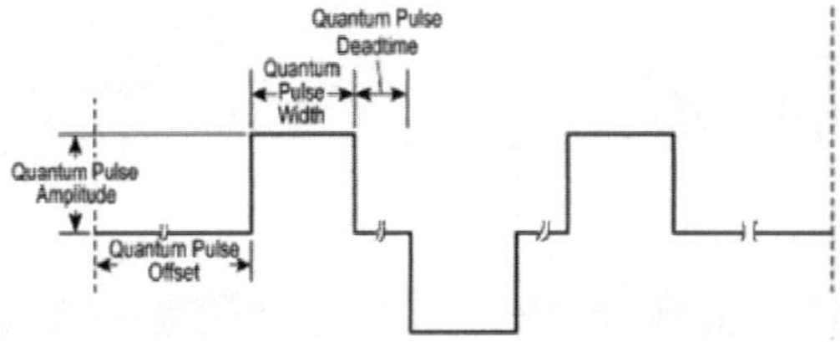


**Figure 4.** Example of Brillouin's "$Q$-Pulse".

example of this pulse design, which Brillouin refers to as a "$Q$-Pulse", is shown in Fig. 4. The pulse width is from ~80 to 1000 ns with a duty cycle of less than 1%. This example shows a pair of pulses with alternating polarity, although same-polarity pulse trains have also been used.

The stimulation power imparted to the core is measured using a circuit shown in Fig. 5. The pulse is generated by a proprietary $Q$-Pulse board and delivered to the core using series and termination resistors that help match the load impedance to that of the pulse board output. Using a high-speed oscilloscope, the voltage across the end of the core nearest the pulse board is measured as well as the voltage across the opposite end of the core across the termination resistor ($Z_{\text{term}}$). The $Z_{\text{term}}$ also acts as a current measuring resistor so the current is calculated as $V_2/Z_{\text{term}}$. The root mean square (rms) voltage across the $Z_{\text{term}}$ is then converted to the rms current.

The voltage across the core is determined using the method shown in Fig. 6. Figure 6a shows the two voltage traces being aligned in a way that minimizes the time difference. This overestimates the power imparted to the core since any phase lag between voltage and current would impart less input power. This voltage difference is shown in the upper plot of Fig. 6b. The current is shown in the middle graph, and the product of these two (power) is shown in the lower plot. It has been shown that the power calculation is essentially the same (within measurement error) whether it is calculated by multiplying the current and voltage plots point by point or by multiplying the calculated rms voltage by the rms current.

In compensation calorimetry, the heater power is varied to keep the core at constant temperature, which generally keeps the inner block at a constant temperature. The difference between the heater power with and without stimulation determines the effect of the stimulation. If this difference is greater than the stimulation that reaches the core, then energy is being produced in the core. Approximately 50 different parameters are collected allowing for calculation of reaction power (the power produced by the process induced by the pulse stimulation). Several calculation methods are possible from these parameters. In Section 2.4, we describe the two methods used.

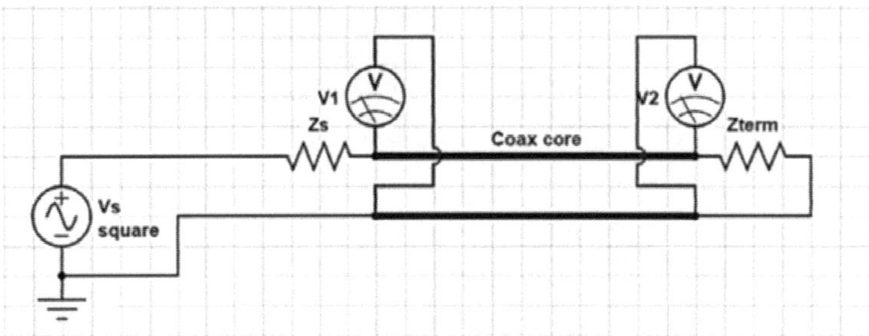


**Figure 5.** Pulse power measurement circuit.

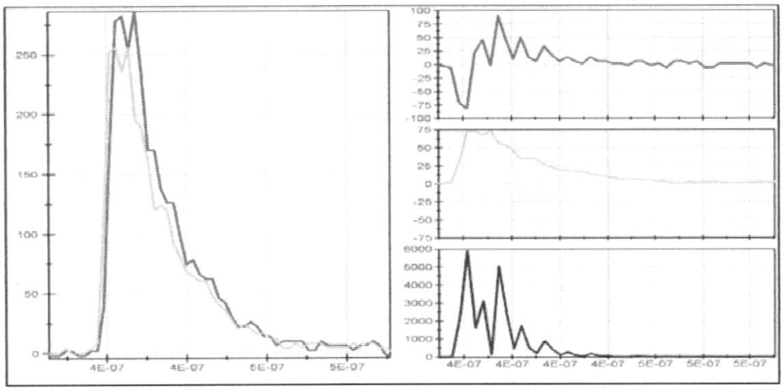

**Figure 6.** Measurement of the pulse power drop across the core: (a) $V_1$ and $V_2$ traces; (b) $V_1 - V_2$, current and power.

## 2.3. Operation

Figure 7 shows a screenshot from the specially designed proprietary automation and data collection computer program used to control and collect results from the IPB reactor/calorimeter system. The program has several panes allowing for control of temperature, pressure, pulse voltage, pulse power, pulse width, and pulse repetition rate and gas composition. The program also collects the heater power, the pulse power at the generator (as well as at the core), all temperatures, water flow rates, and gas pressure. The concentration of hydrogen and oxygen in the argon blanket are collected and measured. In all, ~50 different parameters are collected and stored every 10 s. A sequence file can be used to automatically change any or all of these parameters at specified intervals over a multi-day or multi-week period.

The sheath containing the core is operated with a static fill of hydrogen, helium, or argon gas held at constant pressure up to 10 bar. The temperature of the core is held constant using its embedded heater and thermocouple and controlled from 200 to 600°C. The outer block temperature is held at 25°C by constant-temperature water flowing from a Neslab® chiller.

The power emanating from the $Q$-pulse generator board is held constant as chosen by the program's front panel or the sequence file. Generally, the pulse amplitude (voltage) and pulse width are chosen. The repetition rate is adjusted automatically to maintain the chosen pulse power. Only a minor fraction of this power reaches the core as most of it is lost as heat in the electrical components and the transmission line. Of that reduced power, only a portion of it influences the heater power as explained in Section 2.2. The actual pulse power is measured directly using the methodology presented above.

Operating in power compensation mode, the computer keeps the inner core temperature constant at its set point. When power is imparted from the $Q$-pulse, the heater power is reduced to compensate and maintain a constant temperature. Hence, the core temperature and the inner and outer block temperatures are all held constant when using the same core gas.

First operating in He gas, a sequence was operated from 200 to 600°C in 50°C intervals. At each temperature, a given DC power was applied to the coating on the core. This process was then repeated while applying constant power pulses with varying pulse width at each temperature. Finally, both automated sequences were repeated in hydrogen gas.

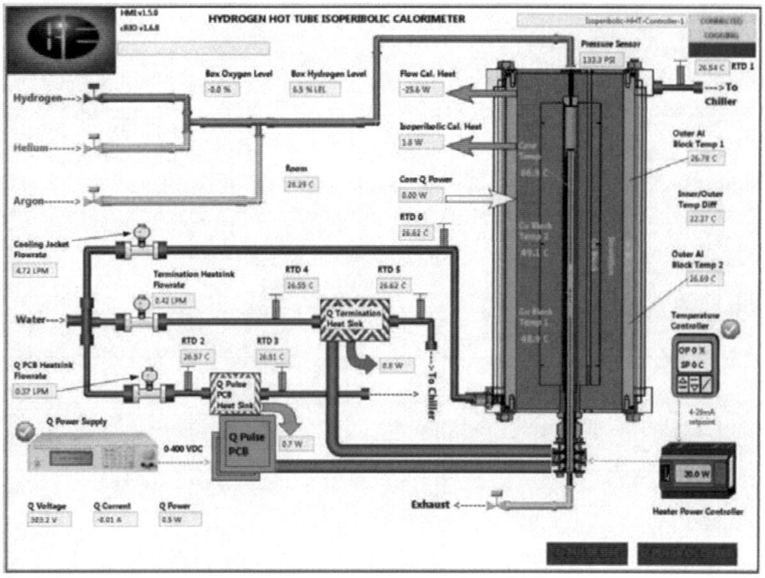


**Figure 7.** Screenshot of the automation and data acquisition computer program in operation.

## 2.4. Analysis

### 2.4.1. *Method A*

In our IPB design, only a fraction of the stimulation power is imparted to the core heater control because the heater/thermocouple combination is only in contact with approximately half of the core's length. The actual fraction imparted to the core is determined by resistively heating the core's coatings using different powers sourced from a well-measured DC power supply and measuring the heater's response at different temperatures. At each temperature, a linear function ($P_{\text{drop}} = mP_{\text{coating}} + b$) is determined between the power imparted to the core's coating via resistive heating and the power reduction in the internal heater necessary to maintain temperature. Representative linear coefficients at different temperatures are shown in Table 1 and represent the input power lost to the environment.

**Table 1.** Correlation of power imparted to the core's internal heater by resistively heating its coating: ($P_{\text{drop}} = mP_{\text{coating}} + b$).

| Temperature (°C) | $m$ | $b$ |
|---|---|---|
| 150 | 0.41 | 0.07 |
| 200 | 0.44 | 0.10 |
| 250 | 0.48 | 0.07 |
| 300 | 0.51 | 0.06 |
| 350 | 0.55 | 0.01 |
| 400 | 0.56 | 0.03 |
| 450 | 0.57 | 0.03 |
| 500 | 0.57 | 0.07 |

The basic calorimetric calculations are shown in Eqs. (1)–(4) when the isoperibolic calorimeter operates in heat flow mode. Heat flow ($Q_{\rm flow}$) is measured using $k_{\rm flow}$, which is determined via calibration and the temperature difference between the inner and outer blocks. Heat loss ($Q_{\rm loss}$) represents the heat loss to air that is not accounted for in ($Q_{\rm flow}$) and is also determined via calibration. The output heat ($Q_{\rm out}$) is the sum of ($Q_{\rm flow}$) and ($Q_{\rm loss}$). The input heat is the sum of power applied to the heater ($Q_{\rm heater}$) and the amount of heat experienced by the heater from the pulse ($Q_{\rm pulse}$). Hence the heat due to the reaction ($Q_{\rm reaction}$) is the difference between the output and input heats.

$$Q_{\rm reaction} = Q_{\rm flow} + Q_{\rm loss} - (Q_{\rm heater} + Q_{\rm pulse}), \tag{1}$$

$$Q_{\rm flow} = k_{\rm flow}(T_{\rm core} - T_{\rm outer}), \tag{2}$$

$$Q_{\rm loss} = k_{\rm loss}(T_{\rm core} - T_{\rm air}), \tag{3}$$

$$Q_{\rm out} = Q_{\rm flow} + Q_{\rm loss}. \tag{4}$$

We use the subscripts to mean operation without $Q$ power and operation with $Q$ power. In power compensation mode, we compare the heater power imparted to the core with and without $Q$ pulses applied. Because $T_{\rm core}, T_{\rm outer}$, and $T_{\rm air}$ are held constant in this mode, $Q_{\rm flow}$ and $Q_{\rm loss}$ are the same with and without $Q$ power. As such, Eq. (4) cannot be used to calculate $Q_{\rm out}$ in power compensation mode. The difference between $Q_{\rm reaction1}$ and $Q_{\rm reaction2}$ is shown in Eq. (5). When $Q$ pulses are not applied, Eq. (6) defines $Q_{\rm pulse}$ and $Q_{\rm pulse}$ to be zero. This simplifies Eq. (5) to that shown in Eq. (7), where $\Delta Q_{\rm heater}$ is the difference between the heater applied with and without $Q$ pulses and $\Delta Q_{\rm out}$ is output power with and without $Q$ power. The empirical determination of $\Delta Q_{\rm out}$ is shown in Eqs. (8)–(10).

$$Q_{\rm reaction2} - Q_{\rm reaction1} = (Q_{\rm flow2} - Q_{\rm flow1}) + (Q_{\rm loss2} - Q_{\rm loss1}) - (Q_{\rm heater2} - Q_{\rm heater1}) - (Q_{\rm pulse2} - Q_{\rm pulse1}). \tag{5}$$

Without $Q$ pulse:

$$Q_{\rm pulse1} = Q_{\rm reaction1} = 0\ {\rm W}, \tag{6}$$

$$Q_{\rm reaction} = (Q_{\rm heater1} - Q_{\rm heater2}) = \Delta Q_{\rm heater} - Q_{\rm pulse} + \Delta Q_{\rm out}. \tag{7}$$

Replacing pulses with DC power through the core to emulate the physical source of the heat, as described in the measurement subsection, allows us to determine the amount of $Q$-pulse power that affects the core heater power when $Q_{\rm reaction} = 0$. Rearranging Eq. (7) where $Q_{\rm heaterDC}$ is the heater power when DC power is applied to the core coating, Eq. (8) allows us to calculate $\Delta Q_{\rm out}$ at different applied DC powers ($Q_{\rm DC}$). Finding the linear fit parameters from the plot of $\Delta Q_{\rm out}$ vs $Q_{\rm DC}$, Eq. (9) shows us the relationship between applied DC power ($Q_{\rm DC}$) and the DC power output to the environment ($\Delta Q_{\rm out}$), which cannot be measured directly.

The same equation can be used to find $\Delta Q_{\rm out}$ with $Q$ power applied substituting ($Q_{\rm pulse}$) for $Q_{\rm DC}$.

$$\Delta Q_{\rm out} = Q_{\rm DC} - (Q_{\rm heater} - Q_{\rm heaterDC}). \tag{8}$$

Since

$$\Delta Q_{\text{out}} = m(Q_{\text{DC}}) + b \quad \text{then} \quad \Delta Q_{\text{out}} = m(Q_{\text{pulse}}) + b. \tag{9}$$

Equation (10) shows the calculation of $Q_{\text{reaction}}$ when operating in power compensation mode where $\Delta Q_{\text{heater}} + \Delta Q_{\text{out}}$ would equal $Q_{\text{pulse}}$ (or $Q_{\text{DC}}$) when $Q_{\text{reaction}} = 0$. Equation (11) defines our effective coefficient of performance for the power compensation mode for our isoperibolic calorimeter system.

$$Q_{\text{reaction}} = \Delta Q_{\text{heater}} - Q_{\text{pulse}} + \Delta Q_{\text{out}}, \tag{10}$$

$$\text{COP} = (\Delta Q_{\text{heater}} + \Delta Q_{\text{out}})/Q_{\text{pulse}} = (\Delta Q_{\text{heater}} + m(Q_{\text{pulse}}) + b)/Q_{\text{pulse}}. \tag{11}$$

### 2.4.2. *Method B*

The second method of analyzing the calorimetry is more direct; instead of calculating the power loss by the calorimeter, it determines the amount of heater power compensation (HPC) for different amounts of DC calibration power $P_{\text{DC}}$ at different temperatures. In fact, this method is analogous to the traditional isoperibolic calorimeter analysis except that it substitutes heater power compensation for the temperature difference, i.e. the calibration curve at each temperature is determined by plotting HPC versus $P_{\text{DC}}$. In order to calculate $Q_{\text{reaction}}$, Method B compares the HPC from the $Q$ pulse experiment to the HPC vs. $P_{\text{DC}}$ calibration curve at the same temperature. Using this DC calibration, the relationship between input power and HPC is determined so that with input pulse power the HPC can be used to back-calculate the power from the pulse imparted into the core.

First, the linear relationship between HPC and DC power ($P_{\text{DC}}$) is found by fitting a linear equation to HPC vs. $P_{\text{DC}}$ when $P_{\text{DC}}$ is varied across the range of $Q_{\text{pulse}}$ powers. These linear coefficients ($M$) are then applied to the HPC measured during $Q_{\text{pulse}}$ stimulation HPC($Q$) to calculate delta output power ($\Delta P_{\text{out}}$) = HPC($Q$)/$M$. $Q_{\text{reaction}}$ is then calculated as shown in Eq. (12), where HPC($Q$) is the actual HPC measured when the pulse is applied. Equation (13) is then used to calculate the coefficient of performance (COP). Alternatively, stimulated power gain (SPG) can be calculated as shown in Eq. (14).

$$Q_{\text{reaction}} = \text{HPC}(Q) - \text{HPC(DC)}, \tag{12}$$

$$\text{COP} = Q_{\text{reaction}}/Q_{\text{pulse}} = (\text{HPC}(Q) - \text{HPC(DC)})/Q_{\text{pulse}}, \tag{13}$$

$$\text{SPG} = \Delta P_{\text{out}}/Q_{\text{pulse}}.$$

**Table 2.** List of linear fit coefficients determined and employed in Method B.

| Temperature (°C) | $M$ |
|---|---|
| 150 | 0.45 |
| 200 | 0.47 |
| 250 | 0.50 |
| 300 | 0.53 |
| 350 | 0.57 |
| 400 | 0.58 |

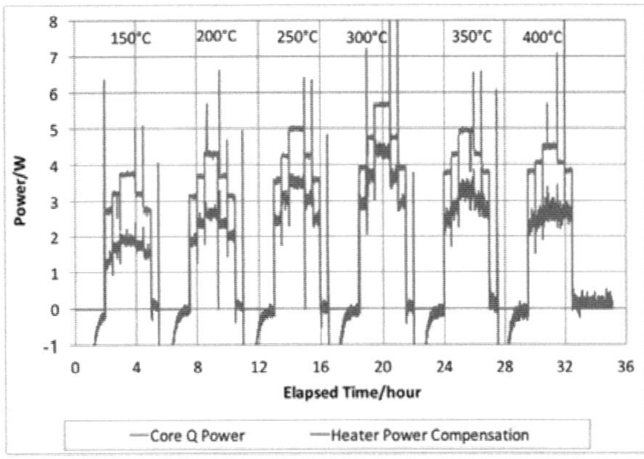


**Figure 8.** Plot of $Q$ pulse power and heater power compensation at constant temperatures from 150 to 400°C.

The linear slope coefficient is similar to the value "$m$" used in Method A, which uses the fit to determine the input power lost to the environment. Method B uses the fit to determine the percentage of input power that interacts with the core's heater and thermocouple. Table 2 shows the values for "$M$", the linear fit coefficient from Method B.

## 3. Results and Discussion

The pulse width was varied from 100 to 300 ns and back to 100 ns while maintaining constant $Q$ power and constant temperature. This was repeated at five different temperatures (held constant) from 150 to 400°C. The reduction in heater power (heater power compensation, HPC) is equal to the $Q$ pulse power that reaches the heater plus $Q_{\text{reaction}}$. Figure 8 shows HPC and $Q$ pulse power at these temperatures while varying the pulse width.

Note that the power compensation amount is very dependent on the pulse length at 300°C. Although the total pulse power from the generator is constant, the pulse power measured at the core does vary with pulse length. Still, the

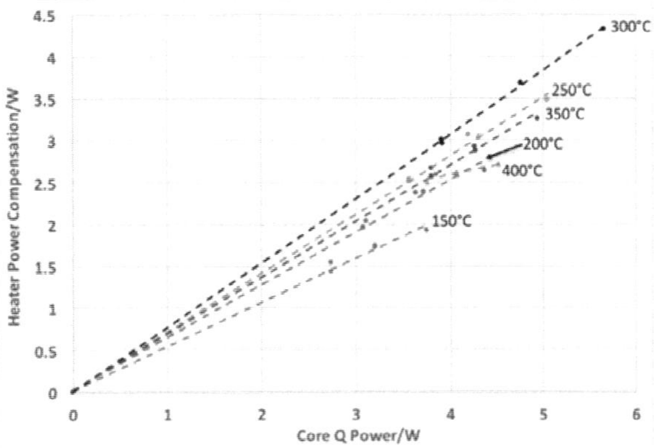


**Figure 9.** Plot of heater power compensation vs. $Q$ pulse power from 150 to 400°C.

| | | | | | | |
|---|---|---|---|---|---|---|
| **EXPERIMENT RUN DETAILS** | | | | | | |
| IPB2 STUDIES ROH, Core 27b Ni/Pd, in H2, CRIO-v167 | | | | | | |
| Operator | Roger H. | Roger H. | Roger H. | Roger H. | Roger H. | Roger H. |
| Date | 09/17/16 | 09/17/16 | 09/17/16 | 09/17/16 | 09/17/16 | 09/17/16 |
| DUT | IPB2 | IPB2 | IPB2 | IPB2 | IPB2 | IPB2 |
| ΔQout calibration | IPB2 28B H2 | IPB2 28B H2 | IPB2 28B H2 | IPB2 28B H2 | IPB2 28B H2 | IPB2 28B H2 |
| **PULSE SYSTEM PARAMETERS** | | | | | | |
| Pulse Width | 100 | 100 | 100 | 100 | 100 | 100 |
| REACTOR GAS | H2 | H2 | H2 | H2 | H2 | H2 |
| Qvoltage, Chroma, VDC | 300 | 300 | 300 | 300 | 300 | 300 |
| Q generator type | Half-H | Half-H | Half-H | Half-H | Half-H | Half-H |
| Core Temp Setting (celsius) | **150** | **200** | **250** | **300** | **350** | **400** |
| Pi-Filter QPOW Setting | 50.00 | 50.00 | 50.00 | 50.00 | 50.00 | 50.00 |
| **COP MEASUREMENT VALUES** | | | | | | |
| Measured pulse power across core oscope (Q_pulse) | **3.75** | **4.31** | **5.00** | **5.66** | **4.94** | **4.51** |
| Heater power - no pulses (Q_heater) | 9.06 | 13.70 | 18.77 | 24.43 | 30.57 | 37.72 |
| Heater power - with pulses (Q_heater) | 7.24 | 11.13 | 15.30 | 20.12 | 27.48 | 34.94 |
| Delta heater power (ΔQ_heater) | **1.83** | **2.58** | **3.47** | **4.31** | **3.08** | **2.77** |
| m (for Q_k equation) | 0.55 | 0.55 | 0.49 | 0.49 | 0.42 | 0.42 |
| b (for Q_k equation) | -0.04 | -0.04 | 0.03 | 0.03 | -0.02 | -0.02 |
| Q_k = m*Q_pulse + b (Q Power dissipated under heat spreader) | 2.02 | 2.32 | 2.46 | 2.78 | 2.04 | 1.87 |
| **COP = (ΔQ_heater + Q_k) / Q_pulse** | **1.02** | **1.14** | **1.19** | **1.25** | **1.04** | **1.03** |

**Figure 10.** Summary of COP calculations from a $Q$-pulse length run similar to that shown in Fig. 9.

magnitude of the power compensation is a greater percentage of the pulse power at 100 ns than at 300 ns. Calculations show that at 300 ns, the $Q_{\mathrm{reaction}}$ is quite small but is of much greater magnitude at 100 ns. Figure 9 shows the effect of temperature on HPC versus pulse power at different temperatures. The summary of the COP results from the data shown in Figs. 8 and 9 calculated using Method A is shown in Fig. 10. Table 3 summarizes the COP results from six such runs.

It is important to note in Table 3 that the runs performed at 300°C showed COP significantly greater than 1.0, while those at 600°C were essentially 1.0 within experimental error. This may be explained as the Pd inner layer totally de-loading its hydrogen as we have seen before at this temperature and the Ni, although retaining hydrogen traverses its Curie point, changing its electrical and chemical properties. Similar results have been seen from more than 50 runs performed over this period.

Method B was used to calculate COP from some more recent runs similar to that shown in Figs. 8 and 9. As shown above, operating above 600°C usually does not yield any reaction heat. Recent runs were operated only up to 400°C. Table 4 summarizes the $Q_{\mathrm{reaction}}$ and SPG calculated from recent runs analyzed using Method B.

There are many more test runs that occurred with Brillouin's IPB hydrogen hot tubes (HHTs), which can be

**Table 3.** Summary of COP calculations from six $Q$-pulse runs similar to that shown in Figs. 8 and 9 up to 600°C.

| Temperature (°C) | Pulse width (ns) | COP | Error ± |
|---|---|---|---|
| 300 | 150 | 1.4 | 0.1 |
| 250 | 150 | 1.4 | 0.1 |
| 300 | 150 | 1.2 | 0.1 |
| 600 | 150 | 1.0 | 0.1 |
| 300 | 300 | 1.3 | 0.1 |
| 600 | 150 | 1.01 | 0.08 |
| 300 | 150 | 1.4 | 0.1 |

**Table 4.** $Q_{\text{reaction}}$ and SPG from recent runs calculated using Method B.

| Temperature (°C) | $Q_{\text{reaction}}$ at 100 ns (W) | $Q_{\text{reaction}}$ at 150 ns (W) | Error ± (W) | SPG at 100 ns | SPG at 150 ns |
|---|---|---|---|---|---|
| 150 | 0.73 | 0.88 | 0.08 | 1.2 | 1.3 |
| 200 | 0.99 | 1.15 | 0.08 | 1.3 | 1.4 |
| 250 | 1.18 | 1.5 | 0.1 | 1.3 | 1.5 |
| 300 | 1.9 | 2.1 | 0.2 | 1.5 | 1.6 |
| 350 | 1.4 | 1.7 | 0.2 | 1.4 | 1.5 |
| 400 | 1.1 | 1.4 | 0.1 | 1.3 | 1.4 |

analyzed using these and other methods but the COP's and SPG's found in those tests are very similar to the runs that were examined and summarized herein.

## 4. Conclusion

LENR can produce thermal power when Ni and other metal-coated tubes are stimulated using fast rise-time pulses. These experiments operated in $H_2$ or He gas from 200 to 600°C. The exact same procedure was performed in each gas. Comparative thermal measurements were performed between heater-only power and heater and pulse power.

These runs were performed in isoperibolic calorimeters operated in power-compensation mode, where the heater adjusts its power to keep the inner and outer temperature difference constant. Over 100 runs were performed on five different Ni-coated cores. Three additional cores were also tested for other experimental purposes. Stimulated power gain values from 1.0 to over 2.0 were measured depending on stimulation conditions. It is important to note that these calculations ignore the heater power necessary to maintain temperature, which can be several times greater than either the stimulation power or power gain. Recent test runs have not averaged above 1.5, although the core's coating composition and metallurgy are still being optimized. The calorimetry is still being optimized. The COP and SPG results presented here ignore the heater power necessary to maintain temperature, and the electrical losses in the pulse generator, both of which can be several times greater than either the stimulation power or power gain.

## Acknowledgements

I would like to acknowledge Dr. Michael McKubre (SRI Emeritus) for his work on the calorimeter design and thank Brillouin Energy scientists Jin Liu, Roger Tong and Mark Garnett for their aid in the calorimetric analysis. I also acknowledge the Brillouin Energy Corp. for their generous support of this work.

## References

[1] R. Godes, R. George, F. Tanzella and M. McKubre, *J. Condensed Matter Nucl. Sci.* **13** (2015) 127.
[2] R. Godes, Drive circuit and method for semiconductor devices, US Patent 8,624,636, 2014.